Deep Learning in Computational Mechanics

Leon Herrmann · Moritz Jokeit · Oliver Weeger ·
Stefan Kollmannsberger

Deep Learning in Computational Mechanics

An Introductory Course

Second Edition

Springer

Leon Herrmann
Bauhaus-Universität Weimar
Weimar, Germany

Moritz Jokeit
Eidgenössische Technische
Hochschule Zürich
Zürich, Switzerland

Oliver Weeger
Technische Universität Darmstadt
Darmstadt, Germany

Stefan Kollmannsberger
Bauhaus-Universität Weimar
Weimar, Germany

ISBN 978-3-031-89528-9 ISBN 978-3-031-89529-6 (eBook)
https://doi.org/10.1007/978-3-031-89529-6

Deep Learning in Computational Mechanics An Introductory Course https://link.springer.com/book/10.1007/978-3-030-76587-3

This Springer imprint is published by the registered company Springer Nature Switzerland AG
The registered company address is: Gewerbestrasse 11, 6330 Cham, Switzerland

To Frida,
may you grow up with artificial intelligence
capable of proper numerical simulation

—Leon

To Mirely,
twenty years are not enough

—Stefan

Preface

This manuscript is a major rework of the first edition of the book *Deep Learning in Computational Mechanics—An Introductory Course* [1]. Alongside comprehensive updates and extensions to the existing chapters, this edition introduces six new chapters to enrich the reader's understanding of the topic. These include Chap. 6 "Machine Learning in Computational Mechanics", Chap. 7 "Material Modeling with Machine Learning", Chap. 8 "Generative Artificial Intelligence", Chap. 9 "Inverse Problems and Deep Learning", a comprehensive review as Chap. 10 "Methodological Overview of Deep Learning in Computational Mechanics", and an outlook as Chap. 11 "The Future of Deep Learning in Computational Mechanics".[1]

Despite the substantial revision, this book is still not (and not intended to be) an expert book on neural networks. Other books, such as [4, 5], are a much better resource for this topic. Instead, this manuscript is a comprehensive introduction to the topic of how learning-based methods, in general, and neural networks, in particular, are used to solve challenges in the field of computational mechanics. To this end, it teaches you the absolute essence of what is needed to solve the simplest possible examples, such as well-known one-dimensional bar and beam problems, or fundamental structural dynamics, thereby enabling you to navigate most of the scientific landscape in neural networks for simulation. Of course, no one would use the presented methodologies to solve such simple tasks. However, the core concepts taught by this book generalize to address more complex problems, for which we provide brief outlooks. Working through this book will put you in the position to do just that: to start your own endeavor in the emerging field of deep learning in computational mechanics.

To improve the reader's understanding of the discussed topics, the book is accompanied by numerous exercises. Besides advancing the understanding, the intention of the exercises is to provide simple implementations of the presented methodologies, such that the reader may directly build on these for their own projects. Pen-and-paper exercises are denoted with (P), while coding exercises are indicated by (C). The implementations of coding exercises and examples within the book are provided

[1] Small parts of the extensions are literal transcripts of [2] and [3].

in Python using the popular PyTorch [6] machine learning library. The accompanying files are available at www.deeplearningincomputationalmechanics.com. In addition, we provide lecture slides as an alternative medium for the presented content—making it easier to use this introductory textbook within the context of a university lecture. Both the exercise files and lecture slides are licensed under a Creative Commons Attribution-NonCommercial-ShareAlike 4.0 International License (CC BY-NC-SA 4.0), such that you may reuse and even adapt the content for your own (non-commercial) purposes as long as appropriate credit is given.

Related textbooks of deep learning (and machine learning) in simulation, which might be valuable to the reader to gain insights from different perspectives are [7] for a broader introduction to machine learning in engineering, [8, 9] for a collection of valuable case studies, [10] for a compilation of state-of-the-art summaries in machine learning from some of the current experts of the field, and [11] for an introductory textbook with a more mathematical viewpoint.

München, Germany — Leon Herrmann
Zürich, Switzerland — Moritz Jokeit
Darmstadt, Germany — Oliver Weeger
Weimar, Germany — Stefan Kollmannsberger

References

1. Kollmannsberger, S., et al. 2021. *Deep learning in computational mechanics: An introductory course*. Studies in Computational Intelligence, vol. 977. Cham: Springer, 2021. ISBN: 978-3-030-76587-3, 978-3-030-76586-6.
2. Herrmann, Leon and Stefan Kollmannsberger. 2024. Deep learning in computational mechanics: A review. In *Computational mechanics*. https://doi.org/10.1007/s00466-023-024 34-4. ISSN: 1432-0924
3. Herrmann, Leon. 2025. *Towards a meaningful integration of neural networks in computational solid mechanics*. Ph.D. thesis. Technische Universität München.
4. Goodfellow, Ian, Yoshua Bengio, and Aaron Courville. 2016. *Deep learning*. MIT Press. ISBN: 0-262-03561-8.
5. Géron, Aurélien. 2023. *Hands-on machine learning with Scikit-Learn, Keras and TensorFlow: Concepts, tools, and techniques to build intelligent systems*, 3rd edn. OCLC: on1346503549. Beijing: O'Reilly Media, Inc., 2023. ISBN: 978-1-09-812597-4.
6. Paszke, Adam, et al. 2019. *PyTorch: An imperative style, high-performance deep learning library*. arXiv:1912.01703 [cs, stat].
7. Brunton, Steven L., and Jose Nathan Kutz. 2022. *Data-driven science and engineering: Machine learning, dynamical systems, and control*. Cambridge, United Kingdom, New York, NY: Cambridge University Press. ISBN: 978-1-00-909848-9.
8. Yagawa, Genki, and Atsuya Oishi. 2021. *Computational mechanics with neural networks*. Lecture Notes on Numerical Methods in Engineering and Sciences. Cham: Springer International Publishing. https://doi.org/10.1007/978-3-030-66111-3. ISBN: 978-3-030-66110-6, 978-3-030-66111-3.
9. Yagawa, G., and Atsuya Oishi. 2023. *Computational mechanics with deep learning: An introduction*. Cham, Switzerland: Springer. ISBN: 978-3-031-11847-0.

10. Rabczuk, Timon, and Klaus-Jürgen Bathe. 2023. Machine learning in modeling and simulation: Methods and applications. Springer International Pu. ISBN: 978-3-031-6643-7.
11. Ray, Deep, Orazio Pinti, and Assad A. Oberai. 2024. *Deep learning and computational physics*. Cham: Springer Nature Switzerland. https://doi.org/10.1007/978-3-031-59345-1. ISBN: 978-3-031-59344-4, 978-3-031-59345-1.

Acknowledgements The authors gratefully acknowledge the funding through the Georg Nemetschek Institut (GNI) under the project DeepMonitor. Oliver Weeger acknowledges the financial support provided by the Deutsche Forschungsgemeinschaft (DFG, German Research Foundation, project number 492770117), the Graduate School of Computational Engineering at TU Darmstadt, and Hessian.AI. Furthermore, a big thank you goes to Thomas Hollowell for updating the exercises and proofreading the manuscript, thereby significantly improving its clarity for students. Also, many thanks to Ralf Denzer and Omid Kianian for valuable comments that improved the mathematical notation and Alexander Henkes for providing generated lattice structures for motivational purposes in the context of generative artificial intelligence. Also, Davide D'Angella shall not be forgotten for aiding in the writing of the first edition of this book [1].

Last and most importantly, a heartfelt gratitude to our beloved families for their relentless support, thus enabling us to live our dreams—without which this book would never have been created.

Competing Interests The authors have no competing interests to declare that are relevant to the content of this manuscript.

Contents

About the Authors

Leon Herrmann has a uniquely diverse background; born in South Africa and growing up in seven different countries. He earned a bachelor's degree in Mechanical Engineering from the Technical University of Denmark (DTU) and a master's degree in Computational Mechanics from the Technical University of Munich (TUM), where he also obtained his doctorate for his work in computational mechanics with neural networks. His primary research focus has been on finite element methods, fracture in composite materials, and combining traditional numerical simulations with modern machine learning techniques.

Moritz Jokeit As a product of the "Mauerfall", Moritz Jokeit grew up in the non-existing town of Bielefeld and the alpine foothills near Rosenheim. Following his bachelor's degree in Civil Engineering, he studied Computational Mechanics at the Technical University of Munich (TUM) and the Polytechnic University of Catalonia (UPC). His passion for deep learning and computational mechanics was transformed into a master thesis that laid the groundwork for this lecture book. After his graduation he continued his research at the Chair of Computational Modeling and Simulation. He is now a doctoral candidate at the Institute for Biomechanics at the ETH Zürich focusing on the mechanics of the spine.

Oliver Weeger is a Full Professor for Cyber-Physical Simulation with the Department of Mechanical Engineering at the Technical University of Darmstadt in Germany. He graduated in Techno-Mathematics from TU Munich in 2011 and obtained his Ph.D. in Mathematics from TU Kaiserslautern in 2015. Before joining TU Darmstadt in 2019, he had been working at the Singapore University of Technology and Design as a Postdoctoral Researcher and Assistant Professor.

His passion for research and education evolves around advanced computational methods, modeling, and optimization approaches for nonlinear, multiscale, and multiphysics problems in engineering. In particular, this includes the fusion of machine learning, classical modeling, and simulation to obtain flexible and yet accurate, reliable and robust predictive models for computational mechanics.

Stefan Kollmannsberger graduated in Civil Engineering in 1998 and worked for several years as heavy underground construction engineer before returning to university to devote himself to computational mechanics. He graduated with a Ph.D. at the Technical University of Munich in 2010, where he enjoyed leading the research group "Simulation in Applied Mechanics" until 2023. Since then, he is full professor at the Bauhaus University in the culturally opulent city of Weimar and heads the Chair of Data Science in Construction. He is dedicated to both teaching and science and uses the content of this lecture book as a basis for an introductory course in the field of artificial intelligence in computational mechanics.

Symbols

$\tilde{\Theta}$	Modified parameters
Θ	Model parameters
Θ^*	Optimal model parameters
A	Cross-sectional area
a	Activation/neural output
$\boldsymbol{A}$	Example matrix
$\boldsymbol{a}$	Layer of activation/neural layer output
$\overline{\boldsymbol{A}}$	Mean matrix
b	Bias
$\boldsymbol{B}$	Mean-centered matrix
$\boldsymbol{B}$	Hessian
$\boldsymbol{b}$	Bias vector
b^*	Optimal bias
$\mathcal{B}$	Subscript to indicate boundary condition error
C	Cost function
C	Courant number
$\boldsymbol{C}$	Constitutive matrix/tensor
$\boldsymbol{c}$	Coefficients of reduced system
$\tilde{\boldsymbol{c}}$	Ground truth vector of wave speeds
c	Heat capacity
c	Compliance
c	Wave speed
$\hat{c}$	Prediction of the wave speed
$\nabla_{\boldsymbol{\Theta}} C$	Gradient of the cost function with respect to the weights and biases (alternatively $\partial C/\partial \boldsymbol{\Theta}$)
$\tilde{C}$	Modified cost function
$\tilde{c}$	Ground truth wave speed
D	Discriminator
D	Dissipation
d	Decoding mapping
d	Damping coefficient

$\mathcal{D}$	Subscript to indicate data-driven
E	Young's modulus
e	Encoding mapping
F	Concentrated force
$\boldsymbol{F}$	Force vector
$\boldsymbol{F}$	Deformation gradient
$\boldsymbol{f}$	Force vector
$\boldsymbol{f}$	Right-hand side of system of differential equations
$\hat{\boldsymbol{f}}$	Prediction of right-hand side of system of differential equations
f	Generic model as a function
$\mathcal{F}$	Subscript to indicate finite element-based loss
f_l	Layer of neural network as a function
f_{NN}	Neural network as a function
G	Generator
$\bar{\boldsymbol{g}}$	Weighted average of gradients in Ada-Grad
$\tilde{\boldsymbol{g}}$	Exponentially decaying average of gradients in RMSProp
g	Dirichlet boundary condition
$\boldsymbol{H}$	Inverse of the Hessian
$\boldsymbol{h}$	Hidden state of recurrent neural network or latent space of an autoencoder
$\hat{\boldsymbol{h}}$	Latent space prediction
h	Neumann boundary condition
$\mathcal{H}$	Hamiltonian
I	Identity
I	Area moment of inertia
$\boldsymbol{I}$	Identity matrix/tensor
$\mathcal{I}$	Subscript to indicate initial condition error
$\boldsymbol{K}$	Filter for a convolutional layer
$\boldsymbol{K}$	Stiffness matrix
$\boldsymbol{K}$	Stiffness matrix
k	Spring stiffness
L	Last layer index
L	Length of the bar
L	Lagrangian
l	Arbitrary layer index
$\mathcal{L}$	Loss component
$\boldsymbol{M}$	Mass matrix
$\boldsymbol{m}$	First statistical moment in Adam
$\tilde{\boldsymbol{m}}$	Corrected first statistical moment in Adam
m	Number of examples/samples
m	Mass
$\mathcal{M}$	Subscript to indicate measurement mismatch
m	Model paraboloid
N	Shape function
$\boldsymbol{n}$	Second statistical moment in Adam

$\tilde{\boldsymbol{n}}$	Corrected second statistical moment in Adam
$\hat{\mathcal{N}}$	Prediction of the non-linear operator of the differential equation
$\mathcal{N}$	Non-linear differential operator
n	Number of features
N_1	Number of spatial discretization points in x_1
N_2	Number of spatial discretization points in x_2
$\hat{\mathcal{N}}^{\mathcal{T}}$	Prediction of the non-linear differential operator in time-dependent partial differential equation
$\mathcal{N}^{\mathcal{T}}$	Non-linear differential operator in time-dependent partial differential equation
N_t	Number of temporal discretization points
$\boldsymbol{P}$	First Piola-Kirchhoff stress tensor
$\boldsymbol{p}$	Inhomogeneous source vector/distributed load vector
$\boldsymbol{p}$	Vector of momentum of degrees of freedom
$\boldsymbol{p}$	Search direction
p	Inhomogeneous source term/distributed load
p	Momentum of a degree of freedom
p	Dropout hyperparameter
$\boldsymbol{R}$	Residuum vector
$\hat{\boldsymbol{r}}$	Prediction of residual vector of the governing differential equation
$\boldsymbol{r}$	Residual vector of the governing differential equation
$\hat{r}$	Prediction of residual of the governing differential equation
$\mathcal{R}$	Subscript to indicate residual of differential equation
r	Residual of the governing differential equation
r_{NN}	Residual as evaluation by the neural network
S	Forward solver
s	Stride length
$\mathcal{T}$	Temporal domain
t	Time
$\tilde{t}$	Temporal coordinate associated with solution $\tilde{u}$
(test)	Superscript to indicate the testing set
(train)	Superscript to indicate the training set
$\boldsymbol{U}$	Singular modes of $\boldsymbol{AA}^*$
$\bar{\boldsymbol{U}}$	Truncated singular modes
$\hat{\boldsymbol{u}}$	Vector of solution prediction
$\tilde{\boldsymbol{U}}$	Economy singular modes
$\tilde{\boldsymbol{u}}$	Ground truth solution vector/measurement associated with coordinate $\tilde{x}$ (and time $\tilde{t}$)
$\boldsymbol{u}$	Solution vector of a differential equation/degrees of freedom
$\hat{u}$	Prediction of the solution u of a differential equation
$\mathcal{U}$	Uniform random distribution
$\tilde{u}$	Ground truth solution/measurement associated with coordinate $\tilde{x}$ (and time $\tilde{t}$)
u	Solution of a differential equation
u_{NN}	Neural network as model of solution

V	Volume fraction
$\boldsymbol{V}$	Singular modes of $\boldsymbol{A}^*\boldsymbol{A}$
$\bar{\boldsymbol{V}}$	Truncated singular modes
$\hat{\boldsymbol{v}}$	Prediction of multiple test functions
$\tilde{\boldsymbol{V}}$	Economy singular modes
$\boldsymbol{v}$	Momentum term in gradient descent with momentum
$\hat{v}$	Prediction of test function
$\mathcal{V}$	Subscript to variational loss
v	Test function
v_{NN}	Neural network as test function
W	Internal energy density/strain energy density
$\boldsymbol{W}$	Weight matrix
$\boldsymbol{W}$	Principal components
$\bar{\boldsymbol{W}}$	Truncated principal components
$\boldsymbol{w}$	Weight vector
$\boldsymbol{w}^*$	Optimal weight vector
$\dot{W}$	Mechanical power
w	Weight
w^*	Optimal weight
$\boldsymbol{X}$	Design matrix containing the features of all samples $\boldsymbol{x}$ or time-series of vectors $\boldsymbol{x}$
$\boldsymbol{X}$	Snapshot matrix
$\overline{\boldsymbol{X}}$	Mean snapshot matrix
$\hat{\boldsymbol{X}}$	Prediction vector of components of the differential equation
$\hat{\boldsymbol{x}}$	Prediction of input vector
$\tilde{\boldsymbol{X}}$	Design matrix associated with target $\tilde{y}$
$\tilde{\boldsymbol{x}}$	Vector input associated with target $\tilde{y}$
$\tilde{\boldsymbol{x}}$	Vector of spatial coordinates associated with vector of solutions $\tilde{\boldsymbol{u}}$
$\boldsymbol{x}$	Input vector
$\boldsymbol{x}$	Vector of spatial coordinates
$\boldsymbol{x}$	Vector of degrees of freedom
$\tilde{x}$	Scalar input associated with target $\tilde{y}$
$\tilde{x}$	Spatial coordinate associated with solution $\tilde{u}$
x	Input
x	Spatial coordinate
x	Degree of freedom
$\hat{\boldsymbol{Y}}$	Time-series prediction
$\hat{\boldsymbol{y}}$	Vector prediction
$\tilde{\boldsymbol{y}}$	Ground truth vector associated with input $\tilde{x}$
$\boldsymbol{y}$ y	Values of a labeled dataset xy
$\hat{y}$	Scalar prediction
$\tilde{y}$	Ground truth scalar associated with input $\tilde{x}$
y	Output
z	Intermediate neural network vector output
z	Intermediate neural network scalar output

z	Prediction before the strong enforcement of the boundary conditions
Γ_D	Dirichlet boundary
Γ_N	Neumann boundary
Δt	Time step size
Δx	Spatial grid size
Ξ	Sparse coefficients in SINDy
Π_e	External energy
Π_i	Internal energy
Π_kin	Kinetic energy
Π_pot	Potential energy
Π_tot	Total energy
Θ	SINDy candidate library
$\boldsymbol{\Sigma}$	Singular values
$\bar{\boldsymbol{\Sigma}}$	Truncated singular values
$\tilde{\boldsymbol{\Sigma}}$	Economy singular values
$\boldsymbol{\Psi}$	Reduced basis
Ω	Spatial domain
Ω	Penalty term
α	Learning rate or step size in the gradient descent algorithm
β	Adams hyperparameters
$\tilde{\boldsymbol{\gamma}}$	Ground truth vector indicator capturing entire domain Ω
γ	Indicator scaling the mass density ρ in the wave equation
γ	Internal variable
$\hat{\gamma}$	Prediction of indicator
$\tilde{\gamma}$	Ground truth indicator
δ	Sensitivity with respect to the cost function and the neurons input
δ	Dirac delta
$\boldsymbol{\varepsilon}$	Strain tensor
ε	Strain component
η	Hyperparameter in gradient descent with momentum defining influence of previous gradients
$\boldsymbol{\kappa}$	Vector of weighting terms
κ	Weighting term
$\boldsymbol{\lambda}$	Vector/set of coefficients of a differential equation
$\hat{\lambda}$	Prediction of the parameters of the differential equation
λ	Coefficients of a differential equation
λ	Lamé’s second parameter
λ	Regularization factor
$\boldsymbol{\mu}$	Cluster center
$\boldsymbol{\mu}$	Mean vector
μ	Lamé’s first parameter
ν	Poisson’s ratio
$\boldsymbol{\xi}$	Random noise acting as input
ρ	Decay rate in RMSProp
ρ	Mass density

$\boldsymbol{\sigma}$	Stress tensor
$\boldsymbol{\sigma}$	Standard deviation vector
σ	Activation function
σ	Stress component
$\tilde{\sigma}$	Ground truth stress associated with coordinate $\tilde{x}$
τ	Breakeven threshold
ψ	Ricker wavelet
ϵ	Noise term
ϵ	AdaGrads stabilization hyperparameter
$\hat{\phi}$	Predicted flux
ϕ	Flux
ϕ	Dissipation potential

Chapter 1
Computational Mechanics Meets Artificial Intelligence

Abstract The integration of artificial intelligence into computational mechanics marks a significant shift in problem-solving approaches for physics and engineering. This chapter outlines the historical development of artificial intelligence from its early stages to applications in computational mechanics. Judging by the growing body of scientific literature, artificial intelligence, and its subtopics form a recent yet important part of research within simulation disciplines. Despite this hype, not only the potential but also the challenges are highlighted through a realistic perspective on the capabilities of neural networks. The chapter serves as an introduction to the book's key motivations, foundational concepts, and scope.

The pursuit of artificial intelligence dates back to the time computers, in a modern sense, were born [26]. Since the emergence of computer science, it has been partially dedicated to finding the key to intelligent machines. Over the last few years, rapid advancements in the field of machine learning have caused quite a sensation. Algorithms suddenly achieved competing results in tasks thought to be the reserve of human intelligence. Notable examples from the last 30 years include

- IBM's Deep Blue chess victory over then world chess champion Garry Kasparov in 1997 [1],
- recommendation systems embedded in major commercial platforms like Netflix, Amazon, and Spotify [2],
- AlexNet winning the ImageNet Large Scale Visual Recognition Challenge 2012 [3], providing the breakthrough for many subsequent state-of-the-art image recognition models, such as GoogLeNet [4], VGG [5], and ResNet [6],
- Google DeepMind's AlphaGo defeating world champion Go player Lee Sedol in 2016 [7], opening the door to success in many other games, including highly complex computer games [8],
- advances in robotics as seen by Boston Dynamics' Learning Robots, OpenAI's Dactyl [9], or even autonomous driving [10–12],

L. Herrmann et al., *Deep Learning in Computational Mechanics*,
https://doi.org/10.1007/978-3-031-89529-6_1

- natural language processing with transformers [13] leading to chatbots such as Google's BERT[1] [14], OpenAI's ChatGPT [15], and Anthropic's Claude launched in 2018, 2022 and 2023 respectively,[2]
- Google DeepMind's AlphaFold, solving the protein folding problem in 2020 [18]—thereby accelerating biological research and potential drug discovery,
- generative image and video tools such as OpenAI's DALL-E [19], Stability AI's Stable Diffusion [20], Midjourney, and OpenAI's Sora AI [21]—relying on generative adversarial networks [22] and diffusion models [23–25].

Notably, this progress and the significance of artificial intelligence has in 2018 been honored by the Turing Award in the informatics community and, in 2024, in a broader sense, has been recognized by the Nobel Committee. Geoffrey Hinton, Yann LeCun, and Yoshua Bengio received the Turing Award in 2018. In 2024, Geoffrey Hinton and John Hopfield were awarded the Nobel Prize in physics for their foundational work on artificial neural networks, and Demis Hassabis, John Jumper, and David Baker were awarded the Nobel Prize in chemistry for their advancements in protein structure prediction (AlphaFold) using deep learning [18]. Figure 1.1 provides a timeline of some of the most important advancements in the history of artificial intelligence.

The reasons for machine learning's successes are the vast availability of data paired with straightforward implementations and powerful graphics hardware. Nowadays, new classes of learning machines are confronted with various complex problems. Whether audio-visual signal processing, weather forecasting, or autonomous driving, data-driven algorithms offer tremendous potential and take computer science a step closer to the ideal of intelligent machines.

The unprecedented success of machine learning has not gone unheeded in other scientific fields. Especially in the presence of adequate data, the application of learning-based algorithms has shown promising results. For instance, image-based medical diagnosis already benefits from outstanding progress in visual object recognition [28]. At the same time, the demand for data constitutes a major obstacle to wider interdisciplinary adoption. Physics and engineering applications often lack sufficient information about the underlying problem. Thus, the use of data-driven approaches seems somewhat naive. Further, engineering sciences have an established, long-standing paradigm of computer-aided problem-solving, which has led to excellent methodologies for many problems. Finding numerical solutions to problems that arise from observing natural or engineered systems describes the essence of computational mechanics. The solution process commences with formalizing the problem in terms of physical quantities. A set of partial differential equations typically governs the resulting mathematical description. Further, computational methods require a discretization of the problem to allow the numerical approximation of continuous variables inside a prescribed domain. Most methods following this scheme have been developed and enhanced over decades [29], hence guaranteeing

[1] Note that BERT is based on an encoder-only transformer architecture, while GPT relies only on the decoder part; see Chap. 8 for more details. This limits BERT to tasks like extractive question answering (i.e., from the input context), while GPTs are generative.

[2] For an open source model, see Meta AI's LlaMa [16, 17] from 2023.

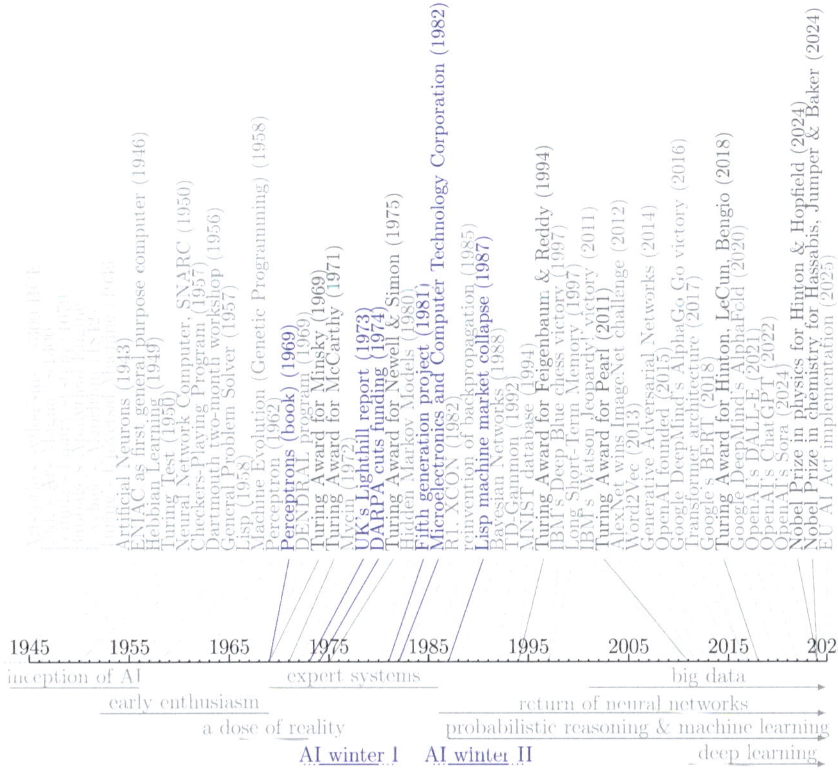

Fig. 1.1 Iconic moments in the history of artificial intelligence. Incidents related to the two artificial intelligence winters (lack of funding and research followed by an overhype) are indicated in blue. Nobel Prizes and Turing Awards are marked in dark gray. The highlighted periods are according to [26]'s introductory chapter, which provides an excellent overview of the history and foundations of artificial intelligence. What will follow next? Artificial general intelligence/superintelligence [27] solving all of humankind's problems, a third artificial intelligence winter, or maybe something in between, such as improved methodologies in computational mechanics?

a high level of robustness and reliability—hindering current machine learning from outperforming classical methods in a general sense.

Nevertheless, the generality and simplicity of learning algorithms have sparked interest in the scientific computing community and inspired early attempts toward the data-driven solution of partial differential equations. Incorporating domain knowledge into the learning model allowed it to compensate for the typical data sparsity in physical problems. The computational limitations at that time restricted the application beyond the scope of canonical examples. However, the availability of enhanced programming environments and potent hardware has encouraged researchers from various disciplines to revisit the proposed ideas over the past years. Current publications demonstrate some potential of deep learning in computational mechanics

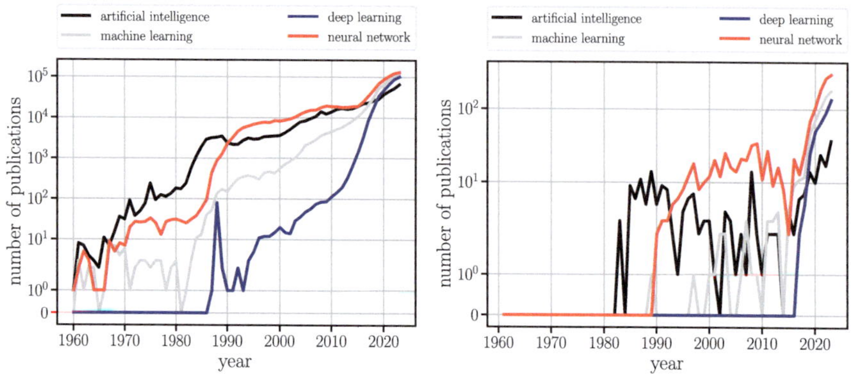

(a) Publications in all fields (b) Publications within computational mechanics

Fig. 1.2 Number of publications concerning artificial intelligence and some of its subtopics since 1960 showing the exponential growth of literature within the field. Illustration inspired by [46] and adapted from [45]

[30–45]. Most approaches deploy neural network architectures to approximate functions, which are difficult to describe with classical methods. Neural networks are an artificial intelligence technology that loosely mimic the brain's neuronal structure and belong to the most successful group of contemporary learning algorithms. Their ability to learn highly non-linear representations is an excellent prerequisite for modeling physical phenomena.

The rapidly increasing interest in deep learning—but also the wider fields of machine learning and artificial intelligence—within computational mechanics is well documented in the scientific literature. By considering the number of publications treating "*Artificial Intelligence*", "*Machine Learning*", "*Deep Learning*", and "*Neural Networks*", the interest can be quantified.[3] Figure 1.2a shows the trend in all journals indexed by Scopus since 1960, while Fig. 1.2b depicts the trend within the computational mechanic's community by considering representative journals[4] at Elsevier and Springer. The trends before 2017 differ slightly, with a steady growth in general but only limited interest within computational mechanics.[5] However, around

[3] The terms have to appear in the title, abstract, or keywords to ensure that the publication's focus is on the corresponding method and does not simply appear as a sidenote within the publication.

[4] The considered journals are *Computer Methods in Applied Mechanics and Engineering, Computers & Mathematics with Applications, Computers & Structures, Computational Mechanics, Engineering with Computers, Journal of Computational Physics*.

[5] Pioneering works exploring neural networks for computational mechanics prior to the current rise of deep learning are compiled in reviews such as [47, 48]; see [49] for a more recent treatment. Contributions across almost all of the discussed categories have already been made before the year 2000. These include data-driven modeling, such as inverse surrogate models [50, 51], physics-informed approaches for solving differential equations [52–54], efforts in simulation substitution, such as constitutive modeling [55, 56] or estimating numerical parameters [57]. Also, efforts to exploit the parallel computation capabilities of neural networks have been made, where more efficient implementations are obtained by constructing networks from discretizations [58–60]. The

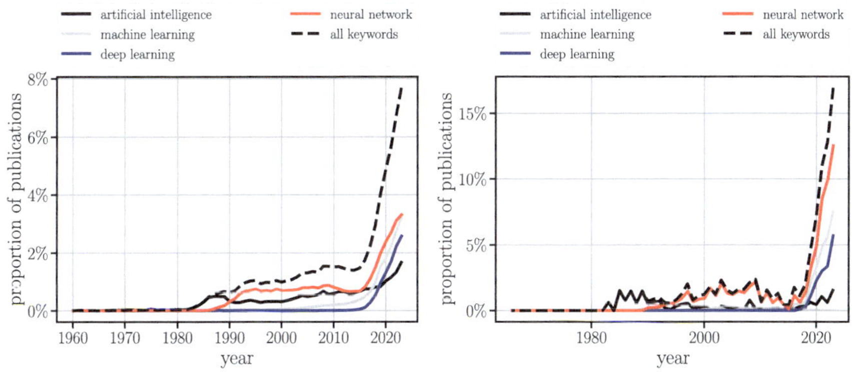

(a) Proportion in all fields (b) Proportion within computational mechanics

Fig. 1.3 Proportion of publications in % treating artificial intelligence and some of its subtopics since 1960. Publications dealing with at least one of the four keywords yield the dashed line

2017, both curves show a shift in trend, namely a vast increase in publications highlighting the interest and potential prospects of artificial intelligence and its subtopics for various applications. The relative importance of artificial intelligence and its subfields is emphasized by considering the proportion of publications within computational mechanics (Fig. 1.3b) and all fields (Fig. 1.3a). In 2023, ~17% of computational mechanics publications considered at least one of the artificial intelligence subtopics, while in the broader scientific field, it is only ~8%. Interestingly, the main contributor within computational mechanics are neural networks.

Despite (or rather because of) the abundance of research content, caution is advised when delving into the field of deep learning in computational mechanics and not letting impressive-looking results fool the reader. To illustrate this point, we want to draw an analogy to state-of-the-art generative machine learning implementations—which are much more sophisticated than all of deep learning within computational mechanics. Using DALL-E 3 [19] integrated within ChatGPT 4 [15] and Stable Diffusion XL [20], we generate an image of Isaac Newton in front of a blackboard with his three laws of motion; see Fig. 1.4. As you can see, the generated images provides believable depictions of Isaac Newton and writing on the blackboard. However, upon closer examination, the writing exposes itself as gibberish and highlights the lack of reliability required for many engineering applications. A further issue arises in

notable exclusions are generative approaches, arising with variational autoencoders [61, 62], and generative adversarial networks [22] in 2014, as well as deep reinforcement learning, which was popularized in the early 2010s [63].

(a) Initial DALL-E-3 prompt: "*Generate an image of Isaac Newton in front of a blackboard on which his three laws are written in mathematical notation and chalk.*"

(b) Follow-up DALL-E-3 prompt to correct the first image: "*The laws on the blackboard are incorrect. Please add the correct formulations. If you are unable to do so, simply focus on the second law, which is F=m*a.*"

(c) Stable Diffusion XL prompt (No Style): "*Generate an image of Isaac Newton in front of a blackboard on which his three laws are written in mathematical notation and chalk.*"

(d) Stable Diffusion XL prompt (Digital Art): "*Generate an image of Isaac Newton in front of a blackboard on which his three laws are written in mathematical notation and chalk.*"

Fig. 1.4 Generated images of Isaac Newton with his three laws of motion using DALL-E-3 [19] and Stable Diffusion XL [20]. Multiple attempts have been made in order to generate the best possible outcome and depict neural networks as favorably as possible

the images from Stable Diffusion XL [20] in which the chalk is clearly a pen (or multiple pens in Fig. 1.4c). An analogy can be drawn to results obtained with neural networks in physics. At first glance, these might seem convincing, but they often already violate fundamental physical laws, such as, for example, the conservation of mass.[6] Furthermore, the degree of accuracy and reliability is simply insufficient, as seen in Fig. 1.5. Three attempts are made at generating basic geometric constellations: (i) a square, (ii) uniformly spaced lines, and (iii) a uniform checkerboard pattern. Only the first task is achieved successfully in Fig. 1.5b, which is, however, accompanied by two failed attempts in Fig. 1.5a, c—thus underlining the robustness issue. With this in mind, one would be skeptical of structures engineered by deep learning. For example, stepping onto an airplane with parts designed and certified by a neural network is currently unimaginable.

This stands in stark contrast to how reliability and robustness are enforced in classical approaches in physics and engineering. Unfortunately, many publications demonstrate how to solve physics problems with neural networks without considering the complete state-of-the-art. Comparisons are only conducted with respect to other neural network-based methods during their performance assessment. However, in many cases, efficient classical solution strategies already exist, which currently are not beaten by novel deep learning-based methods [40, 45, 64]. Thus, deep learning is beaten not only in reliability but also in speed and accuracy. These established methods are all too often neglected in the emerging field of deep learning in computational mechanics (and deep learning in general). Nevertheless, studying these new deep learning-based methods may still provide valuable insights for future methodologies, where meaningful combinations with established classical methods are currently the most promising. To clarify what is possible and how well neural networks perform, this book provides an honest introduction to the most prominent deep learning methodologies employed within computational mechanics. This provides a good overview of the expanding body of literature, with annual growth reaching triple digits within the computational mechanics community alone; see Fig. 1.2b. Note, however, that the current treatment is only a snapshot of the methods available by the year 2024. This could quickly change in the coming years, given the rapidly growing interest in artificial intelligence—making the study of foundational techniques covered in this book even more worthwhile.

Following this introduction, Chap. 2 leads off with the fundamental concepts of machine learning. Subsequently, Chap. 3 elaborates on neural networks and their corresponding algorithms. After reviewing the most important deep learning methodologies, the most prominent topics within deep learning in computational mechanics are treated in detail in the ensuing chapters. This includes a hands-on introduction to

[6] Note that this example is not unflawed. DALL-E 3's intended purpose is not text generation within images. It is capable of doing so to some degree (and it is impressive that it is able to deduce that we are referring to the laws of motion), as it has seen many images containing text, which is, however, not sufficient for it to understand nor reproduce text in images in a senseful manner. However, similarly, most machine learning models in computational mechanics are only trained on data in the hope that they will understand the deeper physics and reproduce these correctly. The underlying problem is that the level of accuracy required in computational mechanics is currently unreachable by deep learning. Text in images is just one example of deficiencies in state-of-the-art generative approaches. Another example is mesh generation. If you are interested in testing the capabilities, try to make DALL-E 3 (or any other image generation implementation) generate a mesh of a rectangular plate with one central circular hole composed of triangular and quadrilateral elements. It is almost impossible to generate anything resembling a proper discretization, no matter how much additional information you provide.

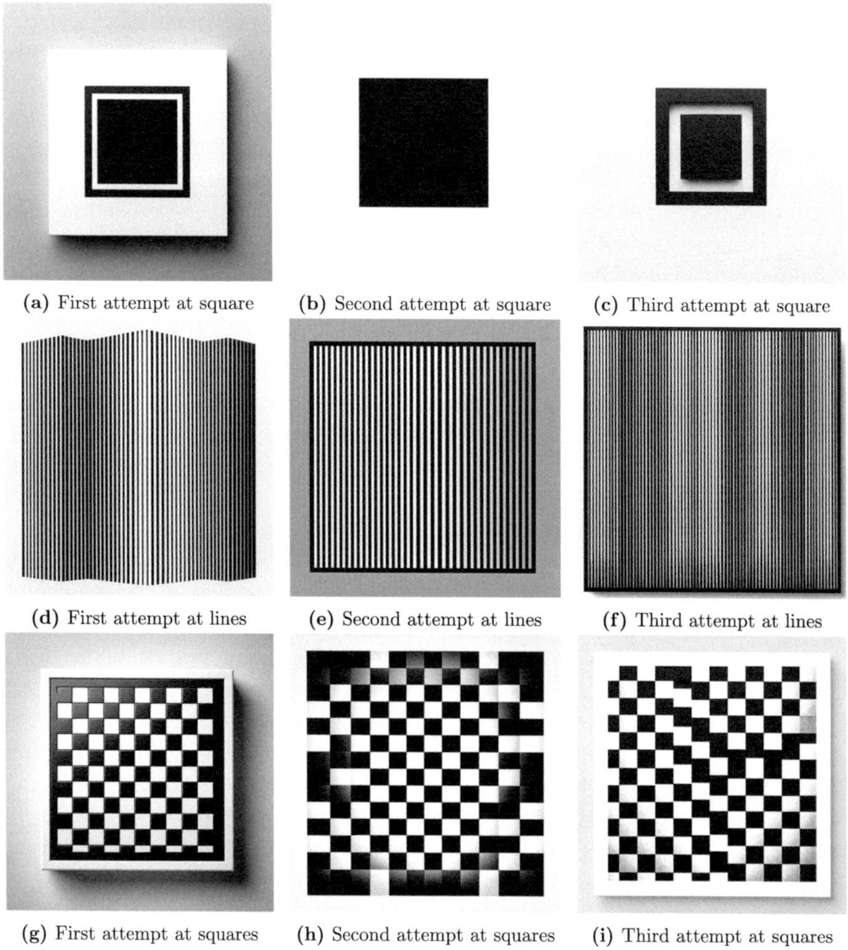

(a) First attempt at square (b) Second attempt at square (c) Third attempt at square
(d) First attempt at lines (e) Second attempt at lines (f) Third attempt at lines
(g) First attempt at squares (h) Second attempt at squares (i) Third attempt at squares

Fig. 1.5 Generated images of basic geometric constellations with increasing complexity using DALL-E-3 [19]. The square was prompted with "*Generate a blank image with only a solid black two-dimensional square in the center.*", the lines with "*Create a square image of uniformly spaced vertical stripes alternating between black and white. These stripes should cover the entire canvas.*", and the squares with "*Create a square image of a uniform checkerboard consisting of black and white squares.*"

physics-informed neural networks in Chap. 4 to solve partial differential equations from mechanics with neural networks and an extension to more advanced topics of physics-informed neural networks in Chap. 5. Before continuing with neural networks, a brief outlook on machine learning methods beyond deep learning is provided on select applications in Chap. 6, which are later combined with neural networks. Chapter 7 continues with material modeling, where data-driven techniques are presented with different neural network architectures and enhanced by physics-informed learning. Subsequently, Chap. 8 discusses generative approaches with which dimensionality reduction, realistic data generation, and anomaly detection can be handled.

Inverse problems are then tackled in Chap. 9, where the main methodologies are demonstrated on the applications of nondestructive testing and topology optimization. Lastly, a broad methodological overview of deep learning in computational mechanics is provided in Chap. 10. The book concludes with Chap. 11, which identifies future research directions and provides guidelines for good scientific practice, a sensitive yet important topic in the area of deep learning in general and particularly when applied to computational mechanics.

Exercises

A prerequisite for most exercises throughout this book is a basic understanding of the Python programming language. Additionally, the machine learning library PyTorch [65] provides the fundamental implementations needed for deep learning. The most important features of PyTorch are:

- Tensor arithmetics, enabling efficient computation and manipulation of multi-dimensional arrays
- Automatic differentiation, facilitating the computation of gradients for neural network training
- Neural networks, through a modular and flexible architecture for building and training diverse models
- GPU acceleration with CUDA,[7, 8] allowing for faster computations by leveraging the power of graphics processing units (GPUs)

All four concepts will be introduced in a tutorial-style manner within the next exercise, followed by two simple examples from solid mechanics.

E.1 Introduction to PyTorch (C)

In PyTorch, a (torch) tensor is a fundamental data structure similar to an array or matrix. It is a multi-dimensional grid of values (all of the same type), which is comparable to the NumPy array from the NumPy [67] library. Hence, most of the functionality available for NumPy arrays is also available for torch tensors. They are also easily converted from and to each other. However, in addition, the torch tensor is integrated with CUDA, allowing for faster processing through seamless GPU parallelization. Furthermore, the torch tensor is equipped with an optional automatic differentiation framework (to be discussed in detail in Sect. 3.4.3 in Chap. 3), which keeps track of all arithmetic operations, thus enabling an in-built differentiation.

---

[7] CUDA stands for compute unified device architecture and is a closed-source parallel computing framework, which is an application programming interface developed by Nvidia for GPUs [66].

[8] For Apple users, the metal performance shaders (MPS) framework similarly provides GPU acceleration for Apple's GPU hardware.

Below, some fundamental operations are provided, which you can try in order to familiarize yourself with the framework.

Tensor Initialization & Manipulation

- Library import is achieved with `import torch`.
- Tensor initialization

```
a = torch.tensor([[0., 2., 4.], [3., 4., 2.]]) # shape (2, 3)
b = torch.zeros(3, 3, 3) # shape (3, 3, 3)
c = torch.linspace(0, 1, 10) # shape (10)
d = torch.ones(1, 3) # shape (1, 3)
```

- Tensor dimensionality is retrieved with `a.shape` or `a.size()`.
- Indexing

```
a[0, 0] # to access first element
a[:, 0] # to access first order tensor in the second dimension
b[0, :, :] # to access first second order tensor
b[-1, -1, -1] # to access last element
a[0, 1:3]     # last two elements of the 0th first order tensor
```

- Fundamental operations can often be called in two ways, with a functional approach, e.g., `torch.flatten(b)`, or in an object-oriented manner, e.g., `b.flatten()`.

```
torch.cat((a, d), 0) # concatenation in dimension 0
torch.transpose(a, 0, 1) # dimensions 0 and 1 are transposed
torch.permute(b, (2, 1, 0)) # dimensions 0 and 2 are permuted
torch.reshape(a, (1, 6)) # or c.view(1, 6) reshape to (1, 6)
torch.unsqueeze(a, 0) # creates an additional empty dimension (1, 2, 3)
torch.squeeze(d, 0) # removes first empty dimension (4)
torch.flatten(b) # flattens b (27)
torch.ones_like(c) # creates a tensor of ones with the shape of c
torch.cos(a) # applies cosine to each entry
```

- Conversion to and from NumPy (imported with `import numpy as np`)

```
a = np.array([2., 3.])
b = torch.from_numpy(a) # convert numpy array to torch tensor
c = b.numpy() # convert to torch tensor from numpy array
```

- Grid creation

```
x = torch.linspace(0, 1, 10)  # shape (10)
y = torch.linspace(0, 2, 5)  # shape (5)
x, y = torch.meshgrid(x, y, indexing='ij')  # grid of shape
    (10, 5)
```

Gradient Computation

- The gradient computation is achieved through automatic differentiation, which relies on the history of all previously performed arithmetic operations. These can be tracked through computation graphs. In PyTorch, computation graphs can be enabled by setting `requires_grad = True`, which tracks all arithmetic operations, thus allowing for automatic differentiation. Only the syntax in PyTorch is explored here. The corresponding theory is explained later in Sect. 3.4.3 in Chap. 3.

```
x.requires_grad = True  # after construction
y = torch.tensor([0.], requires_grad=True)  # during
    construction
```

- To accommodate the computation graph, each tensor contains several member variables, which can be accessed with `x.data` for the `data` variable.

```
data  # values stored in the tensor
requires_grad  # boolean enabling computation graph
grad  # value of the gradient
grad_fn  # function handle storing previous operation
is_leaf  # boolean stating if tensor is a leaf node (see
    Chapter 2)
```

- Consider the array x, which we manipulate with $y = 2x, z = y^2$. The gradients $\partial z/\partial y$ and $\partial z/\partial x$ are obtained by invoking the `backward` function, which computes the Jacobian vector product. The vector $\boldsymbol{v}$ is the input `torch.ones_like(z)`. As the Jacobian $\boldsymbol{J}$ is only non-zero on the diagonal terms (due to the element-wise computations)

$$\boldsymbol{J}^\top \boldsymbol{v} = \begin{pmatrix} \frac{\partial z(y_1)}{\partial y_1} & \dots & \frac{\partial z(y_1)}{\partial y_{10}} \\ \vdots & \ddots & \vdots \\ \frac{\partial z(y_{10})}{\partial y_1} & \dots & \frac{\partial z(y_{10})}{\partial y_{10}} \end{pmatrix} \begin{pmatrix} 1 \\ \vdots \\ 1 \end{pmatrix} = \begin{pmatrix} \frac{\partial z(y_1)}{\partial y_1} & \dots & 0 \\ \vdots & \ddots & \vdots \\ 0 & \dots & \frac{\partial z(y_{10})}{\partial y_{10}} \end{pmatrix} \begin{pmatrix} 1 \\ \vdots \\ 1 \end{pmatrix} = \begin{pmatrix} \frac{\partial z(y_1)}{\partial y_1} \\ \vdots \\ \frac{\partial z(y_{10})}{\partial y_{10}} \end{pmatrix},$$

  this yields the derivatives $\partial z/\partial y$. The derivatives $\partial z/\partial x$ are obtained by the product of the Jacobian between y and x and the computed derivatives $\partial z/\partial y$.
  The variable, not stemming from another manipulation, is x and therefore referred to as the leaf node, i.e., `x.is_leaf` would be `True`. The default behavior is to only store the gradients with respect to the variables at leaf nodes. To counteract this behavior and, for example, store the gradient with respect to y, `y.retain_grad()`

can be specified. In addition, the entire computation graph is deleted after invoking `backward` unless `retain_graph = True` is specified. To perform automatic differentiation on already differentiated quantities, the computation graph must be tracked during differentiation, which is enabled with `create_graph = True`.

```
x = torch.linspace(0., 1., 10, requires_grad=True)
y = 2 * x
z = y ** 2

y.retain_grad()  # otherwise gradient with respect to y will be deleted

z.backward(torch.ones_like(z))

dzdx = x.grad  # gradient of z with respect to x: 8 * x
dzdy = y.grad  # gradient of z with respect to y: 2 * y = 4 * x
```

- The `backward` function is convenient within training loops, where scalar values are derived, thus not requiring input arguments. It operates as part of the `autograd` system, which is the core module for automatic differentiation in PyTorch. For more general-purpose automatic differentiation, direct use of the `autograd` module through the `grad` function can be beneficial. Note the `retain_graph = True` in the first call, which makes it possible to compute a second gradient afterward.

```
from torch.autograd import grad

x = torch.linspace(0., 1., 10, requires_grad = True)
y = 2 * x
z = y ** 2

dz_dx = grad(z, x, torch.ones_like(z), retain_graph=True)[0]
dz_dy = grad(z, y, torch.ones_like(z))[0]
```

- Finally, when the computation graph is no longer needed (e.g., when converting a torch tensor to a NumPy array), the graph can be detached with `z.detach()`.

Neural Networks

- Neural networks will be introduced in detail in Chap. 3. However, to clarify the syntax, neural networks are defined as models that take an input and provide an output. PyTorch contains many in-built models, which can be combined in a modular way to create more complex models. An example is the linear model, which performs a linear transformation.

$$y = \boldsymbol{w} \cdot \boldsymbol{x} + \boldsymbol{b} \tag{1.1}$$

The parameters $\boldsymbol{w}$, $\boldsymbol{b}$ determining the transformation are randomly initialized.

```
model = torch.nn.Linear(2, 3)  # input size: 2 and output
    size: 3
x = torch.tensor([[1., 2.]])  # input
y = model(x)  # output
```

- The parameters $\boldsymbol{w}$, $\boldsymbol{b}$ can be accessed in the following manner.

```
model.weight.data  # returns a tensor of shape (3, 2)
model.bias.data  # returns a tensor of shape (3)
```

GPU Acceleration

- Using torch tensors on a GPU is achieved by specifying the `device` during initialization.[9, 10]

```

a = torch.ones(4, device=device)  # cuda
b = torch.ones(4)  # cpu
```

- or by sending the tensor to the GPU.

```
b = b.to(device)  # cuda
```

- Models (or rather the parameters of the models) are transferable in the same manner.

```
model = torch.nn.Linear(1, 1)
model = model.to(device)
```

- The data can be returned to the central processing unit (CPU) in the same way. Alternatively, one can use `a.cpu()`. This is, for example, a prerequisite when transforming torch tensors to numpy arrays.

E.2 Computing Stress Fields with Tensors (C)

1. Consider a bilinear element, as depicted in Fig. 1.6, with the following shape functions (valid in the range -1 to 1):

---

[9] If you do not have access to a GPU, free GPUs can be accessed through Google collab [68].

[10] As acceleration hardware, this book assumes Nvidia GPUs, which are selected by specifying CUDA with `torch.cuda.is_available()` . For Apple's GPUs, the metal performance shaders (MPS) framework can be selected with `torch.backends.mps.is_available()`, which in this book is not explicitly specified for other use cases of GPUs.

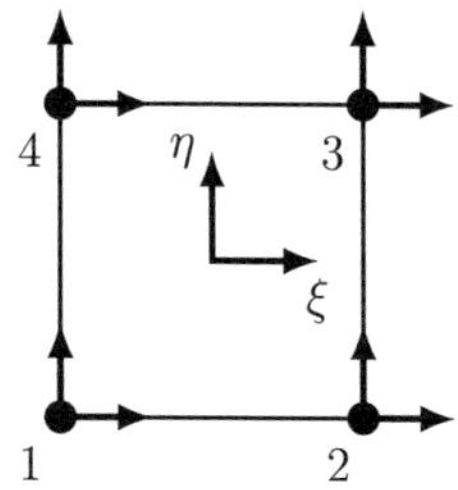

Fig. 1.6 Four node quadrilateral element with bilinear shape functions

$$
\begin{aligned}
N_1(\xi, \eta) &= \frac{1}{4}(1 - \xi)(1 - \eta), \\
N_2(\xi, \eta) &= \frac{1}{4}(1 + \xi)(1 - \eta), \\
N_3(\xi, \eta) &= \frac{1}{4}(1 + \xi)(1 + \eta), \\
N_4(\xi, \eta) &= \frac{1}{4}(1 - \xi)(1 + \eta).
\end{aligned}
$$

A displacement field $\boldsymbol{u}(\xi, \eta)$ can be established with the vector of displacement degrees of freedom $\boldsymbol{u}^i = (u_1^i, u_2^i)^\mathsf{T}$.

$$
\boldsymbol{u}(\xi, \eta) = \sum_{i=1}^{4} N_i(\xi, \eta)\boldsymbol{u}^i
$$

The goal is to implement the displacement field for given nodal displacement degrees of freedom and compute the divergence of the stresses to check the equilibrium. First compute the strains $\boldsymbol{\varepsilon}$ and stresses $\boldsymbol{\sigma}$ using

$$
\begin{aligned}
\boldsymbol{\varepsilon} &= \frac{1}{2}(\nabla \boldsymbol{u} + \nabla \boldsymbol{u}^\mathsf{T}), \\
\boldsymbol{\sigma} &= \boldsymbol{C} : \boldsymbol{\varepsilon},
\end{aligned}
$$

where $\boldsymbol{C}$ is the fourth-order material (constitutive) tensor, defined as

$$
\begin{aligned}
\boldsymbol{C}_{1111} = \boldsymbol{C}_{2222} &= \frac{E}{1 - \nu^2}, \\
\boldsymbol{C}_{1122} = \boldsymbol{C}_{2211} &= \frac{E\nu}{1 - \nu^2}, \\
\boldsymbol{C}_{1212} &= \frac{E}{2(1 + \nu)}, \\
\text{else } \boldsymbol{C}_{ijkl} &= 0.
\end{aligned}
$$

For the double contraction :, the function `torch.tensordot(C, strain)` is helpful. Next compute the stress divergence to determine the out-of-balance body force $\boldsymbol{p}$

$$\nabla \cdot \boldsymbol{\sigma} + \boldsymbol{p} = \boldsymbol{0}. \tag{1.2}$$

Consult the accompanying exercise file `exercise2_1.ipynb` for further guidance. As material properties use $E = 210\,000$, $\nu = 0.3$.
Hint: You can verify your implementation with the analytical derivatives of the shape functions.

2. Now consider the following displacement field

$$u_1(x_1, x_2) = 0, \tag{1.3}$$

$$u_2(x_1, x_2) = -1.7004 \cdot 10^{-7} x_2^2. \tag{1.4}$$

Assuming the material parameters $E = 210\,000$, $\nu = 0.3$, and a gravitational acceleration of $g = 9.81$ (acting in the negative x_2 direction), estimate the material's density ρ. No other load is acting on the specimen. Reuse the solution you obtained from the previous exercise, also provided as `exercise2_2.ipynb`.
Hint: Gravity acts as body load, computed as $p_1 = 0$, $p_2 = -\rho g$.

E.3 Stress Fields from Perceptrons (C)

We now build upon the PyTorch code snippets seen in Exercise E.1 with a simple one input, one output linear perceptron model. Perceptrons were invented in the 1960s and are still relevant for classification problems. Here, the same concept is used to encode a linear relationship and subsequently highlight the automatic differentiation capabilities of PyTorch to differentiate functions.

Now consider a one-dimensional displacement distribution. Here, the displacement $u(x)$ is to be obtained by a linear transformation from the coordinate x, i.e., $u(x) = wx + b$ using

```
model = torch.nn.Linear(1, 1)
```

The model parameters can be set with

```
model.weight.data = torch.tensor([[w]])
model.bias.data = torch.tensor([b])
```

Sample the displacement in the range $x \in [-1, 1]$ and compute the strains ε and stresses σ using the one-dimensional formulas

$$\varepsilon = \nabla u,$$
$$\sigma = E\varepsilon.$$

Use a Young's modulus of $E = 210\,000$ and refer to the file `exercise3.ipynb`.

References

1. Kasparov, Garri. 2018. *Deep thinking: Where machine intelligence ends and human creativity begins*. eng. London: John Murray. ISBN: 978-1-4736-5351-1.
2. Adomavicius, G., and A. Tuzhilin. 2005. Toward the next generation of recommender systems: A survey of the state-of-the-art and possible extensions. *IEEE Transactions on Knowledge and Data Engineering* 17(6): 734–749. ISSN: 1558-2191. https://doi.org/10.1109/TKDE.2005.99.
3. Krizhevsky, Alex, Ilya Sutskever, and Geoffrey E. Hinton. 2017. ImageNet classification with deep convolutional neural networks. *Communications of the ACM* 60(6): 84–90. ISSN: 0001-0782. https://doi.org/10.1145/3065386.
4. Szegedy, Christian et al. 2015. Going deeper with convolutions. In *2015 IEEE Conference on Computer Vision and Pattern Recognition (CVPR)*. Boston, MA, USA: IEEE, June 2015, 1–9. ISBN: 978-1-4673-6964-0. https://doi.org/10.1109/CVPR.2015.7298594.
5. Simonyan, K., and A. Zisserman. 2015. Very deep convolutional networks for large-scale image recognition. en. In *3rd International Conference on Learning Representations (ICLR 2015)*.
6. He, Kaiming et al. 2016. Deep residual learning for image recognition. In *2016 IEEE Conference on Computer Vision and Pattern Recognition (CVPR)*. Las Vegas, NV, USA: IEEE, June 2016, 770–778. ISBN: 978-1-4673-8851-1. https://doi.org/10.1109/CVPR.2016.90.
7. Silver, David et al. 2016. Mastering the game of Go with deep neural networks and tree search. en. *Nature* 529(7587): 484–489. ISSN: 1476-4687. https://doi.org/10.1038/nature16961.
8. Vinyals, Oriol et al. 2019. Grandmaster level in StarCraft II using multi-agent reinforcement learning. en. *Nature* 575(7782): 350–354. ISSN: 1476-4687. https://doi.org/10.1038/s41586-019-1724-z.
9. Andrychowicz, OpenAI: Marcin et al. 2020. Learning dexterous in-hand manipulation. en. *The International Journal of Robotics Research* 39(1): 3–20 (2020). ISSN: 0278-3649, 1741-3176. https://doi.org/10.1177/0278364919887447.
10. Levinson, Jesse et al. 2011. Towards fully autonomous driving: Systems and algorithms. In *2011 IEEE Intelligent Vehicles Symposium (IV)*. ISSN: 1931-0587. June 2011, 163–168. https://doi.org/10.1109/IVS.2011.5940562.
11. Bojarski, Mariusz et al. 2016. *End to End Learning for Self-Driving Cars*. arXiv:1604.07316 [cs]. Apr. 2016. https://doi.org/10.48550/arXiv.1604.07316.
12. Grigorescu, Sorin et al. 2020. A survey of deep learning techniques for autonomous driving. en. *Journal of Field Robotics* 37(3): 362–386. ISSN: 1556-4959, 1556-4967. https://doi.org/10.1002/rob.21918.
13. Vaswani, Ashish et al. 2017. Attention is all you need. In *Advances in Neural Information Processing Systems*, vol. 30. Curran Associates, Inc.
14. Devlin, Jacob et al. 2019. *BERT: Pre-training of Deep Bidirectional Transformers for Language Understanding*. arXiv:1810.04805 [cs]. May 2019. https://doi.org/10.48550/arXiv.1810.04805.
15. OpenAI. *GPT-4 Technical Report*. arXiv:2303.08774 [cs]. Mar. 2023. https://doi.org/10.48550/arXiv.2303.08774.
16. Touvron, Hugo et al. 2023. *Llama 2: Open Foundation and Fine-Tuned Chat Models*. arXiv:2307.09288. July 2023. https://doi.org/10.48550/arXiv.2307.09288.
17. Touvron, Hugo et al. 2023. *LLaMA: Open and Efficient Foundation Language Models*. arXiv:2302.13971. Feb. 2023. https://doi.org/10.48550/arXiv.2302.13971.
18. Jumper, John et al. 2021. Highly accurate protein structure prediction with AlphaFold. en. *Nature* 596(7873): 583–589. ISSN: 1476-4687. https://doi.org/10.1038/s41586-021-03819-2.
19. Shi, Zhan et al. 2020. *Improving Image Captioning with Better Use of Captions*. arXiv:2006.11807 [cs]. June 2020. https://doi.org/10.48550/arXiv.2006.11807.
20. Podell, Dustin et al. 2023. *SDXL: Improving Latent Diffusion Models for High-Resolution Image Synthesis*. arXiv:2307.01952 [cs]. July 2023.
21. Liu, Yixin et al. 2024. *Sora: A Review on Background, Technology, Limitations, and Opportunities of Large Vision Models*. arXiv:2402.17177 [cs]. Apr. 2024. https://doi.org/10.48550/arXiv.2402.17177.

22. Goodfellow, Ian et al. 2014. Generative adversarial nets. In *Advances in Neural Information Processing Systems*, vol. 27. Curran Associates, Inc.
23. Sohl-Dickstein, Jascha et al. 2015. Deep unsupervised learning using nonequilibrium thermodynamics. en. In *Proceedings of the 32nd International Conference on Machine Learning*. PMLR, June 2015, 2256–2265.
24. Ho, Jonathan, Ajay Jain, and Pieter Abbeel. 2020. Denoising diffusion probabilistic models. In *Advances in Neural Information Processing Systems*, vol. 33. Curran Associates, Inc., 6840–6851.
25. Nichol, Alex, and Prafulla Dhariwal. 2021. *Improved Denoising Diffusion Probabilistic Models*. arXiv:2102.09672 [cs, stat]. Feb. 2021. https://doi.org/10.48550/arXiv.2102.09672.
26. Russell, Stuart J., and Peter Norvig. 2022. *Artificial intelligence: A modern approach*. eng. Fourth edition, global edition. Pearson Series in Artificial Intelligence. Harlow: Pearson. ISBN: 978-1-292-40113-3.
27. Nick, Bostrom. 2014. *Superintelligence: Paths, dangers, strategies*. First edition. OCLC: ocn881706835. Oxford: Oxford University Press. ISBN: 978-0-19-967811-2.
28. Lambin, Philippe et al. 2012. Radiomics: Extracting more information from medical images using advanced feature analysis. en. *European Journal of Cancer* 48(4): 441–446. ISSN: 09598049. https://doi.org/10.1016/j.ejca.2011.11.036.
29. Liu, Wing Kam, Shaofan Li, and Harold S. Park. 2022. Eighty years of the finite element method: Birth, evolution, and future. en. *Archives of Computational Methods in Engineering* 29(6): 4431–4453. ISSN: 1886-1784. https://doi.org/10.1007/s11831-022-09740-9.
30. Faller, William E., and Scott J. Schreck. 1996. Neural networks: Applications and opportunities in aeronautics. en. *Progress in Aerospace Sciences* 32(5): 433–456. ISSN: 0376-0421. https://doi.org/10.1016/0376-0421(95)00011-9.
31. Bock, Frederic E., et al. A review of the application of machine learning and data mining approaches in continuum materials mechanics. en. *Frontiers in Materials* 6: 110. ISSN: 2296-8016. https://doi.org/10.3389/fmats.2019.00110.
32. Garnier, Paul et al. 2019. A review on deep reinforcement learning for fluid mechanics. en. In arXiv:1908.04127 [physics] (Aug. 2019). arXiv: 1908.04127.
33. Duraisamy, Karthik, Gianluca Iaccarino, and Heng Xiao. 2019. Turbulence Modeling in the age of data. *Annual Review of Fluid Mechanics* 51(1): 357–377. ISSN: 0066-4189, 1545-4479. https://doi.org/10.1146/annurev-fluid-010518-040547.
34. Baker, Nathan et al. 2019. *Workshop Report on Basic Research Needs for Scientific Machine Learning: Core Technologies for Artificial Intelligence*. en. Tech. rep. 1478744. Feb. 2019, 1478744. https://doi.org/10.2172/1478744.
35. Moosavi, Seyed Mohamad, Kevin Maik Jablonka, and Berend Smit. 2020. The role of machine learning in the understanding and design of materials. en. *Journal of the American Chemical Society* 142(48): 20273–20287. ISSN: 0002-7863, 1520-5126. https://doi.org/10.1021/jacs.0c09105.
36. Brunton, Steven, Bernd Noack, and Petros Koumoutsakos. 2020. Machine Learning for fluid mechanics. en. *Annual Review of Fluid Mechanics* 52(1). tex.ids= Brunton2020a arXiv: 1905.11075, 477–508. ISSN: 0066-4189, 1545-4479. https://doi.org/10.1146/annurev-fluid-010719-060214.
37. Cai, Shengze et al. 2021. Physics-informed neural networks (PINNs) for fluid mechanics: A review. en. *Acta Mechanica Sinica* 37(12): 1727–1738. ISSN: 1614-3116. https://doi.org/10.1007/s10409-021-01148-1.
38. Calzolari, Giovanni, and Wei Liu. 2021. Deep learning to replace, improve, or aid CFD analysis in built environment applications: A review. en. *Building and Environment* 206: 108315. ISSN: 0360-1323. https://doi.org/10.1016/j.buildenv.2021.108315.
39. Adler, Amir, Mauricio Araya-Polo, and Tomaso Poggio. 2021. Deep learning for seismic inverse problems: toward the acceleration of geophysical analysis workflows. *IEEE Signal Processing Magazine* 38(2): 89–119. ISSN: 1558-0792. https://doi.org/10.1109/MSP.2020.3037429.
40. Woldseth, Rebekka V., et al. 2022. On the use of artificial neural networks in topology optimisation. en. *Structural and Multidisciplinary Optimization* 65(10): 294. ISSN: 1615-1488. https://doi.org/10.1007/s00158-022-03347-1.

41. Regenwetter, Lyle, Amin Heyrani Nobari, and Faez Ahmed. 2022. Deep Generative models in engineering design: A review. en. In arXiv:2110.10863 [cs, stat] (Feb. 2022). arXiv: 2110.10863.
42. Shin, Seungyeon, Dongju Shin, and Namwoo Kang. 2023. Topology optimization via machine learning and deep learning: A review. en. *Journal of Computational Design and Engineering* 10(4): 1736–1766. ISSN: 2288-5048. https://doi.org/10.1093/jcde/qwad072.
43. Bishara, Dana, et al. 2023. A state-of-the-art review on machine learning-based multiscale modeling, simulation, homogenization and design of materials. en. *Archives of Computational Methods in Engineering* 30(1): 191–222. ISSN: 1886-1784. https://doi.org/10.1007/s11831-022-09795-8.
44. Rosenkranz, Max et al. 2023. A comparative study on different neural network architectures to model inelasticity. en. *International Journal for Numerical Methods in Engineering* (July 2023), nme.7319. ISSN: 0029-5981, 1097-0207. https://doi.org/10.1002/nme.7319.
45. Herrmann, Leon, and Stefan Kollmannsberger. 2024. Deep learning in computational mechanics: A review. en. *Computational Mechanics*. ISSN: 1432-0924. https://doi.org/10.1007/s00466-023-02434-4.
46. Yagawa, G., and Atsuya Oishi. 2023. *Computational mechanics with deep learning: An introduction*. eng. Cham, Switzerland: Springer. ISBN: 978-3-031-11847-0.
47. Yagawa, G., and H. Okuda. 1996. Neural networks in computational mechanics. en. *Archives of Computational Methods in Engineering* 3(4): 435–512. ISSN: 1886-1784. https://doi.org/10.1007/BF02818935.
48. Waszczyszyn, Zenon, and Leonard Ziemiański. 2001. Neural networks in mechanics of structures and materials – new results and prospects of applications. en. *Computers & Structures* 79(22): 2261–2276. ISSN: 0045-7949. https://doi.org/10.1016/S0045-7949(01)00083-9.
49. Yagawa, Genki, and Atsuya Oishi. 2021. *Computational Mechanics with Neural Networks*. en. Lecture Notes on Numerical Methods in Engineering and Sciences. Cham: Springer. ISBN: 978-3-030-66110-6 978-3-030-66111-3. https://doi.org/10.1007/978-3-030-66111-3.
50. Song, S. J., and L. W. Schmerr. 1992. Ultrasonic flaw classification in weldments using probabilistic neural networks. en. *Journal of Nondestructive Evaluation* 11(2): 69–77. ISSN: 1573-4862. https://doi.org/10.1007/BF00568290.
51. Yagawa, G. et al. 1993. Identification of crack shape hidden in solid by means of neural network and computational mechanics. en. In *Inverse Problems in Engineering Mechanics*, ed. Masataka Tanaka and Huy Duong Bui. International Union of Theoretical and Applied Mechanics. Berlin, Heidelberg: Springer, 213–222. ISBN: 978-3-642-52439-4. https://doi.org/10.1007/978-3-642-52439-4_21.
52. Psichogios, Dimitris C., and Lyle H. Ungar. 1992. A hybrid neural network-first principles approach to process modeling. en. *AIChE Journal* 38(10): 1499–1511. ISSN: 0001-1541, 1547-5905. https://doi.org/10.1002/aic.690381003.
53. Dissanayake, M. W. M. G., and N. Phan-Thien. 1994. Neural-network-based approximations for solving partial differential equations. en. *Communications in Numerical Methods in Engineering* 10(3): 195–201. ISSN: 10698299. https://doi.org/10.1002/cnm.1640100303.
54. Lagaris, I.E., A. Likas, and D.I. Fotiadis. 1998. Artificial neural networks for solving ordinary and partial differential equations. en. In *IEEE Transactions on Neural Networks* 9(5): 987–1000. ISSN: 10459227. https://doi.org/10.1109/72.712178.
55. Theocaris, Pericles S., and P. D. Panagiotopoulos. 1995. Generalised hardening plasticity approximated via anisotropic elasticity: A neural network approach. *Computer Methods in Applied Mechanics and Engineering* 125(1): 123–139. ISSN: 0045-7825. https://doi.org/10.1016/0045-7825(94)00769-J.
56. Furukawa, Tomonari, and Genki Yagawa. 1998. Implicit constitutive modelling for viscoplasticity using neural networks. en. *International Journal for Numerical Methods in Engineering* 43(2): 195–219. ISSN: 0029-5981, 1097-0207. https://doi.org/10.1002/(SICI)1097-0207(19980930)43:2<195::AID-NME418>3.0.CO;2-6.
57. Okuda, Hiroshi et al. 1998. Neural network-based parameter estimation for non-linear finite element analyses. *Engineering Computations* 15(1): 103–138. ISSN: 0264-4401. https://doi.org/10.1108/02644409810200721.

58. Takeuchi, Jun, and Yukio Kosugi. 1994. Neural network representation of finite element method. en. *Neural Networks* 7(2): 389–395. ISSN: 0893-6080. https://doi.org/10.1016/0893-6080(94)90031-0.
59. Yagawa, G., and H. Okuda. 1996. Finite element solutions with feedback network mechanism through direct minimization of energy functionals. en. *International Journal for Numerical Methods in Engineering* 39(5): 867–883. ISSN: 0029-5981, 1097-0207. https://doi.org/10.1002/(SICI)1097-0207(19960315)39:5<867::AID-NME886>3.0.CO;2-Q.
60. Topping, B. H. V., A. I. Khan, and A. Bahreininejad. 1997. Parallel training of neural networks for finite element mesh decomposition. In *Computers & Structures*. Computing in Civil and Structural Engineering 63(4): 693–707. ISSN: 0045-7949. https://doi.org/10.1016/S0045-7949(96)00082-X.
61. Rezende, Danilo Jimenez, Shakir Mohamed, and Daan Wierstra. 2014. *Stochastic Backpropagation and Approximate Inference in Deep Generative Models*. arXiv:1401.4082 [cs, stat]. May 2014. https://doi.org/10.48550/arXiv.1401.4082.
62. Kingma, Diederik P., and Max Welling. 2022. *Auto-Encoding Variational Bayes*. arXiv:1312.6114 [cs, stat]. Dec. 2022. https://doi.org/10.48550/arXiv.1312.6114.
63. Mnih, Volodymyr et al. 2015. Human-level control through deep reinforcement learning. en. *Nature* 518(7540): 529–533. ISSN: 1476-4687. https://doi.org/10.1038/nature14236.
64. McGreivy, Nick, and Ammar Hakim. 2024. Weak baselines and reporting biases lead to overoptimism in machine learning for fluid-related partial differential equations. en. *Nature Machine Intelligence* 6(10): 1256–1269. ISSN: 2522-5839. https://doi.org/10.1038/s42256-024-00897-5.
65. Paszke, Adam et al. 2019. PyTorch: An Imperative Style, High-Performance Deep Learning Library. In arXiv:1912.01703 [cs, stat] (Dec. 2019). arXiv: 1912.01703.
66. Nickolls, John et al. 2008. Scalable Parallel Programming with CUDA: Is CUDA the parallel programming model that application developers have been waiting for? en. *Queue* 6(2): 40–53. ISSN: 1542-7730, 1542-7749. https://doi.org/10.1145/1365490.1365500.
67. Harris, Charles R. et al. 2020. Array programming with NumPy. en. *Nature* 585(7825): 357–362. ISSN: 0028-0836, 1476-4687. https://doi.org/10.1038/s41586-020-2649-2
68. https://colab.google. en.

Chapter 2
Fundamental Concepts of Machine Learning

Abstract Machine learning algorithms are different from conventional algorithms as they improve solutions by means of optimization using experience, i.e., multiple data points. This chapter introduces the foundational principles of machine learning, including data structures, types of learning, optimization techniques, and challenges such as overfitting and regularization. The discussion focuses on fundamental algorithms like linear regression and gradient descent, laying the groundwork for understanding their use in both deep learning and computational mechanics. These concepts are presented in a simple and accessible manner, catering to readers with no experience in machine learning.

2.1 Definition

Nowadays, machine learning is arguably the most successful and widely used technique to address problems that cannot be solved by hand-crafted programs and has exceeded human-level performance in specific tasks [1–3]. In contrast to conventional algorithms following a predefined set of rules, a machine learning algorithm relies on a large amount of data that is observed in nature, hand-crafted by humans, or generated by another algorithm [4]. A more formal definition by Mitchell states that "*a computer program is said to learn from experience E with respect to some class of tasks T and performance measure P, if its performance at tasks T, as measured by P, improves with experience E*" [5]. Taking image recognition as an example, the task T is to classify previously unseen images, the performance measure P corresponds to the amount of correctly classified images, and the experience E includes all images that have been used to train the algorithm. Most machine learning algorithms can be decomposed into the following features: a dataset, a cost function, an optimization procedure, and a parameterized model [6], as depicted in Fig. 2.1. Generally, the cost function defines an optimization criterion by relating the data to the model parameters. Further, the optimization procedure searches for the model parameters that best represent the provided data. The key difference between machine learning and solving an optimization problem is that the optimized model is subsequently used for predictions on previously unseen data.

L. Herrmann et al., *Deep Learning in Computational Mechanics*,
https://doi.org/10.1007/978-3-031-89529-6_2

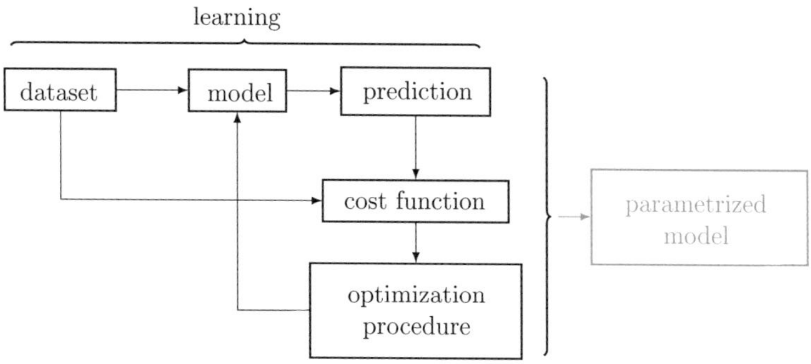

Fig. 2.1 Basic building blocks of most machine learning algorithms

2.2 Data Structure

A machine learning algorithm processes a dataset containing a collection of data points, often referred to as examples. Every example consists of one or more features describing the data point in a quantitative manner. In terms of notation, each example can be written as a vector $\boldsymbol{x}$, where each entry x_j corresponds to a feature of that example. To take advantage of fast implementations of matrix-vector calculus in modern programming languages, all examples are arranged in a so-called design matrix

$$\boldsymbol{X} = \begin{array}{c} \\ \text{example 1} \\ \text{example 2} \\ \vdots \\ \text{example } m \end{array} \begin{array}{c} \begin{array}{cccc} \text{feature 1} & \text{feature 2} & \cdots & \text{feature } n \end{array} \\ \begin{pmatrix} x_{11} & x_{12} & \cdots & x_{1n} \\ x_{21} & x_{22} & \cdots & x_{2n} \\ \vdots & \vdots & \ddots & \vdots \\ x_{m1} & x_{m2} & \cdots & x_{mn} \end{pmatrix} \end{array}.$$

Each row of the matrix represents an example, and each column corresponds to a feature describing the examples [6]. In the case of recognizing gray-scale images, every photo in the dataset is stored as one example vector in the design matrix. Assuming all pictures have a resolution of 50×50 pixels, then every example consists of 2 500 features storing the gray-scale value for each pixel.

It is common practice to split the data into three different subsets, namely into a training set, a validation set, and a test set. The training set, usually containing about $\sim$80% of the initial data, is used for the actual training, i.e., the optimization of weights and biases in the model. The validation set usually contains around $\sim$10% of the dataset and is held back to tune certain user-defined model parameters called hyperparameters, such as, e.g., the number of layers of a neural network. The remaining $\sim$10% are kept for the test set to evaluate the model's performance on unseen data, providing an unbiased evaluation of the model fit. The model cannot be modified by the test set data, as it otherwise would have indirectly been used to tune the model, thereby acting as the validation set. The given percentages are for guidance only since machine learning literature does not prescribe specific figures and leaves it to the practitioner how to concretely subdivide the data [7].

2.3 Types of Learning

Different types of learning exist, which are briefly presented in the following subsections.

2.3.1 *Supervised Learning*

Most problems solved by machine learning algorithms fall into the category of supervised learning [8]. In this context, "*supervised*" indicates that the algorithm is processing a labeled dataset. Thus, next to a design matrix $\tilde{\boldsymbol{X}}$, the dataset comprises a vector $\tilde{\boldsymbol{y}}$ with a label or target $\tilde{y}_i$ for each example $\tilde{\boldsymbol{x}}_i$. For instance, in image classification tasks, each image is paired with a certain category label. The supervised learning algorithm studies the dataset and learns to classify the images into the given categories by comparing its prediction with the given ground truth label.

2.3.2 *Unsupervised Learning*

The goal of unsupervised learning is to find a structure or, more precisely, the probability distribution in the provided data. The data is not labeled, and therefore, no explicit prediction is possible. However, it can be advantageous to apply unsupervised learning to large datasets in order to find inherent structures or repeating patterns in the data. For instance, anomaly detection algorithms are used to identify fraudulent credit card transactions that differ from the usual purchasing behavior of a customer [6].

2.3.3 *Semi-supervised Learning*

As the term suggests, semi-supervised learning combines the two preceding concepts. In cases where only few samples of the data are labeled, unsupervised learning helps to improve the performance of the supervised learning algorithm [6].

2.3.4 *Reinforcement Learning*

The basic idea of reinforcement learning is that an algorithm interacts with an environment to learn a certain decision behavior maximizing the expected average reward [4]. When the decision behavior is modeled by a neural network, the technique is referred to as deep reinforcement learning.[1] In general, reinforcement learning is used for problems involving sequential decision-making in order to fulfill a long-term goal. It is particularly useful when formulating a differentiable cost function is unviable. A group of Google researchers demonstrated the effectiveness of this technique on the example of playing the game of Go [9]. Solely by playing games against itself, the program reached superhuman abilities and was capable of beating the European Go champion.

2.4 Machine Learning Tasks

The following subsections give a small overview of machine learning tasks with a short introduction to suitable algorithms. This list is by far not complete—however, many problems that arise in practice can be related to one of the following categories.

2.4.1 *Regression*

Regression, depicted in Fig. 2.2a, is a supervised learning problem with the goal of predicting a numerical value. Essentially, a regression algorithm outputs a function that maps a given input to an output, usually in the form of a real number. An example is the prediction of house prices based on certain criteria like the area, number of rooms, or the age of the house. A more detailed explanation of linear regression can be found in Sect. 2.5. Further, it is possible to extend the algorithm for the use with higher-order polynomials as well as for the prediction of multiple outputs, known as multivariate regression. Decision tree algorithms and neural networks are also used for regression problems. The latter is introduced in Chap. 3.

2.4.2 *Classification*

Just like regression, classification, shown in Fig. 2.2b, belongs to the category of supervised learning. Instead of a numerical value, their output takes on a discrete form. In other words, a classification algorithm returns a function that assigns a

---

[1] As reinforcement learning is a major field in itself—only partially overlapping with deep learning, the topic is only briefly discussed in Sect. 10.5.

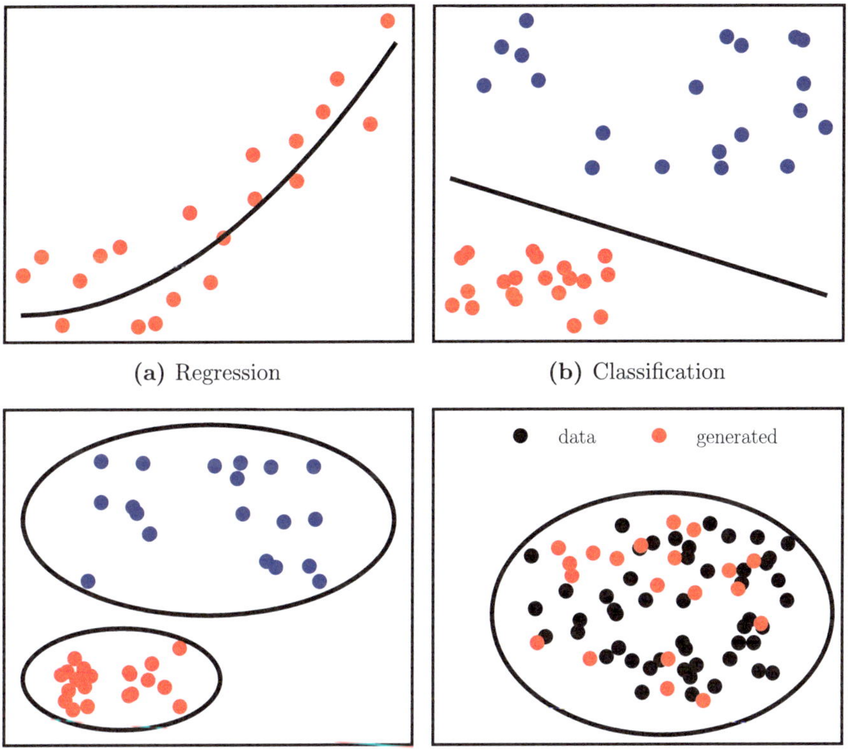

(a) Regression (b) Classification

(c) Clustering (d) Generative modeling

Fig. 2.2 The most prominent machine learning tasks

category to the provided input. Again, the example of classifying the content of an image falls into this class of problems. Even though the name suggests otherwise, logistic regression is a classification algorithm generating a binary output based on the logistic (or sigmoid) function (cf. Chap. 3). Other important algorithms for categorization are support vector machines (see Sect. 6.5) and decision trees like random forest or the more advanced gradient boosting approach. In terms of performance at complex tasks like image recognition, those algorithms have been surpassed by neural networks but are still used for simpler problems [6].

2.4.3 Clustering

Clustering, illustrated in Fig. 2.2c, differs from classification and regression as it is an unsupervised learning task. The algorithm gives feedback about which parts of the data share similarities and, therefore, belong to the same cluster. A popular choice for clustering is the k-means clustering algorithm (to be introduced in Sect. 6.4) that divides the incoming data into k different clusters of examples being close to each other [6].

2.4.4 Generative Modeling

Generative modeling, depicted conceptually in Fig. 2.2d, is a machine learning task focused on generating new data points that resemble a given dataset. Unlike regression and classification, which are concerned with predicting labels or values, generative modeling aims to learn the underlying distribution of the data, allowing it to create new examples that are statistically similar to the original dataset. A key application of generative modeling is the generation of synthetic images, seen in popular tools like DALL-E [10], Stable Diffusion [11], and Midjourney (for image generation), or Sora AI [12] (for video generation), or Shap-E [13] and retrieval-augmented generation [14] (for shape/geometry generation). The most prominent techniques are generative adversarial networks, variational autoencoders, diffusion models, and transformers, all of which will be covered in depth in Chap. 8.

2.5 Linear Regression

To get a better understanding of the way supervised learning works, the simple example of linear regression is introduced. It is well suited to explain concepts that are also applicable to more sophisticated machine learning algorithms. The goal of a regression model is to predict a scalar value $\hat{y} \in \mathbb{R}$ from an input vector $\boldsymbol{x} \in \mathbb{R}^n$. In general, linear regression can be written as

$$\hat{y} = \boldsymbol{w} \cdot \boldsymbol{x} + b = \sum_{j=1}^{n} w_j x_j + b, \tag{2.1}$$

where $\hat{y}$ represents the prediction aimed at matching a target $\tilde{y}$, $\boldsymbol{w}$ is a vector containing the weights, and $\boldsymbol{x}$ denotes the example vector (cf. Fig. 2.3). The constant b is called bias, referring to the case when either the weights or the input is close to zero, the output is biased toward b. Each weight w_j is a coefficient that gets multiplied with the corresponding feature x_j. The magnitude and sign of a weight decide the feature's importance to the prediction y. For instance, if a feature x_j has a corresponding weight w_j which is large in magnitude, small changes in x_j alter the prediction $\hat{y}$ by a great amount. The weights and the bias are the parameters of the model (expressed by $\boldsymbol{\Theta} = \{\boldsymbol{w}, b\}$), which determine how well it performs on the task of predicting target values for new data. In order to get an estimate of the model's future performance, a small part of the labeled data is held back to be used

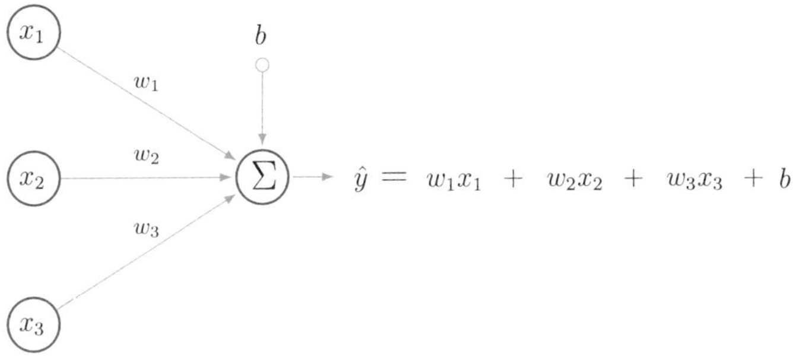

Fig. 2.3 Linear regression for a single example $\boldsymbol{x}$ with three input features

for evaluation. This fraction is called test set while the remaining part used for finding the optimal model parameters and optimal hyperparameters are correspondingly referred to as training set and validation set (cf. Sect. 2.2).

What still needs to be considered is a way to measure the model's performance. A common metric for a data-driven loss $\mathcal{L}_\mathcal{D}$ is the squared error loss. If $\hat{y}$ is the prediction from $\tilde{\boldsymbol{x}}$ and $\tilde{y}$ the ground truth, the squared error loss is simply defined as $(\tilde{y} - \hat{y}(\tilde{\boldsymbol{x}}))^2$. However, this only provides feedback for a single data point $(\boldsymbol{x}_i, y_i)$. To get one measurement for the performance on the whole dataset $\{(\tilde{\boldsymbol{x}}_i, \tilde{y}_i)\}_{i=1}^{m_\mathcal{D}}$, the mean of the squared error (MSE) of all examples $m_\mathcal{D}$ is calculated as follows:

$$\mathcal{L}_\mathcal{D} = \frac{1}{m_\mathcal{D}} \sum_{i=1}^{m_\mathcal{D}} (\tilde{y}_i - \hat{y}_i(\tilde{\boldsymbol{x}}))^2. \tag{2.2}$$

The closer the prediction $\hat{y}_i$ gets to the value of $\tilde{y}_i$, the smaller the error and the better the model's expected performance on examples it has never seen before.

The mean squared error is a popular choice for the cost function C. It has a continuous derivative and naturally penalizes large differences between the true target and the prediction [4]. In the case of linear regression, the squared error loss even leads to a convex optimization problem, meaning the cost function has only one particular minimum.

Now the question is how to find the optimal parameters $\boldsymbol{w}^*$ and b^* that yield a good prediction. This task can be interpreted as an optimization problem with the goal of minimizing the mean squared error. Given Eq. (2.2) and inserting Eq. (2.1), a cost function $C(\boldsymbol{w}, b)$ dependent on the parameters $\boldsymbol{w}$ and b is obtained

$$C(\boldsymbol{w}, b) = \frac{1}{m_\mathcal{D}} \sum_{i=1}^{m_\mathcal{D}} (\tilde{y}_i - (\boldsymbol{w} \cdot \tilde{\boldsymbol{x}}_i + b))^2. \tag{2.3}$$

With the given cost function, it is possible to formulate the optimization problem: Find the model parameters $\boldsymbol{w}$ and b that minimize the cost function $C(\boldsymbol{w}, b)$ or

$$\min_{\boldsymbol{w},b} C(\boldsymbol{w}, b) = \min_{\boldsymbol{w},b} \frac{1}{m_{\mathcal{D}}} \sum_{i=1}^{m_{\mathcal{D}}} (\tilde{y}_i - (\boldsymbol{w}\tilde{\boldsymbol{x}}_i + b))^2. \tag{2.4}$$

Finding the optimal parameters that lead to a good performance of the model can be interpreted as "*learning*".

The general concept that separates machine learning from solving an optimization problem is that the former adapts a model using the training examples and then evaluates it on the test set to emulate the future performance on unseen data. This leads to the important distinction between the training error $\mathcal{L}_{\mathcal{D}}^{(\text{train})}$

$$\mathcal{L}_{\mathcal{D}}^{(\text{train})} = \frac{1}{m_{\mathcal{D}}^{(\text{train})}} \sum_{i=1}^{m_{\mathcal{D}}^{(\text{train})}} (\tilde{y}_i^{(\text{train})} - \hat{y}_i^{(\text{train})})^2 \tag{2.5}$$

and the testing error $\mathcal{L}_{\mathcal{D}}^{(\text{test})}$

$$\mathcal{L}_{\mathcal{D}}^{(\text{test})} = \frac{1}{m_{\mathcal{D}}^{(\text{test})}} \sum_{i=1}^{m_{\mathcal{D}}^{(\text{test})}} (\tilde{y}_i^{(\text{test})} - \hat{y}_i^{(\text{test})})^2. \tag{2.6}$$

To better understand the described concept of linear regression, a one-dimensional linear regression example is given in Fig. 2.4. The goal is to fit the model to the data points by reducing the distance between the regression line and the training examples. Then, as the algorithm is fed with unlabeled data points x_{new}, the fitted line is used for the prediction of $\hat{y}_{\text{new}}$.

In the case of linear regression, it is possible to solve directly for the optimal model parameters w^* and b^* by analytically setting the gradient of the cost function C to zero. This closed-form solution to the minimization problem is also known as the normal equations [6]. These are often expressed in a matrix form to simplify the handling of large datasets. Then, the problem can be written as

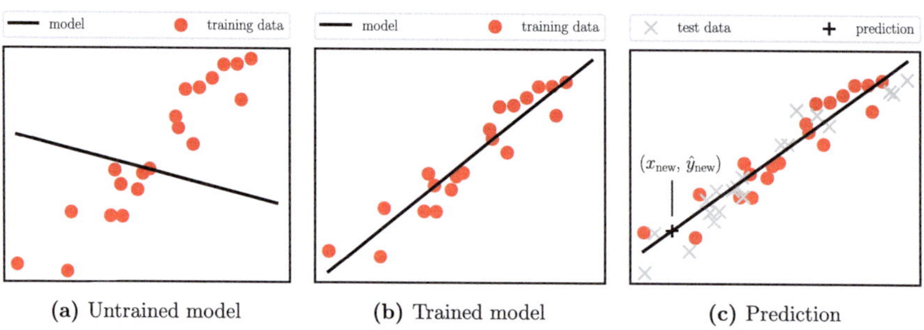

Fig. 2.4 Linear regression example. Left: The initial model does not fit the training data. Center: The model properly fits the training data. Right: The trained model is used to make a prediction $\hat{y}_{\text{new}}$ for an unknown example x_{new}. Figures are inspired by [6]

$$\tilde{X}\boldsymbol{\Theta} = \tilde{\boldsymbol{y}}, \tag{2.7}$$

where $\tilde{\boldsymbol{y}}$ is a vector with the $\tilde{y}$ values of the dataset and $\boldsymbol{\Theta} = (\boldsymbol{w}, b)^{\mathsf{T}}$ consists of the weights and the biases (stored together as a vector) to be determined. The matrix $\tilde{X}$ contains the corresponding features evaluated at each $\tilde{\boldsymbol{x}}$ value. As the dataset is greater than the number of weights and the bias together, the system is overdetermined, and the matrix $\tilde{X}$ is thereby not symmetric and not invertible. The goal is to minimize the residuum vector's magnitude

$$\boldsymbol{R} = \tilde{\boldsymbol{y}} - \tilde{X}\boldsymbol{\Theta}. \tag{2.8}$$

This is done by finding the sum of least squares and minimizing it by setting the derivative with respect to $\boldsymbol{\Theta}$ to zero, which is discussed in further detail in [15]. This results in the following expression for the optimal weights and bias, i.e., the normal equations:

$$\boldsymbol{\Theta}^* = \begin{pmatrix} \boldsymbol{w}^* \\ b^* \end{pmatrix} = (\tilde{X}^{\mathsf{T}}\tilde{X})^{-1}\tilde{X}^{\mathsf{T}}\tilde{\boldsymbol{y}}. \tag{2.9}$$

While this closed-form solution is well known, no closed-form solution is known for many practically relevant problems. In these cases, using a more generally applicable optimization technique is favorable. At the end of this chapter, Sect. 2.8 introduces the gradient descent approach that lays the foundation for many optimization algorithms in modern machine learning problems. Gradient descent techniques allow the iterative search for a minimum, considering a large number of parameters.

2.6 Overfitting Versus Underfitting

The previous section has introduced the test error, in particular, $\mathcal{L}_{\mathcal{D}}^{(\text{test})}$ (Eq. (2.6)), to measure how well a model is expected to perform when confronted with new inputs. The ability to generate good predictions for previously unseen data is called generalization, and the associated error is called the generalization error. Therefore, the test error is also considered to be an estimate of the generalization error.

Supervised learning algorithms like linear regression are based on the idea that a low training error also leads to a small test error. Statistical learning theory provides some assumptions to support this idea [6]:

1. The data contains all the necessary information to solve the problem.
2. Examples in each dataset are independent of each other.
3. The training and test sets are identically distributed, meaning each example is generated with the same probability distribution.

Assuming all given statements are true, the training error ought to be the same as the test error. Since, in reality, the data generating probability distribution is unknown a priori, the test error is usually higher than the training error. Figure 2.5 illustrates

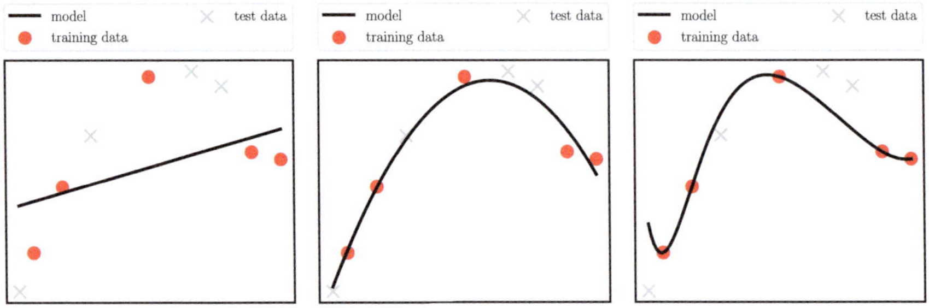

(a) Underfitting (b) Ideal fitting (c) Overfitting

Fig. 2.5 Representation of the underfitting and overfitting problem on the example of a one-dimensional regression. Figures inspired by [4, 6]

the two main reasons. Firstly, the capacity of a model can be too high, meaning the algorithm chooses a complex function that fits the training data perfectly but fails to generalize because it overestimates the importance of noisy data. This behavior is called overfitting (cf. Fig. 2.5c). Secondly, a model with a very low capacity is applied to a highly non-linear problem. For example, if the hypothesis space of a model only contains linear functions, it is not able to represent data that follows quadratic or cubic functions. This case describes underfitting (cf. Fig. 2.5a).

The capacity of a model, more precisely, its ability to fit a wide variety of functions, plays an important role in terms of performance. Choosing the capacity in a way that suits the complexity of the problem or, in other words, finding the balance between underfitting and overfitting is a central challenge in machine learning [6]. The relationship between the training error and the generalization error (estimated by the testing error) with respect to the capacity is depicted in Fig. 2.6. As indicated, the optimal capacity is reached when the generalization error is as low as possible.

A way to tackle the problem of underfitting (cf. Fig. 2.5a) is to increase the set of functions an algorithm is allowed to select. Continuing with the example of linear regression, we assume a simple model with a single input x

$$\hat{y} = wx + b \tag{2.10}$$

as in Eq. (2.15). The model can be easily extended to include polynomials. Adding x^2 as a new feature, a second-order polynomial regression of the following form is obtained:

$$\hat{y} = w_1 x^2 + w_2 x + b, \tag{2.11}$$

where $x_1 = x^2$ and $x_2 = x$. Introducing more polynomials to the model not only increases the number of features x_i but also introduces the number of parameters w_i. Hence, the algorithm has more possibilities to tune and adapt the model to fit the target function appropriately (cf. Fig. 2.5b). Yet, if the polynomial degree and, therefore, the capacity becomes too large, the model starts to overfit [6]. This is shown in Fig. 2.5c for the case of regression with a polynomial degree of $p = 7$

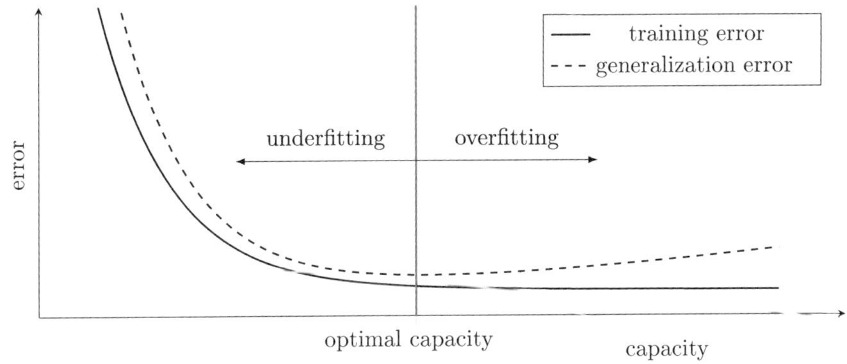

Fig. 2.6 Generalization error (practically estimated by the testing error) and training error in relation with the capacity of a model. Figure inspired by [6, 16, 17]

$$\hat{y} = \sum_{i=1}^{p} \left(w_i x^i\right) + b. \tag{2.12}$$

In the literature, over-, and underfitting are often associated with the statistical terms bias[2] and variance. Both concepts are illustrated by Fig. 2.7. A high bias means the model produces many mistakes on the training set, so it is equivalent to underfitting. Coming from statistics, the term variance describes the model's sensitivity to changes in the dataset. If the variance is high, small deviations in the training data result in a very different model. This behavior is closely related to overfitting [4], as a high sensitivity to the training set, i.e., a large variance, leads to large discrepancies between training and testing errors. A small variance is favored over a small bias, as the quality of the prediction should be reliable, i.e., as consistent as possible.[3] Thus, it is preferable to make a consistent error in terms of the bias, as illustrated by Fig. 2.7.

2.7 Regularization

Overcoming underfitting or overfitting, i.e., finding the appropriate capacity for a model, is one of the most challenging tasks in machine learning. The previous section showed a possibility to deal with underfitting by increasing the number of features x_i. In the same way, overfitting can be reduced when features are removed from the

[2] Not to be confused with the bias model parameter b.

[3] This is equally applicable to human decision-making across industries, where it is preferable to make a consistent error (bias) instead of making noisy decisions, as discussed by Kahneman et al. [18].

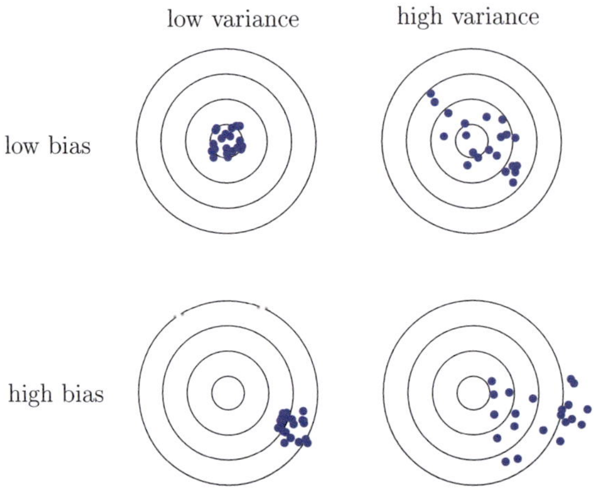

Fig. 2.7 Demonstration of the statistical concepts of variance and bias, in which high bias and low variance are preferred over low bias and high variance. In other words, consistent errors are preferred over unpredictable, i.e., varying errors

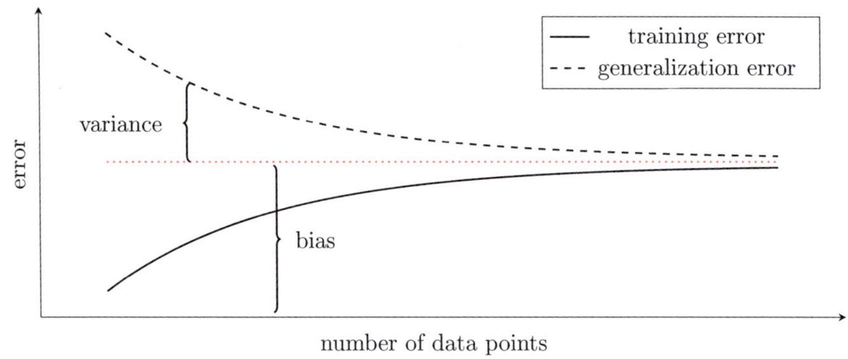

Fig. 2.8 Influence of dataset size on training and generalization error. The larger the amount of available data, the closer both error measurements get. The distance between the training error and the asymptote marked in red can be interpreted as the bias. In the same way, the distance between the generalization error and the asymptote denotes the variance. Figure inspired by [17, 19]

model. However, this option is not very popular since removing features means that information about the problem gets lost. Another countermeasure against overfitting is adding more examples to the training set [6]. In most cases, it is not feasible or simply impossible to gather more data about the problem. Conversely, removing training data can help in the case of underfitting. Yet again, this leads to a loss of information. The influence of the training set size on the error measures is illustrated in Fig. 2.8.

Since the complexity of the underlying problem is usually unknown, it is hardly possible to choose a model with the appropriate capacity beforehand. So, another approach to overcome overfitting is to retain a high capacity and introduce a control mechanism called regularization that prefers the selection of certain functions over

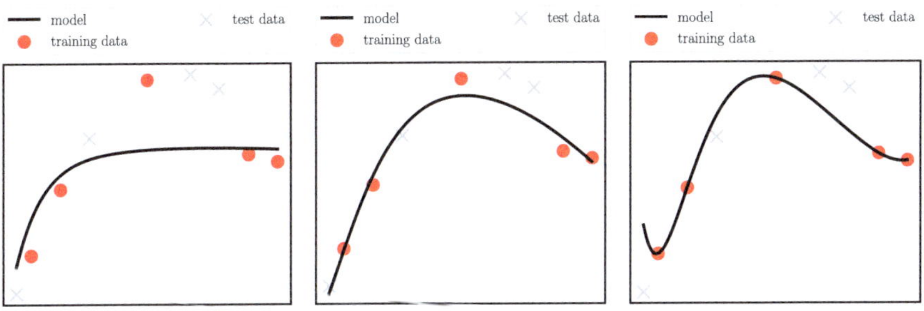

(a) Excessive regularization (b) Proper regularization (c) No regularization ($\lambda \to 0$)

Fig. 2.9 Representation of regularization. Figures inspired by [6, 20]

others. For instance, the cost function for linear regression (cf. Eq. (2.3)) can be modified to include an additional term that penalizes large weights

$$\tilde{C}(\boldsymbol{w}, b) = C(\boldsymbol{w}, b) + \lambda ||\boldsymbol{w}||_1, \tag{2.13}$$

where $||\boldsymbol{w}||_1$ denotes the L^1-norm of the weight vector $\boldsymbol{w}$. Adding a term proportional to the magnitude of the weights forces the algorithm to select a model with smaller weights. In the case of linear or polynomial regression, the weights can be understood as the coefficients determining the slope of the function. For instance, the model depicted in Fig. 2.9c exhibits large oscillations, which indicates an underlying function with large slope coefficients. Hence, penalizing large weights is a way to reduce the slope coefficients and the occurrence of oscillations. The influence of the penalty term is controlled by λ. If λ equals zero, the algorithm simply returns the initial linear regression model (cf. Fig. 2.9c). Contrarily, a large value for λ leads to extremely small weights, which basically eliminates the corresponding features, resulting in a very simple or sparse model (cf. Fig. 2.9a). In practice, this technique called L^1 or LASSO regularization is used for feature selection, a method to remove unimportant features from a model. A similar approach is L^2 or ridge regularization, which adds the square of the L^2-norm of the weights instead of the sum of the absolute values. For linear regression, the regularized cost function can be written as

$$\tilde{C}(\boldsymbol{w}, b) = C(\boldsymbol{w}, b) + \lambda \boldsymbol{w} \cdot \boldsymbol{w}. \tag{2.14}$$

An advantage of L^2 regularization is that the penalty term is differentiable [4]. This gains more importance when used in combination with the gradient-based optimization algorithms, which are introduced in Sect. 2.8.

Apart from model selection and adding a penalty term to the objective function, many other techniques exist to implicitly and explicitly express preferences for a certain solution. A few more are explained in Chapter 3 in the context of neural networks. Overall, these approaches can be summarized under the term regularization. A definition by Goodfellow states: "*Regularization is any modification we make to a learning algorithm that is intended to reduce its generalization error (low variance) but not its training error (low bias)* [6]." Accordingly, regularization is also known as the bias-variance trade-off [4].

When applying L^1 or L^2 regularization to a model, a new parameter λ is added to the objective function. In contrast to weights and bias, λ is not part of the optimization objective and thus has to be set manually by the user. Generally, all external settings controlling the behavior of an algorithm are called hyperparameters [4]. They give the opportunity to further improve the performance by doing a so-called hyperparameter tuning. It is for this reason that a new subset from the training data has to be selected since the test set cannot be used for any parameter optimization. Therefore, the training data is split into a set for training the standard model parameters and a validation set used for tuning the hyperparameters. The importance of a validation set is illustrated by the polynomial regression example in Fig. 2.10. Using the validation data, the polynomial degree $p = 8$ is identified as the best possible hyperparameter (Fig. 2.10a). However, overfitting due to the hyperparameter tuning can be observed when considering the test data. Figure 2.10b with $p = 8$ shows slight oscillations, which lead to a higher mean squared error for the test data than for $p = 3$ in Fig. 2.10c. As model selection and tuning are also prone to overfitting, the test set must remain unused until the final evaluation of the machine learning model to retain an accurate model assessment.

2.8 Optimization Techniques

Optimization is a cornerstone of many machine learning algorithms, playing a critical role in shaping their performance and effectiveness. At its core, the objective of optimization is to minimize or maximize an optimization criterion or objective function $C(\boldsymbol{\Theta})$ by altering $\boldsymbol{\Theta}$ [6]. In the context of machine learning, this function is often referred to as cost function, where $\boldsymbol{\Theta}$ denote the parameters of a model, usually

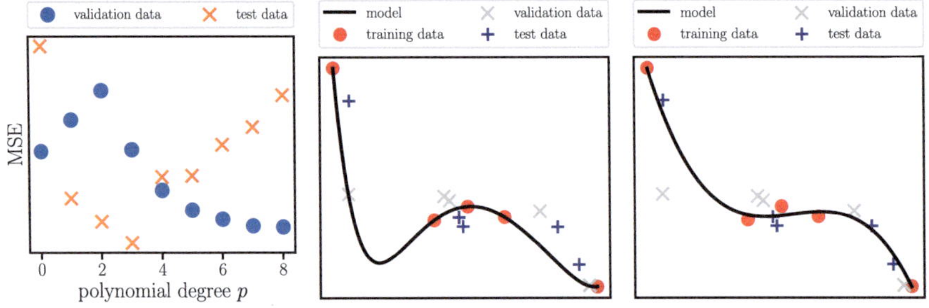

(a) Mean squared error for different polynomial degrees p

(b) $p = 8$

(c) $p = 3$

Fig. 2.10 Overfitting during hyperparameter tuning. Using the validation data, a polynomial degree of $p = 8$ is identified as the best possible hyperparameter. This has, however, led to overfitting, which can be quantified using the test data the weights $\boldsymbol{w}$ and biases b. The ultimate goal is efficiently determining the optimal parameters $\boldsymbol{\Theta}^*$. This challenge has been the focus of extensive research, resulting in various optimization methods. This section will delve into the basics of gradient descent as a foundational concept in Sect. 2.8.1 before exploring two of the most frequently encountered optimization algorithms (including their variations) in deep learning: Adam, discussed in Sect. 2.8.2, and L-BFGS, presented in Sect. 2.8.3.

2.8.1 Gradient Descent

When the optimization criterion is differentiable, a popular choice is the iterative gradient descent algorithm. A gradient descent approach finds a local minimum of a function by taking steps proportional to the negative gradient of the function at a given point. The gradient points in the direction of the steepest ascent. Hence, a small step in the opposite direction leads to a minimization of the function.

The algorithm is commonly used for neural networks but can also be applied to support vector machines or linear regression. In the latter case, the optimization criterion is convex, meaning the function only has one global minimum. For neural networks, the optimization problem is non-convex, and hence, it is more likely to converge to a local minimum [6].

As mentioned in Sect. 2.5, it is possible to analytically solve for the optimal parameters. Nevertheless, it serves as a suitable example to explain the basic steps of gradient descent. To keep things simple, a linear regression model with a scalar input x, a corresponding weight w and a scalar bias b is chosen

$$\hat{y} = wx + b. \tag{2.15}$$

Again, the goal is to find the optimal parameters w^* and b^* that minimize the mean squared error. The resulting cost function can be written as

$$C = \frac{1}{m_{\mathcal{D}}} \sum_{i=1}^{m_{\mathcal{D}}} (\tilde{y}_i - (w\tilde{x}_i + b))^2. \tag{2.16}$$

Now, the cost function's partial derivatives for each parameter have to be calculated

$$\begin{aligned} \frac{\partial C}{\partial w} &= \frac{1}{m_{\mathcal{D}}} \sum_{i=1}^{m_{\mathcal{D}}} -2\tilde{x}_i \left(\tilde{y}_i - (w\tilde{x}_i + b)\right), \\ \frac{\partial C}{\partial b} &= \frac{1}{m_{\mathcal{D}}} \sum_{i=1}^{m_{\mathcal{D}}} -2 \left(\tilde{y}_i - (w\tilde{x}_i + b)\right). \end{aligned} \tag{2.17}$$

If more parameters are involved, it is helpful to rewrite the partial derivatives in the form of the gradient $\nabla_{\boldsymbol{\Theta}} C$. The gradient is nothing more than a vector storing all the partial derivatives with respect to all function parameters. During each iteration step or epoch, to use machine learning terminology, the parameters of the model get updated according to the following rules:

$$
\begin{aligned}
w &\leftarrow w - \alpha \frac{\partial C}{\partial w}, \\
b &\leftarrow b - \alpha \frac{\partial C}{\partial b}.
\end{aligned} \tag{2.18}
$$

In the first step, the parameters are initialized as zero and then updated in each epoch until convergence is reached. The learning rate α controls the step size of each update [4]. A simple Python implementation of a one-dimensional linear regression with gradient descent is provided below, where `w` and `b` correspondingly converge towards 2 and 3 (the parameters defining the linear function `y`).

```
# training data
x = np.linspace(0, 1, 10)
y = 2 * x + 3  # linear function to be approximated

# hyperparameters
lr = 1e-1  # learning rate
epochs = 100  # number of iterations

# initial model
w = 1  # weight
b = 1  # bias

# training loop
for epoch in range(epochs):
    yPred = w * x + b  # prediction
    cost = np.mean((yPred - y) ** 2)
    dcostdw = np.mean(2 * x * (yPred - y))
    dcostdb = np.mean(2 * (yPred - y))
    w -= lr * dcostdw  # gradient descent update
    b -= lr * dcostdb  # gradient descent update
```

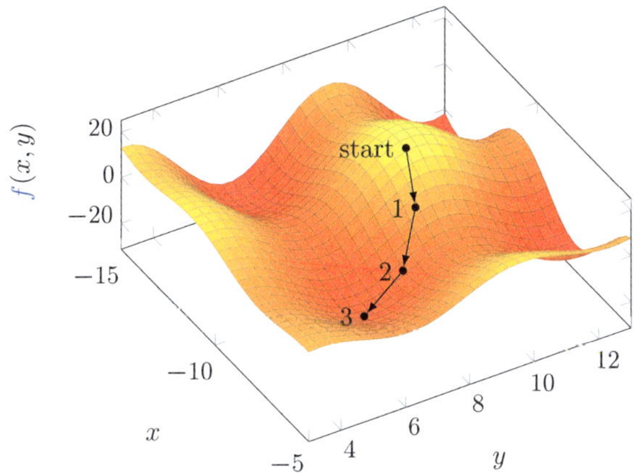

Fig. 2.11 Gradient descent on a surface defined by $f(x, y) = y\sin(x) + x\cos(y)$. Depending on the point of initialization, gradient descent follows the steepest direction toward a minimum of the underlying function

The general algorithm is summarized in Algorithm 1. A graphical representation of gradient descent is provided in Fig. 2.11. It illustrates how the algorithm follows the direction of the steepest descent toward a minimum. The outcome is highly dependent on the initialization of the starting point. The algorithm might descend into another local minimum if the starting point is initialized at a different location. This highlights the difficulties when finding the global minimum in non-convex optimization problems.

Algorithm 1 Gradient descent

Require: dataset $\tilde{\boldsymbol{x}}$, $\tilde{\boldsymbol{y}}$, number of epochs n, step size α, model f
  initialize the model $f(x; \boldsymbol{\Theta})$
  for all n **do**
    Compute the cost function $C(f(\tilde{\boldsymbol{x}}; \boldsymbol{\Theta}), \tilde{\boldsymbol{y}})$
    Compute the gradient $\nabla_{\boldsymbol{\Theta}} C$
    Update the model parameters $\boldsymbol{\Theta} \leftarrow \boldsymbol{\Theta} - \alpha \nabla_{\boldsymbol{\Theta}} C$
  end for

A major drawback of gradient descent is its sensitivity to the choice of the learning rate α. If α is too large, the algorithm starts to oscillate or even fails to converge at all. Conversely, a small α value leads to an extremely low convergence rate [4].

Another issue is that with an increasing number of examples, the computational effort per epoch grows as well because all partial derivatives are evaluated for the entire training set of size $m_{\mathcal{D}}$. Since the whole training set is used, the method described above is often referred to as full-batch gradient descent. Alternatively, the parameter update can be performed instantly after computing the gradient for a single example in the training data. This introduces stochasticity to the algorithm because the gradient of a single example might indicate a substantially different direction than the gradient computed for the complete training batch [6]. However, in practice, stochastic gradient descent—sometimes called online gradient descent often shows better convergence properties, especially when combined with an adaptive learning rate. This is possible due to the stochasticity of the gradient, which enables escapes from local minima in strongly non-convex optimization landscapes. A popular and widely used compromise between full-batch and stochastic gradient descent is mini-batch gradient descent. The idea is to find an approximation of the gradient by evaluating the gradient just for a small sample of the data, a so-called mini batch. Instead of evaluating the gradient for the entire training set, it is replaced by an estimator computed on a sample with a fixed size. Due to the very crude approximation of the gradient by stochastic gradient descent, large oscillations occur during the optimization procedure. By contrast, the full-batch gradient descent approaches the optimum in a straighter manner only depending on the choice of the learning rate, as shown in Fig. 2.12. The mini-batch gradient descent path lies between the two extremes. The differences between the implementation of the three possible optimization strategies are described in Sect. 3.6 in the context of neural networks.

2.8.2 Adam

Although gradient descent is a simple and intuitive optimization algorithm, its high sensitivity to hyperparameters, such as the learning rate, can make it less practical for optimizing complex deep learning models. This limitation led to the development of adaptive optimization algorithms such as Adam [21]. Adam, which stands for adaptive moment estimation, combines the concepts of momentum and adaptive learning rates during gradient descent.

2.8.2.1 Gradient Descent with Momentum

Gradient descent with momentum [22] extends the update rule from Eq. (2.18) by incorporating a momentum term $\boldsymbol{v}_t$, which captures the direction of the previous

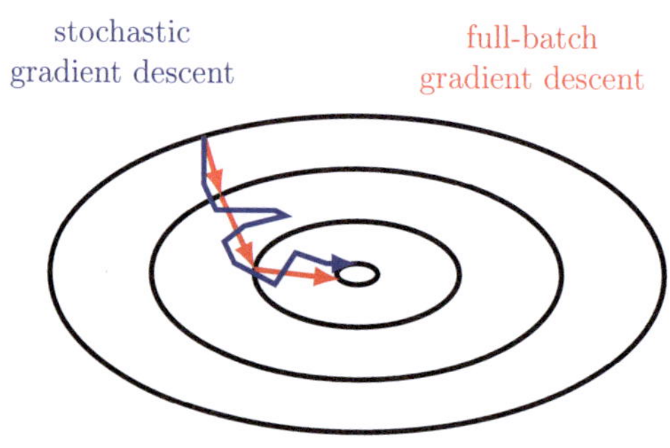

Fig. 2.12 Stochastic gradient descent versus full-batch gradient descent

gradients. The update rule becomes:

$$\boldsymbol{v}_{t+1} = \eta \boldsymbol{v}_t + \nabla_{\boldsymbol{\Theta}} C(\boldsymbol{\Theta}), \tag{2.19}$$

$$\boldsymbol{\Theta}_{t+1} = \boldsymbol{\Theta}_t - \alpha \boldsymbol{v}_{t+1}. \tag{2.20}$$

Here, $\boldsymbol{\Theta} = \{\boldsymbol{w}, b\}$ represents both the weights $\boldsymbol{w}$ and bias b. This formulation introduces an additional hyperparameter η, which controls the exponential decaying influence of the previous gradients in the momentum term $\boldsymbol{v}_t$. Here, $\boldsymbol{v}_t$ is analogous to the velocity towards the solution, and η is analogous to friction slowing that motion in the absence of a gradient.

Incorporating momentum in gradient descent offers two primary advantages. First, it reduces sensitivity to noise, as the momentum term $\boldsymbol{v}_t$ smooths the update across previous gradients. This is particularly beneficial for stochastic gradient descent, where noisy gradient estimates may arise, as seen in Fig. 2.12. Second, momentum improves the likelihood of escaping local minima, as the retained momentum at a local minimum can help facilitate an escape; see Fig. 2.13.

It is worth mentioning a popular extension called the Nesterov accelerated gradient [23]. This method obtains a more accurate estimate of the momentum in Equation (2.19) through a look-ahead approach. Instead of calculating the gradient at the current position, it computes the gradient at the approximate next step $\tilde{\boldsymbol{\Theta}} = \boldsymbol{\Theta}_t - \alpha\eta \boldsymbol{v}_t$ (approximated in terms of momentum). The alternative momentum estimate is given by

$$\boldsymbol{v}_{t+1} = \eta \boldsymbol{v}_t + \nabla_{\tilde{\boldsymbol{\Theta}}} C(\tilde{\boldsymbol{\Theta}}). \tag{2.21}$$

The look-ahead approach in the Nesterov accelerated gradient allows the algorithm to be more aware of the upcoming changes in the objective function, thus making better and more informed updates. This results in faster convergence and improved optimization performance. For an in-depth comparison of gradient descent with momentum and Nesterov accelerated gradient, see [24].

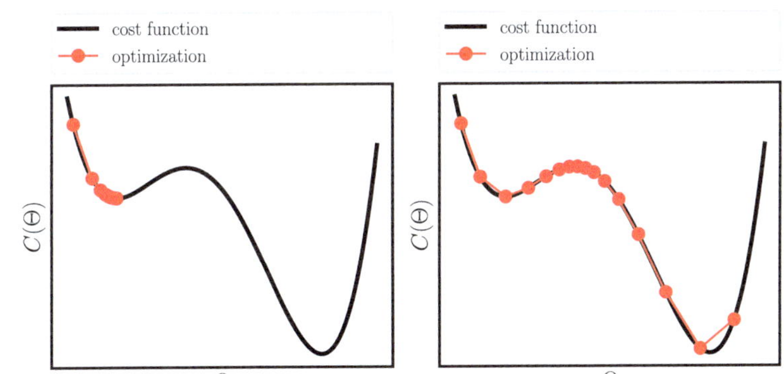

(a) Gradient descent

(b) Gradient descent with momentum

Fig. 2.13 Momentum allows the optimizer to escape from the local minimum

2.8.2.2 AdaGrad

The adaptive gradient algorithm, termed AdaGrad [25], addresses the problem of learning rate decay. Typically, the learning rate is fixed across all parameter updates, which can result in suboptimal convergence for learning rates that are too low—or unstable convergence for learning rates that are too large. This is illustrated in Fig. 2.14, where a too-large learning rate leads to oscillations around the minimum, and a small learning rate leads to extremely slow convergence.

AdaGrad solves this problem by adaptively scaling the learning rate α for each parameter, based on the history $\boldsymbol{\Theta}_t$ of the gradient updates for that parameter. The scaling is performed by the weighted average of the squared gradients $\bar{g}_t^2$, effectively determining the learning rate for each individual weight. The weighted average of the squared gradients is computed with all previous gradients:

$$\bar{g}_t^2 = \sum_{\tau=1}^{t} [\nabla_{\boldsymbol{\Theta}_\tau} C(\boldsymbol{\Theta}_\tau)]^2, \tag{2.22}$$

where the square acts in an element-wise fashion. This leads to the following update rule with an adaptive learning rate:

$$\boldsymbol{\Theta}_{t+1} = \boldsymbol{\Theta}_t - \frac{\alpha}{\sqrt{\bar{g}_t^2 + \epsilon}} \nabla_{\boldsymbol{\Theta}_t} C(\boldsymbol{\Theta}_t). \tag{2.23}$$

Here ϵ is a small numerical stabilization parameter ensuring that the denominator is never zero.

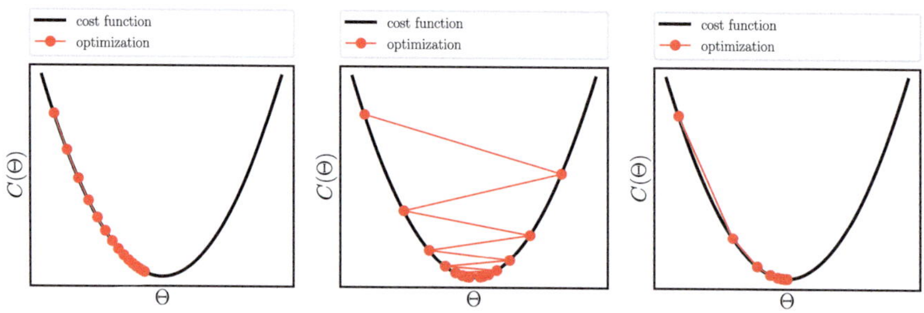

(a) Small learning rate (b) Large learning rate (c) Adaptive learning rate with AdaGrad

Fig. 2.14 Convergence is sensitive to the choice of learning rate. A small learning rate leads to a slow convergence, while a large learning rate leads to oscillations, thus impeding the convergence. Adapting the learning rate during training using AdaGrad mitigates the sensitivity and leads to faster convergence

2.8.2.3 RMSprop

The problem with AdaGrad is a fast reduction in the learning rate due to the accumulation of the squared gradients over the entire training process—and employed in the denominator, which can slow down or even prevent convergence. Therefore, root mean square propagation, commonly referred to as RMSprop, extends the idea of AdaGrad by using a moving average of the squared gradients instead. Practically, this is achieved with an exponentially decaying average $\tilde{g}_t^2$ with a decay hyperparameter ρ, similar to the exponential decaying momentum term in Eq. (2.19) with η.

$$\tilde{g}_t^2 = \rho \tilde{g}_{t-1}^2 + (1-\rho)\left[\nabla_{\boldsymbol{\Theta}_t} C(\boldsymbol{\Theta}_t)\right]^2 \tag{2.24}$$

The parameter update rule, equivalent to Equation (2.23) is then given by

$$\boldsymbol{\Theta}_{t+1} = \boldsymbol{\Theta}_t - \frac{\alpha}{\sqrt{\tilde{g}_t^2 + \epsilon}} \nabla_{\boldsymbol{\Theta}_t} C(\boldsymbol{\Theta}_t). \tag{2.25}$$

Both AdaGrad and RMSprop provide adaptive learning rates for each parameter, addressing the challenges of fixed learning rates. While AdaGrad may suffer from a fast reduction in the learning rate, RMSprop mitigates this issue by using a moving average of the squared gradients, leading to more stable convergence and improved optimization performance.

2.8.2.4 Adam

Integrating the strengths of gradient descent with momentum and RMSprop, the Adam optimization algorithm [21] emerges as a robust and efficient optimization algorithm within deep learning. The momentum is captured in terms of the first statistical moment of the gradients during optimization (compare with Eq. (2.19))

$$\boldsymbol{m}_t = \beta_1 \boldsymbol{m}_{t-1} + (1-\beta_1) \nabla_{\boldsymbol{\Theta}_t} C(\boldsymbol{\Theta}_t), \tag{2.26}$$

while the moving average of the squared gradients from RMSprop is captured in terms of the second statistical moment (compare with Eq. (2.24))

$$\boldsymbol{n}_t = \beta_2 \boldsymbol{n}_{t-1} + (1-\beta_2)\left[\nabla_{\boldsymbol{\Theta}_t} C(\boldsymbol{\Theta}_t)\right]^2. \tag{2.27}$$

Thus, $\boldsymbol{m}_t$ and $\boldsymbol{n}_t$ are estimates of the statistical moments. However, since the moments are initialized with $\boldsymbol{m}_0 = \boldsymbol{n}_0 = \boldsymbol{0}$ at the first iteration, they are biased towards zero. To address this, Adam performs a bias correction, unlike RMSprop. The relation between the estimated moments $\boldsymbol{m}_t, \boldsymbol{n}_t$ and the true moments $\mathbb{E}[\nabla_{\boldsymbol{\Theta}_t} C(\boldsymbol{\Theta}_t)], \mathbb{E}\left[\left(\nabla_{\boldsymbol{\Theta}_t} C(\boldsymbol{\Theta}_t)\right)^2\right]$ is derived in [21], yielding the following correction:

$$\tilde{\boldsymbol{m}}_t = \frac{\boldsymbol{m}_t}{1-\beta_1^t}, \tag{2.28}$$

$$\tilde{\boldsymbol{n}}_t = \frac{\boldsymbol{n}_t}{1-\beta_2^t}. \tag{2.29}$$

This correction aligns with the intuition that the bias mainly impacts the first iterations if neglected, as β_1, β_2—being smaller than one—lose their significance with an increased exponent t, i.e., iterations that have advanced further.

The gradient update is thus performed with the corrected statistical moments

$$\boldsymbol{\Theta}_{t+1} = \boldsymbol{\Theta}_t - \frac{\alpha \tilde{\boldsymbol{m}}_t}{\sqrt{\tilde{\boldsymbol{n}}_t} + \epsilon}, \tag{2.30}$$

where the first statistical moment $\tilde{\boldsymbol{m}}_t$ enforces the momentum with the gradient direction in the nominator, as in Eq. (2.20), and the second statistical moment $\tilde{\boldsymbol{n}}_t$ scales the learning rate as in Eq. (2.25). The algorithm is summarized in Algorithm 2. Finally, an important sidenote is also the choice of hyperparameters, which are typically chosen as $\beta_1 = 0.9$, $\beta_2 = 0.999$, $\epsilon = 10^{-8}$. Without the bias correction from Eqs. (2.28) and (2.29), gradient descent from Sect. 2.8.1 can be recovered with $\beta_1 = 0, \beta_2 = 1$, gradient descent with momentum from Sect. 2.8.2.1 with $0 < \beta_1 < 1, \beta_2 = 1$, and RMSprop from Sect. 2.8.2.3 with $\beta_1 = 0, 0 < \beta_2 < 1$. Note that the learning and decay rates are scaled differently when expressed through the definitions of the Adam formulation.

Algorithm 2 Adam

Require: parameters $\boldsymbol{\Theta}$, cost function $C(\boldsymbol{\Theta})$
  initialize first and second moments $\boldsymbol{m}_0 = 0, \boldsymbol{n}_0 = 0$
  for all t **do**
    Compute gradients $\nabla_{\boldsymbol{\Theta}_t} C(\boldsymbol{\Theta}_t)$
    Update biased first moment $\boldsymbol{m}_t \leftarrow \beta_1 \boldsymbol{m}_{t-1} + (1-\beta_1)\nabla_{\boldsymbol{\Theta}_t} C(\boldsymbol{\Theta}_t)$
    Update biased second moment $\boldsymbol{n}_t \leftarrow \beta_2 \boldsymbol{n}_{t-1} + (1-\beta_2)\left[\nabla_{\boldsymbol{\Theta}_t} C(\boldsymbol{\Theta}_t)\right]^2$
    Correct biased first moment $\tilde{\boldsymbol{m}}_t \leftarrow \boldsymbol{m}_t/(1-\beta_1^t)$
    Correct biased second moment $\tilde{\boldsymbol{n}}_t \leftarrow \boldsymbol{n}_t/(1-\beta_2^t)$
    Update parameters $\boldsymbol{\Theta}_{t+1} \leftarrow \boldsymbol{\Theta}_t - \alpha\tilde{\boldsymbol{m}}_t/(\sqrt{\tilde{\boldsymbol{n}}_t} + \epsilon)$
  end for

Other improved versions of Adam include AdaMax [21] replacing the L^2-norm in the denominator with a max norm; Nadam [26], which combines Adam with Nesterov accelerated gradient; AMSGrad [27] which penalizes excessively large learning rates with an additional term; AdamW [28], which decouples the adaptive learning rate from the optimization step; and RAdam [29], which adjusts the adaptive learning rate with an additional term in order to improve the estimation of the moments in the initial training phase. These extensions can enhance convergence speed, stability, and generalization performance for specific problems. Nonetheless, Adam remains the typical (initial) default choice for most deep learning tasks.

2.8.3 L-BFGS

While Adam is an effective optimization algorithm that relies on first-order gradient information, Newton's methods and quasi-Newton methods, such as the L-BFGS algorithm, offer an alternative approach that leverages second-order gradient information. This typically results in fewer iterations but a higher computational effort per iteration. The fundamental gradient update is identical to the previously discussed optimizers, with the current parameters $\boldsymbol{\Theta}_t$ being updated by a search direction $\boldsymbol{p}_t$ and a step size α_t.

$$\boldsymbol{\Theta}_{t+1} = \boldsymbol{\Theta}_t - \alpha_t \boldsymbol{p}_t \tag{2.31}$$

However, in contrast to the earlier methods, the estimation of the direction $\boldsymbol{p}_t$ is now determined not only by the first-order information of the cost function gradient $\nabla_{\boldsymbol{\Theta}_t} C(\boldsymbol{\Theta}_t)$ but also by its Hessian $\nabla^2_{\boldsymbol{\Theta}_t} C(\boldsymbol{\Theta}_t)$

2.8.3.1 Newton's Method

Newton's method, being the most fundamental second-order gradient optimizer, provides a solid foundation for understanding more advanced optimization techniques. To derive Newton's method, we consider the following truncated Taylor expansion of the cost function at $\boldsymbol{\Theta}$ with the increment $\Delta\boldsymbol{\Theta} = \boldsymbol{p}$.

$$C(\boldsymbol{\Theta} + \boldsymbol{p}) \approx m(\boldsymbol{\Theta} + \boldsymbol{p}) = C(\boldsymbol{\Theta}) + [\nabla_{\boldsymbol{\Theta}} C(\boldsymbol{\Theta})]^{\intercal} \boldsymbol{p} + \frac{1}{2} \boldsymbol{p}^{\intercal} \nabla^2_{\boldsymbol{\Theta}} C(\boldsymbol{\Theta}) \boldsymbol{p} \tag{2.32}$$

Here, the cost function $C(\boldsymbol{\Theta})$ is approximated at $\boldsymbol{\Theta}$ by a model paraboloid $m(\boldsymbol{\Theta})$. The minimum of the paraboloid is found with its stationary point, obtained by setting its derivative to zero and solving for the increment $\boldsymbol{p}$.

$$\nabla_{\boldsymbol{\Theta}} m(\boldsymbol{\Theta} + \boldsymbol{p}) = \nabla_{\boldsymbol{\Theta}} C(\boldsymbol{\Theta}) + \nabla^2_{\boldsymbol{\Theta}} C(\boldsymbol{\Theta}) \boldsymbol{p} = 0 \tag{2.33}$$

$$\boldsymbol{p} = -[\nabla^2_{\boldsymbol{\Theta}} C(\boldsymbol{\Theta})]^{-1} \nabla C(\boldsymbol{\Theta}) \tag{2.34}$$

Using the increment $\boldsymbol{p}$, the following update rule can be formulated:

$$\boldsymbol{\Theta}_{t+1} = \boldsymbol{\Theta}_t - \alpha_t [\nabla^2_{\boldsymbol{\Theta}} C(\boldsymbol{\Theta})]^{-1} \nabla C(\boldsymbol{\Theta}) \tag{2.35}$$

where the direction $\boldsymbol{p}_t$ from Eq. (2.31) is identified as $[\nabla^2_{\boldsymbol{\Theta}} C(\boldsymbol{\Theta})]^{-1} \nabla C(\boldsymbol{\Theta})$. The step size is implicitly defined in Eq. (2.34), where convergence within one step is reached for quadratic cost functions. However, this is generally not the case, and thus an additional step size α_t can be added, which is determined with a line search to find the optimal step given the direction $\boldsymbol{p}_t$. The conceptually simplest line search is the exact line search, where the step size is estimated by inserting the update rule (Eq. (2.35)) into the cost function with a subsequent minimization by finding the stationary point with respect to α_t.

$$\alpha_t = \arg\min_{\alpha_t} C(\boldsymbol{\Theta} + \alpha_t \boldsymbol{p}_t) \tag{2.36}$$

$$\nabla_{\alpha_t} C(\boldsymbol{\Theta} + \alpha_t \boldsymbol{p}_t) = 0 \tag{2.37}$$

More involved methods used in practice are inexact line search methods, such as backtracking line search, the Armijo rule, or the Wolfe conditions. These inexact line search methods strike a balance between computational efficiency and ensuring that the step size choice leads to a sufficient decrease in the cost function.

A significant drawback of Newton's method is the computational complexity involved in computing the inverse Hessian, which is in the order $O(n^3)$ [30]. For reference, the first-order gradient descent methods have a complexity of $O(n)$. Given the potentially large number of parameters in deep learning models, Newton's method becomes impractical due to its computational cost.

2.8.3.2 Quasi-Newton Methods BFGS

The problem of computational complexity is addressed by quasi-Newton methods, which approximate either the Hessian $\boldsymbol{B}$ or the inverse of the Hessian $\boldsymbol{H}$, leading to computationally more efficient methods while retaining the valuable second-order information. Revisiting the quadratic model from Eq. (2.32) with the search direction $\boldsymbol{p}$ and the approximation of the Hessian $\boldsymbol{B}_t = \boldsymbol{B}(\boldsymbol{\Theta}_t)$, we have:

$$m(\boldsymbol{\Theta}_t + \boldsymbol{p}) = C(\boldsymbol{\Theta}_t) + [\nabla_{\boldsymbol{\Theta}_t} C(\boldsymbol{\Theta}_t)]^\top \boldsymbol{p} + \frac{1}{2}\boldsymbol{p}^\top \boldsymbol{B}_t \boldsymbol{p}. \tag{2.38}$$

As previously seen in Eq. (2.35), the model is minimized by

$$\boldsymbol{p}_t = -\boldsymbol{B}_t^{-1} \nabla_{\boldsymbol{\Theta}} C(\boldsymbol{\Theta}_t). \tag{2.39}$$

The Hessian $\boldsymbol{B}$ is approximated using two conditions. First, the Hessian should be positive definite at the minimum, such that

$$\boldsymbol{x}^\top \boldsymbol{B} \boldsymbol{x} > 0 \forall \boldsymbol{x} \in \mathbb{R}^n \ \boldsymbol{0}. \tag{2.40}$$

This ensures that the curvature is positive in all directions and thus locally convex. If the Hessian were not positive definite, negative curvatures would be possible, allowing local maxima and saddle points.

Second it is required that the gradient of the current model $m(\boldsymbol{\Theta}_{t+1})$ should match the gradient of the cost function $C(\boldsymbol{\Theta})$ at the current iteration $t + 1$, i.e., at $\boldsymbol{\Theta}_{t+1}$ and the previous iteration t, i.e., $\boldsymbol{\Theta}_t = \boldsymbol{\Theta}_{t+1} - \alpha_t \boldsymbol{p}_t$, which ensures, that the approximative model m locally matches the cost function C sufficiently well. These conditions can be written as

$$\text{for } t+1\text{: } \nabla_p m(\boldsymbol{\Theta}_{t+1}+\mathbf{0}) = \nabla_{\boldsymbol{\Theta}} C(\boldsymbol{\Theta}_{t+1}), \tag{2.41}$$

$$\text{for } t\text{: } \nabla_p m(\boldsymbol{\Theta}_{t+1}-\alpha_t \boldsymbol{p}_{t+1}) = \nabla_{\boldsymbol{\Theta}} C(\boldsymbol{\Theta}_{t+1}) - \alpha_t \boldsymbol{B}_{t+1}\boldsymbol{p}_t = \nabla_{\boldsymbol{\Theta}} C(\boldsymbol{\Theta}_t). \tag{2.42}$$

Note that the model is derived with respect to the direction $\boldsymbol{p}$, as the parameters $\boldsymbol{\Theta}$, the cost $C(\boldsymbol{\Theta})$, and the Hessian $\boldsymbol{B}_t$ are assumed fixed around the Taylor expansion from Eq. (2.32). By contrast, the true cost function is derived with respect to the weights $\boldsymbol{\Theta}$. It is seen in Eq. (2.41) that the gradients at iteration $t+1$ already match. However, the gradients in the previous iteration t, i.e., Eq. (2.42), can be used to derive the following relation.

$$\alpha_t \boldsymbol{B}_{t+1}\boldsymbol{p}_t = \nabla_{\boldsymbol{\Theta}} C(\boldsymbol{\Theta}_{t+1}) - \nabla_{\boldsymbol{\Theta}} C(\boldsymbol{\Theta}_t) \tag{2.43}$$

The notation is decluttered using the substitutions $\boldsymbol{s}_t = \alpha_t \boldsymbol{p}_t = \boldsymbol{\Theta}_{t+1} - \boldsymbol{\Theta}_t$ and $\boldsymbol{y}_t \nabla_{\boldsymbol{\Theta}} C(\boldsymbol{\Theta}_{t+1}) - \nabla_{\boldsymbol{\Theta}} C(\boldsymbol{\Theta}_t)$, yielding the so-called secant equation

$$\boldsymbol{B}_{t+1}\boldsymbol{s}_t = \boldsymbol{y}_t. \tag{2.44}$$

By multiplication of $\boldsymbol{s}_t^{\mathsf{T}}$, the positive definiteness can be ensured

$$\boldsymbol{s}_t^{\mathsf{T}} \boldsymbol{B}_{t+1}\boldsymbol{s}_t = \boldsymbol{s}_t^{\mathsf{T}} \boldsymbol{y}_t > 0, \tag{2.45}$$

which is thus only the case if

$$\boldsymbol{s}_t^{\mathsf{T}} \boldsymbol{y}_t > 0. \tag{2.46}$$

This condition will not always be satisfied for non-convex functions, wherefore it has to be imposed by constraining the line search of α_t. As the system of equations in the secant equation from Eq. (2.44) only contains n equations, with which $n(n+1)/2$ entries of the symmetric $\boldsymbol{B}(\boldsymbol{\Theta}_t)$ are to be identified, an additional constraint has to be introduced. For this purpose, a Hessian that is closest to the last Hessian approximation helps maintain continuity and stability. Therefore, the difference between the last two Hessians is minimized.

$$\min_{\boldsymbol{B}_{t+1}} ||\boldsymbol{B}_{t+1} - \boldsymbol{B}_t|| \tag{2.47}$$

$$\text{s.t. } \boldsymbol{B}_{t+1} = \boldsymbol{B}_{t+1}^{\mathsf{T}}, \tag{2.48}$$

$$\boldsymbol{B}_{t+1}\boldsymbol{s}_t = \boldsymbol{y}_t \tag{2.49}$$

Using the Frobenius norm enables an easy solution to the minimization problem. This is applied in the DFP method, named after Davidon [31], Fletcher and Powell [32]. Also the BFGS method created by and named after Broyden [33], Fletcher [34], Goldfarb [35], and Shanno [36] uses the Frobenius norm, but approximates the inverse Hessian $\boldsymbol{H}$ instead of the Hessian $\boldsymbol{B}$, such that the inversion in Eq. (2.39) is avoided. The minimization from Eqs. (2.47) and (2.49) can be derived analogously for the inverse Hessian $\boldsymbol{H}_{t+1}$.

$$\min_{\boldsymbol{H}_{t+1}} ||\boldsymbol{H}_{t+1} - \boldsymbol{H}_t|| \tag{2.50}$$

$$\text{s.t. } \boldsymbol{H}_{t+1} = \boldsymbol{H}_{t+1}^{\mathsf{T}}, \tag{2.51}$$

$$\boldsymbol{H}_{t+1}\boldsymbol{s}_t = \boldsymbol{y}_t \tag{2.52}$$

The corresponding update formula for the inverse Hessian is given by

$$\boldsymbol{H}_{t+1} = (\boldsymbol{I} - \rho_t \boldsymbol{s}_t \boldsymbol{y}_t^{\mathsf{T}})\boldsymbol{H}_t(\boldsymbol{I} - \rho_t \boldsymbol{y}_t \boldsymbol{s}_t^{\mathsf{T}}) + \rho_t \boldsymbol{s}_t \boldsymbol{s}_t^{\mathsf{T}} \tag{2.53}$$

$$\text{with } \rho_t = \frac{1}{\boldsymbol{y}_t^{\mathsf{T}} \boldsymbol{s}_t}. \tag{2.54}$$

For the derivation and in-depth explanation of the BFGS and DFP algorithms, see [30]. The algorithm is initialized with $\boldsymbol{H}_0$, which can be computed with finite differences or simply set to the identity matrix, thus neglecting the second-order information for the first iteration. With the update rule of the inverse Hessian, the direction $\boldsymbol{p}_t$ from Eq. (2.39) is computed, which in combination with a line search yields the updated parameters $\boldsymbol{\Theta}_t$. The algorithm is summarized in Algorithm 3. The downside of the BFGS algorithm is that it is very memory intensive, as it has to store and manipulate a generally dense $n \times n$ matrix $\boldsymbol{H}_k$.

Algorithm 3 BFGS adapted from [30]

Require: initial parameters $\boldsymbol{\Theta}_0$, number of epochs, initial inverse Hessian $\boldsymbol{H}_0$
for all t **do**
  Compute search direction $\boldsymbol{p}_t \leftarrow -\boldsymbol{H}_t \nabla_{\boldsymbol{\Theta}} C(\boldsymbol{\Theta}_t)$
  Update parameters (with line search for α_t satisfying Wolfe conditions)
    $\boldsymbol{\Theta}_{t+1} \leftarrow \boldsymbol{\Theta}_t + \alpha_t \boldsymbol{p}_t$
  Compute $\boldsymbol{s}_t \leftarrow \boldsymbol{\Theta}_{t+1} - \boldsymbol{\Theta}_t$
  Compute $\boldsymbol{y}_t \leftarrow \nabla_{\boldsymbol{\Theta}}(\boldsymbol{\Theta}_{t+1}) - \nabla_{\boldsymbol{\Theta}}(\boldsymbol{\Theta}_t)$
  Compute $\rho_t \leftarrow 1/(\boldsymbol{y}_t^{\mathsf{T}} \boldsymbol{s}_t)$
  Update inverse Hessian $\boldsymbol{H}_{t+1} \leftarrow (\boldsymbol{I} - \rho_t \boldsymbol{s}_t \boldsymbol{y}_t^{\mathsf{T}})\boldsymbol{H}_t(\boldsymbol{I} - \rho_t \boldsymbol{y}_t \boldsymbol{s}_t^{\mathsf{T}}) + \rho_t \boldsymbol{s}_t \boldsymbol{s}_t^{\mathsf{T}}$
end for

2.8.3.3 L-BFGS

To circumvent the memory issue for large systems, as they often occur in deep learning, the limited-memory BFGS (L-BFGS) [37] aims to not store the full matrix $\boldsymbol{H}_t$, but to reconstruct it from the most recent vector pairs $\boldsymbol{s}_t$, $\boldsymbol{y}_t$. This is achieved with a recursion formula for $\boldsymbol{H}_k$ using the last m vector pairs $\boldsymbol{s}_t$, $\boldsymbol{y}_t$, which is implemented in a two-loop recursion. Typically, the last 3–20 vector pairs are sufficient. For a description of the recursion formula and two-loop recursion algorithm, see [30] or [37]. The high-level L-BFGS implementation is presented in Algorithm 4 with the difference to Algorithm 3 being the computation of the Hessian.

Algorithm 4 L-BFGS adapted from [30]

Require: initial parameters $\boldsymbol{\Theta}_0$, number of epochs, number of saved states m
  for all t **do**
    Choose $\boldsymbol{H}_t^0$
    Compute search direction with the two-loop recursion
      $\boldsymbol{p}_t \leftarrow -\boldsymbol{H}_t \nabla_{\boldsymbol{\Theta}} C(\boldsymbol{\Theta}_t)$
    Update parameters (with line search for α_t satisfying Wolfe conditions)
      $\boldsymbol{\Theta}_{t+1} \leftarrow \boldsymbol{\Theta}_t + \alpha_t \boldsymbol{p}_t$
    if $t > m$ **then**
      Delete $\boldsymbol{s}_{t-m}, \boldsymbol{s}_{t-m}$
    end if
    Compute and save $\boldsymbol{s}_t \leftarrow \boldsymbol{\Theta}_{t+1} - \boldsymbol{\Theta}_t$
    Compute and save $\boldsymbol{y}_t \leftarrow \nabla_{\boldsymbol{\Theta}}(\boldsymbol{\Theta}_{t+1}) - \nabla_{\boldsymbol{\Theta}}(\boldsymbol{\Theta}_t)$
  end for

Exercises

E.4 Linear Regression (P & C)

1. Evaluate the cost function as defined in Eq. (2.3) using a linear regression with an initialization of the weight $w = 1$ and bias $b = 1$. Use the following data:

$$\tilde{x} = [0, 2, 3],$$
$$\tilde{y} = [1, 6, 7].$$

2. Find the gradient of the cost function with respect to the weight and bias.
3. Find the minimum of the cost function and the corresponding weight and bias directly using the gradients. Then, plot the linear regression.
4. Now repeat the linear regression with the normal equations as in Eq. (2.9) and compare your results with the linear regression from the previous exercise.
5. Find the minimum of the cost function and the corresponding weight and bias by using a gradient descent approach. Use the initial weight of $w = 1$, the initial bias $b = 1$, and a learning rate of $\alpha = 0.1$. Perform three iterations and check if you converge toward the result of the previous exercise.
6. Now, the goal is to implement a linear regression model in Python. A template for the class to be used: `LinearRegression` is provided in `exercise4.ipynb`. The correctness of the implemented class can be verified using the previous hand calculations. The first task is to implement the member function `forward(self, x)`, which takes an input `x` and returns the linear prediction.

7. The next member function to be implemented is `costFunction(self, x, y)`. Here, the cost function is to be found with the input arrays `x` and `y`. The output should be a scalar value.
8. Now the member function `gradient(self, x, y)` has to be added. Here, the inputs `x` and `y` are again arrays that have to be used to determine the scalar gradients of the cost function with respect to the weight and bias. The output is defined as an array of the gradient with respect to the bias and the gradient with respect to the weight.
9. Finally, the member function `train(self, epochs, lr, xTrain, yTrain, xTest, yTest)` has to be implemented. This function approximates the weight and bias that minimize the cost function by using the gradient descent approach. The number of iterations is provided as an input `epochs`, and the learning rate is provided as `lr`. The function has to update the weight and bias after each iteration using the training data `xTrain, yTrain`. The testing data `xTest, yTest` can be used to track the generalization error. The value of the cost function using the training and testing data can be printed on the screen. To reduce the number of outputs for the user, this should only be done for every 10th or 100th iteration.
10. Now, the class can be applied to an example case. Here, a linear regression is applied to random data with a linear tendency. You can try to play around with the sample size, number of epochs, and learning rate.

E.5 Higher-Order Regression (C)

1. The goal is now to extend the linear regression model to a higher-order regression model. A similar class with almost the same member functions is to be implemented. Only slight adjustments are necessary. An outline of the class is provided as `HigherOrderRegression` in `exercise5.ipynb`. The first step is implementing the member function `forward(self, x)`. Here, a higher-order prediction has to be made using the following form,

$$\hat{y} = \sum_{i=1}^{p} (w_i x^i) + b$$

where p is the polynomial degree that is selected during the construction of the class. One thing to note here is that the weight w_i is an array instead of a scalar now.
2. The member function `costFunction(self, x, y)` has to be adjusted accordingly.
3. Now derive the gradients of the cost function with respect to all weights and the bias. The derived quantities can then be inserted in the member function

`gradient(self, x, y)`. The function should return an array with all the gradients with respect to the weights and with respect to the bias. The gradient with respect to the bias should be the last entry.
4. Finally, adjust the member function `train(self, epochs, learningRate, xTrain, yTrain, xTest, yTest)`, such that you ensure that all the weights are adjusted accordingly using gradient descent.
5. Now, the class can be used in an example case defined in `example5.ipynb`. Try to adjust the model parameters, such as the polynomial degree, the number of epochs, and the learning rate. Which phenomena occur when a polynomial degree of 1, 2, or 9 is chosen? How can this phenomenon already be spotted during learning, and how can it be avoided?
6. To avoid overfitting, regularization is now to be added to the higher-order regression class. If the parameter `regularization` is set to a floating-point value, an L^2 regularization is to be enabled. Adjust the member functions `costFunction` and `gradient` accordingly.
7. Apply the regularization to the example in `example5.ipynb` and see if the model is still overfitting. Then, try to vary the regularization parameter. What happens when the parameter goes toward zero, and what happens when it goes toward infinity?

E.6 Adam Optimizer (C)

1. The goal is to implement the Adam optimizer. A basic framework is provided in `exercise6.ipynb`. During the construction of the `AdamOptimizer` class, the hyperparameters and some helper variables are initialized. Specifically, the two statistical moments and the number of performed iterations (to be used as exponent) are stored. The member function `updateParams(self, params, grads)` takes the current parameters `params` and its gradients `grads` and should update the parameters according to the Adam algorithm presented in Algorithm 2.
2. With the implemented Adam optimizer, perform a comparison with gradient descent using the Rosenbrock function $f(x_1, x_2)$ [38] as optimization objective, i.e., as cost function:

$$f(x_1, x_2) = 100\left(x_2 - x_1^2\right)^2 + (1 - x_1)^2 \tag{2.55}$$

The inputs x_1, x_2 are to be adjusted to minimize f. Try different learning rates and initial guesses. Which method is more reliable and which is less sensitive to the choice of hyperparameters?

References

1. Bostrom, Nick. 2014. *Superintelligence: Paths, dangers, strategies*. 1st edn. OCLC: ocn881706835. Oxford: Oxford University Press. ISBN: 978-0-19-967811-2.
2. Tegmark, Max. 2017. *Life 3.0: Being human in the age of artificial intelligence*. 1st edn. New York: Alfred A. Knopf. ISBN: 978-1-101-94659-6 978-1-101-97031-7.
3. Christian, Brian. 2020. *The alignment problem: Machine learning and human values*. 1st edn. New York, NY: W.W. Norton & Company. ISBN: 978-0-393-63582-9.
4. Burkov, Andriy. 2019. *The hundred-page machine learning book*. English. tex.ids= Burkov2019a. Andriy Burkov. ISBN: 978-1-9995795-0-0.
5. Mitchell, Tom M. 1997. *Machine learning*. en. McGraw-Hill series in computer science. tex.ids= Mitchell1997a. New York: McGraw-Hill. ISBN: 978-0-07-042807-2.
6. Goodfellow, Ian, Yoshua, Bengio, and Aaron, Courville. 2016. *Deep learning*. MIT Press. ISBN: 0-262-03561-8.
7. Andrew, Ng. 2020. *Machine learning*. Online course. tex.ids= Ng2020a.
8. Chollet, François. 2018. *Deep learning with Python*. en. OCLC: ocn982650571. Shelter Island, New York: Manning Publications Co. ISBN: 978-1-61729-443-3.
9. Silver, David, et al. 2017. Mastering the game of Go without human knowledge. en. In: *Nature* 550.7676. tex.ids= Silver2017a, pp. 354–359. ISSN: 0028-0836, 1476-4687. https://doi.org/10.1038/nature24270.
10. Shi, Zhan, et al. 2020. *Improving image captioning with better use of captions*. arXiv:2006.11807 [cs]. https://doi.org/10.48550/arXiv.2006.11807.
11. Podell, Dustin, et al. 2023. *SDXL: Improving latent diffusion models for high-resolution image synthesis*. arXiv:2307.01952 [cs].
12. Liu, Yixin, et al. 2024. *Sora: A review on background, technology, limitations, and opportunities of large vision models*. arXiv:2402.17177 [cs]. https://doi.org/10.48550/arXiv.2402.17177.
13. Jun, Heewoo, and Alex, Nichol. 2023. *Shap-E: Generating conditional 3D implicit functions*. arXiv:2305.02463. https://doi.org/10.48550/arXiv.2305.02463.
14. Lewis, Patrick, et al. 2021. *Retrieval-augmented generation for knowledge-intensive NLP tasks*. arXiv:2005.11401. https://doi.org/10.48550/arXiv.2005.11401.
15. Cheney, Ward E., and David R. Kincaid. 2012. *Numerical mathematics and computing*. English. 7th. tex.ids= Cheney2012a. Cengage learning. ISBN: 978-1-133-10371-4.
16. Fortmann-Roe, Scott. 2012. *Understanding the bias-variance tradeoff*.
17. Mehta, Pankaj, et al. 2019. A high-bias, low-variance introduction to Machine Learning for physicists. en. In: *Physics Reports* 810. arXiv: 1803.08823, pp. 1–124. ISSN: 03701573. https://doi.org/10.1016/j.physrep.2019.03.001.
18. Kahneman, Daniel, Olivier, Sibony, and Cass R., Sunstein. 2011. *Noise: A flaw in human judgment*. 1st edn. OCLC: on1249942231. New York: Little, Brown Spark. ISBN: 978-0-316-45140-6 978-0-316-26665-9.
19. Abu-Mostafa, Yaser S., Malik, Magdon-Ismail, and Hsuan-Tien, Lin. 2012. *Learning from data*. English. tex.ids= Abu-Mostafa2012a. S.l.: AMLBook. ISBN: 978-1-60049-006-4.
20. Bishop, Christopher M. 2016. *Pattern recognition and machine learning*. eng. Softcover reprint of the original 1st edn. 2006 (corrected at 8th printing 2009). Information Science and Statistics. New York, NY: Springer New York. ISBN: 978-1-4939-3843-8.
21. Kingma, Diederik P., and Jimmy Ba. 2017. Adam: A method for stochastic optimization. arXiv:1412.6980 [cs].
22. Polyak, B. T. (1964). Some methods of speeding up the convergence of iteration methods. en. In: *USSR Computational Mathematics and Mathematical Physics* 4(5):1–17. ISSN: 0041-5553. https://doi.org/10.1016/0041-5553(64)90137-5.
23. Nesterov, Y. 1983. A method for solving the convex programming problem with convergence rate O(1/k2). *Proceedings of the USSR Academy of Sciences*.
24. Sutskever, Ilya, et al. 2013. On the importance of initialization and momentum in deep learning. In: *Proceedings of the 30th international conference on international conference on machine learning - volume 28*. ICML'13. Atlanta, GA, USA: JMLR.org, pp. III–1139–III–1147.

25. Duchi, John, Elad, Hazan, and Yoram, Singer. 2011. Adaptive subgradient methods for online learning and stochastic optimization. *The Journal of Machine Learning Research* 12.null, pp. 2121–2159. ISSN: 1532-4435.
26. Dozat, T. 2016. Incorporating Nesterov momentum into Adam.
27. Reddi, Sashank J., Satyen, Kale, and Sanjiv, Kumar. 2019. *On the convergence of Adam and beyond*. arXiv:1904.09237 [cs, math, stat] version: 1. https://doi.org/10.48550/arXiv.1904.09237.
28. Loshchilov, Ilya and Frank, Hutter. 2019. *Decoupled weight decay regularization*. arXiv:1711.05101 [cs, math] version: 3. https://doi.org/10.48550/arXiv.1711.05101.
29. Liu, Liyuan, et al. 2021. *On the variance of the adaptive learning rate and beyond*. arXiv:1908.03265 [cs, stat]. https://doi.org/10.48550/arXiv.1908.03265.
30. Nocedal, Jorge and Stephen J., Wright. 2006. *Numerical optimization*. 2nd ed. Springer series in operations research. OCLC: ocm68629100. New York: Springer. ISBN: 978-0-387-30303-1.
31. Davidon, William C. 1991. Variable metric method for minimization. en. SIAM Journal on Optimization 1(1):1–17. ISSN: 1052-6234, 1095-7189. https://doi.org/10.1137/0801001.
32. Fletcher, R., and M. J. D., Powell. 1963. A rapidly convergent descent method for minimization. *The Computer Journal* 6(2):163–168. ISSB:0010-4620, 1460-2067. https://doi.org/10.1093/comjnl/6.2.163.
33. Broyden, C. G. 1970. The convergence of a class of double-rank minimization Algorithms 1. General considerations. *IMA Journal of Applied Mathematics* 6(1):76–90. ISSN: 0272-4960, 1464-3634. https://doi.org/10.1093/imamat/6.1.76.
34. Fletcher, R. 1970. A new approach to variable metric algorithms. *The Computer Journal* 13(3):317–322. ISSN: 0010-4620, 1460-2067. https://doi.org/10.1093/comjnl/13.3.317.
35. Goldfarb, Donald. 1970. A family of variable-metric methods derived by variational means. *Mathematics of Computation* 24(109):23–26. ISSN: 0025-5718, 1088-6842. https://doi.org/10.1090/S0025-5718-1970-0258249-6.
36. Shanno, D. F. 1970. Conditioning of quasi-Newton methods for function minimization. *Mathematics of Computation* 24(111):647–656. ISSN: 0025-5718, 1088-6842. https://doi.org/10.1090/S0025-5718-1970-0274029-X.
37. Nocedal, Jorge. 1980. Updating Quasi-Newton matrices with limited storage. *Mathematics of Computation* 35(151):773–782. ISSN: 0025-5718. https://doi.org/10.2307/2006193.
38. Rosenbrock, H. H. 1960. An automatic method for finding the greatest or least value of a function. *The Computer Journal* 3(3):175–184. ISSN: 0010-4620, 1460-2067. https://doi.org/10.1093/comjnl/3.3.175.

Chapter 3
Neural Networks

Abstract Artificial neural networks (ANNs) are state-of-the-art machine learning architectures modeling neurons and their connections through weights and biases. ANNs serve as universal function approximators, meaning that a sufficiently complex neural network can learn almost any function in any dimension. This flexibility, combined with backpropagation and a learning algorithm, enables one to learn unknown functions with astonishing accuracy. This chapter explains key architectures, such as fully connected, convolutional, and recurrent networks, along with essential concepts like forward propagation, backpropagation, and regularization techniques. Designed for readers new to the field, it provides clear explanations of the basic algorithms as well as examples to demonstrate how these networks can model complex non-linear relationships in physics and mechanics, e.g., deployable as surrogate models.

For supervised learning tasks, artificial neural networks (ANNs) are the most popular state-of-the-art algorithmic architecture [1]. Inspired by the brain's neurons, an early form known as perceptron was created by Frank Rosenblatt in 1958. After the late 1960s, the development stagnated due to the lack of computational power and efficient methods for network training. With the introduction of the backpropagation algorithm in 1986, the learning capabilities of neural networks improved significantly. Nevertheless, their application only became practical in the early 2000s on the hardware available at that time. A major breakthrough leading to more attention was the success of the convolutional neural network known as AlexNet by Krizhevsky et al. [3]. AlexNet won the image recognition challenge ImageNet in 2012 by a large margin [4]. Its success was also due to developments in computer graphics that allowed the exploitation of GPUs for the training process. Since then, artificial neural networks have been applied to solve a great variety of problems. Typical tasks include speech, image, and natural language processing; autonomous driving; playing board and computer games; algorithmic trading; or weather forecasting. Recently, physicists and engineers have also started to investigate how conventional methods can benefit from the capabilities of neural networks [5]. Even though fundamentals in this young area of research have been established, most results of modern neural networks are based on empirical studies and heuristics [2].

This chapter introduces the simplest form of neural networks in Sect. 3.1 and explains their fundamental algorithmic operations such as forward propagation, dif-

L. Herrmann et al., *Deep Learning in Computational Mechanics*,
https://doi.org/10.1007/978-3-031-89529-6_3

ferentiation, and backpropagation in Sects. 3.2 to 3.4 by means of an illustrative example. Necessary details such as possible types of activation functions Sect. 3.5, learning algorithms Sect. 3.6, and regularization Sect. 2.7 are treated in the corresponding chapters. Features of such basic networks are demonstrated at the simple application of approximating a sine function in Sect. 3.8. Section 3.9 is devoted to more advanced neural network architectures that were designed to perform exceptionally well in specific tasks such as image or speech recognition, language translation, or time series forecasting. However, they may also be utilized in computational mechanics, e.g., for infering strains directly from material distributions, see Sect. 3.9.1.2.

3.1 Fully Connected Neural Networks

A neural network can be viewed as parameterized function defining a mapping $y = f_{NN}(x)$. A particularity of neural networks is that they are composed of many nested functions. For instance, three functions f_1, f_2, and f_3 might form the mapping $f_{NN}(x) = f_3(f_2(f_1(x)))$, where each function f_l represents a layer of the network. The information in the form of an input x flows from the input layer through an arbitrary number of "*hidden*" layers to the output layer. Such a network is called a feed-forward neural network. A classic example of a neural network that is not a feed-forward neural network is the recurrent neural network, discussed in Sect. 3.9.6. The number of layers defines the depth, whereas the amount of hidden neurons determines the width of a network. In this context, the term "*deep*" learning refers to neural networks with more than one hidden layer [6].

A simple network architecture is depicted in Fig. 3.1. The circles represent the basic units called "*neurons*", whereas the connections can be interpreted as weights that control the importance of their inputs. If every neuron from the previous layer is connected with each neuron of the next layer, then the neural network is said to be "*fully connected*". An example of an architecture that is not fully connected is the convolutional neural network presented in Sect. 3.9.1.[1]

A single neuron receives a vector of features $\boldsymbol{x}$ and produces a scalar output $a(\boldsymbol{x})$, which serves as an input for the neurons in the next layer [2]. The output of a neuron can be decomposed into two operations. First, the input vector $\boldsymbol{x}$ is transformed into

$$z = \boldsymbol{w} \cdot \boldsymbol{x} + b \tag{3.1}$$

[1] A recurrent neural network is, however, a fully connected neural network, as the subsequent layers are fully connected. To simplify the notation, we abbreviate the fully connected feed-forward neural network architecture illustrated in Fig. 3.1 with "*fully connected neural network*", as often done in the field of deep learning.

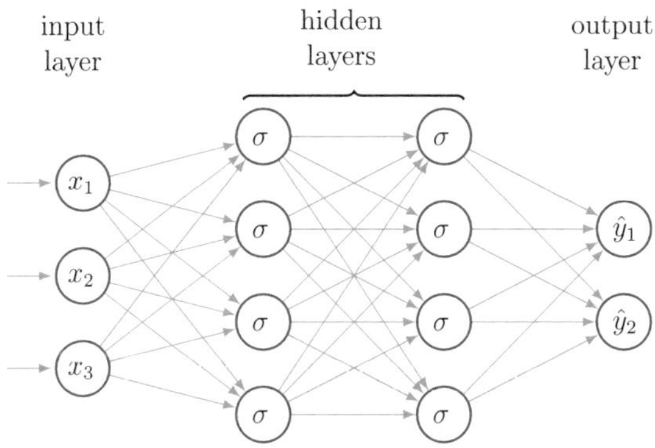

Fig. 3.1 A fully connected feed-forward neural network (commonly shortened to the term "*fully connected neural network*")

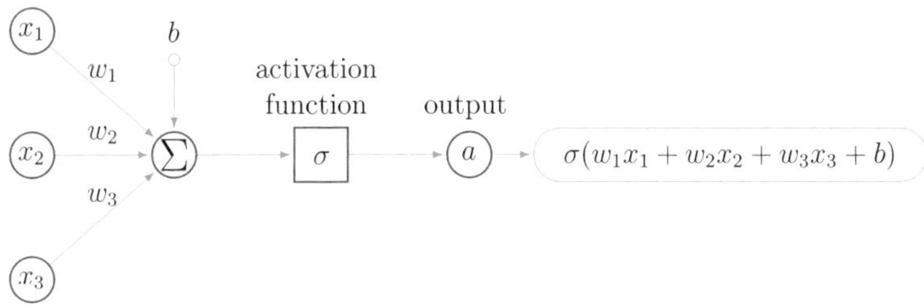

Fig. 3.2 A single neuron with three inputs

with a neuron-specific weight $\boldsymbol{w}$ and bias b. Second, the neuron's output $a(\boldsymbol{x})$ is computed by applying the non-linear function σ to the scalar z resulting from the linear combination in Eq. (3.1) (cf. Fig. 3.2)

$$a(\boldsymbol{x}) = \sigma(z). \tag{3.2}$$

The function σ is called activation function and is usually chosen to be the same for all neurons [6]. Without the use of a non-linear activation function, the model would not be able to represent any non-linearities in the data. The reason is that regardless of how many linear transformations are applied to an input, the output would still remain a linear function of the input. Typical candidates for the activation $\sigma(z)$ are the hyperbolic tangent, the sigmoid function, and the rectified linear unit, in short ReLU (see Fig. 3.3). A more detailed discussion about the properties of the different activation functions is provided in Sect. 3.5.

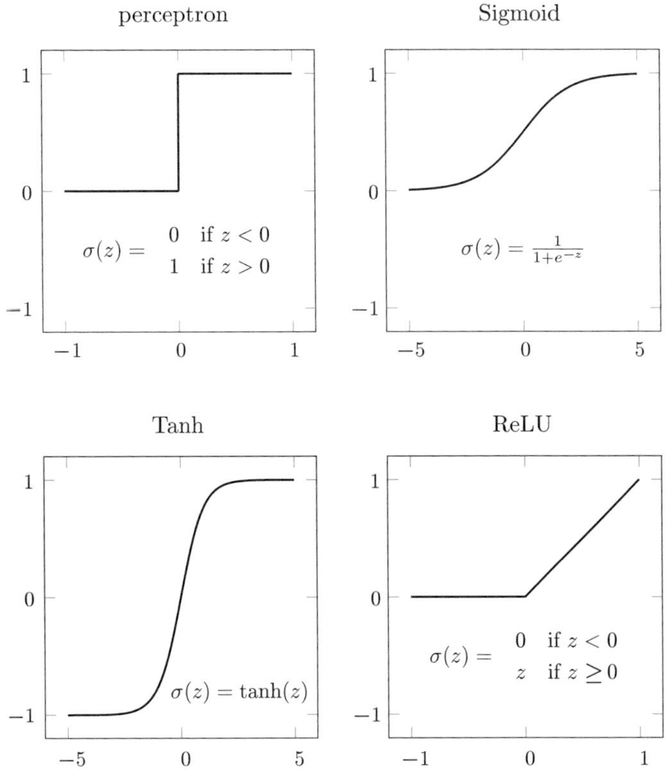

Fig. 3.3 Common activation functions

Depending on the function chosen for the output neurons, fully connected neural networks in their basic form are able to solve either regression or classification tasks. For regression, simply applying a linear function in the output layer leads to a result in the form of a real number. In the case of classification, a common choice is the sigmoid or the softmax function to enforce a binary encoding of the output that can be easily transformed into labels.

A fully connected neural network with one hidden layer is capable of approximating any continuous multi-input/multi-output function with arbitrary precision, given that the hidden layer contains a sufficient amount of neurons [2]. Known as the universal approximation theorem, this hypothesis was formally proven by multiple researchers independently, e.g., in "*Approximation by superpositions of a sigmoidal function*" by Cybenko, to mention one of the first references [7]. A graphical and very intuitive explanation of the universal approximation theorem can be found in chapter four of Nielsen's online book [8]. The explanation's principal idea is that hidden neurons allow the generation of step functions with arbitrary heights and offsets that can be superposed to construct any continuous function [2]. However, the whole concept is rather of theoretical importance. In practice, the usage of deep networks with multiple hidden layers is preferred over a single hidden layer [9–11]. Multi-layer architectures exhibit the same representational power as comparable wide networks with a single hidden layer while being computationally more efficient in the training process, as suggested by [12]. Nevertheless, this topic is controversially discussed and continues to be an area of active research [2].

A fully connected neural network is able to provide a continuous approximation [8] for continuous functions. However, their ability to approximate discontinuous functions is limited. The solution often exhibits oscillations around the discontinuities similar to the Gibbs phenomenon observed in Fourier series approximations of discontinuous functions [13]. In "*Constructive approximation of discontinuous functions by neural networks*", Llanas et al. propose a way to overcome this shortcoming and show an almost uniform approximation of a piece-wise continuous function by a fully connected neural network with a single hidden-layer [14].

Let us now get a first impression of the representational capacities of a neural network. To this end, we generate two-dimensional images by initiating the weights of the neural network randomly without any training and simply inspecting visually how complex the picture looks. Figure 3.4 shows the output of four different neural networks with different numbers of hidden layers using 100 neurons each. The neural network weights are sampled from a uniform distribution $\mathcal{U}$ in the range -1 to 1, while the biases are sampled in the range -10 to 10. The input is a uniform grid in the range $-\frac{1}{2}$ to $\frac{1}{2}$ in both the x_1 and x_2 direction. The most expressive neural network is depicted in Fig. 3.4d and consists of ten hidden layers which lead to 91 301 parameters. Although the function depicted in the picture was generated randomly, the complex patterns already indicate the ability of the neural network to approximate complicated functions. Figure 3.6 further highlights this expressivity by zooming into the image, showcasing the fine-grained nature of the neural network output. For the sake of comparison, Fig. 3.5a, b depict the output of shallower networks using two hidden layers with 300 neurons and five hidden layers with 150 neurons, respectively. Their expressivity is clearly inferior, although the number of parameters used is comparable to the most expressive neural network depicted in Fig. 3.4d. To showcase that shallower networks can also become more expressive, Fig. 3.5c depicts an output from a five-layer network with 800 neurons, which is qualitatively comparable to the output depicted in Fig. 3.4d. However, this choice increases the number of parameters from 91 301 to 2 566 401. Thus, it is typically recommended to use deeper networks, i.e., a larger layer depth, to increase the network capacity. Unfortunately, deeper networks are more difficult to train due to effects such as vanishing gradients [15]. However, remedies exist, which include residual connections [16] or improved activation functions (e.g., ReLU, or leaky ReLU; see Sect. 3.5).

To demonstrate how the high expressivity of neural networks compares to classical approaches, we match the neural network outputs from Fig. 3.4d as closely as possible with three commonly used classical approaches by means of a least square fit. The output is depicted in Fig. 3.7. Figure 3.7a is a piece-wise constant representation, an engineering standard in, e.g., medical images such as CT-scans. Figure 3.7b, c,

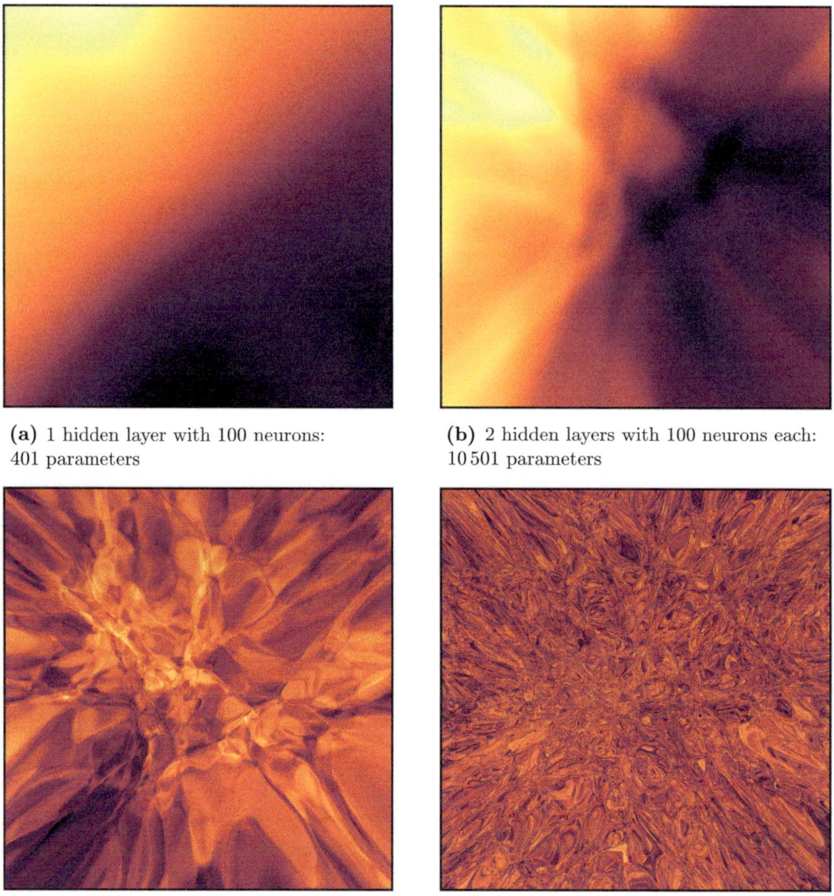

(a) 1 hidden layer with 100 neurons: 401 parameters

(b) 2 hidden layers with 100 neurons each: 10 501 parameters

(c) 5 hidden layers with 100 neurons each: 40 801 parameters

(d) 10 hidden layers with 100 neurons each: 91 301 parameters

Fig. 3.4 Output visualization of a multi-layer neural network with sigmoid activation functions. It has two inputs corresponding to the x and y coordinates of a picture and one output defining the color at the pixel (x, y). The parameters of the network are randomly initialized and are not trained. The code used for generating the pictures was developed by Marquardt [17]

instead are generated using higher-order finite element spaces, which are known to be very expressive, especially for the representation of sufficiently smooth functions. The number of parameters for these three classical representations of functions is chosen to be comparable to the 91 301 parameters used to generate Fig. 3.4d with the previously described neural network. Although the classical approaches capture the overall tendencies, the neural network is able to represent much more fine-grained details. This becomes particularly apparent from the artifacts in the form of squares indicating the boundary of the finite elements in the zoomed-in views in Fig. 3.7a, b.

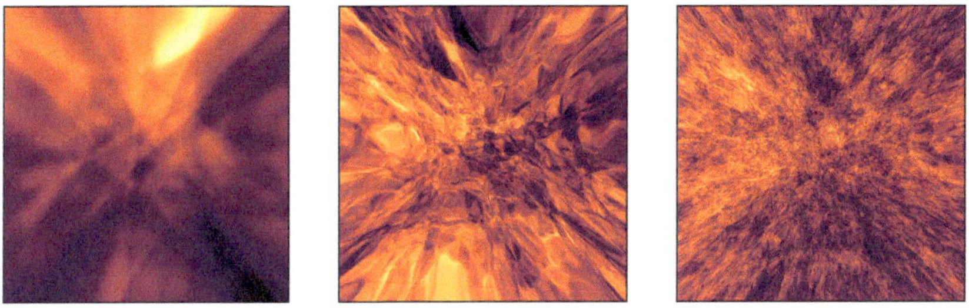

(a) 2 hidden layers with 300 neurons: 91 501 parameters
(b) 5 hidden layers with 150 neurons: 91 201 parameters
(c) 5 hidden layers with 800 neurons: 2 566 401 parameters

Fig. 3.5 Output visualization of a multi-layer neural network with sigmoid activation functions. Instead of increasing the neural network depth as in Fig. 3.4, the effect of the network width is studied, showcasing the inferiority of shallower neural networks with respect to representational capacity

By contrast, the output of the neural networks depicted in Fig. 3.6 is less blurry. The blue and gray boxes illustrate the element sizes for $p = 2$ and $p = 8$ used in Fig. 3.7c, d. No artifacts are visible and the complexity of the zoomed-in view is still very high. While this observation in the eye norm is far from a mathematical proof, it demonstrates that the function spaces that an assembly of classical finite elements is capable of representing are smaller than those that a neural network can capture for the same number of parameters. We termed this feature loosely expressivity above to convey the idea as simply as possible [18].

3.2 Forward Propagation

Before proceeding with the training process of a neural network, it is important to get a better understanding of how the data is propagated from the input to the output layers. A very simple example network with two inputs x_1 and x_2, one hidden layer, and a single output $\hat{y}$ is depicted in Fig. 3.8.

The goal is to compute the prediction $\hat{y}$ of the neural network for one given example

$$\boldsymbol{x} = \begin{pmatrix} x_1 \\ x_2 \end{pmatrix} = \begin{pmatrix} 3 \\ 2 \end{pmatrix}$$

and weights

$$\boldsymbol{W}^{(1)} = \begin{pmatrix} w_{11}^{(1)} & w_{12}^{(1)} \\ w_{21}^{(1)} & w_{22}^{(1)} \end{pmatrix} = \begin{pmatrix} 1 & -3 \\ -2 & 1 \end{pmatrix}, \quad \boldsymbol{w}^{(2)} = \begin{pmatrix} w_1^{(2)} \\ w_2^{(2)} \end{pmatrix} = \begin{pmatrix} 2 \\ -1 \end{pmatrix}.$$

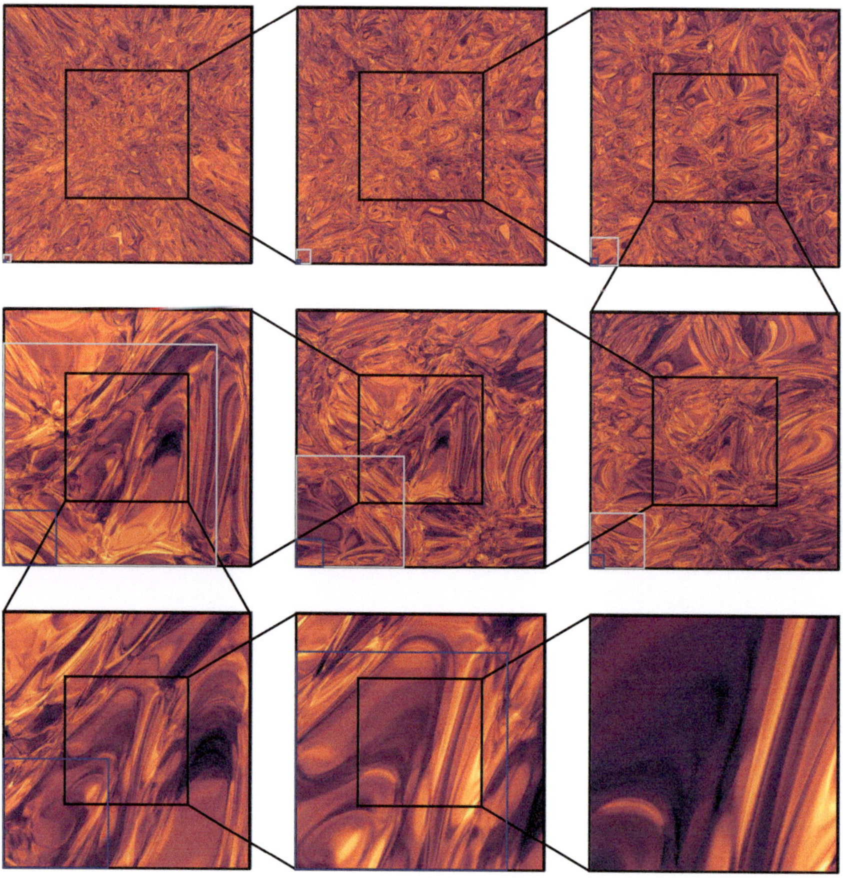

Fig. 3.6 Zoomed in view of the output from Fig. 3.4d relying on 91 301 parameters. The element size of finite element discretizations with a similar number of parameters is depicted in gray and blue. For $p = 2$, 150×150 elements yield 90 601 parameters (blue) and for $p = 8$, 37×37 elements yield 88 209 parameters (gray)

For the sake of simplicity, the biases for each neuron are set to zero and are thus neglected in the following calculations. Furthermore, the activation function for all neurons is chosen to be an element-wise square $\sigma(z_i) = (z_i)^2$. The input vector $\boldsymbol{x}$ is fed to the network and every entry is stored in the corresponding neuron of the input layer without any further modifications. As a first step toward computing the output of the hidden layer, the weighted sum of the inputs is calculated and stored in the variable $\boldsymbol{z}^{(1)}$

$$\boldsymbol{z}^{(1)} = \boldsymbol{W}^{(1)}\boldsymbol{x} = \begin{pmatrix} 1 & -3 \\ -2 & 1 \end{pmatrix} \begin{pmatrix} 3 \\ 2 \end{pmatrix} = \begin{pmatrix} -3 \\ -4 \end{pmatrix}.$$

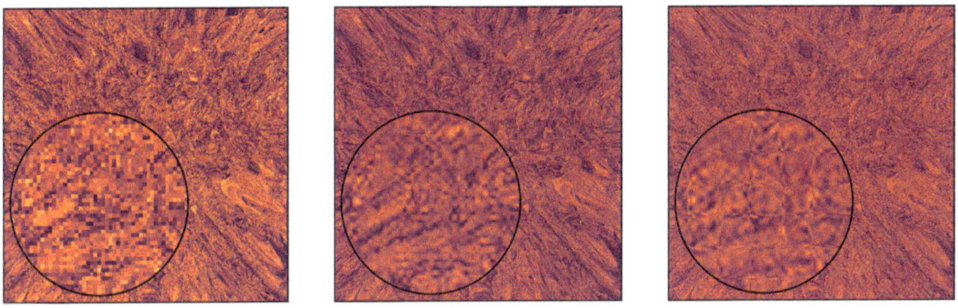

(a) Piece-wise constant ansatz, i.e., pixels with 302 × 302 = 91 204 parameters

(b) Finite element ansatz with integrated Legendre polynomials of degree $p = 2$: 91 301 parameters

(c) Finite element ansatz with integrated Legendre polynomials of degree $p = 8$: 91 301 parameters

Fig. 3.7 Output visualization of classical ansatz spaces. The parameters are set to match the outputs from Fig. 3.4a (top) and Fig. 3.4c (bottom)

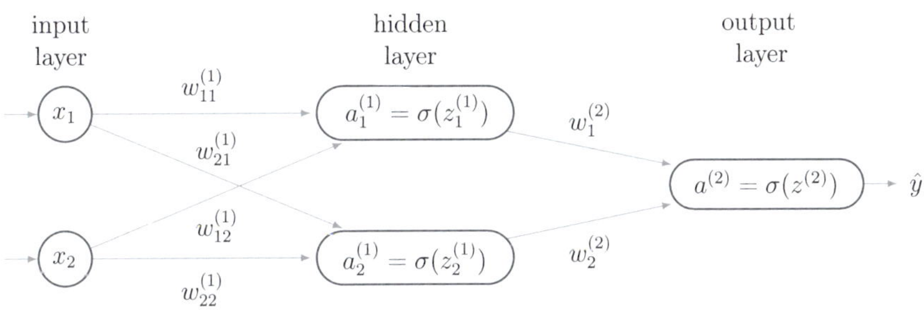

Fig. 3.8 A simple fully connected neural network example

Then, the activation function σ is applied element-wise to the resulting vector $z^{(1)}$ to get the output $\boldsymbol{a}^{(1)}$ of the hidden layer

$$\boldsymbol{a}^{(1)} = \sigma(\boldsymbol{z}^{(1)}) = \begin{pmatrix} \sigma(-3) \\ \sigma(-4) \end{pmatrix} = \begin{pmatrix} 9 \\ 16 \end{pmatrix}.$$

Now, the output $\boldsymbol{a}^{(1)}$ serves as the input for the neurons of the next layer. Again, the linear transformation $\boldsymbol{z}^{(1)}$ is computed first

$$z_1^{(2)} = (\boldsymbol{w}^{(2)}) \cdot a^{(1)} = \begin{pmatrix} 2 & -1 \end{pmatrix} \begin{pmatrix} 9 \\ 16 \end{pmatrix} = 2.$$

Finally, the prediction of the neural network $\hat{y}$ is computed by applying the activation function to $z_1^{(2)}$

$$\hat{y} = a_1^{(2)} = \sigma(z_1^{(2)}) = \sigma(2) = 4.$$

To generalize the computations from the example, the output for the current neuron j in the lth layer is calculated by the following formula [8]:

$$a_j^{(l)} = \sigma(z_j^{(l)}) = \sigma\left(\sum_k w_{jk}^{(l)} a_k^{(l-1)} + b_j^{(l)}\right), \tag{3.3}$$

where $a_k^{(l-1)}$ is the output of the kth neuron from the previous layer $(l-1)$ and $w_{jk}^{(l)}$ is the weight connecting the kth neuron with the current neuron j. After adding the neuron-specific bias $b_j^{(l)}$ to the weighted sum of all outputs k from the previous layer, the non-linear activation function is applied [8].

Using only matrix and vector notation Eq. 3.3 yields [8]

$$\boldsymbol{a}^{(l)} = \sigma(\boldsymbol{z}^{(l)}) = \sigma(\boldsymbol{W}^{(l)}\boldsymbol{a}^{(l-1)} + \boldsymbol{b}^{(l)}). \tag{3.4}$$

The images in Figs. 3.4 and 3.5 were generated in a very similar fashion, namely by forward-propagating arbitrary inputs through a fully connected neural network with random weights and biases. The more complex images were achieved only by adding more hidden layers and neurons to the network architecture.

Implementing a fully connected neural network in Python using PyTorch is straightforward. The following code creates a neural network given an activation function as `activationFunction` and the number of neurons per layer as the list `layerDimensions`.

Listing 3.1 Fully connected neural network

```
# neural network architecture definition
layerDimensions = [1, 10, 10, 1]
activationFunction = torch.nn.ReLU()

# neural network creation
modules = []
for i in range(len(layerDimensions) - 1):
    modules.append(torch.nn.Linear(layerDimensions[i],
                                   layerDimensions[i + 1]))
    modules.append(activationFunction)
model = torch.nn.Sequential(*modules)

# prediction with neural network
x = torch.tensor([[1.], [2.]])
y = model(x)
```

3.3 Differentiation

Up to this point, the neural network was only able to transform an input into an output depending on the randomly initialized set of weights and biases—meaning the network has not utilized any data to learn the optimal parameters $w^\star$ and $b^\star$ that generate the desired output. As for the example of linear regression, learning from data can be formulated as an optimization problem that requires the definition of a suitable cost function. Once again, the mean squared error loss is chosen to quantify the prediction accuracy of the model. The cost function takes on the familiar data-driven form

$$C = \mathcal{L}_\mathcal{D} = \frac{1}{2m_\mathcal{D}} \sum_{i=1}^{m_\mathcal{D}} (\tilde{y}_i - \hat{y}_i)^2, \tag{3.5}$$

but with the addition of the factor $\frac{1}{2}$ to simplify the expression of the derivative needed for later calculations. To find optimal network parameters, the cost function is minimized with the gradient descent method that was introduced in Sect. 2.8. Essential for the iterative parameter update (cf. Eq. (2.18)) is the computation of the gradient of the cost function with respect to the weights and biases.

For simple network architectures, it is possible to compute the corresponding derivatives analytically. Therefore, the network must be expressed as a function of the inputs and the network parameters. Given the example from the previous section (cf. Fig. 3.8), the analytical expression of the network is

$$\hat{y} = \sigma\left(w_1^{(2)}\sigma(w_{11}^{(1)}x_1 + w_{12}^{(1)}x_2) + w_2^{(2)}\sigma(w_{21}^{(1)}x_1 + w_{22}^{(1)}x_2)\right). \tag{3.6}$$

To improve the readability and since the analytical form is valid for all examples, the index i is dropped here. Now, the cost function for a single example of the network from Fig. 3.8 can be written as

$$C = \frac{1}{2}(\tilde{y} - \hat{y})^2 = \frac{1}{2}\left[\tilde{y} - \sigma\left(w_1^{(2)}\sigma(w_{11}^{(1)}x_1 + w_{12}^{(1)}x_2) + w_2^{(2)}\sigma(w_{21}^{(1)}x_1 + w_{22}^{(1)}x_2)\right)\right]^2. \tag{3.7}$$

To compute the analytical expression, e.g., for the derivative $\frac{\partial C}{w_{11}^{(1)}}$, the chain rule is applied to Eq. (3.5) so that

$$\frac{\partial C}{\partial w_{11}^{(1)}} = \frac{\partial C}{\partial \hat{y}} \frac{\partial \hat{y}}{\partial w_{11}^{(1)}}, \tag{3.8}$$

with

$$\frac{\partial C}{\partial \hat{y}} = -(\tilde{y} - \hat{y}). \tag{3.9}$$

Deriving Eq. (3.6) with respect to $w_{11}^{(1)}$ yields

$$\frac{\partial \hat{y}}{\partial w_{11}^{(1)}} = \sigma'\left(w_1^{(2)}\sigma(w_{11}^{(1)}x_1 + w_{12}^{(1)}x_2) + w_2^{(2)}\sigma(w_{21}^{(1)}x_1 + w_{22}^{(1)}x_2)\right) w_1^{(2)}\sigma'(w_{11}^{(1)}x_1 + w_{12}^{(1)}x_2)x_1.$$

Similarly, the remaining derivatives could be computed, but for larger and deeper networks, the analytical differentiation becomes infeasible. In engineering applications, a common approach for the numerical approximation of derivatives is the method of finite differences. However, this approach introduces truncation errors and quickly becomes computationally infeasible for large networks with many parameters [19]. Automatic differentiation, or backpropagation in the case of neural networks, is a more efficient alternative, allowing fast and precise numerical evaluation derivatives with respect to many parameters (including higher-order derivatives). Generally, a computer program is nothing else than the sum of its basic arithmetic operations and elementary function evaluations. Automatic differentiation augments the code during execution by immediately evaluating and storing the derivatives of each basic operation. Then, repeatedly applying the chain rule to the accumulated derivative values allows to compute the derivative of the whole composition [19].

3.4 Backpropagation

3.4.1 *Basic Methodology*

Backpropagation is an efficient technique for determining the partial derivatives of graph-structured functions and in particular neural networks. It belongs to the field of automatic differentiation that deals with the algorithmic computation of derivatives [6]. In the following, the general form of the backpropagation algorithm is derived for fully connected neural networks and then illustrated on the example network from Sect. 3.2 (cf. Fig. 3.8).

If the cost function C is seen as an average over the sum of cost functions C_i of the individual examples i

$$C = \frac{1}{m_{\mathcal{D}}} \sum_{i=1}^{m_{\mathcal{D}}} C_i, \tag{3.10}$$

then the cost C_i for one example i can be written as

$$C_i = \frac{1}{2}(\tilde{y}_i - \hat{y}_i)^2 = \frac{1}{2}(\tilde{y}_i - a_i^{(L)})^2, \tag{3.11}$$

where $a_i^{(L)}$ denotes the output of the last layer L and simultaneously the output $\hat{y}_i$ of the neural network. From here, the computation of the derivatives is only performed for a single example. Thus, the index i in Eq. (3.11) is dropped to improve the readability. As mentioned, the goal of the backpropagation algorithm is to compute the partial derivatives of the cost function with respect to the weights $\frac{\partial C}{\partial w_{jk}^{(l)}}$ and biases $\frac{\partial C}{\partial b_j^{(l)}}$.

Recalling Eq. (3.3), the output of a layer l is defined as

$$a_j^{(l)} = \sigma(z_j^{(l)}), \tag{3.12}$$

where

$$z_j^{(l)} = \sum_k w_{jk}^{(l)} a_k^{(l-1)} + b_j^{(l)} \tag{3.13}$$

is computed using the outputs of the previous layer $a_k^{(l-1)}$.

Now, rewriting $\frac{\partial C}{\partial w_{jk}^{(l)}}$ yields

$$\frac{\partial C}{\partial w_{jk}^{(l)}} = \frac{\partial C}{\partial z_j^{(l)}} \frac{\partial z_j^{(l)}}{\partial w_{jk}^{(l)}} = \delta_j^{(l)} \frac{\partial z_j^{(l)}}{\partial w_{jk}^{(l)}} = \delta_j^{(l)} a_k^{(l-1)}, \tag{3.14}$$

with

$$\frac{\partial z_j^{(l)}}{\partial w_{jk}^{(l)}} = \frac{\partial}{\partial w_{jk}^{(l)}} \sum_h w_{jh}^{(l)} a_h^{(l-1)} + b_j^{(l)} = a_k^{(l-1)}, \tag{3.15}$$

where the index h runs over all neurons from the previous layer $(l-1)$. Similarly, $\frac{\partial C}{\partial b_j^{(l)}}$ is computed as

$$\frac{\partial C}{\partial b_j^{(l)}} = \frac{\partial C}{\partial z_j^{(l)}} \frac{\partial z_j^{(l)}}{\partial b_j^{(l)}} = \delta_j^{(l)} \frac{\partial z_j^{(l)}}{\partial b_j^{(l)}} = \delta_j^{(l)}, \tag{3.16}$$

since

$$\frac{\partial z_j^{(l)}}{\partial b_j^{(l)}} = \frac{\partial}{\partial b_j^{(l)}} \sum_k w_{jk}^{(l)} a_k^{(l-1)} + b_j^{(l)} = 1. \tag{3.17}$$

Equations (3.14) and (3.16) introduce a new variable $\delta_j^{(l)}$ that describes the sensitivity of the cost function toward a change in the neuron's weighted input $z_j^{(l)}$. If $\delta_j^{(l)}$ for each neuron j in layer l is known, the derivative $\frac{\partial C}{\partial w_{jk}^{(l)}}$ can be computed simply by multiplying $\delta_j^{(l)}$ of the jth neuron of the current layer l with the output $a_k^{(l-1)}$ of the kth neuron from the previous layer $(l-1)$. The derivative with respect to the bias $\frac{\partial C}{\partial b_j^{(l)}}$ directly equates to $\delta_j^{(l)}$.

In the next step, an expression for $\delta_j^{(l)}$ in the output layer L is derived

$$\delta_j^{(L)} = \frac{\partial C}{\partial z_j^{(L)}} = \frac{\partial C}{\partial a_j^{(L)}} \frac{\partial a_j^{(L)}}{\partial z_j^{(L)}} = \frac{\partial C}{\partial a_j^{(L)}} \sigma'(z_j^{(L)}) = -(\tilde{y}_j - \sigma(z_j^{(L)}))\sigma'(z_j^{(L)}), \quad (3.18)$$

recalling that $a_j^{(L)} = \sigma(z_j^{(L)})$ and inserting

$$\frac{\partial a_j^{(L)}}{\partial z_j^{(L)}} = \frac{\partial}{\partial z_j^{(L)}} \sigma(z_j^{(L)}) = \sigma'(z_j^{(L)}). \quad (3.19)$$

What is still missing is a general term for $\delta_j^{(L)}$ in an arbitrary layer l. Utilizing the chain rule $\delta_j^{(l)}$ can be expressed in terms of $\delta_k^{(l+1)}$

$$\delta_j^{(l)} = \frac{\partial C}{\partial z_j^{(l)}} = \sum_k \frac{\partial C}{\partial z_k^{(l+1)}} \frac{\partial z_k^{(l+1)}}{\partial z_j^{(l)}} = \sum_k \frac{\partial z_k^{(l+1)}}{\partial z_j^{(l)}} \delta_k^{(l+1)}. \quad (3.20)$$

With $a_j^{(l)} = \sigma(z_j^{(l)})$ and the definition of $z_k^{(l+1)}$ being

$$z_k^{(l+1)} = \sum_j w_{kj}^{(l+1)} a_j^{(l)} + b_k^{(l+1)} = \sum_j w_{kj}^{(l+1)} \sigma(z_j^{(l)}) + b_k^{(l+1)}, \quad (3.21)$$

the derivative with respect to $z_j^{(l)}$ can be calculated as

$$\frac{\partial z_k^{(l+1)}}{\partial z_j^{(l)}} = w_{kj}^{(l+1)} \sigma'(z_j^{(l)}). \quad (3.22)$$

Substituting Eq. (3.22) into Eq. (3.20) finally yields

$$\delta_j^{(l)} = \sum_k w_{kj}^{(l+1)} \delta_k^{(l+1)} \sigma'(z_j^{(l)}). \quad (3.23)$$

With Eqs. (3.14), (3.18), and (3.23), all expressions to calculate the partial derivatives of the simple example network are available. Since the biases in the example are chosen to be zero, the computation of the derivatives has to be executed with respect to the four entries of $\boldsymbol{W}^{(1)}$ and two entries of $\boldsymbol{W}^{(2)}$. This also implies that Eq. 3.16 is not needed. The single data point considered in this example consists of the input vector $\boldsymbol{x} = (3, 2)^\intercal$ and the target $y = 1$. In order to compute the sensitivity for the output neuron $\delta^{(L)}$, the derivative of the activation function $\sigma'(z) = 2z$ and the results for $z_1^{(2)}$ from the previous section are inserted into Eq. (3.18) resulting in

$$\delta_1^{(2)} = -(\tilde{y} - \sigma(z_1^{(2)}))\sigma'(z_1^{(2)}) = -(1 - 4)\sigma'(2) = 12.$$

This intermediate result is then propagated backward to compute $\delta_1^{(1)}$ and $\delta_2^{(1)}$ of the hidden layer using Eq. (3.23)

$$\delta_1^{(1)} = w_1^{(2)} \delta_1^{(2)} \sigma'(z_1^{(1)}) = 2 \cdot 12 \cdot \sigma'(-3) = -144,$$

$$\delta_2^{(1)} = w_2^{(2)} \delta_1^{(2)} \sigma'(z_2^{(1)}) = (-1) \cdot 12 \cdot \sigma'(-4) = 96.$$

Finally, the partial derivatives are calculated with Eq. (3.14) using the previously computed δ-values

$$\frac{\partial C}{\partial w_{11}^{(1)}} = \delta_1^{(1)} x_1 = -144 \cdot 3 = -432,$$

$$\frac{\partial C}{\partial w_{12}^{(1)}} = \delta_1^{(1)} x_2 = -144 \cdot 2 = -288,$$

$$\frac{\partial C}{\partial w_{21}^{(1)}} = \delta_2^{(1)} x_1 = 96 \cdot 3 = 288,$$

$$\frac{\partial C}{\partial w_{22}^{(1)}} = \delta_2^{(1)} x_2 = 96 \cdot 2 = 192,$$

$$\frac{\partial C}{\partial w_1^{(2)}} = \delta_1^{(2)} a_1^{(1)} = 12 \cdot 9 = 108,$$

$$\frac{\partial C}{\partial w_2^{(2)}} = \delta_1^{(2)} a_2^{(1)} = 12 \cdot 16 = 192.$$

The partial derivatives can now be used to update the parameters according to the update rule from Eq. (2.18) in Sect. 2.8:

$$\boldsymbol{W}_{\text{new}}^{(1)} = \boldsymbol{W}^{(1)} - \alpha \nabla_{\Theta} C_{\boldsymbol{W}^{(1)}}, \quad (3.24)$$

$$\boldsymbol{w}_{\text{new}}^{(2)} = \boldsymbol{w}^{(2)} - \alpha \nabla_{\Theta} C_{\boldsymbol{w}^{(2)}}. \quad (3.25)$$

Updating the weights with a learning rate $\alpha = 0.001$ yields

$$\boldsymbol{W}_{\text{new}}^{(1)} = \begin{pmatrix} 1 & -3 \\ -2 & 1 \end{pmatrix} - 0.001 \begin{pmatrix} -432 & -288 \\ 288 & 192 \end{pmatrix} = \begin{pmatrix} 1.432 & -2.712 \\ -2.288 & 0.808 \end{pmatrix} \quad (3.26)$$

and

$$\boldsymbol{w}_{\text{new}}^{(2)} = \begin{pmatrix} 2 \\ -1 \end{pmatrix} - 0.001 \begin{pmatrix} 108 \\ 192 \end{pmatrix} = \begin{pmatrix} 1.892 \\ -1.192 \end{pmatrix}. \quad (3.27)$$

An intuitive explanation of backpropagation can be found in the second chapter of Nielsen's online book "*Neural Networks and Deep Learning*" that was used as a reference for the formal derivation of the presented algorithm [8]. Another illustrative example is described in Christopher Olah's blog post [20].

3.4.2 Input Derivatives

Since neural networks are fully differentiable functions, it is not only possible to compute the derivatives with respect to the network parameters $\boldsymbol{\Theta}$ but also with respect to the input variables, as will be seen for the Sobolev training in Sect. 3.8 or Hamiltonian and Lagrangian neural networks in Sect. 3.9.8.1. Even higher-order derivatives can be evaluated and used to check if a solution satisfies a governing partial differential equation of a given problem. This will be exploited extensively in Chaps. 4 and 5, where neural networks learn to approximate solutions of typical physics and engineering problems governed by partial differential equations. To achieve this, physical solutions (predicted by neural networks) must be differentiated with respect to input quantities, such as space or time.

Computing the derivatives of the network output with respect to the input variables is demonstrated on the example from Sect. 3.2 (cf. Fig. 3.8). To this end, we recall the corresponding analytical expression of the network

$$\hat{y} = \sigma(w_1^{(2)}\sigma(w_{11}^{(1)}x_1 + w_{12}^{(1)}x_2) + w_2^{(2)}\sigma(w_{21}^{(1)}x_1 + w_{22}^{(1)}x_2)). \tag{3.28}$$

For instance, the derivative of the network output $\hat{y}$ with respect to the input variable x_1 takes on the following form:

$$\begin{aligned}\frac{\partial \hat{y}}{\partial x_1} &= \sigma'(w_1^{(2)}\sigma(w_{11}^{(1)}x_1 + w_{12}^{(1)}x_2) + w_2^{(2)}\sigma(w_{21}^{(1)}x_1 + w_{22}^{(1)}x_2)) \\ &\quad \cdot (w_1^{(2)}\sigma'(w_{11}^{(1)}x_1 + w_{12}^{(1)}x_2)w_{11}^{(1)} + w_2^{(2)}\sigma'(w_{21}^{(1)}x_1 + w_{22}^{(1)}x_2)w_{21}^{(1)}).\end{aligned} \tag{3.29}$$

As for the derivatives with respect to the network parameters, an analytical differentiation is not feasible in practice. Here, deep learning frameworks that make use of automatic differentiation and general computation graphs are well-suited for fast and precise numerical evaluation of the desired derivatives.

3.4.3 Generalization with Computation Graphs

The previous two examples of computing derivatives of fully connected neural networks (with respect to parameters or inputs) illustrated the essence of the backpropagation algorithm, which is the sequential application of the chain rule (specifically

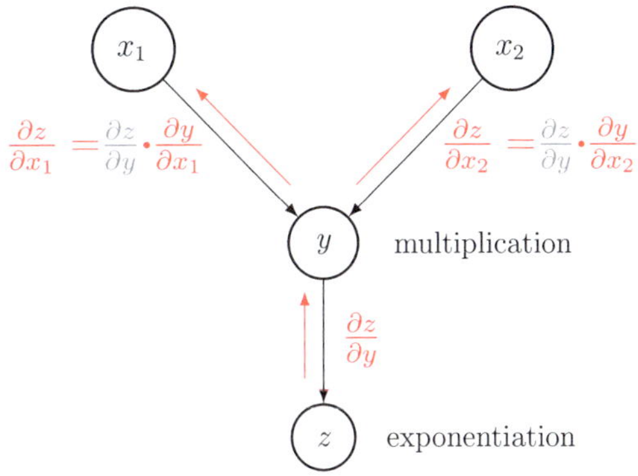

Fig. 3.9 Computation graph constructed from the operations defined in Eqs. (3.30) and (3.31). The elementary operations are indicated by black arrows, while the backpropagation using the chain rule is indicated in red

in reverse mode) to neural network outputs. This is done on the lowest possible level, i.e., the elementary arithmetic operations and functions. In practice (in automatic differentiation frameworks), this is implemented using computation graphs, which keep track of each elementary operation. Each elementary operation is saved as a node with its corresponding value. Consider, for example, the following arithmetic operations

$$y = x_1 \cdot x_2, \tag{3.30}$$

$$z = y^2, \tag{3.31}$$

from which the computation graph in Fig. 3.9 can be constructed. The chain rule can then be applied in a systematic way by propagating backward through the computation graph, as highlighted in red. Thereby, the derivative of z is found with respect to x_1 and x_2.

Taking this concept one step further, a general algorithm can be formulated. Consider the arbitrary node c in Fig. 3.10, which is part of a larger computation graph. It receives the inputs from nodes a and b and propagates to nodes d and e. The gradient information of the cost function C is now backpropagated through the entire graph and reaches node c. The gradient with respect to the value c is obtained by summing all dependencies

$$\frac{\partial C}{\partial c} = \frac{\partial C}{\partial c_d} + \frac{\partial C}{\partial c_e} \tag{3.32}$$

where the derivatives $\frac{\partial C}{\partial c_d}$, $\frac{\partial C}{\partial c_e}$ are obtained by combining the backpropagated gradients and the information of the elementary operation conducted at nodes e and d. This is best explained by considering the information passed backward by node c, which are the gradients $\frac{\partial C}{\partial a_c}$ and $\frac{\partial C}{\partial b_c}$. These are computed with the chain rule:

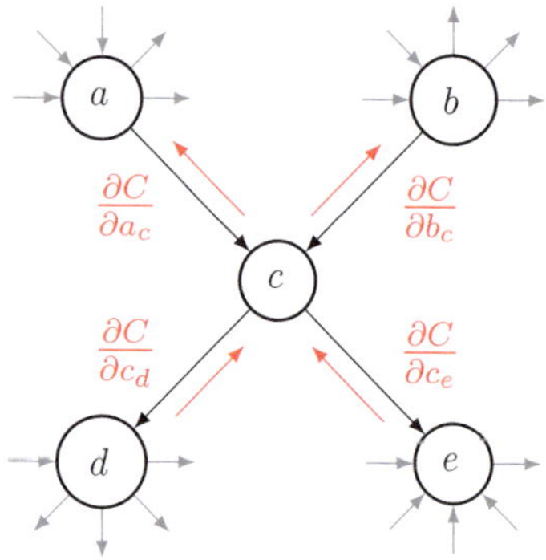

Fig. 3.10 Backpropagation in a general computation graph around node c. The gradient travels backward through the graph, marked in red. The gray arrows indicated connections to the remaining computation graph. Figure inspired from [21]

$$\frac{\partial C}{\partial a_c} = \frac{\partial C}{\partial c}\frac{\partial c}{\partial a_c}, \tag{3.33}$$

$$\frac{\partial C}{\partial b_c} = \frac{\partial C}{\partial c}\frac{\partial c}{\partial b_c}. \tag{3.34}$$

The derivatives $\frac{\partial c}{\partial a_c}$, $\frac{\partial c}{\partial b_c}$ are given through the elementary operation performed at node c. In the case of the previous example illustrated in Fig. 3.9, node y is obtained by a multiplication. The corresponding elementary derivatives are thus given by:

$$\frac{\partial y}{\partial x_1} = x_2, \tag{3.35}$$

$$\frac{\partial y}{\partial x_2} = x_1. \tag{3.36}$$

In practice, the derivatives of elementary arithmetic operations, such as multiplication and addition, and common functions, such as trigonometric and activation functions, are stored in a library, which are then used in the computation graph. Note, also, that any derivatives of any quantity within the computation graph can be obtained, including the input derivatives, as seen in Sect. 3.4.2.

In PyTorch, the automatic differentiation is performed analogously. The computation graph is traversed during backpropagation to compute all derivatives—where default-wise, only the derivatives with respect to the leaf (final) nodes are stored. To illustrate PyTorch's behavior and options, consider the example from Eqs. (3.30) and (3.31), which can be written as:

```
x1 = torch.tensor([1.], requires_grad=True)
x2 = torch.tensor([2.], requires_grad=True)
y = x1 * x2
z = y ** 2
```

As x_1 and x_2 both require gradients, enabled with `requires_grad=True`, a computation graph is constructed. By querying the function handles with `.grad_fn`, the elementary operation performed to arrive at the variable is retrieved. For `y.grad_fn`, `MulBackward` is returned, indicating the multiplication between x_1 and x_2, while `z.grad_fn` returns a `PowBackward` handle.

The backward propagation through the computation graph is performed with

```
z.backward()

x1.grad  # to access gradient dz/dx_1
x2.grad  # to access gradient dz/dx_2
```

which ends at the leaf variables x_1 and x_2. The intermediate derivatives, such as $\frac{\partial z}{\partial y}$, are not stored unless specified prior to the gradient computation. To enable this, one has to specify `y.retain_grad()`. Furthermore, the complete computation graph is deleted after the `.backward()` call. This can be avoided by the `retain_graph =True` option. However, if derivatives such as `x1.grad` or `x2.grad` are used in a subsequent computation, which is to be derived, the option `create_graph=True` has to be enabled additionally. This considers the elementary operations employed during the backpropagation, from which it extends the computation graph.

3.5 Activation Function

As observed in the preceding section, the activation function σ needs to be differentiable in order to train neural networks with gradient-based optimization methods. Inspired by the biological archetype, the first neural networks, known as perceptrons, used a Heaviside step function to imitate an active ($= 1$) or inactive neuron ($= 0$) (see Fig. 3.3). However, the derivative of a step function is zero everywhere except at the jump, which makes it impractical for training networks with gradient-based optimization. Until recently, continuous functions like the hyperbolic tangent or the sigmoid, also known as the logistic function, were popular choices for the activation σ. Unfortunately, with increasing weights, those functions tend to saturate, resulting in very small gradients that prevent the network from further learning [6]. To resolve this problem of "*vanishing gradients*", a new type of activation function was introduced. Rectified linear units or, in short, ReLU (cf. Fig. 3.3) are now used in most feed-forward neural networks [6]. They help to reduce saturation since the function grows linearly for positive inputs. To accommodate negative inputs, a modification of the ReLU called leaky ReLU was introduced. Here negative activation inputs that would result in zero for ReLU are instead multiplied with a small gradient α, as shown in Fig. 3.11. This retains the non-linearity while allowing for negative activations. A disadvantage is the non-differentiable kink at zero. Several activation functions such as the sigmoid linear unit (SiLU, also known as Swish) [22, 23], the exponential linear unit (ELU) [24], the scaled exponential linear unit (SELU) [25], the Gaussian error linear unit (GELU) [26], or the Mish [27] overcome this issue;

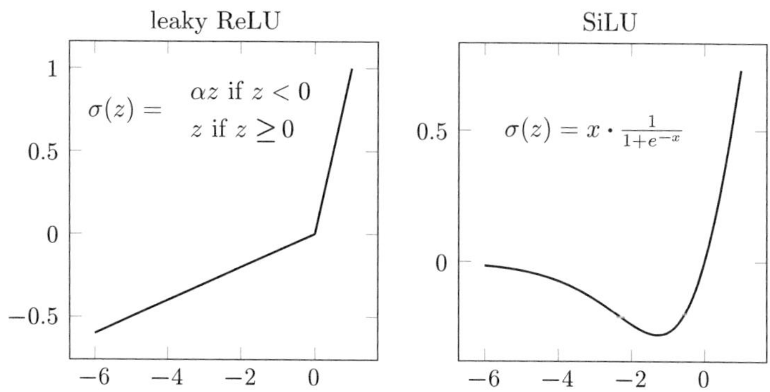

Fig. 3.11 ReLU variations: Leaky ReLU and SiLU activation function

see, e.g., Fig. 3.11 for the SiLU activation function. Further extensions introduce learnable parameters within the activation function [28] to enhance the network's expressivity. In the case of the leaky ReLU, the slope α is updated by the learning algorithm. The modified activation function is therefore called parametric rectified linear unit or, in short, PReLU [29]. Generally, the choice of the activation function has a considerable impact on the performance and especially on the learning rate of a model but usually depends on the specific problem and the experience of the machine learning practitioner [30]. ReLU is, however, often a reasonable starting point if no prior knowledge of the problem is available. For additional information on the selection of the activation function, consult, e.g., [11].

3.6 Learning Algorithm

Within the previous sections, all relevant theory is provided to build a neural network. To summarize, the following aspects are required to train the network for a supervised learning task [31]:

- Input data $\tilde{\boldsymbol{X}}$ and corresponding targets $\tilde{\boldsymbol{y}}$, divided into a training, validation and test set
- Network topology, defined by the input, output, and hidden layers, the corresponding number of neurons, and their connections
- An activation function σ
- A cost function C, which defines the feedback signal used for learning
- A way to compute the gradients with respect to the network parameters, e.g., backpropagation
- An optimizer with hyperparameters, such as the learning rate α, which determines how learning proceeds

Algorithm 5 roughly describes the learning algorithm for a fully connected neural network that regresses a scalar value $\hat{y}$ from a given input $\boldsymbol{x}$.

Algorithm 5 Training a neural network with full-batch gradient descent. The inner loop is only displayed for a better understanding. Normally, the loop over the examples is vectorized for more efficient computations.

Require: training data $\tilde{X}$, targets $\tilde{\boldsymbol{y}}$
  define network architecture (input layer, hidden layers, output layer, activation function) set learning rate α
  initialize weights $\boldsymbol{W}$ and biases $\boldsymbol{b}$
  for all epochs **do**
    for example $i \leftarrow 1$ to $m_{\mathcal{D}}$ **do**
      apply forward propagation: $\hat{y}_i \leftarrow f_{NN}(\boldsymbol{x}_i; \boldsymbol{W}, \boldsymbol{b})$ ▷ cf. Section 3.2
      compute loss: $C_i \leftarrow (\tilde{y}_i - \hat{y}_i)^2$
      apply backpropagation for gradients $\partial C_i / \partial \boldsymbol{W}, \partial C_i / \partial \boldsymbol{b}$ ▷ cf. Section 3.4
    end for
    compute full-batch cost function: $C \leftarrow \frac{1}{m_{\mathcal{D}}} \sum_{i=1}^{m_{\mathcal{D}}} C_i$
    compute full-batch gradients w.r.t. $\boldsymbol{W}$: $\frac{\partial C}{\partial \boldsymbol{W}} \leftarrow \frac{1}{m_{\mathcal{D}}} \sum_{i=1}^{m_{\mathcal{D}}} \frac{\partial C_i}{\partial \boldsymbol{W}}$
    compute full-batch gradients w.r.t. $\boldsymbol{b}$: $\frac{\partial C}{\partial \boldsymbol{b}} \leftarrow \frac{1}{m_{\mathcal{D}}} \sum_{i=1}^{m_{\mathcal{D}}} \frac{\partial C_i}{\partial \boldsymbol{b}}$
    update weights: $\boldsymbol{W} \leftarrow \boldsymbol{W} - \alpha \frac{\partial C}{\partial \boldsymbol{W}}$
    update biases: $\boldsymbol{b} \leftarrow \boldsymbol{b} - \alpha \frac{\partial C}{\partial \boldsymbol{b}}$
  end for

Instead of using full-batch gradient descent, the parameters can be updated with a stochastic or mini-batch gradient descent step. In the case of stochastic gradient descent, the parameter update is executed inside the loop over all examples separately. This means a gradient descent step is taken each time an example is propagated through the network (cf. Algorithm 6).

Algorithm 6 Training a neural network with stochastic gradient descent

---

Require: training data $\tilde{X}$, targets $\tilde{\boldsymbol{y}}$
  define network architecture (input layer, hidden layers, output layer, activation function) set learning rate α
  initialize weights $\boldsymbol{W}$ and biases $\boldsymbol{b}$
  for all epochs **do**
    for example $i \leftarrow 1$ to $m_{\mathcal{D}}$ **do**
      apply forward propagation $\hat{y}_i \leftarrow f_{NN}(\boldsymbol{x}_i; \boldsymbol{W}, \boldsymbol{b})$ ▷ cf. Section 3.2
      compute loss: $C_i \leftarrow (\tilde{y}_i - \hat{y}_i)^2$
      apply backpropagation for gradients $\partial C_i / \partial \boldsymbol{W}, \partial C_i / \partial \boldsymbol{b}$ ▷ cf. Section 3.4
      update weights: $\boldsymbol{W} \leftarrow \boldsymbol{W} - \alpha \frac{\partial C_i}{\partial \boldsymbol{W}}$
      update biases: $\boldsymbol{b} \leftarrow \boldsymbol{b} - \alpha \frac{\partial C_i}{\partial \boldsymbol{b}}$
    end for
  end for

The most popular optimization approach in practice is mini-batch gradient descent (cf. Algorithm 7). Instead of computing the gradients averaged over the whole training set, the gradients are evaluated just for a small part of the training data, the so-called mini-batch. This allows the processing of datasets more efficiently when they are too big to fit the memory as a whole. Apart from that, mini-batch gradient descent is preferred because it is said to introduce a regularizing effect [32] due to the fact that each training minibatch suggests a slightly different direction of best descent, allowing the optimization path to explore more of the parameter space as it searches for optima.

All three approaches can be implemented with the same Python implementation, relying on a neural network model, as the one defined in Listing 3.1. By defining a dataset with `TensorDataset` (for custom datasets, use `Dataset`), a data loader can be defined with `DataLoader`. The data loader's purpose is to load the current (randomized) batch within the training loop. By modifying the batch size to ten, full-batch gradient descent is recovered, while reducing the batch size to one results in stochastic gradient descent. Mini-batch gradient descent is achieved with all intermediate batch sizes.

Listing 3.2 Stochastic, mini-batch and full-batch gradient descent

```python
from torch.utils.data import TensorDataset, DataLoader

# training data
x = torch.linspace(0, 1, 10).unsqueeze(1)
y = 2 * x + 3  # function to be approximated

# hyperparameters
lr = 1e-3
epochs = 1000
batchSize = 2  # 1 for SGD and 10 for full-batch

dataset = TensorDataset(x, y)
dataloader = DataLoader(dataset, batch_size=batchSize,
    shuffle=True)
optimizer = torch.optim.Adam(model.parameters(), lr=lr)

# training loop
for epoch in range(epochs):
    for xBatch, yBatch in dataloader:
        yBatchPred = model(xBatch)  # forward propagation
        cost = 0.5 * torch.mean((yBatchPred - yBatch) **
    2)
        cost.backward()  # backward propagation
        optimizer.step()  # gradient descent update
        optimizer.zero_grad()  # reset gradients
```

Algorithm 7 Training a neural network with mini-batch gradient descent

Require: training data $\tilde{X}$, targets $\tilde{y}$
  define network architecture (input layer, hidden layers, output layer, activation function) set learning rate α
  initialize weights W and biases b
  for all epochs **do**
    shuffle rows of X and y synchronously (optional)
    divide X and y into n batches of size k
    for all batches **do**
      for example $i \leftarrow 1$ to k **do**
        apply forward propagation: $\hat{y}_i \leftarrow f_{NN}(x_i; W, b)$ ▷ cf. Section 3.2
        compute loss: $C_i \leftarrow (\tilde{y}_i - \hat{y}_i)^2$
        apply backpropagation for gradients $\partial C_i / \partial W$, $\partial C_i / \partial b$ ▷ cf. Section 3.4
      end for
      compute mini-batch cost function: $C \leftarrow \frac{1}{k} \sum_{i=1}^{k} C_i$
      compute mini-batch gradient w.r.t. W: $\frac{\partial C}{\partial W} \leftarrow \frac{1}{k} \sum_{i=1}^{k} \frac{\partial C_i}{\partial W}$
      compute mini-batch gradients w.r.t. b : $\frac{\partial C}{\partial b} \leftarrow \frac{1}{k} \sum_{i=1}^{k} \frac{\partial C_i}{\partial b}$
      update weights: $W \leftarrow W - \alpha \frac{\partial C}{\partial W}$
      update biases: $b \leftarrow b - \alpha \frac{\partial C}{\partial b}$
    end for
  end for

3.7 Regularization of Neural Networks

As stated in the previous section on regularization (cf. Sect. 2.7), the task of making an algorithm perform well on new inputs and not only on the training data is one of the biggest challenges in machine learning and is a field of extensive research. Regularization includes all strategies aiming to diminish the test error without increasing the training error. Ideally, they trade a significant reduction of variance for a slightly increased bias. There exist multiple approaches to regularize a machine learning model or, in particular, neural networks. One option is to formulate certain constraints, for example, by directly restricting the parameter values or by adding an extra term to the objective function that constrains the parameters indirectly (through penalization). Some constraints and restrictions can alter an undetermined problem into a determined one; others prefer simpler models for better generalization properties. The idea of choosing the simplest hypothesis among competing explanations stems back to the fourteenth century and is known as Occam's razor. Overall, these constraints and restrictions are often designed to encode prior knowledge about the problem and, if chosen properly, can help to reduce the generalization error [6]. Furthermore, Sect. 2.7 introduced the three cases of underfitting (or high bias), overfitting (or high variance), and an ideal model capacity that matches the complexity of the underlying problem.

Most problems tackled by deep learning algorithms, like image recognition or audio sequences, are too complex to be modeled precisely. However, experience has shown that building a large model with an appropriate regularization mechanism yields the best results in terms of minimizing the generalization error [6].

3.7.1 Early Stopping

When plotting the training and validation error for each iteration, it is often observed that while the training error declines steadily, the test error begins to increase again (cf. Fig. 3.12). This phenomenon typically occurs when the model has sufficient capacity to overfit the problem. The principal idea of early stopping is to halt the training process as soon as the validation error rises and the model enters the overfitting regime (see point A in Fig. 3.12) [2]. In this way, the fitting of particular features of the training samples can be avoided. It is essential to monitor the error on a validation set and not the test set because the number of training steps of the gradient-based algorithm has now become a hyperparameter of the model. Every time the validation error decreases, the weights and biases of the model are stored. The algorithm terminates when the validation error has not improved over a predefined number of iteration steps (see point B in Fig. 3.12). Finally, the model parameters are returned at the point of the lowest validation error.

Due to its simplicity and effectiveness, early stopping is a very popular regularization method often applied in practice. It can be seen as a hyperparameter selection algorithm determining the ideal number of training steps. The only additional costs

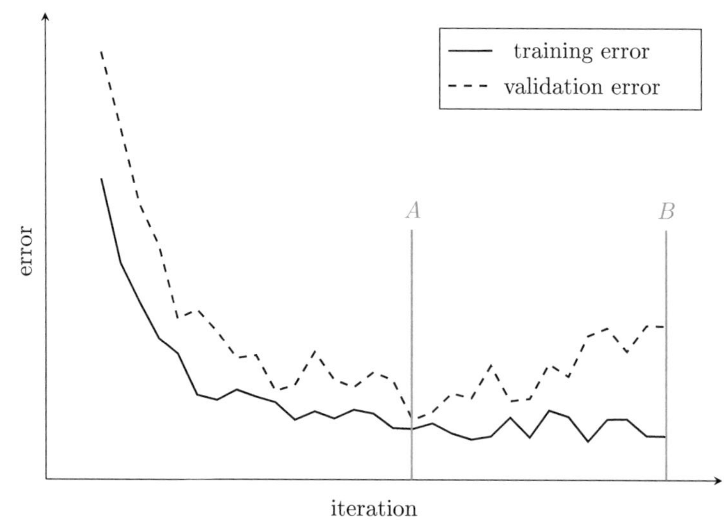

Fig. 3.12 Early stopping: at point A, the validation error reaches its minimum. At point B, the optimization routine is terminated since the validation error has not shown any improvement for the prescribed number of iterations

are the evaluations of the validation set[2] after each epoch and the memory used to store the parameters. Conversely, the computational costs are often reduced significantly since the execution of unnecessary training steps is prevented.

3.7.2 L^1 and L^2 Regularization

The two main representatives from the family of parameter norm penalties have already been introduced in Sect. 2.8, namely L^1 and L^2 regularization. The methods have briefly been described for linear regression, and it is straightforward to extend them to regularize neural networks. Both regularization techniques aim to limit the capacity of a model by penalizing the parameters $\boldsymbol{\Theta}$ of the model with the help of a penalty term $\Omega(\boldsymbol{\Theta})$ that is added to the cost function C [6]. In a general form, this can be expressed as

$$\tilde{C} = C + \lambda\Omega, \tag{3.37}$$

where $\tilde{C}$ is the regularized and C the unregularized cost function. The coefficient λ is a hyperparameter weighting the relative contribution of the penalty term Ω. If $\lambda = 0$, no regularization is applied, and larger values for λ result in a more regularized model. In the case of L^1 regularization, the cost function takes the form

$$\tilde{C} = C + \lambda||\boldsymbol{w}||_1, \tag{3.38}$$

and

$$\tilde{C} = C + \frac{\lambda}{2}\boldsymbol{w} \cdot \boldsymbol{w} \tag{3.39}$$

for L^2 regularization, respectively, where C can be any cost function, e.g., the mean squared error.

How these terms influence the training process can be shown when deriving the update rules of the gradient descent algorithm for the regularized cost function $\tilde{C}$ [6]. Beginning with computing the partial derivatives of Eq. (3.38)

$$\frac{\partial\tilde{C}}{\partial\boldsymbol{w}} = \frac{\partial C}{\partial\boldsymbol{w}} + \lambda\,\text{sign}(\boldsymbol{w}), \tag{3.40}$$

where $\text{sign}(\boldsymbol{w})$ is applied element-wise and Eq. (3.39)

$$\frac{\partial\tilde{C}}{\partial\boldsymbol{w}} = \frac{\partial C}{\partial\boldsymbol{w}} + \lambda\boldsymbol{w}, \tag{3.41}$$

---

[2] These evaluations are relatively cheap in comparison to the computational effort of the gradient computation for the training data. No gradients have to be computed, and in turn, no computational graph has to be constructed for the validation data.

the learning rule for the weights takes on the updated form

$$\boldsymbol{w} \rightarrow \boldsymbol{w}' = \boldsymbol{w} - \alpha\lambda\,\mathrm{sign}(\boldsymbol{w}) - \alpha\frac{\partial C}{\partial \boldsymbol{w}} \tag{3.42}$$

for L^1 regularization and

$$\boldsymbol{w} \rightarrow \boldsymbol{w}' = \boldsymbol{w}\,(1 - \alpha\lambda) - \alpha\frac{\partial C}{\partial \boldsymbol{w}}, \tag{3.43}$$

for L^2 regularization, respectively. In comparison, the update rule for an unregularized cost function is defined as

$$\boldsymbol{w} \rightarrow \boldsymbol{w}' = \boldsymbol{w} - \alpha\frac{\partial C}{\partial \boldsymbol{w}}. \tag{3.44}$$

Depending on the choice of the norm, different solutions are preferred [8]. L^1 regularization encourages the selection of a few connections of high importance while the other weights are forced toward zero. Looking at Eq. (3.42), the weights always shrink by a constant amount and eventually tend toward zero, leading to a sparse set of weights. By contrast, L^2 regularization shrinks the weights proportional to w, so for small weights, the reduction is much smaller compared to L^1 regularization and, thus, is rather seen as weight decay. One intuitive explanation for why smaller weights are preferred is that they reduce the sensitivity toward changes in the inputs. If the weights are large, even a small variation of the input can drastically alter the output.

The given formulas only consider the weights because regularizing the biases can have unwanted effects, such as severe underfitting. Theoretically, the hyperparameter λ could be chosen individually for each layer, but usually, it is set equally for the whole network to reduce the search space during hyperparameter tuning [6].

Overall, regularization with parameter norm penalties can effectively reduce the generalization error of neural networks and can, therefore, improve their performance on a multitude of tasks [8]. Nevertheless, the effectiveness is usually demonstrated empirically rather than by a mathematical proof.

3.7.3 Ensemble and Bagging Methods

Instead of altering the gradient flow through penalty terms, ensemble [33, 34] and bagging [35] methods [6, 8] combine predictions of multiple models, i.e., an ensemble of models, to reduce the variance of the model. Thereby, overall accuracy and robustness are increased. Multiple neural networks are trained independently on different subsets of the training data with different initializations, i.e., their corresponding outputs will vary. To select the preferred output, a voting or averaging scheme is applied to the results. Assuming all networks overfit the data in a different way, taking an average over the outputs can help to prevent this type of overfitting. The key difference between ensemble and bagging methods is that in ensemble methods, the output of different machine learning models is averaged, while for bagging methods, the same neural network architecture is considered.

3.7.4 Dropout

Dropout regularization [36] takes the idea of ensemble and bagging methods further by altering the network itself instead of relying on training multiple networks or modifying the cost function [8]. Before initializing the usual training workflow, a random selection of half the neurons in hidden layers are temporarily dropped, while input and output neurons stay untouched. Then, forward- and backpropagation, as well as the parameter update, are executed. Importantly, dropout works with stochastic gradient-based methods in which the cycle is only applied to one mini-batch of the training data (see Sect. 2.8). Then, for the next mini-batch, the dropped neurons are restored, and the procedure is repeated, now deactivating another random subset of hidden neurons (cf. Fig. 3.13). Thus, the weights and biases are learned, with only half of the hidden neurons activated for each mini-batch. Conversely, the entire set of hidden neurons is active when the network is eventually used for making predictions. To compensate for that, the outgoing weights of every hidden neuron are halved. Other quotas than dropping 50% of hidden neurons are also possible and are usually defined with the hyperparameter $p \in [0, 1]$.

Dropout acts in a similar way as ensemble methods from Sect. 3.7.3, imitating it without the extensive computational effort related to training multiple networks. Dropping a set of neurons during each mini-batch update is similar to training different neural networks. The dropout method can be considered as averaging the effects of a large number of multiple neural networks and thus may reduce overfitting at a reduced cost compared to ensemble methods. This encourages redundant representations in the network and prevents a single neuron from carrying a large fraction

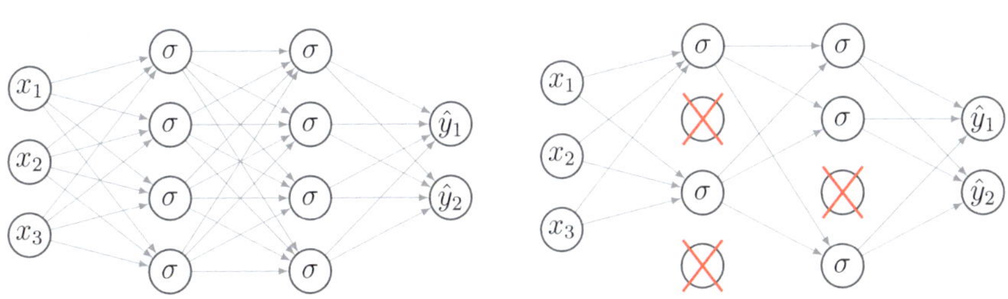

(a) Standard neural network (b) Neural network with dropout

Fig. 3.13 Dropout. Illustration inspired by [8]

of the network's inference capability since every neuron is sometimes randomly deactivated.

3.7.5 Data Augmentation

The best way to increase the generalization abilities of a model without interfering with the model itself is to train it on more data. In most cases, obtaining more training data is not feasible, but sometimes, the creation of "*fake*" data can be an option. A good example is the task of image classification, where a high-dimensional input is mapped to a single classifier. This implies that the model has to be invariant to a wide range of transformations. For the example of image recognition, it is fairly easy to introduce small variations like translation, rotation, or scaling to the input images; see Fig. 3.14.

Another data augmentation technique is the inclusion of noise [6], as seen in Fig. 3.14d. Most classification and regression tasks should still be solvable, even when a small amount of random noise is added to the input data. However, neural networks do not seem to be very robust when dealing with noisy inputs. A possibility to increase the robustness is to actually train the network with noise-injected data.

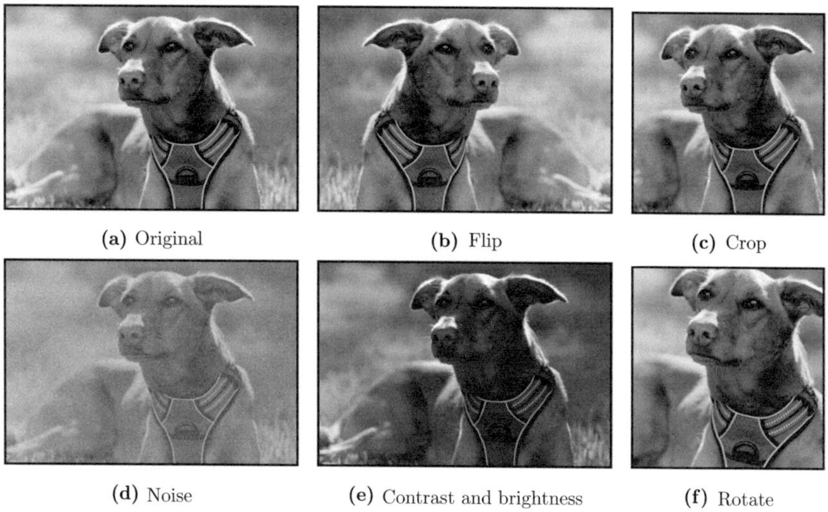

Fig. 3.14 Data augmentation with possible transformations (of Kurt)

3.8 Approximating the Sine Function

As stated in Sect. 3.1, fully connected neural networks with at least one hidden layer are capable of approximating any continuous function with arbitrary precision given a sufficient number of hidden neurons. On the example of the sine function, this section investigates some of the introduced characteristics of machine learning and neural networks. The sine is a suitable choice since it is a continuous function, and its periodicity offers enough complexity to serve as an academic example. All following results are generated using Python and the PyTorch framework.

The basic architecture of the fully connected network is depicted in Fig. 3.15. It consists of one input unit, two hidden layers with 50 hidden neurons each as well as one output unit. The goal is to approximate

$$f(x) = \sin(2\pi x), x \in [-1, 1]$$

in the interval $[-1, 1]$. The neural network approximation can be written as

$$f_{NN} = \boldsymbol{w}_3^{\top} \sigma\left(\boldsymbol{W}_2\left(\sigma(\boldsymbol{w}_1 \cdot \boldsymbol{x} + \boldsymbol{b}_1)\right) + \boldsymbol{b}_2\right) + b_3 = \hat{y}, \tag{3.45}$$

with the hyperbolic tangent $\sigma(z_j) = \tanh(z_j)$ serving as the activation function for the hidden neurons.

In PyTorch, such a network can be implemented using the `torch.nn.Sequential` class—which is the most convenient way to construct feed-forward neural networks:

```
modules = []
modules.append(torch.nn.Linear(1, 50))
modules.append(torch.nn.Tanh())
modules.append(torch.nn.Linear(50, 50))
```

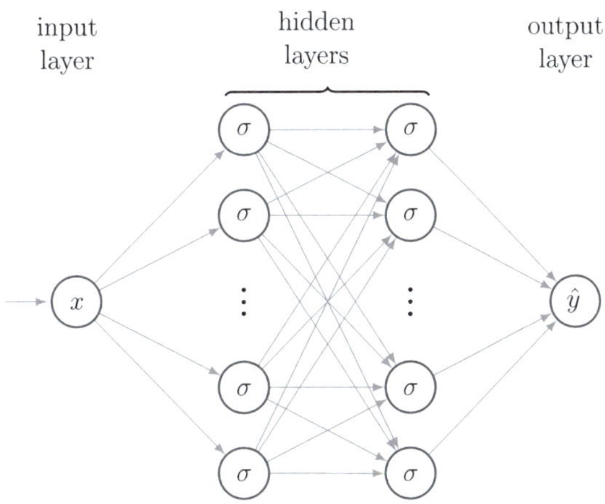

Fig. 3.15 Fully connected neural network architecture used for approximating the sine function. The network takes x as an input and outputs the corresponding $\hat{y}$ coordinate. Each of the two hidden layers consists of 50 neurons with Tanh activation functions. The output neuron uses a linear activation

```
modules.append(torch.nn.Tanh())
modules.append(torch.nn.Linear(50, 1))
# non-linearity is not appended after last linear layer

model = torch.nn.Sequential(*modules)
```

Fully connected feed-forward layers are defined with `torch.nn.Linear`, which creates a linear transformation given the input and output size. The prediction from an input `x` is performed by calling the neural network model with `model(x)` or `model.forward(x)`.

The training set containing 40 samples is generated by computing $\tilde{y} = \sin(2\pi x) + \epsilon$ for a uniform random distribution of values x across the interval $[-1, 1]$. By adding noise with the term ϵ, noisy measurements often occurring in real-world applications are simulated. The noise term is computed as

$$\epsilon = 0.1 \cdot \mathcal{U}(-1, 1),$$

where $\mathcal{U}(-1, 1)$ denotes values drawn from a uniform random distribution. Similarly, a validation set is generated by sampling 40 randomly chosen points in the interval $[-1, 1]$. The test set is simply represented by the analytical solution of the sine function.

This is accomplished with the following code.

```
x = torch.rand((40, 1)) * 2 - 1  # 40 points in interval -1 and 1
noise = torch.rand(x.shape) * 0.2 - 0.1
y = torch.sin(2 * torch.pi * x) + noise
```

Since the prediction of the sine function is a regression task, the mean squared error loss is chosen as the cost function C

$$C = \mathcal{L}_{\mathcal{D}}^{(\text{train})} = \frac{1}{m_{\mathcal{D}}^{(\text{train})}} \sum_{i=1}^{m_{\mathcal{D}}^{(\text{train})}} (\tilde{y}_i^{(\text{train})} - \hat{y}_i^{(\text{train})})^2. \tag{3.46}$$

In Python the cost function is defined as:

```
def costFunction(y, yPred):
    return torch.mean((y - yPred) ** 2)
```

which can be derived using the `.backward()` function.

```
yPred = model(x)
cost = costFunction(y, yPred)
cost.backward()
```

This is possible as long as the quantity to be derived is a scalar and part of the computation graph. As explained in Sect. 3.4.3, the derivatives with respect to the leaf variables can be accessed at the leaf variables with `.grad`. In this case, the leaf variables are the neural network weights and biases, accessible with `model.parameters()`.

Using the cost function C and the respective gradient $\nabla_\Theta C$, the loss on the training data is minimized by applying full-batch gradient descent, in particular, the commonly used optimizer Adam with a learning rate of $\alpha = 0.001$. The computation of the gradient is handled by PyTorch, which provides a very fast automatic differentiation algorithm to compute the partial derivatives with respect to all parameters of the network.

The training algorithm is implemented with the following code snippet, which first defines the optimizer and subsequently performs the full-batch training loop from Algorithm 5.

```
optimizer = torch.optim.Adam(model.parameters(), lr=0.001)

for epoch in range(epochs):
    optimizer.zero_grad()  # set gradients to zero
    yPred = model(x)
    cost = costFunction(y, yPred)
    cost.backward()
    optimizer.step()
```

Here, it is essential to reset the gradients of the leaf variables in each epoch, as the gradients obtained with `cost.backward()` act cumulatively on `.grad` of the leaf variables. Note that the entire data `(x,y)` is used in the training loop. To track the generalization capabilities, a second prediction and cost evaluation—unused during the parameter update, i.e., `optimizer.step()`—has to be added.

The weights of the network are initialized by drawing samples from a uniform distribution within certain bounds. The bounds $[-l, l]$ are defined as $l = \sqrt{\frac{1}{n_{\text{in}}}}$ with n_{in} being the number of input units in the weight tensor $\boldsymbol{w}^{(l)}$. The initialization of the biases $\boldsymbol{b}^{(l)}$ follows the same scheme. As already indicated in Sect. 2.8, the choice of initial parameters can have a severe impact on the results since the gradient-based algorithm may descend into a completely different minimum of the cost function. For a detailed and interactive explanation of how the weight initialization influences the convergence of deep neural network training, the reader is referred to the excellent blog post by Katanforoosh and Kunin [37].

First results at different epochs of the training process are shown in Fig. 3.16. While the number of training iterations is still low (cf. Fig. 3.16a), the model struggles to fit the training data and the underlying sine function. As soon as the network has learned the parameters for a sufficient amount of time (cf. Fig. 3.16b), it fits the training data and also generalizes well on the test data across the sine. If the training continues for too long (cf. Fig. 3.17a), the model starts to fit every example in the training set. This also includes the noisy data points resulting in overfitting and bad predictions for the test data.

As discussed in Sect. 3.7.2, a way to resolve the problem of overfitting is to use parameter norm penalties like L^2 regularization, which adds a term to the cost

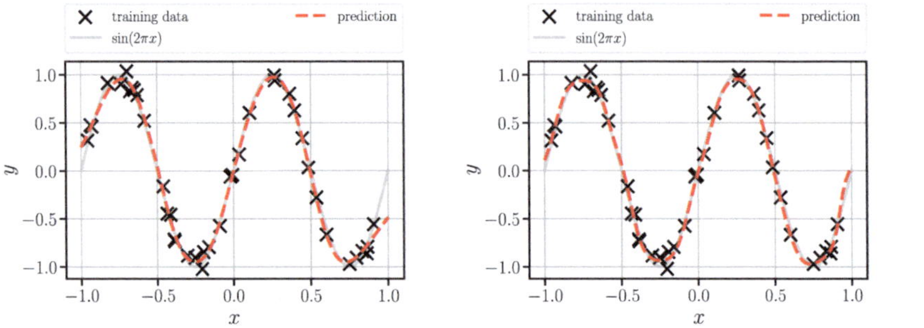

(a) Neural network still showing signs of underfitting after 500 training epochs

(b) Neural network properly fitting the underlying sine function after 5 000 iterations

Fig. 3.16 Predictions of the example network at different training epochs

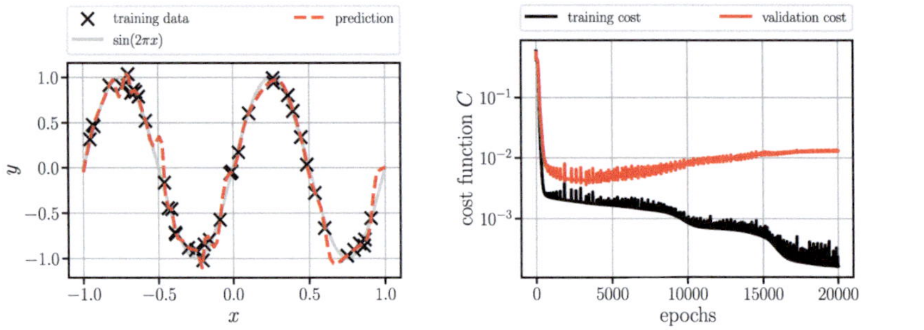

(a) Neural network prediction after 20 000 epochs of training and no regularization. Prediction shows clear signs of overfitting

(b) Training and validation error plotted for each training iteration. The spikes are caused by the Adam optimizer adapting the learning rate

Fig. 3.17 The example network trying to predict the sine function after 20 000 training iterations

function penalizing large weights. The influence of the penalty term is controlled by the hyperparameter λ. The plots in Fig. 3.18 show the results with applied L^2 regularization for different values of λ. For large values of λ, the model exhibits underfitting, meaning it fails to fit the training data as well as the test data (cf. Fig. 3.18b). However, when the ideal value for λ is found, the network is able to fit the data points of the training set, while at the same time a generalization to the test data is achieved (cf. Fig. 3.18a). If no regularization ($\lambda = 0$) is applied, the model shows the typical signs of overfitting (cf. Fig. 3.17a). In that case, the function passes through almost every noisy data point in the training data but fails to represent the underlying sine wave.

Another problem becomes apparent when considering the derivatives of the prediction. Despite achieving a decent approximation of the sine function in Fig. 3.16b, its derivative, depicted in Fig. 3.19, is less accurate. Inaccurate gradients of predictions are a prominent and important challenge of neural networks, as will similarly be seen in numerical simulation examples in Chap. 5. Better predictions of the derivative

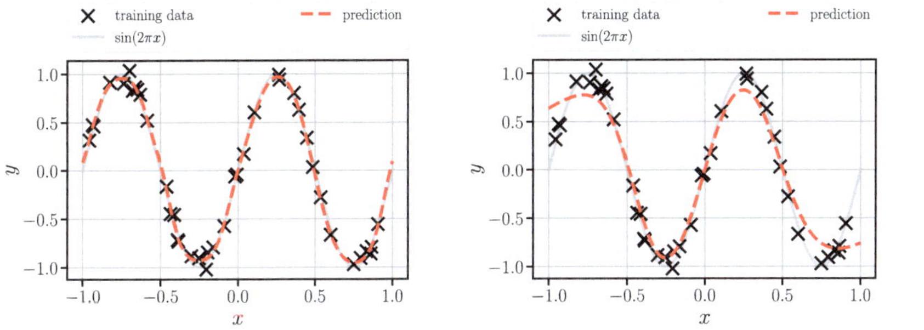

(a) Prediction after 20 000 epochs of training and L^2 regularization ($\lambda = 1 \times 10^{-5}$)

(b) Prediction after 20 000 epochs of training and L^2 regularization ($\lambda = 1 \times 10^{-2}$)

Fig. 3.18 Regularization of sine example

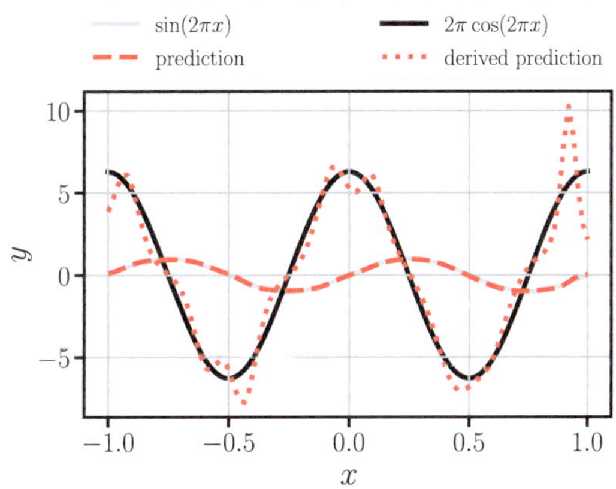

Fig. 3.19 Prediction from Fig. 3.16b and its derivative

can, however, be amended by Sobolev training [38]. In contrast to only considering the function of interest during training, derivative information of the function is supplied to the neural network during training through the cost function. Using derivative information with neural networks requires differentiation with respect to the input, which can readily be achieved with backpropagation from Sect. 3.4, as was discussed in detail in Sect. 3.4.2.

In the case of the sine function example, labeled data of the sine and its derivative, i.e., cosine, could be employed. Specifically, we consider 20 random data points of sine and 20 random data points of its derivative for training instead of the 40 random data points of only the sine function. The resulting prediction and derivative after 5 000 epochs are provided in Fig. 3.20, in which the quality of the derivative has improved over Fig. 3.19 without a worse approximation of the sine function. Furthermore, the second derivative, absent in the training dataset, suffers from fewer artifacts (see Fig. 3.21) than the first derivative obtained without Sobolev training (see Fig. 3.19).

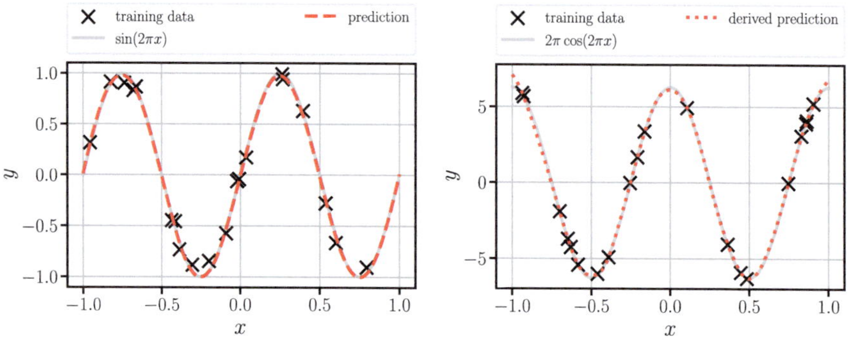

(a) Prediction (b) Derivative of prediction

Fig. 3.20 Predictions after 5000 epochs of Sobolev training using 20 random datapoints of $y = \sin(2\pi x)$ and 20 of its derivative $2\pi \cos(2\pi x)$

Fig. 3.21 Second derivative of prediction after 5 000 epochs of Sobolev training; seen in Fig. 3.20

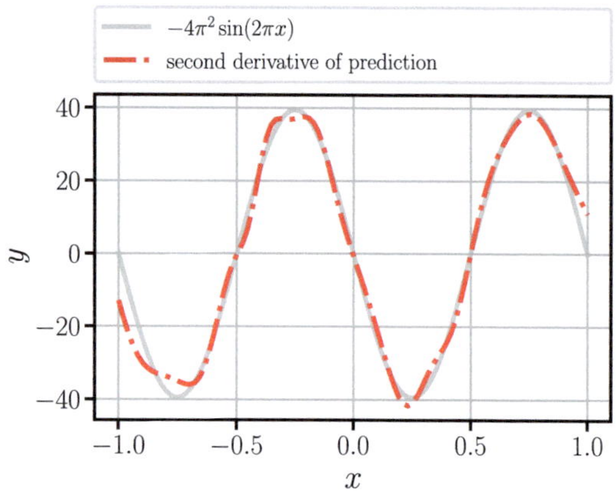

Investigating the seemingly simple example of approximating the sine function reveals that the results are sensitive to a great variety of factors. All the different kinds of hyperparameters and possible settings determined by the user have a great influence on the output and the learning speed of the model. For example, choosing a different activation function, a different learning rate, or even another optimizer can drastically change the outcome. Figure 3.22 displays a few representative results that demonstrate the influence of some of the aforementioned settings and hyperparameters.

3.9 Advanced Architectures

This section introduces more advanced network architectures that have been very successful in their respective field of application. Adapted to a particular data structure

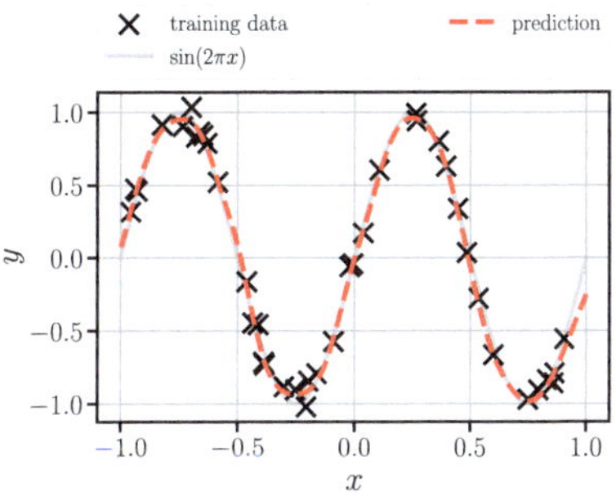

(a) Reference neural network with Sigmoid activations properly fitting the underlying sine function after 5 000 iterations. The network parameters are learned by applying the Adam optimizer with a learning rate of $\alpha = 0.001$

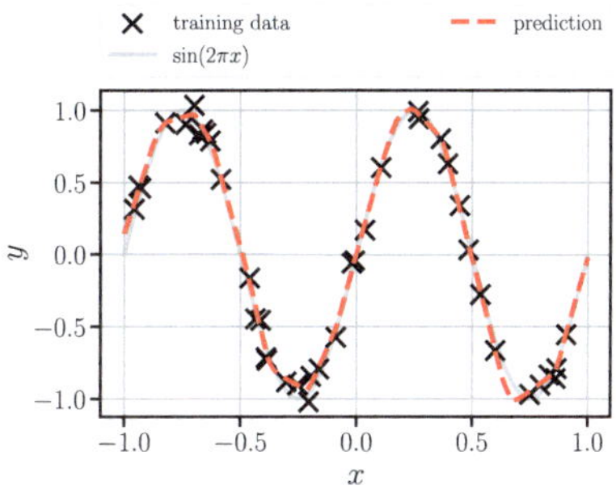

(b) Prediction generated using ReLU as the activation function. With ReLU activations the network already starts to overfit at 5 000 epochs

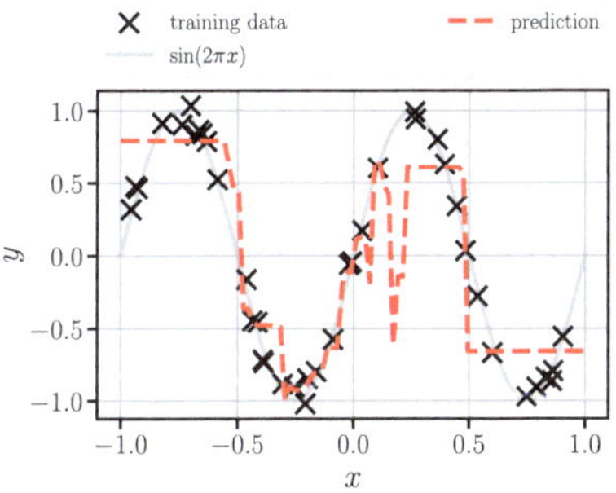

(c) Prediction after 5 000 epochs with learning rate $\alpha = 0.1$. This example shows that a large learning rate can prevent the network from learning

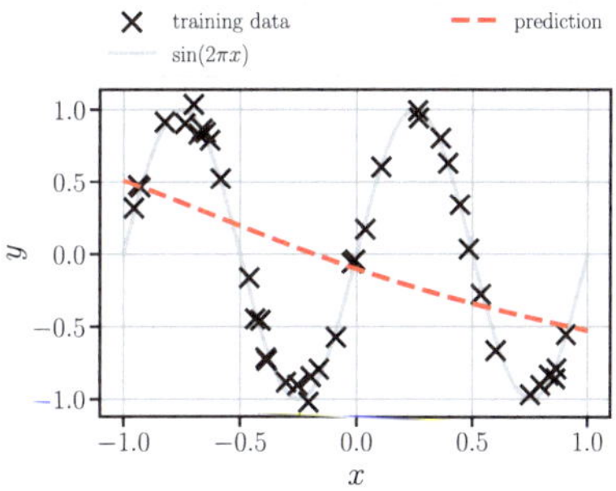

(d) Prediction after 5 000 epochs using stochastic gradient descent (SGD) instead of the Adam optimizer. A drawback of SGD is the slower convergence rate, since the parameters are updated after the evaluation of each single example (cf. Section 3.6)

Fig. 3.22 Parameter study for the example network predicting the sine function. Besides the parameter in question, all parameters remain the same as in the reference example on the top left

or task, they achieve superior results compared to fully connected feed-forward neural networks. There are numerous variants, and every other scientific publication on neural networks proposes yet another variant of a network architecture. Usually these networks are derived or form combinations of the architectures discussed next. The simpler variants with direct applications to computational mechanics are explained in detail (e.g. for Convolutional Neural Networks, Sect. 3.9.1, or Lagrangian or Hamiltonian Neural Networks. Section 3.9.8) including an application to Computational Mechanics. Other network architectures, such as Generative Network architectures (Chap. 8), are briefly meantioned in this section and picked up in later chapters or sections using concrete examples. More complicated, yet important network architectures are meantioned as well, but only their general idea and functionality is laid out, and the interested reader is pointed to corresponding recent literature.

3.9.1 Convolutional Neural Networks

3.9.1.1 Methodology

Convolutional neural networks (CNNs) were originally designed for image processing [39]. Their distinct architecture allows them to account for the spatial structure of a picture and thus makes convolutional neural networks the preferred choice for image classification tasks [8]. When recognizing an object in an image, it can be assumed that features describing the object are found in close proximity to each other. Furthermore, the exact location in the image is not important for the identification of an object. Mostly, only the relative position of features to each other matters. Convolutional neural networks exploit these two realities about physical objects, namely locality and translational invariance. This is achieved through the specialized architecture, referred to as the inductive bias. Like an ordinary neural network, a convolutional neural network consists of neurons that have learnable weights and biases. What differs is the arrangement of the processed data and the mathematical operations employed in the hidden layers, which are separated into convolutional and pooling operations. The input layer is no longer a vector but has the shape of a matrix to account for the grid-like structure of pixels in an image. For instance, a convolutional neural network is processing a square image of 7×7 pixels where each pixel corresponds to a gray-scale value. Then, a hidden neuron in the convolutional layer is connected with a small region of the input pixels, e.g., a 3×3 window, called the local receptive field of the hidden neuron (see Fig. 3.23).

The size of the local receptive field and the distance between the fields, also referred to as stride length s, define the output neurons of the convolutional layer. In the example shown in Fig. 3.23, the local receptive field of size 3×3 moves over the input layer with a stride length of one $s = 1$. Since the input layer consists of 7×7 pixels, the corresponding convolutional layer leads to 5×5 output neurons. Together with the size of the local receptive field, the stride length is a hyperparameter of the convolutional network. A specialty of convolutional neural networks is that all hidden

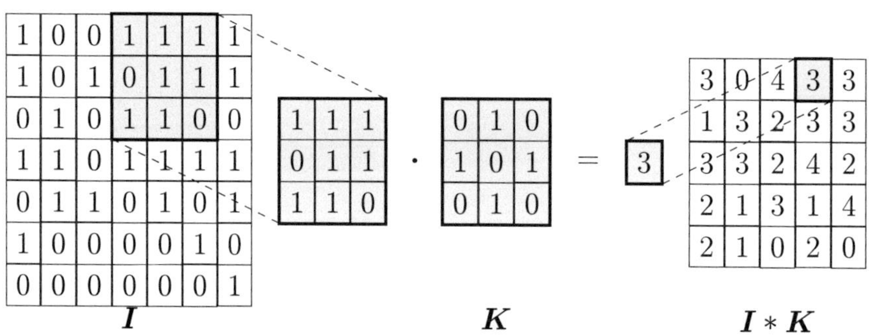

Fig. 3.23 Convolution layer: the local receptive field or filter $\boldsymbol{K}$ of size 3×3 with a stride length of one $s = 1$ is applied to the input layer $\boldsymbol{I}$

$$K = \begin{pmatrix} 0 & 0 & 0 \\ 0 & 1 & 0 \\ 0 & 0 & 0 \end{pmatrix}, s = 1$$

(**a**) Identity filter

$$K = \frac{1}{16}\begin{pmatrix} 1 & 2 & 1 \\ 2 & 4 & 2 \\ 1 & 2 & 1 \end{pmatrix}, s = 1$$

(**b**) Gaussian filter

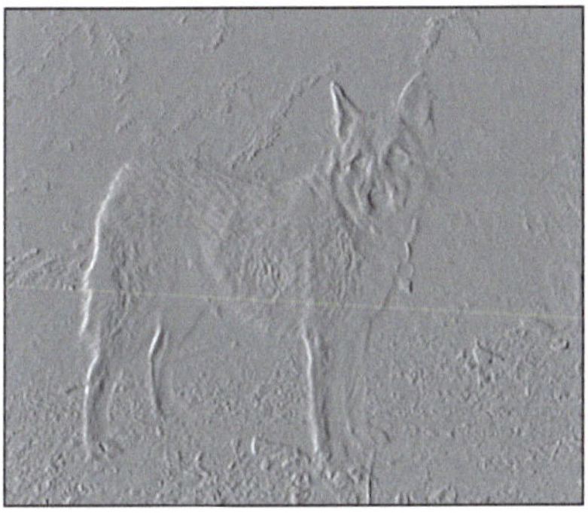

$$K = \begin{pmatrix} -1 & 0 & 1 \\ -2 & 0 & 2 \\ -1 & 0 & 1 \end{pmatrix}, s = 1$$

(**c**) Sobel filter in x-direction

$$K = \frac{1}{9}\begin{pmatrix} 1 & 1 & 1 \\ 1 & 1 & 1 \\ 1 & 1 & 1 \end{pmatrix}, s = 3$$

(**d**) Averaging filter

Fig. 3.24 Different convolutional filters $\boldsymbol{K}$ with stride lengths s applied to the image (of Lucy) depicted in (**a**). In a convolutional neural network the filters $\boldsymbol{K}$ are learnable

neurons in the convolutional layer share the same weights and bias, often referred to as the filter $\boldsymbol{K}$. As a result, all neurons detect the same feature at different locations in the input image, which implies the characteristic of translational invariance (see Fig. 3.24 for examplary convolutional filters applied to the image in Fig. 3.24a). The output of the convolution operation is called the feature map. A convolutional layer normally consists of more than one feature map in order to detect multiple localized features. Every feature map is generally followed by the application of a pooling operation that condenses the output of the convolution (see Fig. 3.25). For example, each unit in the pooling layer summarizes a region of 2×2 neurons in the previous layer. A common pooling technique is max-pooling, where the maximum value of

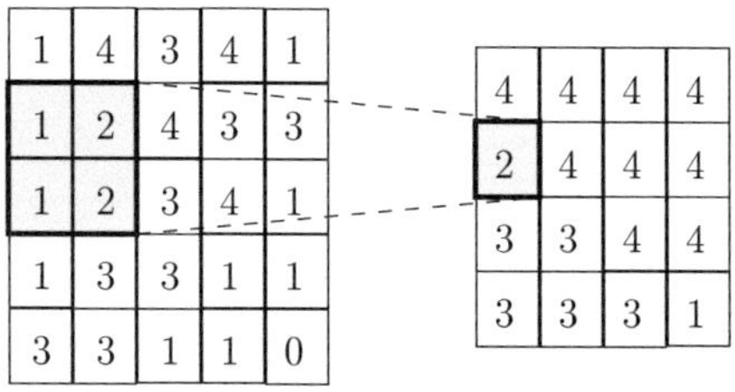

Fig. 3.25 Max-pooling with a 2 × 2 filter and stride length of one ($\max\{1, 2, 1, 2\} = 2$)

the respective region is taken. Pooling can be interpreted as a query that checks if a feature has been detected somewhere while neglecting the exact position in the image.

In the end, a fully connected layer can be added to perform the classification task. The arrangement of multiple convolutional and pooling layers allows very deep and expressive network architectures, such as exemplified in Fig. 3.26. Here, the connection to fully connected neural networks is established, where a key difference lies in the data being processed: matrices $\boldsymbol{x}_i \in \mathbb{R}^{I \times I}$ in the case of convolutional networks as opposed to scalars $x_i \in \mathbb{R}$ for fully connected networks. Consequently, there is a distinction in terminology, where the layer depth is denoted in neurons for the fully connected network and in channels for the convolutional network.

The second key difference is the convolutional operation, which can be viewed as a method of weight sharing. Sharing weights and biases in combination with the pooling operation reduces the amount of trainable parameters significantly. Thus, convolutional networks learn much faster than fully connected neural networks with comparable expressive power [8]. The backpropagation algorithm for calculating the gradients only needs minor adaptations to work with convolution and pooling operations.

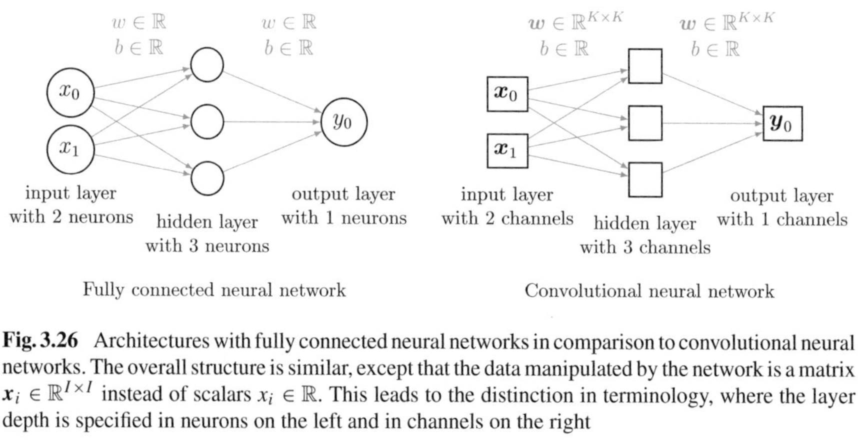

Fig. 3.26 Architectures with fully connected neural networks in comparison to convolutional neural networks. The overall structure is similar, except that the data manipulated by the network is a matrix $\boldsymbol{x}_i \in \mathbb{R}^{I \times I}$ instead of scalars $x_i \in \mathbb{R}$. This leads to the distinction in terminology, where the layer depth is specified in neurons on the left and in channels on the right

A useful formula for the computation of the output dimension O after the convolution applied to an input of size I is the following:

$$O = \left\lfloor \frac{I - K + 2P}{s} \right\rfloor + 1 \tag{3.47}$$

Here K is the filter size, s the stride length, and P the padding size. Padding is a tool to avoid a change in dimension by increasing the dimension of the input. In the case of zero-padding, this is in the form of rows and columns with zeros added at the boundaries of a two-dimensional input. In the previously presented examples, no padding was used, i.e., $P = 0$.

In PyTorch, a convolutional neural network is implemented with `torch.nn.Conv2d`.[3] An example of a network consisting of two convolutional layers is presented below.

```
class CNN(torch.nn.Module):
    def __init__(self, inputImages, hiddenImages, outputImages):
        super(CNN, self).__init__()  # inheritance from torch.nn.Module
        self.cnn1 = torch.nn.Conv2d(inputImages, hiddenImages,
                                    kernel_size=3, stride=1, padding=1)
        self.cnn2 = torch.nn.Conv2d(hiddenImages, outputImages,
                                    kernel_size=3, stride=1, padding=1)
        self.activation = torch.nn.ReLU()

    def forward(self, x):
        y = self.activation(self.cnn1(x))
        y = self.cnn2(y)
        return y
```

A training example relying on the implemented convolutional neural network in which a two-dimensional paraboloid is to be learned is given in the following code snippet. Again, only the prediction and cost function computation are given, which can be combined with the gradient descent implementation from Listing 3.2.

[3] Use `torch.nn.Conv2d` for a network acting on two-dimensional images. For one-dimensional grids, use `torch.nn.Conv1d`, and for three-dimensional voxel data, use `torch.nn.Conv3d`.

```
imageSize = 10
hiddenImages = 10

# training data
x = torch.linspace(-1, 1, imageSize)
y = torch.linspace(-1, 1, imageSize)
x, y = torch.meshgrid(x, y, indexing='ij')
z = x ** 2 * y ** 2

x = torch.randn((1, 1, imageSize, imageSize))
y = z.unsqueeze(0).unsqueeze(0)

model = CNN(1, hiddenImages, 1)

yPred = model(x)
cost = torch.mean((yPred - y) ** 2)
```

3.9.1.2 Example: Learning Strain Distributions

To further familiarize the reader with deep learning and, in particular, some of the more advanced architectures, a second example beyond the sine function approximation from Sect. 3.8 is considered. Specifically, strain distributions in heterogeneous media are to be learned, i.e., a surrogate model is to be built using more advanced neural network architectures. For this purpose, convolutional neural networks from Sect. 3.9.1 are utilized within a U-net architecture, which are further elaborated with Fig. 8.7 in Chap. 8.

In the following example, we consider a fiber matrix material, illustrated by the cross-section in Fig. 3.27. The matrix material (white) is characterized by $E_f = 3$ GPa, and $\nu_f = 0.4$, while the fibers (red) are defined in terms of $E_f = 85$ GPa, and $\nu_f = 0.22$ [40]. Different specimen are generated and simulated with the finite element method under plane strain conditions, which yields a dataset consisting of the material distribution $E(x_1, x_2)$ and the corresponding strain distributions $\varepsilon_{11}(x_1, x_2), \varepsilon_{22}(x_1, x_2), \varepsilon_{12}(x_1, x_2)$. The information is stored in two-dimensional 32×32 arrays, i.e., images; see Fig. 3.28c. The aim of the surrogate model is to predict the spatial strain distributions given speicific loading and boundary conditions, but variable distributions of material, which are defined in terms of fiber E_f, ν_f and matrix E_m, ν_m material.

First, a suitable neural network has to be chosen. Three neural networks are considered: (NN 1) a U-net with five convolutional layers and 94 978 parameters, (NN 2) a feedforward convolutional neural network with four layers and 38 405 parameters, (NN 3) the same U-net as NN 1 followed by a three-layer feedforward convolutional neural network, which relies on 154 471 parameters. To assess the potential of each network, each is trained on a single data sample. The resulting

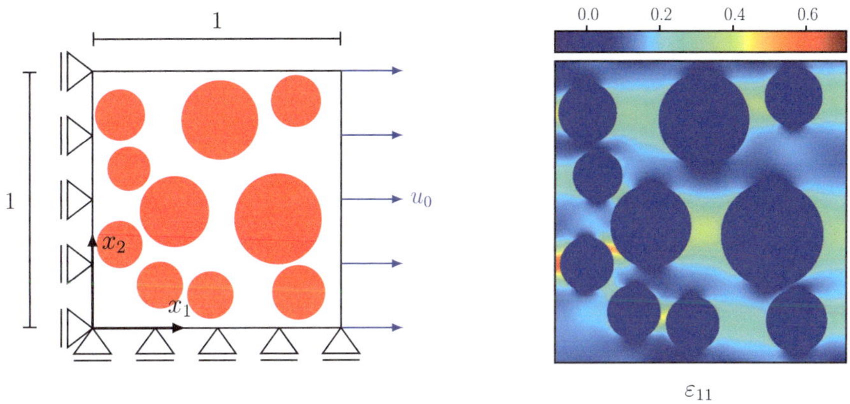

Fig. 3.27 Fiber matrix domain $\Omega \in [0, 1] \times [0, 1]$ m, where the red circles are cross-sections of the fibers, and the white background is the matrix material. The domain is loaded under uni-axial tension with the prescribed displacement $u_0 = 0.1$ m

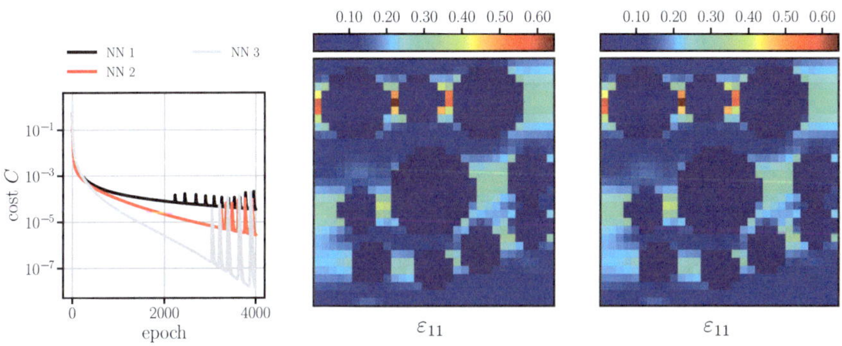

(a) Learning histories of the three considered neural networks on a single datapoint

(b) Prediction of the training datapoint using NN 2

(c) Ground truth of the training datapoint

Fig. 3.28 Selection of neural network architecture by training on a single datapoint

learning histories of these networks after hyperparameter tuning are provided in Fig. 3.28a. NN 3 achieves the lowest cost function and is therefore selected.

Next, the number of training samples is increased, and a validation set is employed to determine the generalization error. A good practice is to slowly and iteratively increase the training set size $m_{\mathcal{D}}$, e.g., in multiples of two, and adjust the network architecture and hyperparameters in case of convergence issues. Figure 3.29 illustrates the training histories for correspondingly $m_{\mathcal{D}} = 4$, $m_{\mathcal{D}} = 16$, and $m_{\mathcal{D}} = 64$, where the training error increases for larger training sets, but more importantly, the validation error, i.e., the generalization error, decreases. Using regularization, such as L^2 regularization in Fig. 3.29b, further pushes the validation error closer to the training error. This is particularly noticeable for $m_{\mathcal{D}} = 4$ and $m_{\mathcal{D}} = 16$, where

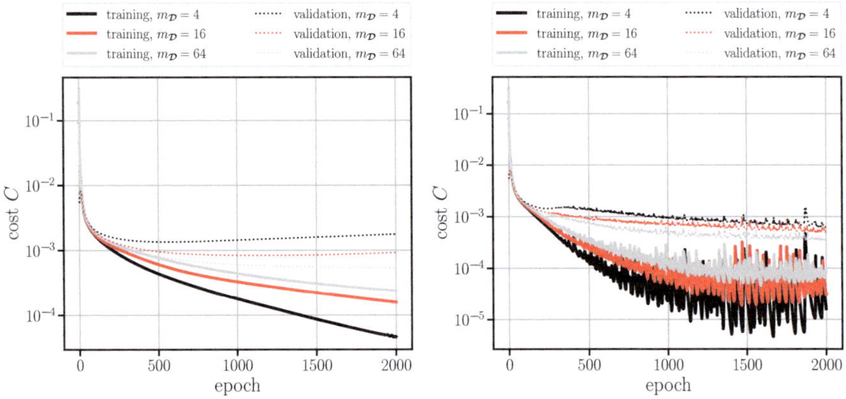

(a) Without regularization **(b)** With L^2 regularization relying on $\lambda = 10^{-3}$

Fig. 3.29 Learning histories of NN 3 with a larger number of training samples $m_{\mathcal{D}}$

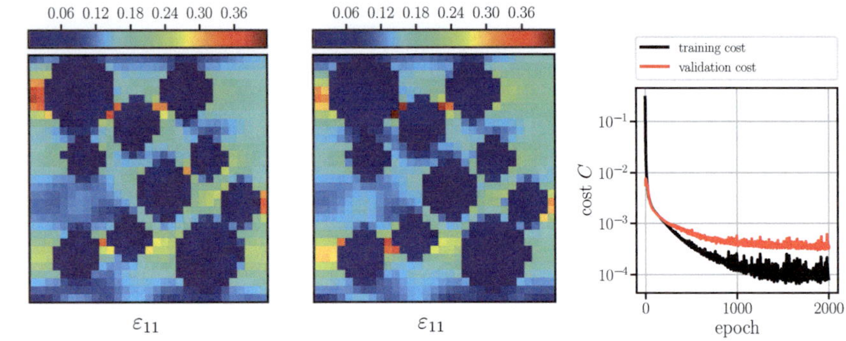

(a) Prediction of validation sample **(b)** Ground truth of validation sample **(c)** Learning history

Fig. 3.30 Surrogate model for predicting strains from material distribution using $m_{\mathcal{D}} = 128$ training samples

overfitting is more prominent when foregoing regularization; see Fig. 3.29a. The corresponding improvements in terms of the best achieved validation error are 219% for $m_{\mathcal{D}} = 4$, 170% for $m_{\mathcal{D}} = 16$, and 161% for $m_{\mathcal{D}} = 64$.

Figure 3.30a provides a prediction of the validation set using the final model with $m_{\mathcal{D}} = 128$ training samples, which further decreases the validation error by 111% in comparison to the best model attained in Fig. 3.29b. For reference, the corresponding learning history is depicted in Fig. 3.30c. More technical details are provided in the context of Exercise E.12. The prediction from Fig. 3.30a can be compared to the ground truth in Fig. 3.30b, from which minor errors are noticeable by visual inspection. However, the prediction time of one sample ($\sim 1.5 \cdot 10^{-3}$ s) is roughly 37 time faster than a comparable finite element simulation with 32×32

elements at polynomial degree $p = 1$. Furthermore, the prediction time stays constant for multiple parallel predictions until the limits of the GPU memory are reached, which is not possible for classical methods such as the finite element method. These speedups are even more pronounced for more expensive simulations beyond linear elasticity. This is useful in time-critical applications, a domain where reduced order or surrogate models are usually applied. But it is not worthwhile in cases where only one prediction is necessary, since the cost of acquiring the training data can not not be neglected. To account herefore and asses if such a surrogate model offers a more cost-efficient alternative, the breakeven threshold τ [41, 42] can be considered

$$\tau = \frac{T_{\text{data}} + T_{\text{train}}}{T_{\text{simulation}} - T_{\text{surrogate}}}, \tag{3.48}$$

where T_{data} is the time for data collection, T_{train} the time for training, $T_{\text{simulation}}$ the time for a single simulation (at a similar accuracy level, i.e., not necessarily the accuracy during data collection [43]), and $T_{\text{surrogate}}$ the time for a single evaluation of the surrogate model. Only when the surrogate is applied at least τ times is it worth training a corresponding surrogate model. In the example at hand, the data was collected with a finite element discretization of 64×64 elements with polynomial degree $p = 3$, resulting in a simulation time of ~0.88 s per sample. The 128 training samples and 32 validation samples result in a total data collection time of $T_{\text{data}} \approx 140$ s. Considering the effort of a single training for 4000 epochs from Fig. 3.30 being about $T_{\text{train}} \approx 36$ s (here the tuning effort is neglected), and the speedup $T_{\text{simulation}} - T_{\text{surrogate}} \approx 55.5 \cdot 10^{-3} - 1.5 \cdot 10^{-3}$ s $= 54 \cdot 10^{-3}$ s, the breakeven threshold is determined as $\tau \approx 3\,300$ with Eq. (3.48). Thus, the obtained surrogate model would need to be used at least 3 300 times before making the training worthwhile. Even if this is achieved, the accuracy and reliability of the predictions are not guaranteed [42, 44, 45]. The data requirements and the reliability are partially addressed in the upcoming chapters using physics-informed neural networks, or problem-specific remedies, in material modeling and inverse problems.

3.9.2 Graph Neural Networks

While convolutional neural networks have been highly successful in image processing tasks, they are limited in their ability to handle unstructured data, such as point clouds or meshes. Such data can be represented as graphs, enabling their manipulation through graph neural networks. Graph neural networks have emerged as a powerful tool for learning with graph-structured data, allowing to capture complex patterns and (topological) relationships within the data.

The most general approach is the message-passing neural network framework [46], which encompasses various graph-based neural network models. A popular and specialized instance of message-passing neural networks is the graph convolutional neural network [47, 48], which has been widely used in numerous applications, such

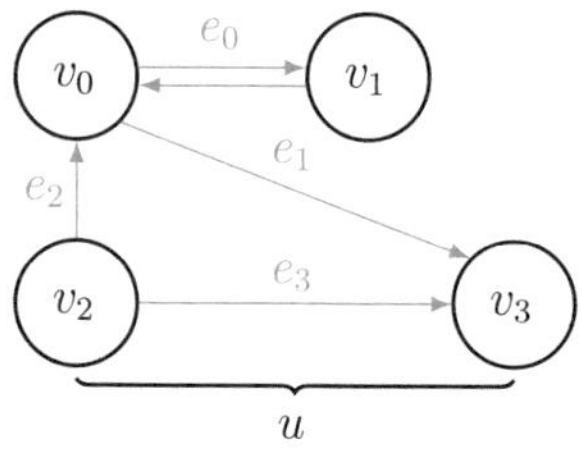

Fig. 3.31 An example graph with nodes $\boldsymbol{v}$, edges $\boldsymbol{e}$ and a global attribute u

as molecular fingerprints for drug discovery [49], citation networks [47], matrix completion for recommender systems [50], traffic forecasting [51]. For aplications in computational mechanics, see, e.g., the mesh-based simulations of flags in the wind presented in [52, 53].

To illustrate the message-passing approach, consider a directed graph, such as the one in Fig. 3.31. The graph is defined in terms of node features $\boldsymbol{v} = \{v_0, v_1, \dots\}$, edge features $\boldsymbol{e} = \{e_0, e_1, \dots\}$ and a global attribute u. Message-passing neural networks recursively update the nodes and edges by passing messages between adjacent nodes, exploiting the following invariant:

$$v_i^s \xrightarrow{e_i} v_i^r \tag{3.49}$$

Each edge e_i has a sender node v_i^s and a receiver node v_i^r. Note that this also holds for undirected graphs, where each edge can be treated as bidirectional and thus has two receivers and senders, as illustrated in Fig. 3.31 with edge e_0. This edge invariance allows to formulate an algorithm independent of the overall graph structure, with the edge being treated as the lowest level.

First, all edge features are updated using an edge neural network $e_i' = f^e(e_i, v_i^s, v_i^r, u; \boldsymbol{\Theta}_e)$, considering the sender and receiver nodes, and the global attribute. Next, all updated features of edges $\boldsymbol{e}_i'$ connecting to a node v_i are aggregated to a scalar $\bar{e}_i' = \rho^{e\rightarrow v}(\boldsymbol{e}_i')$. Typical aggregation functions ρ include summation, averaging, or maximum. This step reduces the variable number of edge connections to a scalar, ensuring that the output is always of the same size, which then can be considered by a second neural network updating the node features $v_i' = f^v(\bar{e}_i', v_i, u; \boldsymbol{\Theta}_v)$.

Finally, additional aggregations can be performed over all edges $\rho^{e\rightarrow u}(\boldsymbol{e})$ and all nodes $\rho^{v\rightarrow u}(\boldsymbol{v})$, yielding $\bar{e}', \bar{v}'$, which are used in a third neural network to update the global attribute $u' = f^u(\bar{e}', \bar{v}', u; \boldsymbol{\Theta}_u)$. The algorithm is summarized in Algorithm 8 and describes a single graph block, which can be applied recursively, allowing nodes to pass messages further. The message can only propagate through the graph up to the recursion depth, i.e., the number of graph blocks. The formulation offers great flexibility in choosing which graph quantities to update and which to use as input to the neural networks. This is explored in detail in [46].

Graph-convolutional neural networks are a specialized instance of message-passing neural networks designed for learning on graph-structured data. While they share some similarities with message-passing neural networks, such as updating node fea-

Algorithm 8 Graph block in a message passing neural network

Require: Graph consisting of nodes $\boldsymbol{v}$, edges $\boldsymbol{e}$ and a global attribute u.
for all edges e_i **do**
  Update edges: $e_i' = f^e(e_i, v_i^s, v_i^r, u; \boldsymbol{\Theta}_e)$
end for
for all nodes v_i **do**
  Find all edges connecting to node v_i: $\boldsymbol{e}_i'$
  Aggregate adjacent edges: $\bar{e}_i' = \rho^{e\rightarrow v}(\boldsymbol{e}_i')$
  Update nodes: $v_i' = f^v(\bar{e}_i', v_i, u; \boldsymbol{\Theta}_v)$
end for
Aggregate all edges $\boldsymbol{e}$: $\bar{e}' = \rho^{e\rightarrow u}(\boldsymbol{e})$
Aggregate all nodes $\boldsymbol{v}$: $\bar{v}' = \rho^{v\rightarrow u}(\boldsymbol{v})$
Update global attribute: $u' = f^u(\bar{e}', \bar{v}', u; \boldsymbol{\Theta}_u)$

tures based on information from neighboring nodes, graph-convolutional neural networks employ a more specific message-passing mechanism. In graph-convolutional neural networks, the neighborhood aggregation step uses a convolution-like operation to combine features from the nodes and their neighbors. The main advantage of graph-convolutional neural networks over the more general message-passing neural networks is their simplicity and ease of implementation, which has led to their popularity in graph-based learning. However, this simplicity can also be a limitation, and the broader message-passing neural networks framework may be more suitable for certain problems that require more flexibility in the choice of message-passing mechanisms, for example when messages should not only be passed along a certain topology defined by edges.

3.9.3 Autoencoder

Autoencoders encode the input into a smaller latent vector and subsequently decode the latent vector to reconstruct the original input. Decoding randomly sampled latent vectors results in new outputs that resemble the original data. The sampling step transforms the autoencoder into a variational autoencoder [54, 55]. Auteoncoders are covered in more detail in Chap. 8.

3.9.4 Generative Adversarial Networks

Generative adversarial networks (GANs) consist of two neural networks. A generator network generates data, while a discriminator network tries to distinguish between generator outputs and a real dataset. During training, the two networks compete. This leads to a generator that is able to produce new samples resembling the original dataset and a discriminator, that can indicate if a sample lies within the expectation.

More details and applications on generative adversarial networks are provided in Chap. 8.

3.9.5 Bayesian Neural Networks

Up to this point, the presented neural networks do not quantify the uncertainty of their predictions and tend to behave overconfidently even when their outputs are incorrect [56, 57]. Bayesian neural networks [58–60] address this issue by providing a probabilistic framework for modeling uncertainty. The goal of a Bayesian neural network is to find the posterior distribution of the weights $\boldsymbol{\Theta}$ of a network $f(x; \boldsymbol{\Theta})$ with respect to the data D: $p(\boldsymbol{\Theta}|D)$.

During inference, the posterior is used to assess the uncertainty of a prediction $\hat{y}(x)$: $p(\hat{y}|x, D)$. Practically, this can be achieved by sampling n different sets of weights $\boldsymbol{\Theta}_i$ from the posterior distribution $p(\boldsymbol{\Theta}|D)$, yielding n predictions $\hat{y}_i = f(x; \boldsymbol{\Theta}_i)$. For regression, these predictions are used to compute the average $\hat{y}$ and covariance $\hat{\Sigma}_y$:

$$\hat{y} = \frac{1}{n}\sum_{i}^{n} \hat{y}_i \tag{3.50}$$

$$\hat{\Sigma}_y = \frac{1}{n-1}\sum_{i}^{n} (\hat{y}_i - \hat{y})(y_i - \hat{y})^{\mathsf{T}} \tag{3.51}$$

The covariance expresses the uncertainty of the prediction. The challenge in Bayesian neural networks lies in obtaining the posterior $p(\boldsymbol{\Theta}|D)$. To this end, Bayes' theorem is used:

$$p(\boldsymbol{\Theta}|D) = \frac{p(D|\boldsymbol{\Theta})p(\boldsymbol{\Theta})}{p(D)} = \frac{p(D|\boldsymbol{\Theta})p(\boldsymbol{\Theta})}{\int_{\boldsymbol{\Theta}} p(D|\boldsymbol{\Theta})p(D)d\boldsymbol{\Theta}} \tag{3.52}$$

Here, $p(D|\boldsymbol{\Theta})$ is the likelihood, representing the probability of observing the data D given the weights $\boldsymbol{\Theta}$, $p(\boldsymbol{\Theta})$ is the prior distribution of the weights. Typically, the prior distribution $p(\boldsymbol{\Theta})$ is modeled by a Gaussian distribution with a zero mean and a relatively large standard deviation, whereas the likelihood $p(D|\boldsymbol{\Theta})$ is approximated with a Gaussian with the mean predicted by a neural network $\hat{\mu} = f(x; \boldsymbol{\Theta}_\mu)$ and either a constant standard deviation σ or an additional neural network prediction $\hat{\sigma} = f(x; \boldsymbol{\Theta}_\sigma)$, which allows the model to learn the noise structure in the data.

The denominator $p(D)$ in Eq. (3.52) is the marginal likelihood, also called the model evidence. The marginal likelihood is the probability of observing the data D given the chosen model (i.e., the neural network) for any possible parameter. Thus, the parameters are marginalized out via the integration in the denominator. This integration is, however, intractable, as all possible weights would have to be considered.

Two main approaches exist to circumvent the evaluation of the integral: Markov chain Monte Carlo [61] and variational inference [62]. In this overview, we will focus on variational inference, where the posterior is approximated by a distribution $p(\boldsymbol{\Theta}|D) \approx q_\phi(\boldsymbol{\Theta})$ parametrized by ϕ. The assumed distribution can then be fit to the true prior by minimizing the Kullback-Leibler divergence KL [63]:

$$\min_\phi \mathrm{KL}(p(\boldsymbol{\Theta}|D), q_\phi(\boldsymbol{\Theta})) = \min_\phi \int_{\boldsymbol{\Theta}} q_\phi(\boldsymbol{\Theta}) \log\left(\frac{q_\phi(\boldsymbol{\Theta})}{p(\boldsymbol{\Theta}|D)}\right) d\boldsymbol{\Theta} \tag{3.53}$$

Thus, the integration in the denominator of Eq. (3.52) has been replaced by an optimization. The optimization can be conducted in combination with the neural network training by maximizing the so-called "*evidence lower bound*" (ELBO) with respect to both the weights $\boldsymbol{\Theta}$ and the variational parameters ϕ:

$$\max_{\boldsymbol{\Theta},\phi} \sum_i^N \log(p(\hat{y}(x_i)|x_i, \boldsymbol{\Theta})) - \mathrm{KL}(p(\boldsymbol{\Theta}|D), q_\phi(\boldsymbol{\Theta})) \tag{3.54}$$

As the likelihood $p(D|\boldsymbol{\Theta})$ is assumed to be Gaussian distributed with constant variance, the maximization of the log-likelihood term is equivalent to the minimization of the mean squared error; see [64]. The optimization is conducted with stochastic variational inference [65], which is equivalent to stochastic gradient descent utilized for variational inference [66].

This is only one form of Bayesian neural networks. Many more exist, such as those that utilize Markov Chain Monte Carlo methods [61], ensemble techniques [67], or dropout [68] as a Bayesian approximation. However, the common ground is that they provide a probabilistic framework to model uncertainty in neural network predictions by approximating the posterior distribution of the weights using various techniques, such as variational inference, to account for the intractable nature of the true posterior distribution. For an excellent introductory tutorial with an accompanying code, the reader is referred to [66].

3.9.6 Recurrent Neural Networks

Consider a state of a system that is changing over time. Conventional neural networks are only able to evaluate a current state of a system described by the fixed-size input vector. Information from previous states can neither persist nor be passed on to later stages. Recurrent neural networks (RNNs) address this issue by adding loops that allow the persistence and propagation of information over several states of a system each of which may be interpreted as a snaphsot of the system at a certain point in time. The recurrent neural network achieves learning changes of the system from one point in time to the next by propagating a hidden state $\boldsymbol{h}_t$ repeatedly. A recurrent neural network can be seen as multiple copies of the same network, each forwarding

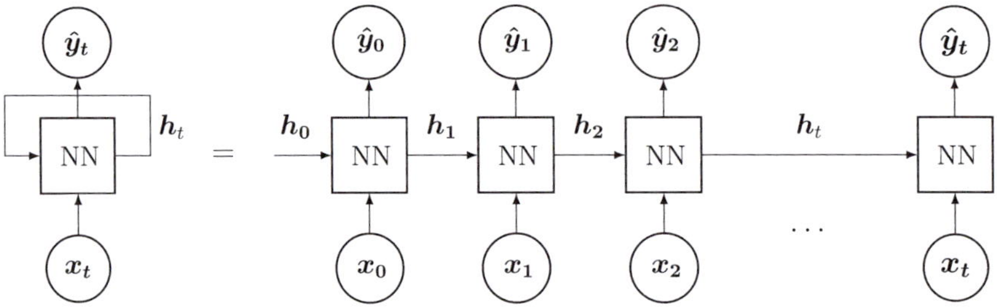

Fig. 3.32 An unrolled recurrent neural network. Adapted from [69] with permission

a message in the form of the hidden state $\boldsymbol{h}_t$ to their respective successor [69]. The sketch of an unrolled recurrent neural network is depicted in Fig. 3.32. As shown, recurrent neural networks are chain-like structures that take in sequences of data $\boldsymbol{x}_0, \boldsymbol{x}_1, \ldots, \boldsymbol{x}_t$ and generate a sequential output $\hat{\boldsymbol{y}}_0, \hat{\boldsymbol{y}}_1, \ldots, \hat{\boldsymbol{y}}_t$. Sequential data often occurs in tasks like speech recognition, language modeling, and translation, making recurrent neural networks the preferred choice for these kinds of problems. They may also be used to describe physical systems in computational mechanics that evolve during time.

In PyTorch, a recurrent neural network can be defined with `torch.nn.RNN`, which maps from an input $\boldsymbol{x}_t$ to the hidden state $\boldsymbol{h}_t$, as[4]

$$\boldsymbol{h}_t = \sigma(\boldsymbol{W}_x \boldsymbol{x}_t + \boldsymbol{b}_x + \boldsymbol{W}_h \boldsymbol{h}_{t-1} + \boldsymbol{b}_h). \tag{3.55}$$

It automatically handles the unrolling of the recurrent unit for a given sequence length (determined by the first dimension of the input `x`), yielding all hidden states. These are subsequently mapped to the output $\boldsymbol{y}_t$ via a linear transformation. More complex transformations can be employed. Also, multiple recurrent layers can be used, e.g., two, as in the code snippet below, where the recurrent unit is given as

$$\boldsymbol{h}_t^{(1)} = \sigma(\boldsymbol{W}_x^{(1)} \boldsymbol{x}_t + \boldsymbol{b}_x^{(1)} + \boldsymbol{W}_h^{(1)} \boldsymbol{h}_{t-1}^{(1)} + \boldsymbol{b}_h^{(1)}), \tag{3.56}$$

$$\boldsymbol{h}_t^{(2)} = \sigma(\boldsymbol{W}_x^{(2)} \boldsymbol{h}_t^{(1)} + \boldsymbol{b}_x^{(2)} + \boldsymbol{W}_h^{(2)} \boldsymbol{h}_{t-1}^{(2)} + \boldsymbol{b}_h^{(2)}). \tag{3.57}$$

Note that this leads to two hidden states $\boldsymbol{h}_t^{(1)}$ and $\boldsymbol{h}_t^{(2)}$, which evolve during the unrolling process.

[4] In PyTorch, two bias vectors $\boldsymbol{b}_x$, $\boldsymbol{b}_h$ are used despite their linear dependence. This implementational choice makes the framework compatible with cuDNN [70], a GPU-accelerated deep learning library.

```python
class RNN(torch.nn.Module):
    def __init__(self, inputSize, hiddenStateSize, outputSize):
        super(RNN, self).__init__()  # inheritance from torch.nn.Module
        self.rnn = torch.nn.RNN(inputSize, hiddenStateSize,
                                nonlinearity='relu', num_layers=2)
        self.linear = torch.nn.Linear(hiddenStateSize, outputSize)

    def forward(self, x):
        h, _ = self.rnn(x)  # unrolls the RNN
        y = self.linear(h)  # transforms hidden layer to output layer
        return y
```

The `torch.nn.RNN` implementation is easy to use, as unrolling is automatically taken care of, but limits customization. Implementing a recurrent neural network on a lower level using `torch.nn.RNNCell` provides more design freedom. The same recurrent neural network from above can be implemented in the following manner.

```python
class RNN(torch.nn.Module):
    def __init__(self, inputSize, hiddenStateSize, outputSize):
        super(RNN, self).__init__()  # inheritance from torch.nn.Module
        self.rnnCell1 = torch.nn.RNNCell(inputSize, hiddenStateSize,
                                         nonlinearity='relu')
        self.rnnCell2 = torch.nn.RNNCell(hiddenStateSize,
   hiddenStateSize,
                                         nonlinearity='relu')
        self.linear = torch.nn.Linear(hiddenStateSize, outputSize)

    def forward(self, x):
        h1 = torch.zeros(x.shape[1], self.rnnCell1.hidden_size,
                         dtype=torch.float)
        h2 = torch.zeros(x.shape[1], self.rnnCell2.hidden_size,
                         dtype=torch.float)

        outputs = []
        for t in range(x.shape[0]):
            h1 = self.rnnCell1(x[t, :, :], h1)
            h2 = self.rnnCell2(h1, h2)
            y = self.linear(h2)
            outputs.append(y)
        outputs = torch.stack(outputs, dim=0)
        return outputs
```

Either implementation can be employed in the following training scenario, where a cubic relation in time is to be learned. Only the prediction and cost function computation are shown, as the remaining training loop is identical to the previously discussed training loops (see Listing 3.2). Different sample trajectories are generated with different initial conditions `t0`.

```
sequenceLength = 8
inputSize = 1  # and output size
hiddenStateSize = 500
numberOfSequences = 3  # or number of samples

# training data
t0 = torch.linspace(0, 1, numberOfSequences)
t = torch.linspace(0, 1, sequenceLength + 1)
t, t0 = torch.meshgrid(t, t0, indexing='ij')
y = (t0 + t) ** 3

y = y.unsqueeze(-1)
x = y[:-1]  # (sequenceLength, numberOfSequences, inputSize)
y = y[1:]  # (sequenceLength, numberOfSequences, outputSize)

model = RNN(inputSize, hiddenStateSize, inputSize)

yPred = model(x)  # prediction
cost = torch.mean((yPred - y) ** 2)  # cost
```

One shortcoming of standard recurrent neural networks is that they are only able to connect very recent information with the current task. An illustrative example is the prediction of words in text sequences. For instance, the final word in the sentence "*the color of a lemon is yellow*" can directly be derived from the preceding words. Often, the relevant information needed for the prediction is found much earlier in the sequence. In the text excerpt, "*We used too many lemons for our lemonade... We did not like the lemonade, it was too sour*", the gap between the information and the place of prediction is much larger, and recurrent neural networks usually fail to learn such long-term dependencies [71]. This problem can be solved by introducing a special implementation of recurrent neural networks, called long short-term memory networks (LSTMs) [72] that often achieve outstanding results in the aforementioned tasks. The key concept of LSTMs is their cell state $\boldsymbol{c}_t$, acting as long-term memory, where information has to pass several gates before it gets forwarded. Three different gates control and protect the cell state by deciding if the information is added $\boldsymbol{i}_t$, kept $\boldsymbol{o}_t$, or discarded $\boldsymbol{f}_t$. The cell state is updated with

$$\boldsymbol{c}_t = \boldsymbol{c}_{t-1} \odot \boldsymbol{f}_t + \boldsymbol{i}_t \odot \sigma(\boldsymbol{W}_{c_x} \boldsymbol{x}_t + \boldsymbol{W}_{c_h} \boldsymbol{h}_{t-1} + \boldsymbol{b}_c), \tag{3.58}$$

where $\odot$ denotes an element-wise multiplication, i.e., the Hadamard product. The next hidden state is then computed with

$$\boldsymbol{h}_t = \sigma(\boldsymbol{c}_t) \odot \boldsymbol{o}_t \tag{3.59}$$

from which the output $\boldsymbol{y}_t$ can be computed through additional layers. The three gates act on the input $\boldsymbol{x}_t$ and previous hidden state $\boldsymbol{h}_{t-1}$ in the following manner:

$$\text{the input gate: } \boldsymbol{i}_t = \sigma(\boldsymbol{W}_{i_x}\boldsymbol{x}_t + \boldsymbol{W}_{i_h}\boldsymbol{h}_{t-1} + \boldsymbol{b}_i), \tag{3.60}$$

$$\text{the output gate: } \boldsymbol{o}_t = \sigma(\boldsymbol{W}_{o_x}\boldsymbol{x}_t + \boldsymbol{W}_{o_h}\boldsymbol{h}_{t-1} + \boldsymbol{b}_o), \tag{3.61}$$

$$\text{the forget gate: } \boldsymbol{f}_t = \sigma(\boldsymbol{W}_{f_x}\boldsymbol{x}_t + \boldsymbol{W}_{f_h}\boldsymbol{h}_{t-1} + \boldsymbol{b}_f), \tag{3.62}$$

where all weights $\boldsymbol{W}$ and biases $\boldsymbol{b}$ are learnable. This enables the LSTM to either forget previous cell states $\boldsymbol{c}_{t-1}$ with the forget gate $\boldsymbol{f}_t$, add new information via the input gate $\boldsymbol{i}_t$ and finally filter the cell state $\boldsymbol{c}_t$ with the output gate $\boldsymbol{o}_t$. In this way, LSTMs are capable of remembering information and storing long-term dependencies. Since recurrent neural networks employ a sequential structure, the backpropagation algorithm is adapted to take into account the temporal component, resulting in a method called "*backpropagation through time*" [69]. Another common remedy for long-term dependencies are gated recurrent units (GRU) [73].

The concept of recurrent neural networks, i.e., employing a temporally evolving hidden state, can be combined with other advanced architectures, such as convolutional, graph, Bayesian networks, or even autoencoders. For more in-depth content about recurrent neural networks, the reader is referred to Chapter 10 of Goodfellow's "*Deep Learning*" book [6] and Christopher Olah's blog post "*Understanding LSTMs*" [69]. Applications of recurrent neural networks in computational mechanics are discussed in Sects. 10.1.1.2, 10.1.1.2.

3.9.7 Transformer

Another prominent network architecture propelled to fame through chatbots like BERT [74] or ChatGPT [75] are transformers [76]. Transformers utilize self-attention mechanisms to process input sequences, capturing dependencies between elements regardless of their distance in the sequence—thus acting as an alternative to recurrent neural networks and their modern variants. The capability of handling long input sequences is enhanced by positional encoding, which helps the model retain the order of the sequence data. Transformers and their applications to computational mechancis are explored in more detail in Sect. 8.6.

3.9.8 Physics-Inspired Architectures for Dynamics

In mechanics, Hamiltonian mechanics (Sect. 3.9.8.2) and the Lagrangian framework (Sect. 3.9.8.4) are prominent methods to describe dynamical systems. In both cases, the quantities of interest are derived from a single scalar function through differentiation, namely the Hamiltonian $\mathcal{H}$ or the Lagrangian L. This is is a particularly attractive setting for deep learning, as a potentially high-dimensional system can be derived from a scalar output from a neural network. Corresponding developments are Hamiltonian neural networks [77] (Sect. 3.9.8.3) and Lagrangian neural networks [78–80] (Sect. 3.9.8.5). Aside from the architectural advantage, these networks generalize better than standard neural networks, as they implicity obey the structure of Hamiltonian mechanics and Lagrangians. Both concepts are introduced by means of a simple one-dimensional mass-spring model from structural dynamics (Sect. 3.9.8.1). For a more in-depth but equally accessible treatment, we recommend Sam Greydanus' blog posts on the topic [81, 82] (from which we have taken strong inspiration for this section).

3.9.8.1 One-Dimensional Structural Dynamics Problem

We consider the one-dimensional mass-spring model, illustrated in Fig. 3.33. The system can be described by the ordinary differential equation

$$m\ddot{x}(t) + kx(t) = 0 \tag{3.63}$$

with the homogeneous solution (i.e., without external loading)

$$x(t) = A\sin(\omega t + \phi) \tag{3.64}$$

where the natural frequency is given as $\omega = \sqrt{k/m}$. The parameters A and ϕ are determined with the initial conditions $x(0), \dot{x}(0)$. In the following, we consider an initial position of $x(0) = 0$ and an initial velocity of $\dot{x}(0) = 1$. Furthermore, we choose the model parameters $k = 10$ and $m = 1$.

3.9.8.2 Hamiltonian Mechanics

In Hamiltonian mechanics, systems are described by coordinate pairs, namely $[\boldsymbol{x}(t), \boldsymbol{p}(t)]$, where $\boldsymbol{x}(t)$ describes the position and $\boldsymbol{p}(t) = m\dot{\boldsymbol{x}}(t)$ the momentum. The momentum consists of the mass m and velocity $\dot{\boldsymbol{x}}(t)$. The scalar Hamiltonian $\mathcal{H}(\boldsymbol{x}(t), \boldsymbol{p}(t))$ represents the system's total energy Π_{tot}, and is defined as

$$\frac{d\boldsymbol{x}}{dt} = \frac{\partial \mathcal{H}}{\partial \boldsymbol{p}}, \tag{3.65}$$

$$\frac{d\boldsymbol{p}}{dt} = -\frac{\partial \mathcal{H}}{\partial \boldsymbol{x}}. \tag{3.66}$$

Given the Hamiltonian, the time evolution of $\boldsymbol{x}$ and $\boldsymbol{p}$ can be recovered through integration over the time interval $[t, t + \Delta t]$

$$\boldsymbol{x}(t + \Delta t) = \boldsymbol{x}(t) + \int_t^{t+\Delta t} \left.\frac{\partial \mathcal{H}}{\partial \boldsymbol{p}}\right|_\tau d\tau, \tag{3.67}$$

$$\boldsymbol{p}(t + \Delta t) = \boldsymbol{p}(t) - \int_t^{t+\Delta t} \left.\frac{\partial \mathcal{H}}{\partial \boldsymbol{x}}\right|_\tau d\tau, \tag{3.68}$$

which can be discretized by a forward Euler scheme

$$\boldsymbol{x}(t + \Delta t) = \boldsymbol{x}(t) + \left.\frac{\partial \mathcal{H}}{\partial \boldsymbol{p}}\right|_t \Delta t, \tag{3.69}$$

$$\boldsymbol{p}(t + \Delta t) = \boldsymbol{p}(t) - \left.\frac{\partial \mathcal{H}}{\partial \boldsymbol{x}}\right|_t \Delta t. \tag{3.70}$$

Thus, Hamiltonian mechanics enable the prediction of the next position $\boldsymbol{x}(t + \Delta t)$ and momentum $\boldsymbol{p}(t + \Delta t)$ from the current position $\boldsymbol{x}(t)$, the momentum $\boldsymbol{p}(t)$ and the scalar Hamiltonian $\mathcal{H}(\boldsymbol{x}(t), \boldsymbol{p}(t))$. For the mass-spring model from Fig. 3.33, the Hamiltonian is known as the system's total energy Π_{tot} composed of the kinetic Π_{kin} and potential energy Π_{tot}

$$\mathcal{H} = \Pi_{\text{tot}} = \Pi_{\text{kin}} + \Pi_{\text{tot}} = \frac{1}{2}m\dot{x}^2 + \frac{1}{2}kx^2. \tag{3.71}$$

However, the Hamiltonian may not be known for more complex or partially undefined systems. Hamiltonian neural networks can predict unseen dynamics of a system by learning the Hamiltonian by observing the dynamics.

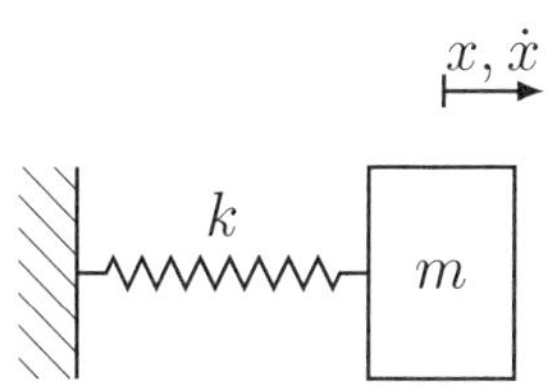

Fig. 3.33 One-dimensional mass-spring model from structural dynamics

3.9.8.3 Hamiltonian Neural Networks

Assuming, we have access to data triplets consisting of positions, velocities, and accelerations, i.e., $\{\tilde{\boldsymbol{x}}(t_i), \tilde{\dot{\boldsymbol{x}}}(t_i), \tilde{\ddot{\boldsymbol{x}}}(t_i)\}_{i=1}^{m_\mathcal{D}}$ (from which $\tilde{\boldsymbol{p}}(t_i) = m\tilde{\dot{\boldsymbol{x}}}(t_i)$ and $\tilde{\dot{\boldsymbol{p}}}(t_i) = m\tilde{\ddot{\boldsymbol{x}}}(t_i)$ can directly be computed), a mapping to the Hamiltonian can be established as

$$\hat{\mathcal{H}} = \mathcal{H}(\boldsymbol{x}(t), \boldsymbol{p}(t); \boldsymbol{\Theta}) \tag{3.72}$$

using a neural network with parameters $\boldsymbol{\Theta}$. The quantities $d\hat{\boldsymbol{x}}/dt$, and $d\hat{\boldsymbol{p}}/dt$ can be retrieved from the estimated Hamiltonian $\hat{\mathcal{H}}$ through Eqs. (3.65) and (3.66). By comparing the estimated derivatives with the labeled data, a cost function can be formulated as

$$\begin{aligned} C = \frac{1}{m_\mathcal{D}} \sum_{i=1}^{m_\mathcal{D}} \Bigg(& \left\| \frac{d\tilde{\boldsymbol{x}}(t_i)}{dt} - \frac{\partial \hat{\mathcal{H}}(\tilde{\boldsymbol{x}}(t_i), \tilde{\boldsymbol{p}}(t_i); \boldsymbol{\Theta})}{\partial \tilde{\boldsymbol{p}}(t_i)} \right\|^2 + \\ & \left\| \frac{d\tilde{\boldsymbol{p}}(t_i)}{dt} + \frac{\partial \hat{\mathcal{H}}(\tilde{\boldsymbol{x}}(t_i), \tilde{\boldsymbol{p}}(t_i); \boldsymbol{\Theta})}{\partial \tilde{\boldsymbol{x}}(t_i)} \right\|^2 \Bigg), \end{aligned} \tag{3.73}$$

which is minimized by adjusting $\boldsymbol{\Theta}$ through gradient-based optimization.

In PyTorch, the training dataset `trainingData` can be defined in the following manner, relying on the analytical solution from Eq. (3.64).

```
x = lambda t: A * np.sin(omega * t + phi)
dxdt = lambda t: omega * A * np.cos(omega * t + phi)
ddxdtt = lambda t: - omega ** 2 * A * np.sin(omega * t + phi)

tTraining = np.linspace(0, tmaxTraining, numberOfTrainingSamples)
trainingData = torch.vstack((
torch.from_numpy(x(tTraining)),
m * torch.from_numpy(dxdt(tTraining)),
m * torch.from_numpy(ddxdtt(tTraining)))
).to(torch.float32).T
```

As only $\boldsymbol{x}$ and $\boldsymbol{p}$ serve as inputs to the neural network, a model input `modelInputTraining` is defined from the training dataset. Importantly, the possibility of differentiating with respect to `modelInputTraining` must be ensured.

Listing 3.3 Conversion of training data to neural network input

```
modelInputTraining = trainingData[:, :2]
modelInputTraining.requires_grad = True
```

Using the model input `modelInputTraining` to predict the Hamiltonian `HPred`, the derivative quantities $d\hat{\mathbf{x}}/dt$ and $d\hat{\mathbf{p}}/dt$ (cf. Eqs. (3.65) and (3.66)) are computed with automatic differentiation (using the `getDerivative` function, which will be introduced in Chap. 4 in Listing 4.1). With $d\hat{\mathbf{x}}/dt$ and $d\hat{\mathbf{p}}/dt$, the cost function from Eq. (3.73) is obtained.

```
HPred = model(modelInputTraining)

dHPreddx = getDerivative(HPred, modelInputTraining, 1)
cost = torch.mean((dHPreddx[:, 0] + trainingData[:, 2]) **
    2 +
(dHPreddx[:, 1] - trainingData[:, 1] / m) ** 2)
```

This computation is embedded in the gradient-based optimization loop, previously seen in Listing 3.2. Once the neural network is trained, the system's dynamics can be identified from a starting point using the forward Euler scheme from Eqs. (3.69) and (3.70).

Listing 3.4 Forward Euler scheme to predict the dynamics of the system, cf. Eqs. (3.69) and (3.70)

```
xPrediction = torch.zeros((2, numberOfTimeSteps + 1),
requires_grad=False)
xPrediction[:, 0] = trainingData[0, :2] # initial
    conditions

for i in range(numberOfTimeSteps):
    currentx = xPrediction[:, i].unsqueeze(0).detach()
    currentx.requires_grad = True
    HPred = model(currentx)

    dHPreddx = getDerivative(HPred, currentx, 1)
    xPrediction[:, i + 1] = (xPrediction[:, i] +
    dt * torch.tensor([dHPreddx[0, 1], -dHPreddx[0, 0]])
    ).detach()
```

Predictions relying on a training with ten data points, equidistantly distributed in the first three-quarters of the oscillations period, are presented in Fig. 3.34a–c. These include the transient response (Fig. 3.34a), the evolution in the position-momentum space, i.e., (x, p) (Fig. 3.34b), and the energy evolution (Fig. 3.34c). These results are compared to a conventional neural network trained in a more straightforward manner. Specifically, a direct mapping to the derivative quantities (without a Hamiltonian) is considered.

$$\begin{pmatrix} \hat{\dot{x}} \\ \hat{\dot{p}} \end{pmatrix} = f(x, p; \boldsymbol{\Theta}) \tag{3.74}$$

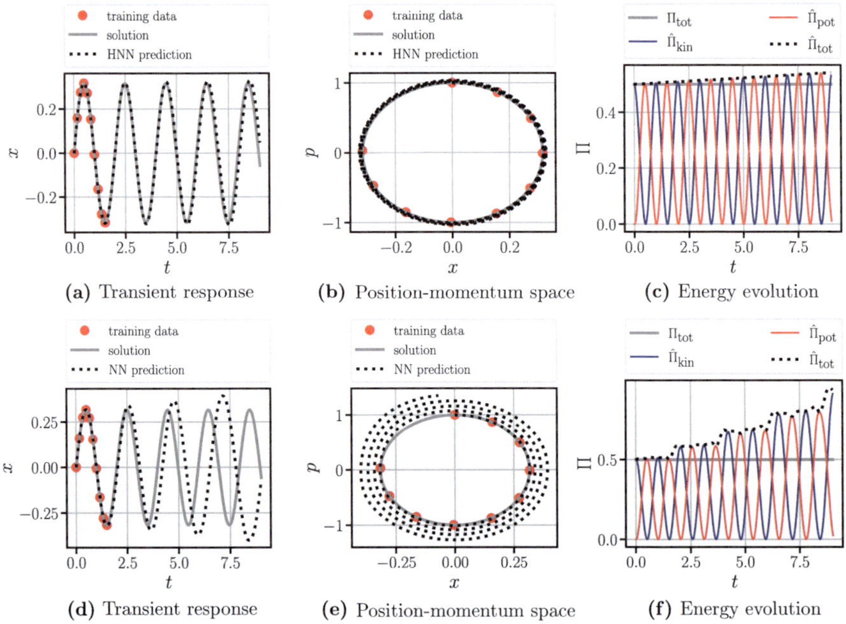

Fig. 3.34 Comparison of Hamiltonian neural network (HNN) predictions (top) and conventional neural network (NN) predictions (bottom)

The neural network f is trained with a standard data-driven cost function

$$C = \frac{1}{m_{\mathcal{D}}} \sum_{i=1}^{m_{\mathcal{D}}} \left(\|\tilde{\dot{\boldsymbol{x}}} - \hat{\dot{\boldsymbol{x}}}(\tilde{\boldsymbol{x}}(t_i), \tilde{\boldsymbol{x}}(t_i); \boldsymbol{\Theta})\|^2 + \|\tilde{\dot{\boldsymbol{p}}} - \hat{\dot{\boldsymbol{p}}}(\tilde{\boldsymbol{x}}(t_i), \tilde{\boldsymbol{p}}(t_i); \boldsymbol{\Theta})\|^2 \right). \quad (3.75)$$

As seen by comparing the analogous results from Fig. 3.34d–f with Fig. 3.34a–c, the Hamiltonian neural network's generalization is superior. However, note that the Hamiltonian neural network also does not conserve the total energy; see Fig. 3.34c. This error is however introduced by the forward Euler integration scheme from Eqs. (3.69) and (3.70) and can be avoided with energy-conserving time-stepping schemes, such as the trapezoidal rule or symplectic Euler. This effect can be isolated by predicting the Hamiltonian from the solution from Eq. (3.64) instead of evolving the dynamics from an initial condition through the time integration from Eqs. (3.67) and (3.68). Figure 3.35 illustrates the corresponding predicted Hamitlonian, which oscillates by about ~0.057%, but clearly does not grow. By contrast, energy accumulation would persist for the conventional neural network in Fig. 3.34f with a better time-stepping scheme, as the bulk of the error arises from the neural network's prediction errors. Thus, the Hamiltonian neural network generalizes better, because it learns to conserve a scalar quantity (related to the enery) from the observations.

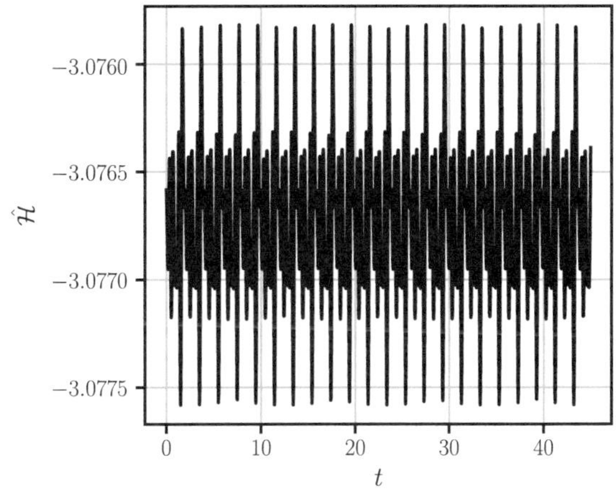

Fig. 3.35 Hamiltonian prediction from the true solution from Eq. (3.64)

Moreover, it is worth pointing out that the predicted Hamiltonian $\hat{\mathcal{H}}$ from Fig. 3.35 does not approximate the (constant) analytical Hamiltonian defined by Eq. (3.71) (which should be $\frac{1}{2}$). As only the derivatives of quantities are considered by the cost function in Eq. (3.73), the learned Hamiltonian can be arbitrarily shifted and is thereby not unique, i.e., will vary for a different neural network initialization. This also makes physical sense, as energies are defined up to an arbitrary constant—only energy differences are physically meaningful—so shifting the Hamiltonian does not affect the physical predictions.

In Fig. 3.34, both neural networks consisted of two hidden layers with 50 neurons each, the hyperbolic tangent as activation function, and 5 000 epochs of training with Adam and a learning rate of $\alpha = 10^{-3}$. For further details; see Exercise E.13.

3.9.8.4 Euler-Lagrange Equation

The Lagrangian framework is an alternative approach to analyze dynamical systems. The Lagrangian L offers a more general formulation than the Hamiltonian in terms of coordinates. The Lagrangian represents the system's energy. For the mass-spring system in Fig. 3.33, the Lagrangian is defined as

$$L = \Pi_{\text{kin}} - \Pi_{\text{tot}} = \frac{1}{2} m\dot{x}^2 - \frac{1}{2} kx^2. \tag{3.76}$$

The Lagrangian obeys the Euler-Lagrange equation

$$\frac{d}{dt}\frac{\partial L}{\partial \dot{\boldsymbol{x}}} = \frac{\partial L}{\partial \boldsymbol{x}}, \tag{3.77}$$

which ensures the principle of least action. Analogously to the Hamiltonian neural networks, Lagrangian neural networks predict a scalar, namely the Lagrangian $\hat{L}$, while relying on Eq. (3.77) to formulate a cost function.

3.9.8.5 Lagrangian Neural Networks

By parametrizing the Lagrangian with a neural network

$$\hat{L} = L(\boldsymbol{x}, \dot{\boldsymbol{x}}; \boldsymbol{\Theta}), \tag{3.78}$$

the quantity of interest $\ddot{\boldsymbol{x}}$ can be recovered by rewriting Eq. (3.77). This allows the direct computation of $\ddot{\boldsymbol{x}}$ from the parametrized Lagrangian $\hat{L}$. To this end, the time derivative in Eq. (3.77) is first expanded via the chain rule (assuming no time dependence of L on t)

$$\frac{d}{dt}\frac{\partial L}{\partial \dot{\boldsymbol{x}}} = \frac{\partial^2 L}{\partial \boldsymbol{x} \partial \dot{\boldsymbol{x}}}\dot{\boldsymbol{x}} + \frac{\partial^2 L}{\partial \dot{\boldsymbol{x}}^2}\ddot{\boldsymbol{x}} = \frac{\partial L}{\partial \boldsymbol{x}}. \tag{3.79}$$

Solving for $\ddot{\boldsymbol{x}}$ yields

$$\ddot{\boldsymbol{x}} = \left(\frac{\partial^2 L}{\partial \dot{\boldsymbol{x}}^2}\right)^{-1} \left(\frac{\partial L}{\partial \boldsymbol{x}} - \frac{\partial^2 L}{\partial \boldsymbol{x} \partial \dot{\boldsymbol{x}}}\dot{\boldsymbol{x}}\right). \tag{3.80}$$

Again relying on a dataset consisting of $\{\tilde{\boldsymbol{x}}(t_i), \tilde{\dot{\boldsymbol{x}}}(t_i), \tilde{\ddot{\boldsymbol{x}}}(t_i)\}_{i=1}^{m_\mathcal{D}}$, a cost function can be established as

$$C = \frac{1}{m_\mathcal{D}} \sum_{i=1}^{m_\mathcal{D}} \left[\left\| \tilde{\ddot{\boldsymbol{x}}}(t_i) - \hat{\ddot{\boldsymbol{x}}}\big(\hat{L}; \tilde{\boldsymbol{x}}(t_i), \tilde{\dot{\boldsymbol{x}}}(t_i)\big) \right\|^2 \right], \tag{3.81}$$

where $\hat{\ddot{\boldsymbol{x}}}$ is defined according to Eq. (3.80):

$$\hat{\ddot{\boldsymbol{x}}}\big(\hat{L}; \tilde{\boldsymbol{x}}(t_i), \tilde{\dot{\boldsymbol{x}}}(t_i)\big) = \\ \left(\frac{\partial^2 \hat{L}(\tilde{\boldsymbol{x}}(t_i), \tilde{\dot{\boldsymbol{x}}}(t_i); \boldsymbol{\Theta})}{\partial \tilde{\dot{\boldsymbol{x}}}(t_i)^2}\right)^{-1} \left(\frac{\partial \hat{L}(\tilde{\boldsymbol{x}}(t_i), \tilde{\dot{\boldsymbol{x}}}(t_i); \boldsymbol{\Theta})}{\partial \tilde{\boldsymbol{x}}(t_i)} - \frac{\partial^2 \hat{L}(\tilde{\boldsymbol{x}}(t_i), \tilde{\dot{\boldsymbol{x}}}(t_i); \boldsymbol{\Theta})}{\partial \tilde{\boldsymbol{x}}(t_i) \partial \tilde{\dot{\boldsymbol{x}}}(t_i)} \tilde{\dot{\boldsymbol{x}}}(t_i)\right). \tag{3.82}$$

Implementation-wise, the differences to the Hamiltonian neural network are minor. The training dataset is modified slightly to contain the direct derivatives of x instead of the terms premultiplied by the mass m.

```
trainingData = torch.vstack((
torch.from_numpy(x(tTraining)),
torch.from_numpy(dxdt(tTraining)),
torch.from_numpy(ddxdtt(tTraining))
).to(torch.float32).T
```

The model input from Listing 3.3 is reused in the mapping to the Lagrangian `LPred`, from which the acceleration `ddxttPred` is computed with Eq. (3.82). This enables the cost computation according to Eq. (3.81).

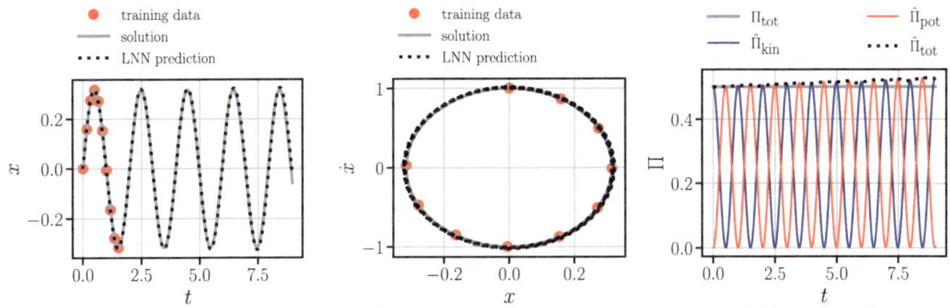

(a) Transient response (b) Position-velocity space (c) Energy evolution

Fig. 3.36 Predictions of Lagrangian neural network (LNN) that can be compared to the Hamiltonian neural network and conventional neural network predictions from Fig. 3.34

```
LPred = model(modelInputTraining)

dLPreddx = getDerivative(LPred, modelInputTraining, 1)
Hessian = getDerivative(dLPreddx[:, 0], modelInputTraining, 1)[:, 1]
ddxttPred = (1. / dLPreddx[:, 1]) * (dLPreddx[:, 0] -
Hessian * modelInputTraining[:, 1])
cost = torch.mean((trainingData[:, 2] - ddxttPred) ** 2)
```

Similar to Sect. 3.9.8.3, the dynamics can be predicted from the acceleration $\ddot{x}$, that can be derived from L with Eq. (3.82), using the forward Euler scheme from Listing 3.4. Analogously, as in Fig. 3.34, the predictions of the Lagrangian neural network are provided in Fig. 3.36. The results are similar to those achieved with the Hamiltonian neural network. Again the energy accumulation is caused by the integration scheme. The same network architecture, dataset, and training algorithm were employed, apart from an extended training consisting of 10 000 epochs.[5] Further details can be found in Sect. E.13.

Similar to the Hamiltonian, the learned Lagrangian is not unique either. Figure 3.37 illustrates three distinct time-dependent Lagrangians obtained by different neural network initializations. These are compared to the analytical Lagrangian from Eq. (3.76), which is not recoverable by the Lagrangian neural network.

Lastly, Hamiltonian and Lagrangian neural networks are compared directly by considering a longer time interval, namely 45 s, i.e., five times longer than in Figs. 3.34 and 3.36. More accurate integration schemes from `scipy.integrate.solve_ivp` are employed, avoiding spurious energy accumulations. As seen by Fig. 3.38, no spurious energies accumulate over time for both networks. Thus, both networks reliably learn the dynamics of the system in Fig. 3.33 from only ten data points. By contrast, the conventional neural network suffers from a severe energy accumulation despite the improved integration scheme. In conclusion, much more accurate predictions for unseen data, i.e., the generalization gain, comes from the implicit enforcement of Hamiltonian mechanics and the Lagrangian framework.

[5] In our experience, the method was more challenging to train than the Hamiltonian neural network.

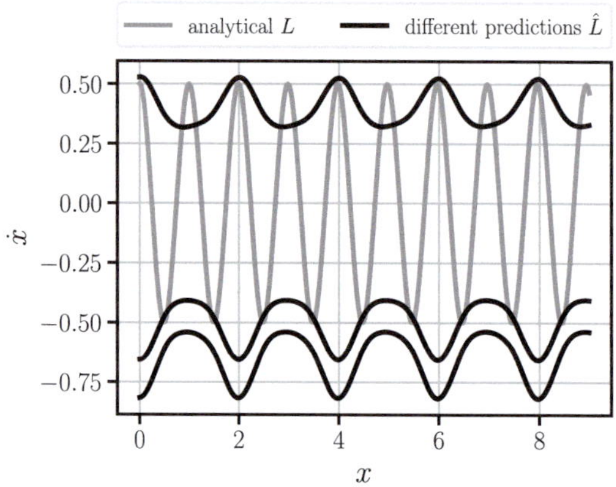

Fig. 3.37 Evolution of the Lagrangian. The Lagrangian neural network results in different Lagrangians $\hat{L}$ for different neural network initializations. These do not correspond to the analytical Lagrangian L from Eq. (3.76)

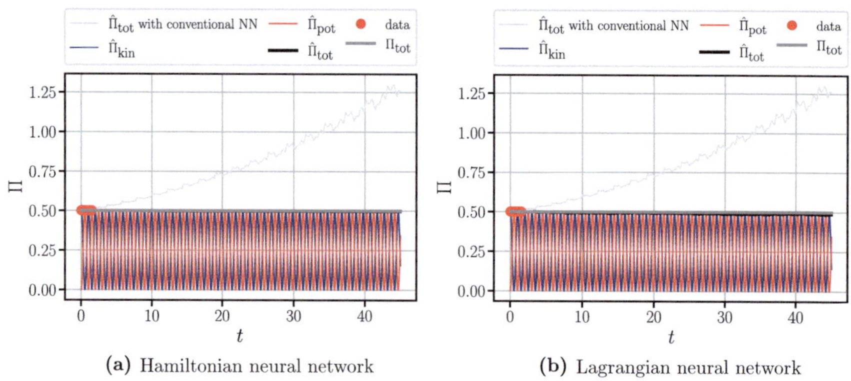

Fig. 3.38 Long-term energy evolution with a suitable time-stepping scheme of Hamiltonian and Lagrangian neural network, illustrating the superior generalization capabilities. The conventional neural network (NN) suffers from a severe spurious energy increase due to errors in the predictions. The considered time interval is five times as large than in Figs. 3.34c, f, and 3.36c

A shortcoming of both approaches lies in their sole applicability to dynamical systems through the Lagrangian and the Hamiltonian.[6] A different and more general perspective, using differential equations, is achieved by physics-informed neural networks, which will be discussed in the upcoming Chaps. 4 and 5.

[6] Although the concepts were discussed only for energy-conserving systems, extensions to dissipative systems have been proposed [83–85].

Exercises

E.7 Approximating the Sine Function (C)

This exercise revisits the example of a simple, fully connected feed-forward neural network approximating the sine function, which was introduced in the manuscript in Sect. 3.8. The goal of this exercise is to deepen the understanding of neural networks by showing the influence of different hyperparameters and familiarizing the reader with PyTorch.

1. Change the neural network architecture to a single hidden layer with two neurons and run the script `exercise 7 .ipynb`. What do you observe?
2. Now set the network to two hidden layers with 50 neurons each. What phenomena can be observed?
3. Name ways to overcome this problem.
4. Take a look at the cost function history. At which epoch should the algorithm stop to prevent the model from overfitting?
5. Set the learning rate to $\alpha = 0.1$. What do you observe?
6. Now reset the learning rate to $\alpha = 0.001$ and apply L^2-regularization by setting the value for $\lambda > 0$. Find a value where the model fits the sine curve without overfitting.
7. Increase the regularization parameter λ until underfitting occurs and plot the result.
8. Change the model to use a rectified linear unit (ReLU) activation function and plot the result.
9. Why does the neural network not learn the sine function properly at the boundaries (and beyond the boundaries)?

E.8 Neural Network Derivatives (P)

1. Express the output $\hat{y}$ of the neural network depicted in Fig. 3.39 as a function of the inputs x_1, x_2 and the network parameters, the weights

$$\boldsymbol{W}^{(1)} = \begin{pmatrix} w_{11}^{(1)} & w_{12}^{(1)} \\ w_{21}^{(1)} & w_{22}^{(1)} \end{pmatrix}, \quad \boldsymbol{w}^{(2)} = \begin{pmatrix} w_1^{(2)} \\ w_2^{(2)} \end{pmatrix}$$

and biases

$$\boldsymbol{b}^{(1)} = \begin{pmatrix} b_1^{(1)} \\ b_2^{(1)} \end{pmatrix}, \quad b^{(2)}$$

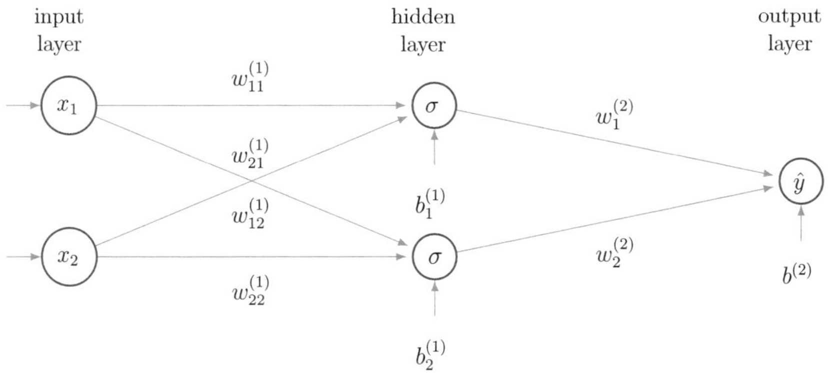

Fig. 3.39 A simple fully connected neural network example

2. Compute all partial derivatives $\frac{\partial \hat{y}}{\partial w_{jk}^{(l)}}$ and $\frac{\partial \hat{y}}{\partial b_j^{(l)}}$ assuming the activation function σ is given as $\sigma = (\cdot)^2$.
3. The cost function is defined as $C = (\tilde{y} - \hat{y})^2$. Compute the partial derivatives of the cost function C with respect to the weights and biases.
   Hint: $\frac{\partial C}{\partial w_{jk}^{(l)}} = \frac{\partial C}{\partial \hat{y}} \frac{\partial \hat{y}}{\partial w_{jk}^{(l)}}$

E.9 Neural Network from Scratch (C)

1. The goal is to implement a fully connected feed-forward neural network from scratch (relying only on the NumPy library). To this end, a class structure of a neural network is provided as `neuralNetwork` in `exercise 9 .ipynb`. Besides a constructor, the class consists of the following member functions:

   - `forward(self, x)` for the forward propagation, i.e., the prediction,
   - `backward(self, y)` for the backward propagation, i.e., the gradient computation
   - `zero_grad(self)` to reset the gradients (before the next forward/backward propagation)
   - `step(self, lr)` to update the parameters with gradient descent

   Implement these four member functions and compare them to the outputs of a corresponding implementation in PyTorch provided as `neuralNetworkTorch`.

2. For the prediction `forward(self, x)`, use Eq. (3.4). The forward propagation can be verified with the provided PyTorch neural network `neuralNetworkTorch`, which uses the same weights and biases.
3. For the backward propagation `backward(self, y)`, Eqs. (3.14), (3.16), (3.18), and (3.23) are helpful. Start by implementing the backward propagation for only a single sample. Subsequently, extend the implementations for an arbitrary number of samples. To this end, recall the definitions provided in Eqs. (3.10) and (3.11). Construct a loop over all samples and add the contributions to the corresponding gradients. The intermediate states, needed for the next layers, can be stored in the helper variables `currentWeightGradients` and `currentBiasGradients`. Remember to divide the gradients by the number of samples $m_{\mathcal{D}}$, as defined by the cost function C in Eq. (3.10). In both steps, the correctness of the implementation can be checked with the PyTorch neural network `neuralNetworkTorch`. When verifying the extended version for arbitrarily many samples, make sure to increase the number of samples to at least two: `x = np.expand_dims(np.linspace (0, 1, 2), 1)`.
4. Resetting the gradients in `zero_grad(self)` can be performed in the same way as in the constructor.
5. The gradient descent update should be performed according to Eq. (2.18) relying on the learning rate `lr`.
6. Apply the neural network to learn the function $f(x) = x^2$ in the range from zero to one. The driver file is provided in `exercise 9 .ipynb`.
7. Next, change the function to a more challenging one. Specifically, consider $f(x) = \sin(2\pi x)^2$. Learning this function is almost impossible. Change the optimizer to the Adam optimizer with the flag `selectOptimizer = 'Adam'`, which will make use of the Adam implementation from Sect. E.6. In combination with the Glorot (Xavier) initialization `torch.nn.init.xavier_uniform_` [15] (in the constructor of `neuralNetwork`), the function can be learned efficiently. Try disabling the Glorot initialization with the flag `xavierInitialization=True` in the constructor for both the gradient descent and Adam optimization.
8. Finally, compare your implementation with the provided PyTorch implementation. Which one is faster?

E.10 Neural Network Expressivity (C)

Check the script provided in `exercise 10 .ipynb` to visualize the potential representational capacity of a neural network. Here, you are able to change the number of neurons in the hidden layers and the number of layers to reproduce the Figs. 3.3 and 3.5 inspired by [17]. At the same time, you can check the number of trainable parameters with `summary` from the `torchsummary` package. Alternatively, compute the number analytically. Come up with a formula assuming two input neurons, one output neuron, and N_d hidden layers with N_w neurons each. What is the number of biases N_b and the number of weights N_w? Verify your result with the output from `summary`.

E.11 Convolutional Neural Network (P & C)

1. Given the following single one-dimensional input $\boldsymbol{x}_1$ and two sets of weights $\boldsymbol{w}_1^{(1)}$, $\boldsymbol{w}_2^{(1)}$, compute the two one-dimensional outputs $\hat{\boldsymbol{y}}_1$, $\hat{\boldsymbol{y}}_2$ assuming a stride of $s = 1$ and no padding.

$$\boldsymbol{x}_1 = [1, 4, 6, 8, 2]$$
$$\boldsymbol{w}_1^{(1)} = [1, 0, 1]$$
$$\boldsymbol{w}_2^{(1)} = [0, 0, 2]$$

2. Repeat the previous exercise with a stride of $s = 2$.
3. Now two inputs $\boldsymbol{x}_1$ and $\boldsymbol{x}_2$ are provided and again two outputs are desired $\hat{\boldsymbol{y}}_1$ and $\hat{\boldsymbol{y}}_2$. Thus four sets of weights $\boldsymbol{w}_{11}$, $\boldsymbol{w}_{12}$, $\boldsymbol{w}_{21}$ and $\boldsymbol{w}_{22}$ are required to perform the convolution. Use a stride of $s = 2$ to compute the outputs $\hat{\boldsymbol{y}}_1$ and $\hat{\boldsymbol{y}}_2$.

$$\boldsymbol{x}_1 = [3, 2, 1, 5, 2]$$
$$\boldsymbol{x}_2 = [5, 2, 0, 8, 0]$$
$$\boldsymbol{w}_{11} = [2, 3, 0]$$
$$\boldsymbol{w}_{12} = [1, 1, 1]$$
$$\boldsymbol{w}_{21} = [0, 2, 0]$$
$$\boldsymbol{w}_{22} = [0, 2, 2]$$

4. To familiarize yourself with PyTorch's syntax for convolutional neural networks and the effect that a single convolutional filter can have on an image, check the code provided in `exercise 11 .ipynb`. The code reproduces the images presented in Fig. 3.24.

E.12 Learning Strain Distributions (C)

This exercise revisits the example of learning strain distributions with more advanced neural network architectures, described in Sect. 3.9.1.2. The exercise's aim is to understand how to handle larger networks and more data-intensive problems in PyTorch by relying on a more professional workflow. Training and validation data and their split is handled by PyTorch's inbuilt `DataSet` and `DataLoader` from `torch.utils.data`. Information on the designed neural network architecture is obtainable through `summary` from `torchsummary`. TensorBoard acts as a visualization toolkit to monitor the training process, which significantly eases hyperparameter tuning. All of these tools can be found implemented for the strain distribution example in `exercise 12 .ipynb` and the helper file `NeuralNetwork.py`.

1. Perform a training of a single example by running the file `exercise 12 . ipynb`. Next, activate the TensorBoard options `writeGraph`, `writeHistogram`, `writeLearningHistory`, `writePredictions`, and check the corresponding syntax enabling the TensorBoard features. After rerunning the training, TensorBoard can be invoked with `tensorboard --logdir=logs` in the terminal while being located in the same directory as the exercise files. This will provide a link to the form http://localhost:6006 (the port number 6006 will vary), which, when opened, provides access to the TensorBoard interface. Before proceeding, familiarize yourself with the interface and make sure to understand the `scalars`, `images`, `graphs`, and `histograms` tabs. If something is unclear, consult the implementations in `exercise 12 .ipynb`.
2. Using TensorBoard (with `writeHistogram` and `writePredictions` deactivated), compare the three neural network implementations, accessible by changing `selectNN`. For a better comparison of the learning histories in TensorBoard, make sure to toggle the y-axis log scale in the `scalars` menu. Adjust the hyperparameters to boost performance.
3. Select the best performing neural network and corresponding hyperparameters and start increasing the number of training samples with `numberOfTrainingSamples`. In case of convergence issues, readjust the hyperparameters or increase the network capacity.
4. As the goal is to reduce the validation error as much as possible, introduce a L^2-regularization through the parameter `weightDecay`. What happens to the training and validation costs? Readjust the other hyperparameters if necessary. Can you beat the surrogate model from Sect. 3.9.1.2 in terms of validation cost?

E.13 Hamiltonian & Lagrangian Neural Networks (C)

The aim of this exercise is to implement a Hamiltonian and a Lagrangian neural network, described in Sect. 3.9.8 and reproduce the results of the mass-spring model seen in Fig. 3.33. To this end, three files are provided: `exercise 13 _1.ipynb` for the Hamiltonian neural network, `exercise 13 _2.ipynb` for the Lagrangian neural network, and `exercise 13 _3.ipynb` for a conventional neural network predicting the dynamics of a system. In all three files, the cost function has to be implemented within the training loop. For the implementation, consult Eqs. (3.73), (3.75), and (3.81). After incorporating the cost computation, train and compare the three approaches. Adapt the hyperparameters if necessary. How do the methods behave for longer prediction times (change the `tmaxValidation` and `numberOfTimeSteps` for the forward Euler scheme), with larger training datasets (change `tmaxTraining` and `numberOfTrainingSamples`), with different neural network architectures (change `neurons` and `layers`), or with different model parameters (change `k` and `m`)? If you want to isolate the energy accumulation error of the predictions, consider a different time-stepping scheme.

References

1. Marquardt, Florian. 2019. Machine Learning for Physicists. Lecture. tex.ids= Marquardt2019a.
2. Mehta, Pankaj, et al. 2019. A high-bias, low-variance introduction to Machine Learning for physicists. en. *Physics Reports* 810. arXiv: 1803.08823, 1–124. ISSN: 03701573. https://doi.org/10.1016/j.physrep.2019.03.001.
3. Krizhevsky, Alex, Ilya Sutskever, and Geoffrey E Hinton. 2012. ImageNet classification with deep convolutional neural networks. In *Advances in neural information processing systems*, vol. 25. Curran Associates, Inc.
4. Demuth, Howard B., et al. 2014. *Neural Network Design*. 2nd. tex.ids= Demuth2014a. Stillwater, OK, USA: Martin Hagan. ISBN: 978-0-9717321-1-7.
5. Carleo, Giuseppe, et al. 2019. Machine learning and the physical sciences. en. *Reviews of Modern Physics* 91(4). tex.ids= Carleo2019a arXiv: 1903.10563, 045002. ISSN: 0034-6861, 1539-0756. https://doi.org/10.1103/RevModPhys.91.045002.
6. Goodfellow, Ian, Yoshua Bengio, and Aaron Courville. 2016. *Deep learning*. MIT Press. ISBN: 0-262-03561-8.
7. Cybenko, G. 1989. Approximation by superpositions of a sigmoidal function. en. *Mathematics of Control, Signals and Systems* 2(4): 303–314. ISSN: 1435-568X. https://doi.org/10.1007/BF02551274.
8. Nielsen, Michael A. 2015. *Neural networks and deep learning*. en. tex.ids= Nielsen2015a. Determination Press.
9. Eldan, Ronen, and Ohad Shamir. 2016. The power of depth for feedforward neural networks. en. In *Conference on learning theory*. PMLR, 907–940.
10. Perekrestenko, Dmytro, et al. 2018. *The universal approximation power of finite-width deep ReLU networks*. arXiv:1806.01528 [cs, stat]. https://doi.org/10.48550/arXiv.1806.01528.
11. Géron, Aurélien. 2023. *Hands-on machine learning with Scikit-Learn, Keras and TensorFlow: Concepts, tools, and techniques to build intelligent systems*. Third edition. OCLC: on1346503549. Beijing: O'Reilly Media, Inc. ISBN: 978-1-09-812597-4.
12. Mhaskar, Hrushikesh, Qianli Liao, and Tomaso Poggio. 2016. Learning Functions: When Is Deep Better Than Shallow. en. In arXiv:1603.00988 [cs]. tex.ids= Mhaskar2016a arXiv: 1603.00988.
13. Hewitt, Edwin, and Robert E. Hewitt. 1979. The Gibbs-Wilbraham phenomenon: An episode in fourier analysis. en. In *Archive for History of Exact Sciences* 21(2). tex.ids= Hewitt1979a, 129–160. ISSN: 0003-9519, 1432-0657. https://doi.org/10.1007/BF00330404.
14. Llanas, B., S. Lantarón, and F. J. Sáinz. 2008. Constructive approximation of discontinuous functions by neural networks. en. *Neural Processing Letters* 27(3): 209–226. ISSN: 1370-4621, 1573-773X. https://doi.org/10.1007/s11063-007-9070-9.
15. Glorot, Xavier, and Yoshua Bengio. 2010. Understanding the difficulty of training deep feedforward neural networks. en. In *Proceedings of the Thirteenth International Conference on Artificial Intelligence and Statistics*. tex.ids= Glorot2010a ISSN: 1938-7228 section: Machine Learning, 249–256.
16. He, Kaiming, et al. 2016. Deep residual learning for image recognition. In *2016 IEEE Conference on Computer Vision and Pattern Recognition (CVPR)*. Las Vegas, NV, USA: IEEE, 770–778. ISBN: 978-1-4673-8851-1. https://doi.org/10.1109/CVPR.2016.90.
17. Marquardt, Florian. 2017. *Visualize the output of a multilayer network*. tex.ids= Marquardt2017a.

18. Schwab, Christoph, and Jakob Zech. 2023. Deep learning in high dimension: Neural network expression rates for analytic functions in $L^2(\mathbb{R}^d \gamma_d)$. en. *SIAM/ASA Journal on Uncertainty Quantification* 11(1): 199–234. ISSN: 2166-2525. https://doi.org/10.1137/21M1462738.
19. Baydin, Atılım Gunes, et al. 2018. Automatic differentiation in machine learning: A survey. en. *Journal of Machine Learning Research* 43.
20. Olah, Christopher. 2015. *Calculus on Computational Graphs: Backpropagation.*
21. Russell, Stuart J., and Peter Norvig. 2022. *Artificial intelligence: A modern approach.* eng. Fourth edition, global edition. Pearson Series in Artificial Intelligence. Harlow: Pearson. ISBN: 978-1-292-40113-3.
22. Ramachandran, Prajit, Barret Zoph, and Quoc V. Le. 2017. *Searching for Activation Functions.* arXiv:1710.05941 [cs]
23. Elfwing, Stefan, Eiji Uchibe, and Kenji Doya. 2018. Sigmoid-weighted linear units for neural network function approximation in reinforcement learning. *Neural Networks.* Special issue on deep reinforcement learning 107: 3–11. ISSN: 0893-6080. https://doi.org/10.1016/j.neunet.2017.12.012.
24. Clevert, Djork-Arné, Thomas Unterthiner, and Sepp Hochreiter. 2016. *Fast and Accurate Deep Network Learning by Exponential Linear Units (ELUs).* arXiv:1511.07289 [cs]. https://doi.org/10.48550/arXiv.1511.07289.
25. Klambauer, Günter, et al. 2017. Self-normalizing neural networks. In *Proceedings of the 31st International Conference on Neural Information Processing Systems.* NIPS'17. Red Hook, NY, USA: Curran Associates Inc., 972–981. ISBN: 978-1-5108-6096-4.
26. Hendrycks, Dan, and Kevin Gimpel. 2023. *Gaussian Error Linear Units (GELUs).* arXiv:1606.08415 [cs]. https://doi.org/10.48550/arXiv.1606.08415.
27. Misra, Diganta. 2020. *Mish: A Self Regularized Non-Monotonic Activation Function.* arXiv:1908.08681 [cs, stat]. https://doi.org/10.48550/arXiv.1908.08681.
28. Jagtap, Ameya D., and George Em Karniadakis. 2020. Adaptive activation functions accelerate convergence in deep and physics-informed neural networks. *Journal of Computational Physics* 404: 109136. arXiv: 1906.01170, ISSN: 00219991. https://doi.org/10.1016/j.jcp.2019.109136.
29. He, Kaiming, et al. 2015. *Delving Deep into Rectifiers: Surpassing Human-Level Performance on ImageNet Classification.* arXiv:1502.01852 [cs] version: 1. Feb. 2015. https://doi.org/10.48550/arXiv.1502.01852.
30. Ng, Andrew. 2020. Machine Learning. Online course. tex.ids= Ng2020a.
31. Chollet, François. 2018. *Deep learning with Python.* en. OCLC: ocn982650571. Shelter Island, New York: Manning Publications Co. ISBN: 978-1-61729-443-3.
32. Wilson, D. Randall, and Tony R. Martinez. 2003. The general inefficiency of batch training for gradient descent learning. en. *Neural Networks* 16(10): 1429–1451. ISSN: 08936080. https://doi.org/10.1016/S0893-6080(03)00138-2.
33. Opitz, David, and Richard Maclin. 1999. Popular ensemble methods: An empirical study. *Journal of Artificial Intelligence Research* 11(1): 169–198. ISSN: 1076-9757.
34. Wan, Shaohua, and Hua Yang. 2013. Comparison among methods of ensemble learning. In *Proceedings of the 2013 International Symposium on Biometrics and Security Technologies.* ISBAST '13. USA: IEEE Computer Society, 286–290. ISBN: 978-0-7695-5010-7. https://doi.org/10.1109/ISBAST.2013.50.
35. Breiman, Leo. 1996. Bagging predictors. en. *Machine Learning* 24(2): 123–140. ISSN: 1573-0565. https://doi.org/10.1007/BF00058655.
36. Srivastava, Nitish, et al. 2014. Dropout: A simple way to prevent neural networks from overfitting. *Journal of Machine Learning Research* 15(56): 1929–1958. ISSN: 1533-7928.
37. Katanforoosh, Kian, and Daniel Kunin. 2019. *Initializing neural networks.*
38. Czarnecki, Wojciech M., et al. 2017. Sobolev training for neural networks. In *Advances in Neural Information Processing Systems*, vol. 30. Curran Associates, Inc.
39. Lecun, Y., et al. 1998. Gradient-based learning applied to document recognition. *Proceedings of the IEEE* 86(11). Conference Name: Proceedings of the IEEE, 2278–2324. ISSN: 1558-2256. https://doi.org/10.1109/5.726791.

40. Herrmann, Leon, et al. 2022. The influence of the fiber–matrix microstructure on the energy release rate of off-axis tunnel cracks in laminates. *Composites Part B: Engineering* 247: 110338. ISSN: 1359-8368. https://doi.org/10.1016/j.compositesb.2022.110338.
41. Keshavarzzadeh, Vahid, et al. 2021. Image-based multiresolution topology optimization using deep disjunctive normal shape model. en. *Computer-Aided Design* 130: 102947. ISSN: 00104485. https://doi.org/10.1016/j.cad.2020.102947.
42. Woldseth, Rebekka V., et al. 2022. On the use of artificial neural networks in topology optimisation. en. *Structural and Multidisciplinary Optimization* 65(10): 294. ISSN: 1615-1488. https://doi.org/10.1007/s00158-022-03347-1.
43. McGreivy, Nick, and Ammar Hakim. 2024. Weak baselines and reporting biases lead to overoptimism in machine learning for fluid-related partial differential equations. en. *Nature Machine Intelligence* 6(10): 1256–1269. ISSN: 2522-5839. https://doi.org/10.1038/s42256-024-00897-5.
44. Heaven, Douglas. 2019. Why deep-learning AIs are so easy to fool. en. In *Nature* 574(7777): 163–166. ISSN: 0028-0836, 1476-4687. https://doi.org/10.1038/d41586-019-03013-5.
45. Herrmann, Leon, and Stefan Kollmannsberger. 2024. Deep learning in computational mechanics: A review. en. *Computational Mechanics*. ISSN: 1432-0924. https://doi.org/10.1007/s00466-023-02434-4.
46. Battaglia, Peter W., et al. 2018. *Relational inductive biases, deep learning, and graph networks*. arXiv:1806.01261 [cs, stat]. https://doi.org/10.48550/arXiv.1806.01261.
47. Kipf, Thomas N., and Max Welling. 2017. *Semi-Supervised Classification with Graph Convolutional Networks*. arXiv:1609.02907 [cs, stat]. https://doi.org/10.48550/arXiv.1609.02907.
48. Monti, Federico, et al. 2018. *Dual-Primal Graph Convolutional Networks*. arXiv:1806.00770 [cs, stat]. https://doi.org/10.48550/arXiv.1806.00770.
49. Duvenaud, David K, et al. 2015. Convolutional networks on graphs for learning molecular fingerprints. In *Advances in Neural Information Processing Systems*, vol. 28. Curran Associates, Inc.
50. van den Berg, Rianne, Thomas N. Kipf, and Max Welling. 2017. *Graph Convolutional Matrix Completion*. arXiv:1706.02263 [cs, stat]. https://doi.org/10.48550/arXiv.1706.02263.
51. Li, Yaguang, et al. 2018. *Diffusion Convolutional Recurrent Neural Network: Data-Driven Traffic Forecasting*. arXiv:1707.01926 [cs, stat]. https://doi.org/10.48550/arXiv.1707.01926.
52. Sanchez-Gonzalez, Alvaro, et al. 2020. *Learning to Simulate Complex Physics with Graph Networks*. https://doi.org/10.48550/arXiv.2002.09405.
53. Pfaff, Tobias, et al. 2021. *Learning Mesh-Based Simulation with Graph Networks*. https://doi.org/10.48550/arXiv.2010.03409.
54. Rezende, Danilo Jimenez, Shakir Mohamed, and Daan Wierstra. 2014. *Stochastic Backpropagation and Approximate Inference in Deep Generative Models*. arXiv:1401.4082 [cs, stat]. https://doi.org/10.48550/arXiv.1401.4082.
55. Kingma, Diederik P., and Max Welling. 2022. *Auto-Encoding Variational Bayes*. arXiv:1312.6114 [cs, stat]. https://doi.org/10.48550/arXiv.1312.6114.
56. Guo, Chuan, et al. 2017. On calibration of modern neural networks. In *Proceedings of the 34th International Conference on Machine Learning - Volume 70*. ICML'17. Sydney, NSW, Australia: JMLR.org, 1321–1330.
57. Nixon, Jeremy, et al. 2020. *Measuring Calibration in Deep Learning*. arXiv:1904.01685 [cs, stat]. https://doi.org/10.48550/arXiv.1904.01685.
58. Lampinen, Jouko, and Aki Vehtari. 2001. Bayesian approach for neural networks—review and case studies. en. *Neural Networks* 14(3): 257–274. ISSN: 0893-6080. https://doi.org/10.1016/S0893-6080(00)00098-8.
59. Titterington, D. M. 2004. Bayesian methods for neural networks and related models. *Statistical Science* 19(1): 128–139. ISSN: 0883-4237, 2168-8745. https://doi.org/10.1214/088342304000000099.
60. Goan, Ethan, and Clinton Fookes. 2020. Bayesian neural networks: An introduction and survey. en. In *Case Studies in Applied Bayesian Data Science: CIRM Jean-Morlet Chair, Fall 2018*, ed. Kerrie L. Mengersen, Pierre Pudlo, and Christian P. Robert. Lecture Notes in Mathematics.

Cham: Springer, 45–87. ISBN: 978-3-030-42553-1. https://doi.org/10.1007/978-3-030-42553-1_3.
61. Hastings, W. K. 1970. Monte Carlo sampling methods using Markov chains and their applications. *Biometrika* 57(1): 97–109. ISSN: 0006-3444. https://doi.org/10.2307/2334940.
62. Blei, David M., Alp Kucukelbir, and Jon D. McAuliffe. 2017. Variational inference: A review for statisticians. en. *Journal of the American Statistical Association* 112(518): 859–877. ISSN: 0162-1459, 1537-274X. https://doi.org/10.1080/01621459.2017.1285773.
63. Kullback, S., and R. A. Leibler. 1951. On information and sufficiency. en. In *The Annals of Mathematical Statistics* 22(1): 79–86. ISSN: 0003-4851. https://doi.org/10.1214/aoms/1177729694.
64. Bishop, Christopher M. 2016. *Pattern Recognition and Machine Learning*. eng. Softcover reprint of the original 1st edition 2006 (corrected at 8th printing 2009). Information Science and Statistics. New York, NY: Springer New York. ISBN: 978-1-4939-3843-8.
65. Graves, Alex. 2011. Practical variational inference for neural networks. In *Advances in Neural Information Processing Systems*, vol. 24. Curran Associates, Inc.
66. Jospin, Laurent Valentin, et al. 2022. Hands-on Bayesian neural networks—a tutorial for deep learning users. *IEEE Computational Intelligence Magazine* 17(2): 29–48. ISSN: 1556-603X, 1556-6048. https://doi.org/10.1109/MCI.2022.3155327.
67. Lakshminarayanan, Balaji, Alexander Pritzel, and Charles Blundell. 2017. Simple and scalable predictive uncertainty estimation using deep ensembles. In *Proceedings of the 31st International Conference on Neural Information Processing Systems*. NIPS'17. Red Hook, NY, USA: Curran Associates Inc., 6405–6416. ISBN: 978-1-5108-6096-4.
68. Gal, Yarin, and Zoubin Ghahramani. 2016. Dropout as a Bayesian approximation: Representing model uncertainty in deep learning. In *Proceedings of the 33rd International Conference on International Conference on Machine Learning - Volume 48*. ICML'16. New York, NY, USA: JMLR.org, 1050–1059.
69. Olah, Christopher. 2015. *Understanding LSTM Networks*.
70. Chetlur, Sharan, et al. 2014. *cuDNN: Efficient Primitives for Deep Learning*. arXiv:1410.0759 [cs]. https://doi.org/10.48550/arXiv.1410.0759.
71. Hochreiter, Sepp. 1991. Untersuchungen zu dynamischen neuronalen Netzen. In *Diploma, Technische Universität München* 91(1).
72. Hochreiter, Sepp, and Jürgen Schmidhuber. 1997. Long short-term memory. In *Neural Computation* 9(8): 1735–1780. ISSN: 0899-7667. https://doi.org/10.1162/neco.1997.9.8.1735.
73. Ballakur, Amulya Arun, and Arti Arya. 2020. Empirical evaluation of gated recurrent neural network architectures in aviation delay prediction. In *2020 5th International Conference on Computing, Communication and Security (ICCCS)*, 1–7. https://doi.org/10.1109/ICCCS49678.2020.9276855.
74. Devlin, Jacob, et al. 2019. *BERT: Pre-training of Deep Bidirectional Transformers for Language Understanding*. arXiv:1810.04805 [cs]. https://doi.org/10.48550/arXiv.1810.04805.
75. OpenAI. 2023. *GPT-4 Technical Report*. arXiv:2303.08774 [cs]. https://doi.org/10.48550/arXiv.2303.08774.
76. Vaswani, Ashish, et al. 2017. Attention is all you need. In: *Advances in neural information processing systems*, vol. 30. Curran Associates, Inc.
77. Greydanus, Sam, Misko Dzamba, and Jason Yosinski. 2019. *Hamiltonian Neural Networks*. arXiv:1906.01563 [cs]. https://doi.org/10.48550/arXiv.1906.01563.
78. Lutter, Michael, Kim Listmann, and Jan Peters. 2019. Deep Lagrangian networks for end-to-end learning of energy-based control for under-actuated systems. In *2019 IEEE/RSJ International Conference on Intelligent Robots and Systems (IROS)*. ISSN: 2153-0866. 7718–7725. https://doi.org/10.1109/IROS40897.2019.8968268.
79. Lutter, Michael, Christian Ritter, and Jan Peters. 2019. *Deep Lagrangian Networks: Using Physics as Model Prior for Deep Learning*. arXiv:1907.04490 [cs, eess, stat]. https://doi.org/10.48550/arXiv.1907.04490.
80. Cranmer, Miles, et al. 2020. *Lagrangian Neural Networks*. arXiv:2003.04630 [physics, stat]. https://doi.org/10.48550/arXiv.2003.04630.
81. Greydanus, Sam, Misko Dzamba, and Jason Yosinski. 2019. *Hamiltonian Neural Networks*.

82. Greydanus, Sam, Miles Cranmer, and Stephan Hoyer. 2020. *Lagrangian Neural Networks*.
83. Zhong, Yaofeng Desmond, Biswadip Dey, and Amit Chakraborty. 2020. *Dissipative SymODEN: Encoding Hamiltonian Dynamics with Dissipation and Control into Deep Learning*. arXiv:2002.08860 [cs, eess, stat]. https://doi.org/10.48550/arXiv.2002.08860.
84. Sosanya, Andrew, and Sam Greydanus. 2022. *Dissipative Hamiltonian Neural Networks: Learning Dissipative and Conservative Dynamics Separately*. arXiv:2201.10085 [cs]. https://doi.org/10.48550/arXiv.2201.10085.
85. Eidnes, Sølve, et al. 2023. Pseudo-Hamiltonian neural networks with state-dependent external forces. *Physica D: Nonlinear Phenomena* 446: 133673. ISSN: 0167-2789. https://doi.org/10.1016/j.physd.2023.133673.

Chapter 4
Introduction to Physics-Informed Neural Networks

Abstract Physics-informed neural networks (PINNs) are used for problems where data is scarce. In such cases, neural networks approximate the unknown solution to a differential equation that governs the physical problem by minimizing the residual. Interestingly, PINNs can also be used to discover differential equations themselves. In this chapter, PINNs are illustrated with basic one-dimensional and two-dimensional examples, ranging from a static bar, a plate in membrane action, to the transient temperature evolution using the non-linear heat equation. These examples are accompanied by simple codes, allowing minor augmentations to solve other differential equations in arbitrary dimensions.

As seen in Sect. 3.9.1.2, generating an accurate surrogate model of even a simple physical system usually requires a large amount of data. However, data acquisition from experiments or simulations is often infeasible, as not all data is observable, or too costly in terms of memory, time, or time to analysis. With this in mind, Raissi et al. proposed an approach that augments surrogate models with existing knowledge about the underlying physics of a problem [20]. In many cases, the governing equations or empirically determined rules defining the problem are known a priori. For instance, an incompressible flow has to satisfy the law of conservation of mass. By incorporating this information, the solution space is drastically reduced, and, as a result, less training data is needed to learn the solution.

The idea of adding prior knowledge to a machine learning algorithm (to solve differential equations) is not completely new. The studies by [20] were inspired by papers of Psichogios and Ungar [1], Lagaris et al. [2], and more recent developments by Kondor [22], Hirn et al. [3] and Mallat [4]. Nevertheless, Raissi et al. popularized the idea and extended existing concepts with fundamentally new approaches like discrete time-stepping schemes.[1] The methodology was demonstrated on a

[1] This discrete time-stepping scheme is briefly discussed in Paragraph 10.1.2.2 in Chap. 10. The first version of this book [5] covered the discrete time-stepping approach in-depth. We, however, realized that the approach creates more confusion than benefit in an introductory textbook due to its complexity and lack of usefulness. Although it is based on the Runge-Kutta method, it does not outperform the Runge-Kutta method in terms of accuracy and computation time.

L. Herrmann et al., *Deep Learning in Computational Mechanics*,
https://doi.org/10.1007/978-3-031-89529-6_4

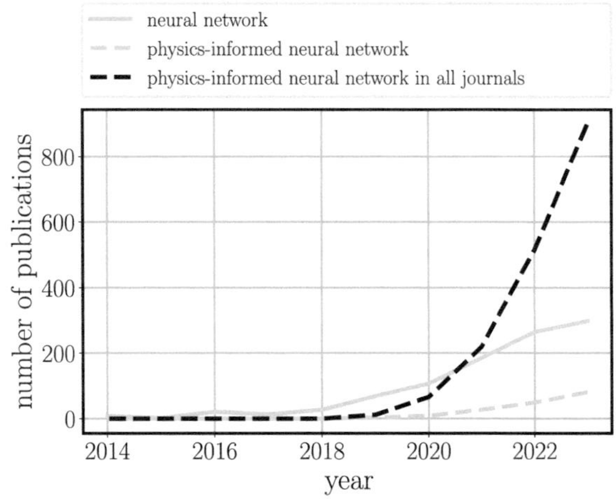

Fig. 4.1 Number of publications concerning physics-informed neural networks within computational mechanics compared to publications on neural networks within computational mechanics (relying on the same selection of journals as in Fig. 4.1b in Chap. 1). The dashed black line depicts the number of publications within all journals on physics-informed neural networks

variety of examples that are of interest in a physics and engineering context [20].[2]

The paper "*Physics-informed neural networks: A deep learning framework for solving forward and inverse problems involving non-linear partial differential equations*" [20] by Raissi et al. is highly referenced and marks a starting point for further research on physics-informed surrogate models—supported by the trend in publications (see Fig. 4.1) concerning physics-informed neural networks since 2017, the date of publication of [20].

Although the publication [20] has caused a drastic increase in literature concerned with the method—with ∼27% of neural network-related computational mechanics publications treating physics-informed neural networks in the year 2023—caution must be advised. Currently, physics-informed neural networks are neither capable of achieving faster nor more accurate solutions to partial differential equations—in comparison to classical methods, except for highly specialized tasks. Physics-informed neural networks are employed to solve a single problem and do not serve as surrogate models employable in fast online predictions. Each acquisition of a solution is accompanied by an expensive training procedure. The main reason behind the lack of accuracy is that current tools are inefficient in exploring the highly non-linear neural network solution spaces, ending in unsatisfactory local minima compared to solutions obtained with established classical methods.

Nevertheless, an argument for investigating the solution procedure for forward and inverse problems with physics-informed neural networks can still be made.[3] Sup-

[2] The accompanying code was written in Python and utilizes the popular GPU-accelerated machine learning framework Tensorflow [6]. Additionally, the code is publicly available on GitHub, allowing others to explore physics-informed neural networks and contribute to their development [7].

[3] The forward problem is concerned with solving the partial differential equations, such as, e.g., achieved by the surrogate model for learning strains in Sect. 3.9.1.2 (i.e., approximating the solutions to the two-dimensional elasticity equations). By contrast, for the inverse problem, the differential

pose one succeeds at learning a single accurate solution. In that case, only a relatively small step has to be taken to go from one solution to multiple solutions, i.e., training a surrogate model based only on a differential equation and thus not relying on any data—yielding a universal discretization technology for any differential equation. However, this seems unfeasible with the current state of deep learning training techniques.[4] At the moment, physics-informed neural networks are mainly applicable in the niche domain of solving differential equations in the forward or inverse setting with considerable prior knowledge of the solution in terms of measurements. In other words, physics-informed neural networks are helpful in incorporating measurement data in the simulation.

Despite its shortcomings, the contribution [20] has spawned a creative spirit within computational mechanics, which now embraces neural networks beyond supervised learning. Therefore, physics-informed neural networks are worthwhile to understand and can provide valuable insights into what to do and what not to do with deep learning in computational mechanics. Thus, the subject is discussed extensively in the current and subsequent chapter, covering the most important physics-informed neural network concepts available in the current literature.

4.1 Overview

The general idea of physics-informed neural networks is to solve problems where only limited data is available, e.g., noisy measurements from an experiment. To compensate for the data scarcity, the algorithm is enriched with physical laws governing the problem at hand. Typically, those laws are described by parameterized non-linear partial differential equations of the form

$$\mathcal{N}[u; \lambda] = 0, \;\; x \in \Omega, \;\; t \in \mathcal{T}. \tag{4.1}$$

Here, the latent solution $u(t, x)$ depends on time $t \in \mathcal{T} = [0, T]$ and a spatial variable $x \in \Omega$, where Ω refers to a space in $\mathbb{R}^D$. Furthermore, $\mathcal{N}[u; \lambda]$ represents a non-linear differential operator with coefficients λ, potentially composed of gradients with respect to space and time. This description covers a wide variety of problems ranging from advection-diffusion reaction of chemical or biological systems to the governing equations of continuum mechanics.

The following sections distinguish between two different use cases for physics-informed neural networks, namely data-driven inference and data-driven identification of partial differential equations. The first case addresses forward problems where

equation (model identification) or its coefficients (parameter identification) are to be identified, as, e.g., seen in the context of the Hamiltonian and Lagrangian neural networks in Sect. 3.9.8, where models were discovered.

[4] Major breakthroughs in optimization algorithms for neural networks might make physics-informed neural networks feasible.

the coefficients λ are known and the hidden solution $u(t, x)$ is computed based on initial and boundary data. The second case solves an inverse problem. Given scattered data of the solution $u(t, x)$, the goal is to identify the coefficients λ—or potentially even the non-linear operator $\mathcal{N}$—of the partial differential equation.

For stationary or transient problems, the physics-informed neural network computes the underlying partial differential equation. Therefore, a classical fully connected neural network approximates the solution $u(t, x)$ and the physics-enriched part evaluates the corresponding partial derivatives forming the left-hand side of Eq. (4.1), i.e., the residual of the differential equation. Both the network approximating $u(t, x)$ as well as the whole physics-informed neural network depend on the same set of parameters. Those shared weights and biases are trained by minimizing a cost function, which consists of multiple mean squared error losses. The cost function in its general form can be written as

$$C = \mathcal{L}_{\mathcal{D}} + \mathcal{L}_{\mathcal{R}}. \tag{4.2}$$

The first term, here denoted as $\mathcal{L}_{\mathcal{D}}$, computes the error of the approximation $u(t, x)$ at known data points. In the case of a forward problem, this term entails data representing the boundary and initial conditions, while for inverse problems, the solution at different points inside the domain is provided. The other term in the cost function, $\mathcal{L}_{\mathcal{R}}$, enforces the partial differential equation on a large set of randomly chosen collocation points inside the domain. In particular, this term penalizes the error of the approximation of the residual of the partial differential equation at every collocation point.

Generally, there are two ways to enforce boundary conditions of a forward problem. In case of a weak enforcement, the term $\mathcal{L}_{\mathcal{D}}$ from Eq. (4.2) entails multiple terms that enforce the solution at the boundary. Alternatively, regular and constant boundary conditions can be enforced in a strong sense. To this end, the output of the network is adapted so that the predicted solution automatically satisfies the boundary condition for any given neural network output. This simplifies the cost function since the boundary terms can be neglected. As with other deep learning algorithms, the optimal network parameters are found by minimizing the cost function (cf. Eq. (4.2)). In the context of physics-informed neural networks, gradient-based optimization algorithms like Adam or L-BFGS are often used [8, 20].

Even though the proposed approach is not guaranteed to converge to a global minimum, and thus, an accurate solution $u(x)$, [7, 20] showed empirically that the method achieves accurate results for several simple problems. Assuming that the partial differential equation along with its boundary and initial conditions has a unique solution and is well-posed, the physics-informed neural network is able to approximate the solution for simple problems at the collocation points and interpolates it in between. Further requirements are a network architecture with adequate representational power, and a sufficient number of collocation points $m_{\mathcal{R}}$. In addition to the collocation-based method, [20] proposed a discrete-time model for both forward and inverse problems. This approach makes use of a discrete time-stepping scheme to predict the solution $u(x)$ at time t^{n+1} from limited snapshot data at time t^n. An advantage of the discrete method is that it omits the use of temporally distributed collocation points since only a solution for the initial state at t^n and boundary conditions must be provided. For more details; see Paragraph 10.1.2.2 in Chap. 10.

The following sections explain the different types of physics-informed neural networks on forward (cf. Sect. 4.2) and inverse problems (cf. Sect. 4.3). After beginning with introductory examples of one- and two-dimensional linear elasticity, the subsequent section illustrates the continuous formulation with a non-linear time-dependent heat problem.

4.2 Data-Driven Inference

In general, the problem of inference can be phrased as: find the solution $u(t, x)$ for given coefficients λ. Since all values of λ are known, Eq. (4.1) simplifies to

$$\mathcal{N}[u] = 0, \quad x \in \Omega, \quad t \in \mathcal{T}. \tag{4.3}$$

4.2.1 *One-Dimensional Static Model*

A very simple application of the physics-informed neural network is the one-dimensional linear elastic static bar (see Fig. 4.2), which is governed by an ordinary differential equation and the corresponding boundary conditions. A solution in an equilibrium state is sought. Therefore, the differential equation only applies to the spatial domain $\Omega \in \mathbb{R}^1$. The differential equation operator is then given as $\mathcal{N}[u] = \frac{d}{dx}(EA\frac{du}{dx}) + p$. This includes an inhomogeneous source term p, which defines a distributed load. The system can thereby be expressed as

$$\frac{d}{dx}\left(EA\frac{du}{dx}\right) + p = 0 \qquad \text{on } \Omega, \tag{4.4}$$

$$EA\frac{du}{dx} = F \qquad \text{on } \Gamma_N, \tag{4.5}$$

$$u = g \qquad \text{on } \Gamma_D. \tag{4.6}$$

The Neumann and the Dirichlet boundaries are represented by Γ_N and Γ_D, respectively, where F denotes a concentrated load on Γ_N, and g prescribes a displacement

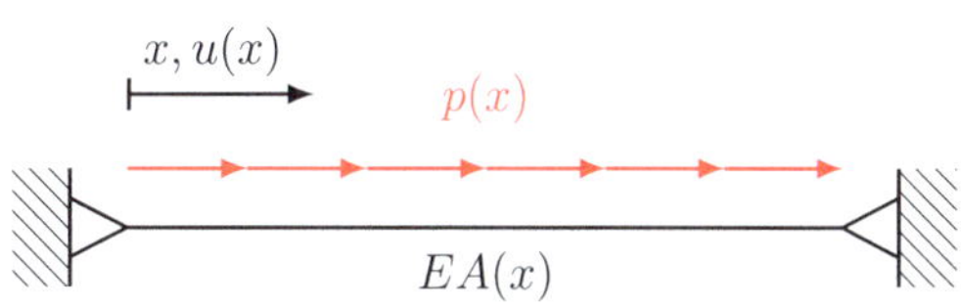

Fig. 4.2 Example static bar configuration with homogeneous Dirichlet boundary conditions at both ends

on Γ_D. The Young's modulus $E(x)$ and the cross-sectional area $A(x)$ may vary with respect to x.

In the physics-informed neural network, a deep neural network is used to approximate the unknown solution $u(x)$. So far, this approach does not differ from the regression approaches introduced in Chaps. 2 and 3. However, with the exception of the Physics-Inspired Architectures for Dynamics introduced in Sect. 3.9.8, these models rely on a tremendous amount of labeled training data. Additionally, the unknown solution must lie within the space of solutions spanned by the data. By contrast, and instead of having labeled training data in the whole domain, the solution for the example at hand is only known at the boundaries (i.e., at $x = 0$ and $x = 1$). Additionally, Eq. (4.4) is forced to be satisfied at each collocation point inside the domain. To this end, the network is extended to compute the left-hand-side, i.e., the residual of Eq. (4.4)

$$r = \frac{d}{dx}\left(EA\frac{du}{dx}\right) + p. \tag{4.7}$$

Since a neural network is fully differentiable, it is not only possible to compute the derivatives of u with respect to the parameters necessary for training. Likewise, the automatic differentiation capabilities of libraries such as PyTorch or TensorFlow allow the fast computation of derivatives of u with respect to the input variable x (cf. Sect. 3.4.2).

All together, this extended network architecture can be interpreted as a physics-informed neural network with outputs $\hat{u} = u_{NN}(x; \boldsymbol{\Theta})$ and $\hat{r} = r_{NN}(x; \boldsymbol{\Theta})$. As depicted in Fig. 4.3, the first part simply approximates the solution $u(x)$ with a fully connected neural network. The second part represents the computation of Eq. (4.7), $r(x)$ using the corresponding first-order d_x and second-order derivatives d_{xx} of u with respect to x, and the identity I of u. Note that both networks $u_{NN}(x; \boldsymbol{\Theta})$ and $r_{NN}(x; \boldsymbol{\Theta})$ depend on the same set of parameters $\boldsymbol{\Theta}$, namely the weights and biases of the first network u_{NN}. The dashed connections in Fig. 4.3 are simply introduced to visualize the composition of r and do not contain trainable network parameters.

To better understand how a physics-informed neural network is implemented, a specific example is introduced and illustrated in Fig. 4.2. Let us define a domain $\Omega = [0, 1]$ with $\Gamma_D = \{x \mid x = 0, x = 1\}$ and $\Gamma_N = \emptyset$. The material parameters are set to $EA = 1$. A solution for the displacement is chosen as

$$u(x) = \sin(2\pi x), \tag{4.8}$$

which, after insertion into the differential equation, Eq. (4.4), results in the distributed load

$$p(x) = 4\pi^2 \sin(2\pi x). \tag{4.9}$$

Furthermore, the following Dirichlet boundary conditions apply

$$u(0) = u(1) = 0. \tag{4.10}$$

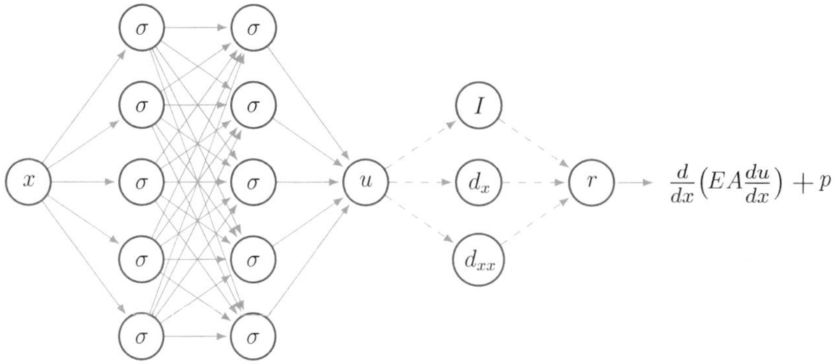

Fig. 4.3 Conceptual physics-informed neural network for the static bar equation. The left part shows the fully connected neural network and the right part represents the physics-informed neural network. The dashed lines denote non-trainable weights

The approach of pre-selecting a solution and computing the remaining quantities to determine the underlying problem is referred to as the method of manufactured solutions. It is commonly used to verify numerical solvers on sufficiently complex examples [9].

With this information, it is now possible to define a cost function. In case of the stationary physics-informed neural network, this cost function is assembled from two mean squared error losses, as given in Eq. (4.2)

$$C = \mathcal{L}_{\mathcal{B}} + \mathcal{L}_{\mathcal{R}}, \tag{4.11}$$

where

$$\mathcal{L}_{\mathcal{B}} = \frac{1}{m_{\mathcal{B}}} \sum_{i=1}^{m_{\mathcal{B}}} \left(\hat{u}\left(\tilde{x}_{\mathcal{B}}^{i}\right) - \tilde{u}_{\mathcal{B}}^{i}\right)^{2}, \tag{4.12}$$

and

$$\mathcal{L}_{\mathcal{R}} = \frac{1}{m_{\mathcal{R}}} \sum_{i=1}^{m_{\mathcal{R}}} \left(\hat{r}\left(x_{\mathcal{R}}^{i}\right)\right)^{2}. \tag{4.13}$$

The solution on the boundary is represented by $m_{\mathcal{B}}$ labeled data points $\{\tilde{x}_{\mathcal{B}}^{i}, \tilde{u}_{\mathcal{B}}^{i}\}_{i=1}^{m_{\mathcal{B}}}$. The boundary loss $\mathcal{L}_{\mathcal{B}}$ is computed by comparing the approximation u_{NN} with the labels $\tilde{u}_{\mathcal{B}}$ of the training data. The error of the approximation at known data points from Eq. (4.2) is given by $\mathcal{L}_{\mathcal{D}} = \mathcal{L}_{\mathcal{B}}$, as only the boundary data points are known. Note also that here, the boundary loss $\mathcal{L}_{\mathcal{B}}$ solely includes Dirichlet boundary conditions, as only these were defined in the example. However, the application of Neumann boundary conditions works similarly and is discussed further in Sect. 4.2.3.

In order to enforce Eq. (4.3) in the whole spatial domain, a set of $m_{\mathcal{R}}$ collocation points $\{x_{\mathcal{R}}^{i}\}_{i=1}^{m_{\mathcal{R}}}$ is generated using a uniform distribution throughout the one-dimensional domain. The corresponding loss $\mathcal{L}_{\mathcal{R}}$ from Eq. (4.13) is computed as the mean squared error of the prediction of r_{NN} at all collocation points. The physics-informed neural network can now be trained by minimizing the cost function given in Eq. (4.11). Instead of a stochastic gradient descent procedure, the second-order L-BFGS optimizer can be employed [10]. For more complex problems, a combination of Adam and L-BFGS is typical.

The physics-informed network is implemented in PyTorch. At first, a neural network is created to predict the displacement $u(x)$. To keep matters as simple as possible, the network only has a single hidden layer with the dimension `hiddenDim`. More involved networks as introduced in Listing 3.1, are likely a better chocie, especially for more complex examples. The input and output dimensions are correspondingly given as `inputDim` and `outputDim`. As previously explained in Sect. 3.8, the linear transformations from one layer to the next are defined by `torch.nn.Linear` and the non-linear activation function is chosen as the hyperbolic tangent, `torch.nn.Tanh`. There is no activation function in the output layer, meaning that the range of the outputs is not constrained.

```
model = torch.nn.Sequential(torch.nn.Linear(inputDim,
    hiddenDim),
                            torch.nn.Tanh(),
                            torch.nn.Linear(hiddenDim,
    outputDim))
```

With the neural network, the first part of the physics-informed neural network illustrated in Fig. 4.3 is defined. A displacement prediction $\hat{u}$ at $x = 0.5$ can be made with the following code.

```
uPred = model(torch.tensor([0.5]))
```

The second part of the network requires the computation of the derivatives with respect to the input x. This is achieved with PyTorch's built-in automatic differentiation (see Sect. 3.4.3), which creates and retains a computational graph to store all information necessary for calculating the gradients.

Listing 4.1 Wrapped automatic differentiation to compute derivatives

```
def getDerivative(y, x):
    dydx = grad(y, x, torch.ones_like(y),
    create_graph=True,
    retain_graph=True)[0]
    return dydx
```

Finally, $r(x)$ from Eq. (4.7) is computed with the following code.

```
def r(model, x, EA, p):
    u = model(x)
    dudx = getDerivative(u, x)
    dEAdudxx = getDerivative(EA(x) * dudx, x)
    r = dEAdudxx + p(x)
    return r
```

Now, to calculate the loss function of the differential equation $\mathcal{L}_{\mathcal{R}}$, the example values are inserted into the model.

```
x = torch.linspace(0, 1, 10, requires_grad=True).unsqueeze(1)
EA = lambda x: 1 + 0 * x
p = lambda x: 4 * torch.pi ** 2 * torch.sin(2 * torch.pi * x)
rPred = r(model, x, EA, p)
lossR = torch.sum(rPred ** 2)
```

Combining $\mathcal{L}_{\mathcal{R}}$ with the loss function of the boundary conditions $\mathcal{L}_{\mathcal{B}}$ allows to compute the cost function C from Eq. (4.11). The implementation for this example is shown in the code snippet below. Note that only the Dirichlet boundary conditions are imposed here. Neumann boundary conditions can be imposed in a similar way with the small addition of computing the derivatives relying on the `getDerivative` function from Listing 4.1. This is described in more detail in the upcoming Sect. 4.2.3.

```
u0 = 0
u1 = 0

u0Pred = model(torch.tensor([0.]))
u1Pred = model(torch.tensor([1.]))
lossB = (u0Pred - u0) ** 2 + (u1Pred - u1) ** 2
```

It is now possible to compute and minimize the cost function C via a stochastic gradient descent approach or more sophisticated methods such as the L-BFGS optimizer. When the cost function is sufficiently small, an accurate prediction of the displacement $u(x)$ is to be expected. The entire training procedure is summarized in Algorithm 9 which also concludes the basic presentation of physics-informed neural networks.

Algorithm 9 Training a physics-informed neural network for the static solution of the problem described in Eq. (4.4)

Require: training data for boundary condition $\{\tilde{x}^i_{\mathcal{B}}, \tilde{u}^i_{\mathcal{B}}\}_{i=1}^{m_{\mathcal{B}}}$
generate $m_{\mathcal{R}}$ collocation points with a uniform distribution $\{x^i_{\mathcal{R}}\}_{i=1}^{m_{\mathcal{R}}}$
define network architecture (input, output, hidden layers, hidden neurons)
initialize network parameters $\boldsymbol{\Theta}$: weights $\{\boldsymbol{W}^l\}_{l=1}^L$ and biases $\{\boldsymbol{b}^l\}_{l=1}^L$ for all layers L
set hyperparameters for optimizer (epochs, learning rate $\alpha, \dots$)

for all epochs **do**
  $\hat{\boldsymbol{u}}_{\mathcal{B}} \leftarrow u_{NN}(\tilde{\boldsymbol{x}}_{\mathcal{B}}; \boldsymbol{\Theta})$
  $\hat{\boldsymbol{r}} \leftarrow r_{NN}(\boldsymbol{x}_{\mathcal{R}}; \boldsymbol{\Theta})$
  compute $\mathcal{L}_{\mathcal{B}}, \mathcal{L}_{\mathcal{R}}$ ▷ cf. Equation (4.12) and (4.13)
  compute cost function: $C \leftarrow \mathcal{L}_{\mathcal{B}} + \mathcal{L}_{\mathcal{R}}$
  update parameters: $\boldsymbol{\Theta} \leftarrow \boldsymbol{\Theta} - \alpha \frac{\partial C}{\partial \boldsymbol{\Theta}}$ ▷ Adam or L-BFGS
end for

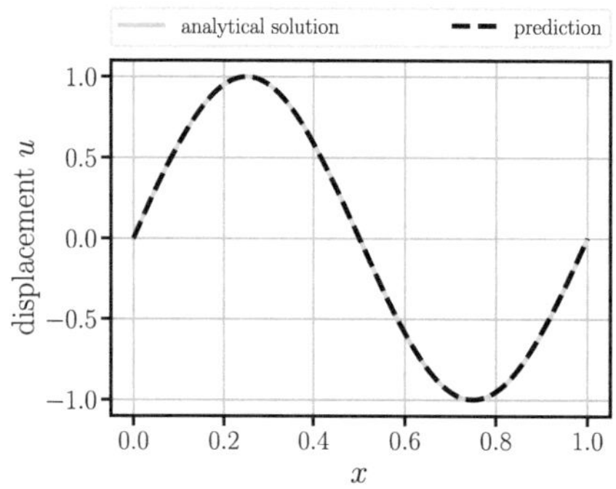

(a) Displacements

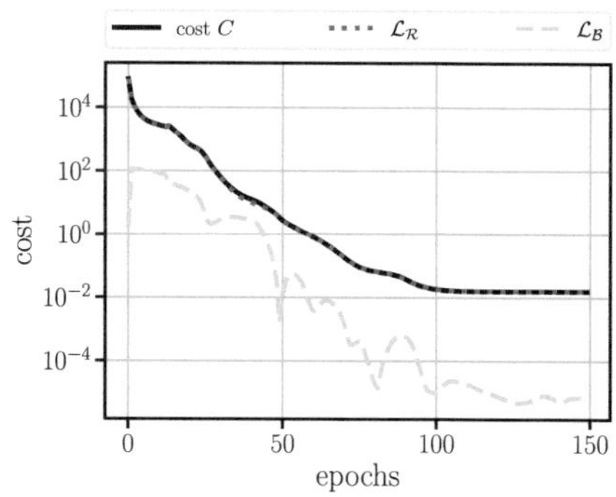

(b) Cost function history

Fig. 4.4 Linear elastic static bar example. Left: the displacements $u(x)$ computed with a physics-informed neural network and the analytic solution are compared. Right: the training history is illustrated by showing the cost function for each epoch

Alternatively, it is possible to use a strong enforcement of the Dirichlet boundary conditions to simplify the minimization problem. For that, the neural network's output must be adapted to automatically satisfy the boundary conditions for any given input. Then, the mean squared error loss of the Dirichlet boundary conditions can be dropped. The strong enforcement of boundary conditions is demonstrated on a two-dimensional plate in membrane action in Sect. 4.2.2 and a transient heat problem by Sect. 4.2.3.

The static bar example is now used to build a physics-informed neural network and subsequently trained with L-BFGS (see Exercise E.14 for the implementation). The corresponding results are presented in Fig. 4.4. Here, the boundary conditions are enforced via the cost function. The plot on the left shows that the estimated and analytic displacements coincide. Additionally, the training history is illustrated on the right, showing the cost function for every training epoch. Note the difference in the order of magnitudes between the cost function of the differential equation and the cost function of the boundary conditions. Adjusting this difference with a weighting factor on the boundary condition term is possible, yet difficult to determine a priori. Automatic schemes exist, as will be discussed in Sects. 4.2.2 and 5.2.3. A simpler alternative for this example would be to strongly enforce the Dirichlet boundary conditions, which leads to an unconstrained optimization problem, as only the differential equation loss has to be minimized.

4.2.2 Two-Dimensional Static Model

The physics-informed neural network methodology can be extended to higher dimensional problems in a straightforward manner. To this end, we consider a two-dimensional static model—a plate in membrane action [11] (see Fig. 4.5). As with the static bar from Sect. 4.2.1, the solution is independent of time t. However, instead of considering an ordinary differential equation with a scalar solution quantity u, a partial differential equation with a vector-valued solution $\boldsymbol{u} = (u_1, u_2)^\intercal$ is considered. The differential equation operator indirectly acts on the displacement field $\boldsymbol{u}$ through the stress field $\boldsymbol{\sigma}(\boldsymbol{u})$: $\mathcal{N}[\boldsymbol{u}] = \nabla \cdot \boldsymbol{\sigma}(\boldsymbol{u}) + \boldsymbol{p}$. Here, $\boldsymbol{p}$ represents a distributed load within the plate. The differential equation defined on the two-dimensional domain $\Omega \subset \mathbb{R}^2$ is

$$\nabla \cdot \boldsymbol{\sigma}(x_1, x_2) + \boldsymbol{p}(x_1, x_2) = \boldsymbol{0} \qquad \text{on } \Omega. \tag{4.14}$$

The stress $\boldsymbol{\sigma}$ can be computed with the strains $\boldsymbol{\varepsilon}$ obtained from the displacements $\boldsymbol{u}$. Specifically, we model the plate in membrane action under plane stress. Thus, the constitutive relation is described as

$$\boldsymbol{\sigma} = \begin{pmatrix} \sigma_{11} \\ \sigma_{22} \\ \sigma_{12} \end{pmatrix} = \boldsymbol{C}\boldsymbol{\varepsilon} = \frac{E}{1-\nu^2} \begin{pmatrix} 1 & \nu & 0 \\ \nu & 1 & 0 \\ 0 & 0 & 1-\nu \end{pmatrix} \boldsymbol{\varepsilon}, \tag{4.15}$$

where $\boldsymbol{C}$ is the plane stress constitutive matrix in Voigt notation. The strains $\boldsymbol{\varepsilon}$ in turn are obtained by deriving the displacement field $\boldsymbol{u} = (u_1, u_2)^\intercal$:

$$\boldsymbol{\varepsilon}(x_1, x_2) = \frac{1}{2}\left(\nabla \boldsymbol{u}(x_1, x_2) + \nabla \boldsymbol{u}(x_1, x_2)^\intercal\right), \tag{4.16}$$

which is equivalent to

$$\boldsymbol{\varepsilon} = \begin{pmatrix} \varepsilon_{11} \\ \varepsilon_{22} \\ \varepsilon_{12} \end{pmatrix} = \begin{pmatrix} \frac{\partial u_1}{\partial x_1} \\ \frac{\partial u_2}{\partial x_2} \\ \frac{1}{2}\left(\frac{\partial u_1}{\partial x_2} + \frac{\partial u_2}{\partial u_1}\right) \end{pmatrix}. \tag{4.17}$$

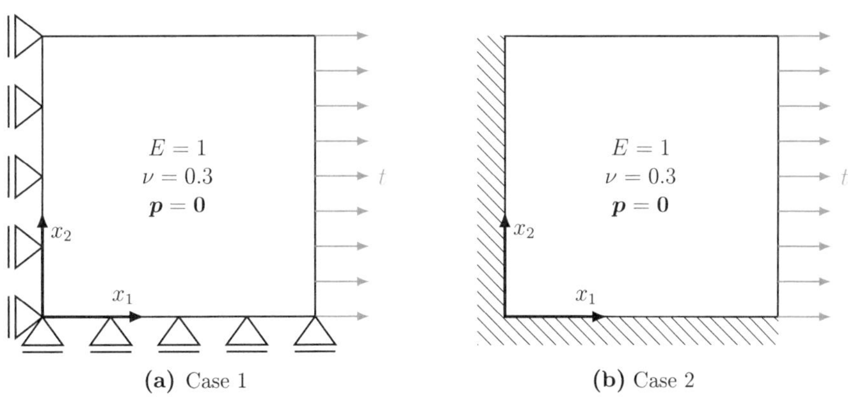

Fig. 4.5 Two square plates of size $L \times L$ with different boundary conditions

In addition to the differential equation, boundary conditions must be formulated to make the problem well-posed. For this purpose, we consider two specific cases illustrated in Fig. 4.5. Both cases consider a square plate of size $L \times L$. Case 1 is a uni-axial tension test with a simple analytical solution.

$$\boldsymbol{u}^{\text{Analytical}} = \left(\frac{x_1 t}{E}, -\frac{x_2 \nu t}{E} \right)^{\mathsf{T}} \tag{4.18}$$

Case 2 introduces additional displacement constraints at the left and bottom edges, which restrict the vertical contraction. This introduces a stress singularity at the bottom right corner. Hence, learning the solution of case 2 is more challenging. The boundary conditions of the two cases are summarized in Table 4.1. These consist of Dirichlet boundary conditions acting directly on the displacements and Neumann boundary conditions applied to the stresses. The stresses are connected to the displacements through the displacement's first derivative.

Specifically a Young's modulus of $E = 1$ and a Poisson's ratio of $\nu = 0.3$ are considered. The dimension of the plate is given as $L = 1$, and no distributed load is assumed, i.e., $\boldsymbol{p} = \boldsymbol{0}$, while the traction at the right edge is set to $t = 1$.

With the governing equations established, we can formulate the physics-informed neural network. Analogously to the one-dimensional case from Sect. 4.2.1, a neural network serves as predictor of the solution field $\hat{\boldsymbol{u}} = u_{NN}(x_1, x_2; \boldsymbol{\Theta})$. The cost function C is again established with the residual of the differential equation $\mathcal{L}_{\mathcal{R}}$ and the boundary loss $\mathcal{L}_{\mathcal{B}}$ (without additional data, i.e., $\mathcal{L}_{\mathcal{D}} = \mathcal{L}_{\mathcal{B}}$):

$$C = \mathcal{L}_{\mathcal{R}} + \mathcal{L}_{\mathcal{B}}. \tag{4.19}$$

Using Eqs. (4.14)–(4.17), the residual of the differential equation can be formulated as

$$\boldsymbol{r} = \nabla \cdot \boldsymbol{\sigma}(\boldsymbol{u}) + \boldsymbol{p}. \tag{4.20}$$

Thus, the physics-informed neural network computes the residual $\hat{\boldsymbol{r}} = r_{NN}(x_1, x_2; \boldsymbol{\Theta})$ from the predicted displacements $\hat{\boldsymbol{u}}$, as illustrated in Fig. 4.6. Note that the residual $\boldsymbol{r} = (r_1, r_2)^{\mathsf{T}}$ is now a two-dimensional vector instead of a scalar, as in the one-dimensional bar example from Sect. 4.2.1. The loss is thus expressed via a sum over the two components evaluated at $m_{\mathcal{R}}$ collocation points $\{x^i_{\mathcal{R}_1}, x^i_{\mathcal{R}_2}\}_{i=1}^{m_{\mathcal{R}}}$.

Table 4.1 Boundary conditions of the two square plate cases shown in Fig. 4.5

	Case 1	Case 2
Bottom edge	$u_2(x_1, 0) = 0, \sigma_{12}(x_1, 0) = 0$	$u_2(x_1, 0) = 0, u_1(x_1, 0) = 0$
Right edge	$u_1(0, x_2) = 0, \sigma_{12}(0, x_2) = 0$	$u_1(0, x_2) = 0, u_2(0, x_2) = 0$
Top edge	$\sigma_{22}(x_1, L) = 0, \sigma_{12}(x_1, L) = 0$	$\sigma_{22}(x_1, L) = 0, \sigma_{12}(x_1, L) = 0$
Left edge	$\sigma_{11}(L, x_2) = t, \sigma_{12}(L, x_2) = 0$	$\sigma_{11}(L, x_2) = t, \sigma_{12}(L, x_2) = 0$

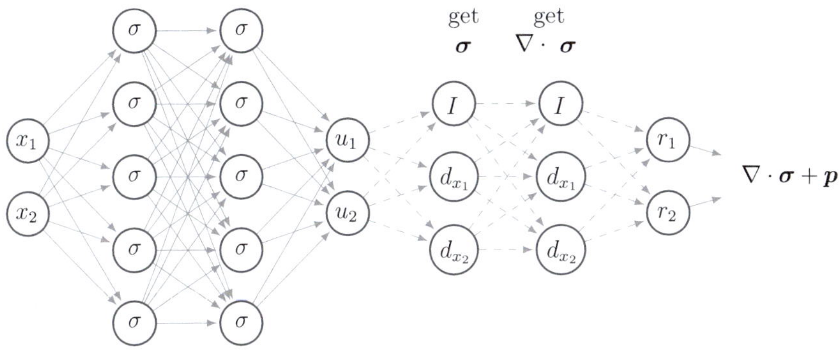

Fig. 4.6 Conceptual physics-informed neural network for the static plate in membrane action. The left part shows the fully connected neural network and the right part represents the physics-informed neural network. The dashed lines denote non-trainable weights

$$\mathcal{L}_{\mathcal{R}} = \frac{1}{m_{\mathcal{R}}} \sum_{i=1}^{m_{\mathcal{R}}} \sum_{j=1}^{2} \left(\hat{r}_j(x^i_{\mathcal{R}_1}, x^i_{\mathcal{R}_2})\right)^2 \tag{4.21}$$

Although the principal concepts remain the same, the corresponding Python code becomes more tedious due to the additional dimension in the input and output. The prediction is performed with the following code snippet.

```
x1 = torch.linspace(0, Lx1, Nx1)
x2 = torch.linspace(0, Lx2, Nx2)
x1, x2 = torch.meshgrid(x1, x2, indexing='ij')
x1.requires_grad = True
x2.requires_grad = True
modelInput = torch.cat((x1.reshape(-1, 1), x2.reshape(-1,
    1)), 1)

uPred = model(modelInput)
```

A uniformly spaced grid is created with the `torch.meshgrid` function, and only afterward are the operations tracked with the computation graph, activated with `.requires_grad = True`. The `modelInput` is a concatenation of the two coordinates. To calculate the differential equation's residual, first, the strains and, subsequently, the stresses have to be computed. To this end, a function computing the strains is established, which relies on the same `getDerivative` function from Listing 4.1.

```
def getStrains(u, x1, x2, Nx1, Nx2):
    strain = torch.zeros((3, Nx1, Nx2))
    strain[0] = getDerivative(u[:, 0].reshape(Nx1, Nx2), x1, 1)
    strain[1] = getDerivative(u[:, 1].reshape(Nx1, Nx2), x2, 1)
    strain[2] = 0.5 * (getDerivative(u[:, 0].reshape(Nx1, Nx2), x2, 1)
```

```
                    + getDerivative(u[:, 1].reshape(Nx1, Nx2), x1, 1))
    return strain
```

The residual `lossPDE` can thus be computed in the following manner.

```
strain = getStrains(uPred, x1, x2, Nx1, Nx2)
stress = torch.tensordot(C, strain, dims=1)

dsig11dx1 = getDerivative(stress[0], x1, 1)
dsig12dx2 = getDerivative(stress[2], x2, 1)
dsig21dx1 = getDerivative(stress[2], x1, 1)
dsig22dx2 = getDerivative(stress[1], x2, 1)

lossPDE = 0  # equilibrium computation
lossPDE += (torch.sum((dsig11dx1 + dsig12dx2 + p1(x1, x2)) ** 2)
            / (Nx1 * Nx2))  # divide by number of collocation points
lossPDE += (torch.sum((dsig21dx1 + dsig22dx2 + p2(x1, x2)) ** 2)
            / (Nx1 * Nx2))  # divide by number of collocation points
```

Here `C` is the constitutive matrix in Voigt notation, which in plane stress is specified as:

```
C = torch.zeros((3, 3))
C[0, 0] = 1
C[0, 1] = nu
C[1, 0] = nu
C[1, 1] = 1
C[2, 2] = 1 - nu
C *= E / (1 - nu ** 2)
```

The boundary loss is also more tedious due to the now eight boundary terms, as listed in Table 4.1. Additionally, the mixture of Dirichlet and Neumann boundary conditions and the physics-informed loss (Eq. (4.21)) pose difficulties during training, as imbalances between the different loss components may arise and prohibit proper convergence. A remedy is to employ a strong enforcement of the boundary conditions, i.e., modifying the neural networks output $z = (z_1, z_2)^\intercal$, such that the boundary conditions are satisfied a priori. This is straightforward for Dirichlet boundary conditions on geometrically simple domains. For more complicated geometries; see [12] and for Neumann boundary conditions; see [13], as will be discussed in more detail in the paragraph on boundary conditions in Sect. 10.1.2.1.

For the two considered cases, the following non-unique enforcements could be applied.

$$\text{case 1: } \boldsymbol{u} = \begin{pmatrix} x_1 z_1 \\ x_2 z_2 \end{pmatrix}, \qquad \text{case 2: } \boldsymbol{u} = \begin{pmatrix} x_1 x_2 z_1 \\ x_1 x_2 z_2 \end{pmatrix}. \tag{4.22}$$

In case 1, evaluating u_1 at the left edge, i.e., inserting $x_1 = 0$ will always yield zero, thus fulfilling the first Dirichlet boundary condition. Correspondingly, the component

u_2 always evaluates to zero at the bottom boundary. In case 2, the entire displacement field $\boldsymbol{u}$ evaluates to zero at the left and bottom edges. Note that the choice of neural network output manipulation is not unique. An alternative could, for example, be composed of multiplications of trigonometric functions or higher-order polynomial terms. Important is, however, that the prediction beyond the prescribed boundary condition and inside the domain Ω is non-zero, as this could make it impossible for the network to learn the correct solution. Strongly enforcing boundary conditions will be revisited in Sect. 5.2.3.

In Python, the enforcements can be defined with `lambda` functions. Here, we specifically consider case 1.

```
strongEnforcement1 = lambda z1, x1, x2: x1 * z1
strongEnforcement2 = lambda z2, x1, x2: x2 * z2
```

These can then be applied to the neural network output in order to enforce the Dirichlet boundary conditions.

```
z = model(modelInput)
u1 = strongEnforcement1(z[:, 0], x1[:, 0], x2[:, 1]).
    unsqueeze(1)
u2 = strongEnforcement2(z[:, 1], x1[:, 0], x2[:, 1]).
    unsqueeze(1)
uPred = torch.cat((u1, u2), 1)
```

As the Dirichlet boundary conditions are fulfilled a priori, only the Neumann boundary conditions have to be considered within the boundary loss $\mathcal{L}_{\mathcal{B}}$. These consist of the traction boundary conditions, consisting of the stress components $\sigma_{11}, \sigma_{22}, \sigma_{12}$ and can be formulated as

$$
\begin{aligned}
\mathcal{L}_{\mathcal{B}} = & \frac{1}{m_{\mathcal{B}_{11}}} \sum_{i=1}^{m_{\mathcal{B}_{11}}} \left(\sigma_{11}[\hat{\boldsymbol{u}}(\tilde{x}^i_{\mathcal{B}_1}, \tilde{x}^i_{\mathcal{B}_2})] - \tilde{\sigma}^i_{\mathcal{B}_{11}}\right)^2 \\
& + \frac{1}{m_{\mathcal{B}_{22}}} \sum_{i=1}^{m_{\mathcal{B}_{22}}} \left(\sigma_{22}[\hat{\boldsymbol{u}}(\tilde{x}^i_{\mathcal{B}_1}, \tilde{x}^i_{\mathcal{B}_2})] - \tilde{\sigma}^i_{\mathcal{B}_{22}}\right)^2 \\
& + \frac{1}{m_{\mathcal{B}_{12}}} \sum_{i=1}^{m_{\mathcal{B}_{12}}} \left(\sigma_{12}[\hat{\boldsymbol{u}}(\tilde{x}^i_{\mathcal{B}_1}, \tilde{x}^i_{\mathcal{B}_2})] - \tilde{\sigma}^i_{\mathcal{B}_{12}}\right)^2,
\end{aligned}
\tag{4.23}
$$

where $\tilde{x}^i_{\mathcal{B}_1}, \tilde{x}^i_{\mathcal{B}_2}$ are the $m_{\mathcal{B}_{11}} + m_{\mathcal{B}_{22}} + m_{\mathcal{B}_{12}}$ sampled coordinates of the boundaries under consideration with labeled values of either $\tilde{\sigma}^i_{\mathcal{B}_{11}}, \tilde{\sigma}^i_{\mathcal{B}_{22}}$ or $\tilde{\sigma}^i_{\mathcal{B}_{12}}$.

The corresponding implementation of the boundary loss composed of the Neumann boundary condition is given below. Again we consider case 1.

```
lossBC = 0
# inhomogeneous Neumann boundary condition at right edge
lossBC += torch.sum((stress[0][-1, :] - force) ** 2) / Nx1

```

```
# normal stress at top edge
lossBC += torch.sum((stress[1][:, -1] - 0) ** 2) / Nx1
# shear stress at top edge
lossBC += torch.sum((stress[2][:, -1] - 0) ** 2) / Nx1
# shear stress at bottom edge
lossBC += torch.sum((stress[2][:, 0] - 0) ** 2) / Nx1
# shear stress at left edge
lossBC += torch.sum((stress[2][0, :] - 0) ** 2) / Nx2
# shear stress at right edge
lossBC += torch.sum((stress[2][-1, :] - 0) ** 2) / Nx2
```

Despite the strong enforcement of the Dirichlet boundary conditions, the problem of imbalances between the residual loss $\mathcal{L}_{\mathcal{R}}$ and the boundary loss $\mathcal{L}_{\mathcal{B}}$ persists. A remedy also discussed in the paragraph on boundary conditions in Sect. 10.1.2.1 and elaborated in Sect. 5.2.3, is to introduce an automatic weighting of the loss components

$$C = \kappa_{\mathcal{R}}\mathcal{L}_{\mathcal{R}} + \kappa_{\mathcal{B}}\mathcal{L}_{\mathcal{B}}, \tag{4.24}$$

where $\kappa_{\mathcal{R}}$ and $\kappa_{\mathcal{B}}$ are learnable parameters $\boldsymbol{\kappa} = \{\kappa_{\mathcal{R}}, \kappa_{\mathcal{B}}\}$. However, instead of minimizing the cost function C with respect to these (where the solution for both would be $-\infty$), the aim is to maximize the cost function. This can be framed as a minimax optimization

$$\min_{\boldsymbol{\Theta}} \max_{\boldsymbol{\kappa}} C \tag{4.25}$$

This ensures, that the more difficult loss term ($\mathcal{L}_{\mathcal{R}}$ or $\mathcal{L}_{\mathcal{B}}$) always receives additional attention during training. This technique is related to attention utilized by transformers, seen in Sect. 3.9.7.

In PyTorch, additional parameters `weights` can be added to the `optimizer` in order to consider them during training, i.e., make them trainable.

```
weights = torch.tensor([1., 1.], requires_grad=True)
optimizer.add_param_group({'params': weights})
optimizer.param_groups[-1]['lr'] = lr  # different learning rate can be specified
```

During the training, the `cost` is defined as Eq. (4.24) using the trainable `weights`. To ensure a maximization instead of a minimization with respect to these `weights`, the gradients—accessed with `.grad`—are simply multiplied with `-1`.

```
# within the training loop
cost = weights[0] * lossPDE + weights[1] * lossBC
cost.backward()
weights.grad *= -1
optimizer.step()
```

The reader is referred to Exercise E.17 for the complete code.

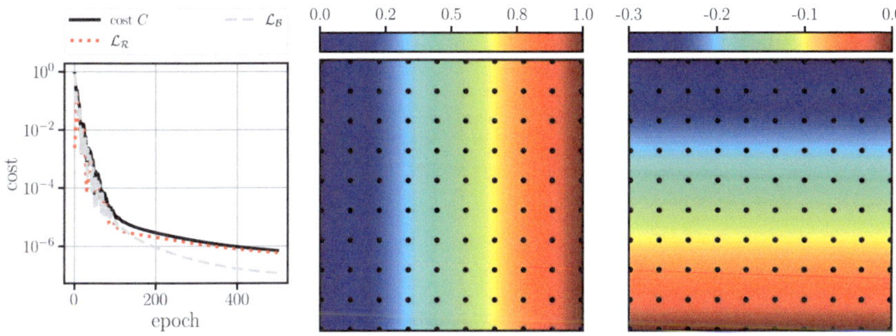

(a) Training history (b) $\hat{u}_1$ (c) $\hat{u}_2$

Fig. 4.7 Displacement prediction of case 1 from Fig. 4.5 using a physics-informed neural network. The 10×10 collocation points are indicated by the black circles

Using the presented formulation, the two cases from Fig. 4.5 are simulated with the physics-informed neural network and compared with results obtained by the finite element method [14]. For both cases, a neural network with two input and two output neurons, three hidden layers with 20 neurons each, and SiLU activation functions[5] (except on the output layer) are employed. This yields 942 trainable parameters.

Case 1

The predicted displacements after 500 epochs are depicted in Fig. 4.7b, c. As seen by the training history provided in Fig. 4.7a, the cost function C is minimized relatively quickly and would continue to decrease with further training. This is also to be expected, as the solution is relatively trivial. In fact, due to the strong enforcement of the boundary conditions from Eq. (4.22), the neural network simply has to output t/E for u_1 and $\nu t/E$ for u_2 in the entire domain (compare with the analytical solution from Eq. (4.18)). Thus, the network has to output a constant for each of its two outputs. The fast convergence is, therefore, not as impressive as it might seem at first sight, espeically considering that the problem is actually linear.

The analytical solution from Eq. (4.18) can also be used to assess the accuracy. Figure 4.8 shows the squared pointwise errors for both the physics-informed neural network and the finite element method in the x_1 and x_2 directions. The physics-informed neural network relies on 10×10 collocation points, while the finite element method uses 5×5 elements with integrated Legendre polynomials of degree up to $p = 2$, corresponding to 242 degrees of freedom. The finite element method achieves machine accuracy, while the neural network's errors are in the order of 10^{-7}.

[5] We use a smooth version of the ReLU, as smooth functions perform better with physics-informed neural networks, due to the additional differentiations for the residual computation. For example, a ReLU with the kink could easily cause discontinuities in the strain field. At the same time, we retain most of the advantages of the ReLU activation function.

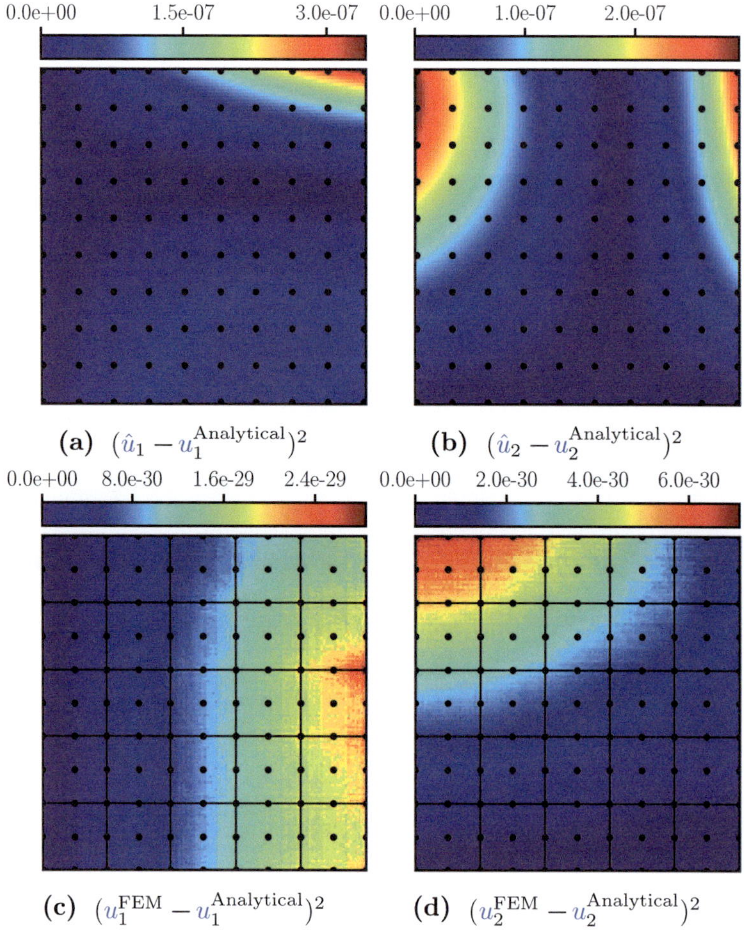

Fig. 4.8 Pointwise errors of physics-informed neural network and the finite element method obtained by comparison to the analytical solution from Eq. (4.18). The physics-informed neural network utilizes 10×10 collocation points, while the finite element method relies on 5×5 elements with integrated Legendre polynomials of up to degree $p = 2$

The physics-informed neural network also has inferior computational times. For the employed finite element discretization, assembly and solving time take $\sim 9.1 \cdot 10^{-3}$ s, whereas the physics-informed neural network requires $\sim 1.2 \cdot 10^{-2}$ s for each epoch when using 10×10 collocation points. Considering that about 500 epochs were necessary, the network is about three orders of magnitude slower.

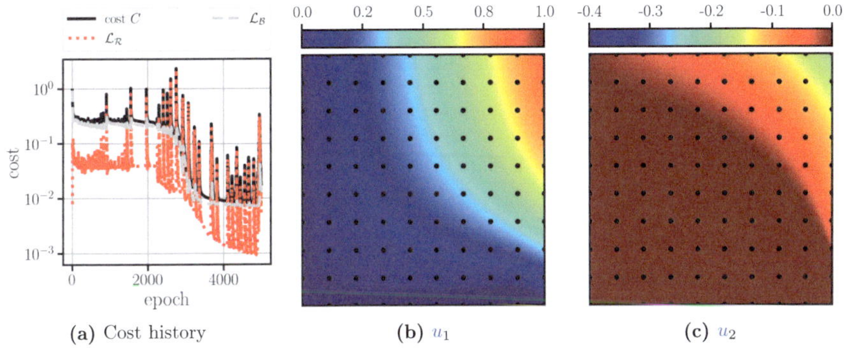

Fig. 4.9 Displacement prediction of case 2 from Fig. 4.5 using a physics-informed neural network. The 10×10 collocation points are indicated by the black circles

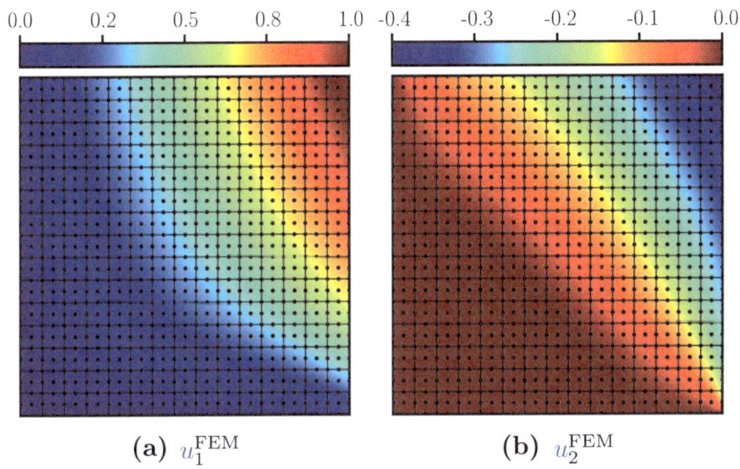

Fig. 4.10 Finite element reference solution of case 2 from Fig. 4.5 relying on 15×15 elements with polynomial degree $p = 2$

Case 2

The physics-informed neural network results for case 2 are presented in Fig. 4.9, which resemble the finite element reference solution from Fig. 4.10 using 15×15 elements with polynomial degree $p = 2$. Due to the stress singularity in the bottom right corner, the physics-informed neural network struggles. As seen from the training history in Fig. 4.9a, the result improves only slowly after about 3 000 epochs. The finite element simulation takes $\sim 8.3 \cdot 10^{-2}$ s for constructing and solving the system of equations. A single epoch of the network training is faster at $\sim 1.4 \cdot 10^{-2}$ s but would require convergence within six epochs in order to beat the finite element

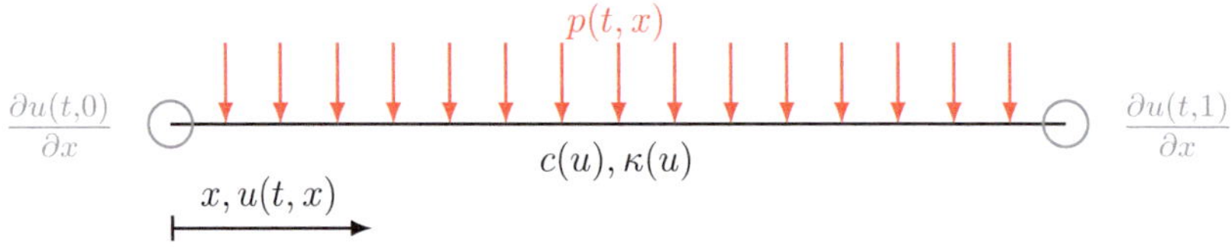

Fig. 4.11 Example configuration for the one-dimensional heat equation with homogeneous Neumann boundary conditions at both ends

method. However, a worse solution—distinguishable already by visual inspection—is reached even after 5 000 epochs.

4.2.3 Continuous-Time Model

An initial boundary value problem described by a one-dimensional non-linear heat equation (see Fig. 4.11) is introduced to demonstrate the application of physics-informed neural networks to transient problems. The evolution of temperature $u(t, x)$ as a function of time t and space x is described by a partial differential equation and boundary conditions of the following form

$$c\frac{\partial u}{\partial t} - \frac{\partial}{\partial x}\left(\kappa\frac{\partial u}{\partial x}\right) - p = 0 \qquad \text{on } \mathcal{T} \times \Omega, \tag{4.26}$$

$$\kappa\frac{\partial u}{\partial x} = h \qquad \text{on } \mathcal{T} \times \Gamma_N, \tag{4.27}$$

$$u = g \qquad \text{on } \mathcal{T} \times \Gamma_D, \tag{4.28}$$

$$u(x, 0) = u_0 \qquad \text{on } \Omega. \tag{4.29}$$

Here, $\Omega \subset \mathbb{R}^1$ represents the one-dimensional spatial domain, whereas $\mathcal{T}$ denotes the temporal domain. Together they form the space-time domain $\mathcal{T} \times \Omega$ with Neumann and Dirichlet boundaries represented by Γ_N and Γ_D, respectively. In engineering applications the heat capacity $c(u)$ or the thermal conductivity $\kappa(u)$ are often temperature-dependent which introduces a non-linearity to Eq. (4.3).

The specific problem investigated in the following and illustrated in Fig. 4.11 is defined on the space-time domain $\mathcal{T} \times \Omega = t \in [0, 0.5] \times x \in [0, 1]$ with $\Gamma_D = \emptyset$ and $\Gamma_N = \{x \mid x = 0,\ x = 1\}$. In particular, the problem is subject to homogeneous Neumann boundary conditions

$$\frac{\partial u(t, 0)}{\partial x} = \frac{\partial u(t, 1)}{\partial x} = 0, \tag{4.30}$$

and the initial condition

$$u(0, x) = u_0. \tag{4.31}$$

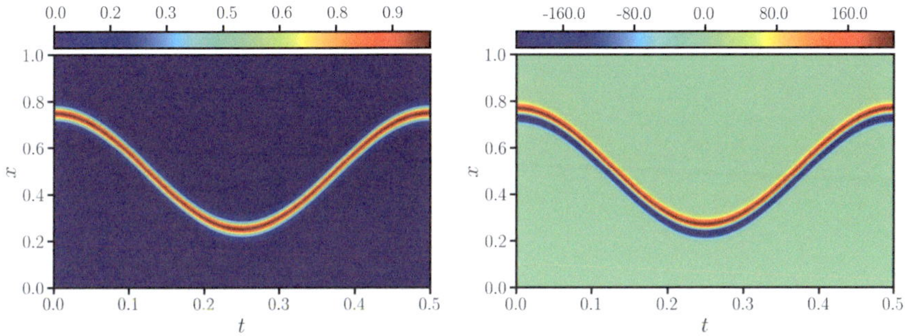

(a) Manufactured solution u (b) Manufactured heat flux ϕ

Fig. 4.12 Manufactured solution u and corresponding flux ϕ for the one-dimensional heat transfer problem

Both the heat capacity $c(u)$ and the thermal conductivity $\kappa(u)$ are temperature-dependent and defined as (inspired by [15])

$$c(u) = 1/2000\, u^2 + 500, \tag{4.32}$$

$$\kappa(u) = 1/100\, u + 7. \tag{4.33}$$

In order to verify the network predictions, a manufactured solution of the following form is proposed

$$u = \exp\left(-\frac{(x-s)^2}{2\sigma^2}\right), \tag{4.34}$$

where $\sigma = 0.02$. Conventionally, such Gaussian bell formulations are used to model laser-induced heat sources[6] p [17] which in this case is traveling along a path s

$$s(t) = \frac{1}{4}\cos\left(\frac{2\pi t}{t_{\max}}\right) + \frac{1}{2}. \tag{4.35}$$

Here, the Gaussian bell term is chosen as the manufactured solution to the problem instead.[7] Plugging Eq. (4.34) into Eq. (4.26) yields

$$p = \frac{\kappa u}{\sigma^2} + u\frac{x-s}{\sigma^2}\left[c\frac{\partial s}{\partial t} - \frac{x-s}{\sigma^2}\left(\kappa + u\frac{\partial \kappa}{\partial u}\right)\right]. \tag{4.36}$$

The manufactured solution from Eq. (4.34) should by no means be interpreted as a realistic example. It simply serves as a fast and direct way to verify the implementa-

[6] Laser-induced heat sources are for example employed in powder bed fusion processes, a popular additive manufacturing technique [16].

[7] Note that the selected, manufactured solution only satisfies the homogeneous Neumann boundary conditions from Eq. (4.30) approximately, which is, however, sufficient for the example at hand.

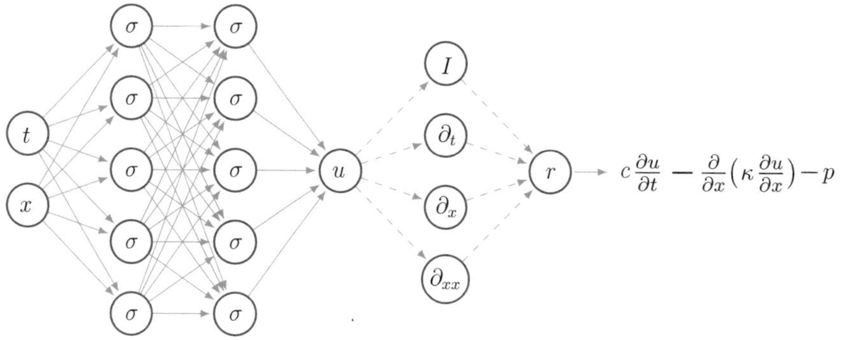

Fig. 4.13 Conceptual physics-informed neural network for the heat equation. The left side shows the fully connected neural network and the right side represents the physics-informed part. The dashed lines denote non-trainable weights

tion and validate the predictions of the trained physics-informed neural network. The manufactured solution u and the corresponding heat flux ϕ are depicted throughout space and time in Fig. 4.12.

The implementation that generated the following results is based on the code accompanying [7, 20], but was adapted to employ the PyTorch framework. Like in the static example, a fully connected neural network $u_{NN}(t, x; \boldsymbol{\Theta})$ is used to predict the unknown displacement field $u(x, t))$ specified in Eqs. (4.26), (4.31), (4.30). The inputs of the network are the temporal variable t and the spatial variable x. The function `buildModel` (cf. Sect. 4.2.1) determines the layer and neuron architecture of the fully connected neural network u_{NN} which approximates the temperature distribution $u(t, x)$. To define the output of the physics-informed neural network $r_{NN}(t, x; \boldsymbol{\Theta})$ (cf. Fig. 4.13), the residual of Eq. (4.26) is

$$r = c\frac{\partial u}{\partial t} - \frac{\partial}{\partial x}\left(\kappa\frac{\partial u}{\partial x}\right) - p = c\frac{\partial u}{\partial t} - \frac{\partial \kappa}{\partial u}\frac{\partial u}{\partial x}\frac{\partial u}{\partial x} - \kappa\frac{\partial^2 u}{\partial x^2} - p. \tag{4.37}$$

The code corresponding to Eq. (4.37) is given below,

```
def modelR(t, x):
    u = modelU(t, x)
    dudt = getDerivative(u, t, 1)
    dudx = getDerivative(u, x, 1)
    ddudxx = getDerivative(u, x, 2)
    k = 0.01 * u + 7
    dkdu = 0.01
    c = 0.0005 * u ** 2 + 500
    p = sourceTerm(t,x)
    r = c * dudt - dkdu * dudx * dudx - k * ddudxx - p
    return r
```

wherein line 10 forms the core of the implmementation of Eq. (4.37). Before that, `self.modelU(t, x)` calls the neural network $u_{NN}(t, x; \boldsymbol{\Theta})$ to return the predicted temperature `u` at t and x. Furthermore, the previously introduced function `getDerivative` from Listing 4.1 is extended to recursively compute the necessary first- and second-order partial derivatives.

Listing 4.2 Wrapped automatic differentiation to compute higher-order derivatives

```
def getDerivative(y, x, n):
    if n == 0:
        return y
    else:
        dydx = grad(y, x, torch.ones_like(y), create_graph=True,
                    retain_graph=True, allow_unused=True)[0]
        return getDerivative(dydx, x, n - 1)
```

The source term p is computed with `sourceTerm(t,x)` according to Eq. (4.36).

```
def sourceTerm(t, x):
    tMax = 0.5
    sigma = 0.02
    uMax = 1
    s = 0.25 * torch.cos(2 * torch.pi * t / tMax) + 0.5
    dsdt = -0.5 * torch.sin(2 * torch.pi * t / tMax) * torch.pi / tMax
    uAnalytical = uMax * torch.exp(-(x - s) ** 2 / (2 * sigma ** 2))
    k = 0.01 * uAnalytical + 7
    dkdu = 0.01
    c = 0.0005 * uAnalytical ** 2 + 500
    s = k * uAnalytical / sigma ** 2
        + uAnalytical * (x - s) / sigma ** 2 * (c * dsdt
        - (x - s) / (sigma ** 2) * (k + uAnalytical * dkdu))
    return s
```

In order to train the network, the following cost function is defined

$$C = \mathcal{L}_{\mathcal{I}} + \mathcal{L}_{\mathcal{B}} + \mathcal{L}_{\mathcal{R}} \tag{4.38}$$

Here, the term $\mathcal{L}_{\mathcal{I}}$

$$\mathcal{L}_{\mathcal{I}} = \frac{1}{m_{\mathcal{I}}} \sum_{i=1}^{m_{\mathcal{I}}} \left(\hat{u}(0, \tilde{x}_{\mathcal{I}}^i; \boldsymbol{\Theta}) - \tilde{u}_{\mathcal{I}}^i \right)^2, \tag{4.39}$$

enforces the initial condition (cf. Eq. (4.31)) by penalizing the error between the network prediction $\hat{u} = u_{NN}(0, \tilde{x}_{\mathcal{I}}^i; \boldsymbol{\Theta})$ and the initial values $\{\tilde{x}_{\mathcal{I}}^i, \tilde{u}_{\mathcal{I}}^i\}_{i=1}^{m_{\mathcal{I}}} = \{\tilde{x}_{\mathcal{I}}^i, u_0^i(\tilde{x}_{\mathcal{I}}^i)\}_{i=1}^{m_{\mathcal{I}}}$ at $m_{\mathcal{I}}$ points randomly drawn from a uniform distribution. The term $\mathcal{L}_{\mathcal{B}}$

$$\mathcal{L}_{\mathcal{B}} = \frac{1}{m_{\mathcal{B}_0}} \sum_{i=1}^{m_{\mathcal{B}_0}} \left(\frac{\partial}{\partial \tilde{x}} \hat{u} \left(\tilde{t}^i_{\mathcal{B}_0}, 0; \boldsymbol{\Theta} \right) \right)^2 + \frac{1}{m_{\mathcal{B}_1}} \sum_{i=1}^{m_{\mathcal{B}_1}} \left(\frac{\partial}{\partial \tilde{x}} \hat{u} \left(\tilde{t}^i_{\mathcal{B}_1}, 1; \boldsymbol{\Theta} \right) \right)^2, \tag{4.40}$$

enforces the Neumann boundary condition according to Eq. (4.30) at correspondingly $m_{\mathcal{B}_0}$ and $m_{\mathcal{B}_1}$ random samples $\{\tilde{t}_{\mathcal{B}_0}, \tilde{x}^i_{\mathcal{B}_0}\}_{i=1}^{m_{\mathcal{B}_0}}$, $\{\tilde{t}_{\mathcal{B}_1}, \tilde{x}^i_{\mathcal{B}_1}\}_{i=1}^{m_{\mathcal{B}_1}}$ on each boundary $x = 0$ and $x = 1$. Finally, adding the error of the residual $\mathcal{L}_{\mathcal{R}}$

$$\mathcal{L}_{\mathcal{R}} = \frac{1}{m_{\mathcal{R}}} \sum_{i=1}^{m_{\mathcal{R}}} \left(\hat{r} \left(\tilde{t}^i_{\mathcal{R}}, \tilde{x}^i_{\mathcal{R}} \right) \right)^2 \tag{4.41}$$

ensures that the solution satisfies the governing equation from Eq. (4.26). As proposed by [20], the $m_{\mathcal{R}}$ collocation points $\{\tilde{t}_{\mathcal{R}}, \tilde{x}^i_{\mathcal{R}}\}_{i=1}^{m_{\mathcal{R}}}$ are generated by a Latin-hypercube sampling technique [18] instead of uniformly distributed collocation points as in the previous two examples. Following Eqs. (4.39), (4.41), the cost function is implemented as follows.

```
def costFunction(t0, x0, u0, tBC0, xBC0, tBC1, xBC1, t, x):
    uICPred = modelU(t0, x0)
    uBC0Pred = modelU(tBC0, xBC0)
    uBC1Pred = modelU(tBC1, xBC1)
    duBC0Preddx = getDerivative(uBC0Pred, xBC0, 1)
    duBC1Preddx = getDerivative(uBC1Pred, xBC1, 1)
    rPred = modelR(t, x)
    lossIC = torch.mean((u0Pred - u0) ** 2)
    lossBC = (torch.mean((uBC0Pred) ** 2)
             + torch.mean((uBC1Pred) ** 2))
    lossPDE = torch.mean(rPred ** 2)
    return lossIC, lossBC, lossPDE
```

Subsequently, a full-batch gradient-based optimization procedure searches for the optimal network parameters $\boldsymbol{\Theta}^*$ by minimizing the cost function in Eq. (4.38). Here, the Adam optimizer is initially employed during the optimization, followed by the L-BFGS method. This combination was adapted from [7] and has empirically been observed to be the most robust approach throughout this study.

Algorithm 10 summarizes the introduced training procedure employed for a continuous prediction of the temperature distribution $u(t, x)$. Figure 4.14a shows the history of the cost function terms.

After the algorithm terminated, the network $u_{NN}(t, x; \boldsymbol{\Theta})$ with trained parameters $\boldsymbol{\Theta}$ is used to predict the continuous temperature distribution over the whole domain $\mathcal{T} \times \Omega = [0, 0.5] \times [0, 1]$. The result, along with a comparison between the prediction and the manufactured solution, is presented in Fig. 4.15.

When optimizing a cost function that consists of multiple terms, it is vital to balance the influence of each term. Figure 4.14b shows an example where no balancing is applied. The term $\mathcal{L}_{\mathcal{R}}$ is magnitudes larger than the two other terms $\mathcal{L}_{\mathcal{I}}$ and $\mathcal{L}_{\mathcal{B}}$. Here, the initial and boundary conditions are underrepresented in the cost function,

Algorithm 10 Training a physics-informed neural network for the continuous solution of the problem described in Eq. (4.26)

Require: training data for initial condition $\{0, \tilde{x}^i_{\mathcal{I}}, \tilde{u}^i_{\mathcal{I}}\}_{i=1}^{m_{\mathcal{I}}}$
Require: training data for boundary condition $\{\tilde{t}^i_{\mathcal{B}}, \tilde{x}^i_{\mathcal{B}}, \tilde{u}^i_{\mathcal{B}}\}_{i=1}^{m_{\mathcal{B}}}$
  generate $m_{\mathcal{R}}$ collocation points with Latin-hypercube sampling $\{t^i_{\mathcal{R}}, x^i_{\mathcal{R}}\}_{i=1}^{m_{\mathcal{R}}}$
  define network architecture (input, output, hidden layers, hidden neurons)
  initialize network parameters $\boldsymbol{\Theta}$: weights $\{\boldsymbol{W}^l\}_{l=1}^{L}$ and biases $\{\boldsymbol{b}^l\}_{l=1}^{L}$ for all layers L
  set hyperparameters for Adam optimizer (Adam epochs, learning rate $\alpha, \ldots$)
  set hyperparameters for L-BFGS optimizer (L-BFGS epochs, convergence criterion, ...)

  procedure TRAIN
    compute $\{\hat{u}(0, \tilde{x}^i_{\mathcal{I}}; \boldsymbol{\Theta})\}_{i=1}^{m_{\mathcal{I}}}$
    compute $\{\frac{\partial}{\partial x}\hat{u}(\tilde{t}^i_{\mathcal{B}}, \tilde{x}^i_{\mathcal{B}}; \boldsymbol{\Theta})\}_{i=1}^{m_{\mathcal{B}}}$
    compute $\{\hat{r}(t^i_{\mathcal{R}}, x^i_{\mathcal{R}}; \boldsymbol{\Theta})\}_{i=1}^{m_{\mathcal{R}}}$
    compute $\mathcal{L}_{\mathcal{I}}, \mathcal{L}_{\mathcal{B}}, \mathcal{L}_{\mathcal{R}}$ ▷ cf. Equation (4.30) to (4.41)
    evaluate cost function: $C \leftarrow \mathcal{L}_{\mathcal{I}} + \mathcal{L}_{\mathcal{B}} + \mathcal{L}_{\mathcal{R}}$
    update parameters: $\boldsymbol{\Theta} \leftarrow \boldsymbol{\Theta} - \alpha \frac{\partial C}{\partial \boldsymbol{\Theta}}$ ▷ Adam or L-BFGS
  end procedure

  for all Adam-epochs **do**
    run TRAIN with Adam optimizer
  end for
  for all L-BFGS-epochs **do**
    run TRAIN with L-BFGS optimizer
  end for

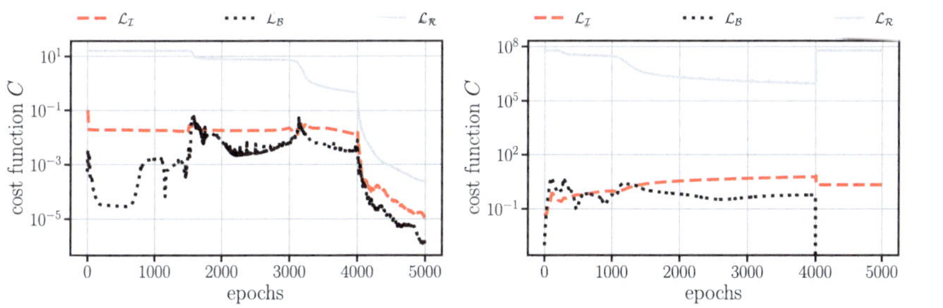

(a) Balanced cost function terms plotted over the number of training iterations

(b) Unbalanced cost function terms plotted over the number of training iterations

Fig. 4.14 Balanced and unbalanced cost function history for 4 000 Adam epochs and 1 000 L-BFGS epochs

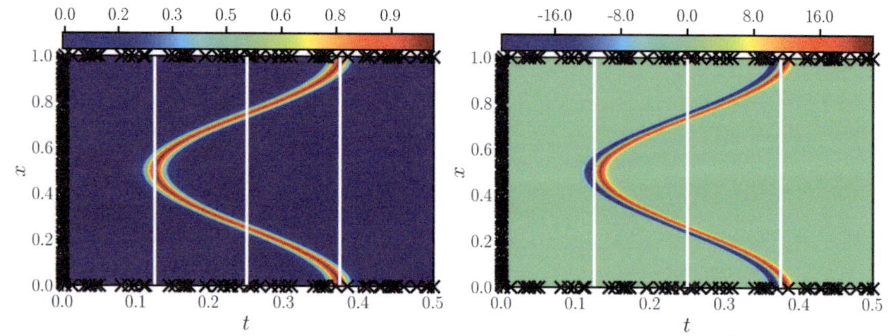

(a) temperature $\hat{u}(x,t)$ (b) heat flux $\hat{\phi}(x,t)$

Fig. 4.15 Prediction of the temperature distribution u and corresponding heat flux ϕ. The crosses indicated the initial condition and boundary condition collocation points. The three white lines are slices in time at $t = 0.125$ s, $t = 0.25$ s, and $t = 0.375$ s at which the analytical solution (Eq. (4.34)) is compared to in Fig. 4.16.

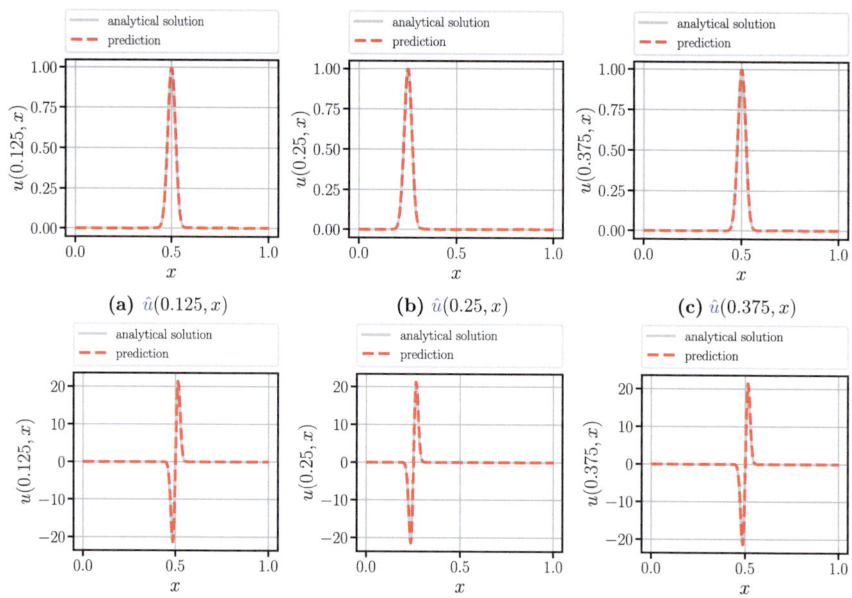

(a) $\hat{u}(0.125, x)$ (b) $\hat{u}(0.25, x)$ (c) $\hat{u}(0.375, x)$

(d) $\hat{\phi}(0.125, x)$ (e) $\hat{\phi}(0.25, x)$ (f) $\phi(0.375, x)$

Fig. 4.16 Comparison of temperature and heat flux at three time slices with analytical solution from Eq. (4.34) and the algorithm does not converge to a correct solution (cf. Fig. 4.14b). Nabian and Meidani addressed this issue by introducing weight factors adjusting the relative importance of each term in the cost function [19]. In the example at hand, an empirically determined factor of 5×10^{-4} must be added to the term `lossPDE` in `costFunction` to gain satisfactory results (cf. Figs. 4.14a and 4.15).

```
lossPDE = torch.mean((5e-4 * rPred) ** 2)
```

This is an alternative to the automatic adjustment of the weighting, as performed in Sect. 4.2.2. Here, the weighting factor is a constant hyperparameter obtained through hyperparameter tuning. The advantage is that the balancing is achieved from the first iteration onward. The methods could, however, also be combined by preselecting a weighting, which may be adapted by the minimax optimization described in Sect. 4.2.2.

4.3 Data-Driven Identification

System identification from sparse data is another class of problems often encountered in physics and engineering applications. The task of system identification can be phrased as: find the parameters λ that best describe the observed data. The main idea is to use the non-linear partial differential equation, Eq. (4.1), together with a solution $u(t, x)$ to solve the inverse problem of identifying the parameters λ. Note that the entire Chap. 9 is dedicated to this class of problems, so that this section can be viewed as a simple preliminary introduction to this topic in the framework of physics-informed neural networks.

A neural network is used to predict the parameter λ instead of the solution $u(t, x)$. The remaining steps are almost identical to the data-driven inference. The partial differential equation is represented by the gradients of the solution u and parameter λ with respect to the input parameters x and t. These are used to compute the mean squared error loss of the partial differential equation $\mathcal{L}_{\mathcal{R}}$, boundary conditions $\mathcal{L}_{\mathcal{B}}$ and initial conditions $\mathcal{L}_{\mathcal{I}}$. The sum of the mean squared error losses is then used as a cost function C as in Eq. (4.2). If the boundary or initial conditions are independent of the parameter λ, their corresponding loss terms can be neglected. Finally, the cost function is minimized by the procedure previously outlined in Sect. 4.2.

4.3.1 One-Dimensional Static Model

To illustrate the data-driven identification using a physics-informed neural network, the example presented in Sect. 4.2.1 is revisited. However, now it is assumed that the displacement $u(x)$ is known while the cross-sectional properties $EA(x)$ are to be determined. As before, the domain $\Omega = [0, 1]$ is investigated. To increase the complexity of the problem, a cubic variation in the cross-sectional properties is

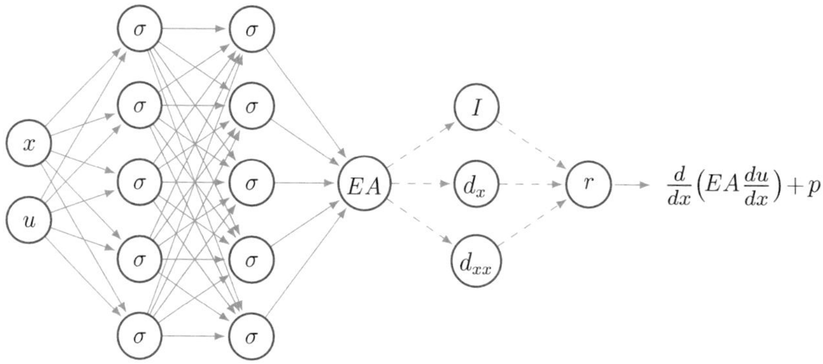

Fig. 4.17 Conceptual physics-informed neural network for data-driven identification of the static bar equation. The left part shows the fully connected neural network and the right part represents the physics-informed neural network. The dashed lines denote non-trainable weights

defined as

$$EA(x) = x^3 - x^2 + 1. \tag{4.42}$$

Again, the manufactured solution $u(x)$ is chosen as

$$u(x) = \sin(2\pi x), \tag{4.43}$$

which leads to the following distributed load $p(x)$ after insertion into the differential equation from Eq. (4.4)

$$p(x) = -2(3x^2 - 2x)\pi\cos(2\pi x) + 4(x^3 - x^2 + 1)\pi^2\sin(2\pi x). \tag{4.44}$$

The chosen solution fulfills the following boundary conditions

$$u(0) = u(1) = 0. \tag{4.45}$$

However, the boundary conditions do not influence the learning of the model parameters, as the model parameters are not part of the boundary conditions. Therefore, the boundary loss $\mathcal{L}_{\mathcal{B}}$ can be omitted.

The network architecture to solve this inverse problem, illustrated in Fig. 4.17, is similar to the one presented in Fig. 4.3 in Sect. 4.2.1. The difference lies in the inputs being the coordinates x as well as the displacements u. Additionally, the left fully connected neural network predicts the stiffness EA instead of the displacements u. The differential equation loss is identical to Eq. (4.13), while the boundary loss is omitted.

Given this information, an inverse physics-informed neural network can be constructed to predict the cross-sectional properties $EA(x)$ of the bar. This is shown

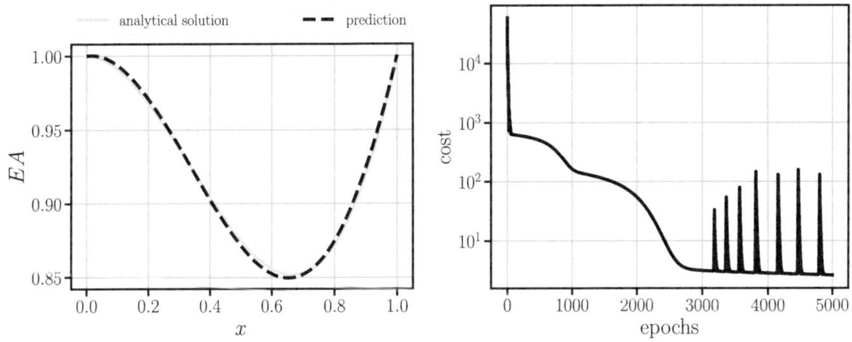

Fig. 4.18 Linear elastic static bar example. Left: the cross-sectional properties $EA(x)$ computed with a physics-informed neural network and the analytic expression are compared. Right: the training history is illustrated by showing the cost function for each epoch

in Fig. 4.18. Here, it is seen that the varying cross-section is approximated well. Additionally, the training history is shown along with the cost function.

Note that variations of this inverse network exist. Raissi et al. [20] extend a neural network with a trainable parameter EA. Their neural network uses coordinates x as inputs and displacements u as output, similar to the forward-driven approach. The cost function is then defined as the sum of the mean squared error between the measurements $\hat{u}$ and the predictions u and the physics-informed loss. This is not necessary in the presented example, as we know the solution u in the full domain, i.e., we have full domain knowledge. Tartakovsky et al. [21] use two neural networks. The first network predicts the displacement u from the coordinates x, and the second network predicts the stiffness EA from the predicted displacements u. Another variation is to remove the displacement u as input (cf. Fig. 4.17), as u does not provide the neural network with any information when learning a single inverse solution. Its removal might, however, decrease the expressivity of the neural network.

Exercises

E.14 Physics-Informed Neural Network for a Static Bar (P & C)

A bar in the domain $x \in [0, 3/2]$ is given with the varying cross-sectional properties

$$EA(x) = x^2 + 1.$$

The goal is to produce a manufactured solution to later verify the implementation of a physics-informed neural network. The following displacement solution is prescribed

$$u(x) = 1 - \cos(3\pi x).$$

1. Compute the distributed load $p(x)$ from the differential equation given in the manuscript. Additionally, state the two homogeneous boundary conditions.
2. Now, the goal is to implement a physics-informed neural network with PyTorch for a static bar. For this purpose, a template has been prepared as `exercise 14 .ipynb`. As a first step, the previously identified distributed load $p(x)$ is to be incorporated as a lambda function `distLoad`, thus completing the problem setup.
3. Next, the key helper functions of the physics-informed neural network have to be implemented. The first is the `getDisplacements(model, x)` function, which using a neural network `model` and coordinates `x` predicts the displacements at `x`. No strong enforcement is to be used. Thus, the function simply returns the neural network output.
4. Furthermore, the loss terms have to be computed with `getLossTerms(x, xB, u, uB, EA, distLoad, uBLabel)`. This includes both the residual of the differential equation $\mathcal{L}_{\mathcal{R}}$ and the boundary condition loss $\mathcal{L}_{\mathcal{B}}$. The residual is computed with the coordinates `x`, and the corresponding displacement predictions `u`, axial stiffnesses `EA`, and distributed load `distLoad` (all evaluated at `x`). The boundary conditions are defined with `uBLabel`, which is a list containing a list per boundary condition. Each list contains three values: the value of the boundary condition, the degree of differentiation (so 0 for a Dirichlet and 1 for a Neumann boundary condition), and the coordinate. This is the required information for the boundary loss computation. Furthermore, the displacement prediction `uB` at the boundary coordinates `xB` is passed to `getLossTerms`. The coordinate `xB` seems redundant as it is already contained in `uBLabel`, but it makes differentiation of `uB` with respect to `xB` easier.
5. With the computation of the loss terms, the cost function can be established. Implement `getCostFunction(lossTerms)`, where `lossTerms` is a list containing the two loss terms. The two terms should be weighted equally, i.e., with one.
6. The last step is completing the training loop. In each epoch, a closure function `closure()` has to be defined, which will be used by the optimizer. The function's purpose is to compute the gradients $\partial C / \partial \boldsymbol{\Theta}$ and return the cost C. Thus, the displacements are to be predicted at `x` and `xB` in order to compute the loss terms with `getLossTerms` and the cost function with `getCostFunction`. Lastly, the backward propagation has to be performed with `cost.backward(retain_graph=True)` .[8] Furthermore, remember to reset the gradients before every call with `optimizer.zero_grad()`. Lastly, it must be noted that `closure()` does not take any input arguments. The variables needed inside are simply captured from the outer (local) scope in which the function `closure()` is defined. Such a function, which accesses variables from its enclos-

[8] `retain_graph=True` is not necessary for the Adam optimizer, but for optimizers performing multiple calls to the closure function, as, e.g., the L-BFGS optimizer.

ing scope, is called a closure function.
Hint: Take inspiration from the prediction and cost computation provided above the closure definition.

7. Run the `exercise 14 .ipynb` file and try changing the hyperparameters. What happens if you increase or decrease the learning rate or the number of epochs or introduce a weighting factor for the boundary conditions?
8. Now try to reproduce the example from the manuscript in Sect. 4.2.1 by replacing the problem hyperparameters in the `exercise 14 .ipynb` file. Find suitable model parameters to estimate the solution quickly and precisely.

E.15 Data-Driven Identification Using Physics-Informed Neural Networks for a Static Bar (P & C)

You are given the following information on a static bar defined in the domain $x \in [0, 1]$

$$
\begin{aligned}
EA(x) &= 1 - u(x), \\
p(x) &= \sin(x), \\
u(x) &= 1 - \sqrt{2\sin(1)x - 2\sin(x) + 1}.
\end{aligned}
$$

1. What is an important difference between the given data and the examples provided in the manuscript?
2. The goal is to implement an inverse physics-informed neural network for a static bar. This is to be done with the help of the `exercise 15 .ipynb` file, which is a modified version of the exercise file of Exercise E.14. The changes are indicated in red. The first step is to incorporate the problem data above in the code.
3. Next the loss term computation with `getLossTerms(x, u, EA, distLoad)` has to be implemented. Given the stiffness prediction `EA`, the analytical displacements `u`, and the distributed load `distLoad` evaluated at `x`, compute the differential equation residual. No boundary condition loss is necessary, leading to only a single loss term.
4. Perform the inversion in order to recover the stiffness distribution. How do the hyperparameters affect the inversion?
5. Reproduce the example from the manuscript in Sect. 4.3 by replacing the hyperparameters in the `exercise 15 .ipynb` file. Find good model parameters to estimate the solution as fast and accurately as possible.

E.16 Normalization and Weighting (C)

1. The influence of the scaling of the solution u, the scaling of the differential equation, and the weighting of the different loss terms, i.e., the residual and boundary terms, are investigated. For this, the example bar from Exercise E.14 is revisited. Further, the file `exercise16.ipynb` is provided. The file is almost identical to the implementation established in Exercise E.14. The difference lies in the scaling factors `scaleODE`, which scales the differential equation terms, and `scaleSolution`, which scales the solution u. Both are initially set to 1. The goal is to investigate the influence of the scaling of the differential equation and the solution. Try increasing and decreasing these factors. What do you observe? If the given training hyperparameters inhibit convergence of the minimization procedure, change them. You can adjust the number of epochs and the learning rate or change the optimizer. Additionally, you can even try to use multiple training blocks, for example, first using an Adam optimizer and then an L-BFGS optimizer. What can you conclude about the scaled cases? Moreover, which scaling is more critical?
2. Now, a weighting factor is introduced on the boundary condition loss, `boundaryLossWeight`. Initially, it is set to 1. Try varying this weighting factor while keeping the remaining hyperparameters constant. You can track the cost C after 500 epochs and fill out the following table:

`boundaryLossWeight`	10^{-3}	10^{-2}	10^{-1}	10^{0}	10^{1}	10^{2}	10^{2}	10^{3}
cost								

Which value yields the best convergence? This is a classical case of manual hyperparameter tuning.
Hint: Note that results might vary due to the random initialization. It is best to run the cases multiple times and average the results.

E.17 Physics-Informed Neural Network for a Plate in Membrane Action (C)

The implementation of a plate in membrane action with a physics-informed neural network is to be studied. To this end, we consider case 1 from Fig. 4.5a, i.e., a uniaxial tension test. A plate of size 1×1 under plane stress with material properties $E = 1, \nu = 0.3$ and a traction of $t = 1$ on the right edge is studied. The boundary conditions are listed in Table 4.1.

1. The implementation is prepared in `exercise17.ipynb`, where the first task is to implement the `getDisplacements(model, input, strongEnforcement1, strongEnforcement2)` function. A strong enforcement is to be applied after querying the output of the neural network `model`. Using Eq. (4.22), the boundary conditions are enforced with lambda functions:

```
strongEnforcement1 = lambda z1, x1, x2: x1 * z1
strongEnforcement2 = lambda z2, x1, x2: x2 * z2
```

   The coordinates are stored in `input` in a concatenated manner.
   Hint: The strong enforcement is to be applied to each displacement component and subsequently concatenated with `torch.cat`, yielding the final displacement field.
2. In order to compute the residual, a helper function for the strain computation, is required. The strains are to be returned in Voigt notation with `getStrains(u, x1, x2, N1, N2)`. For this, consider the predicted displacements `u` at coordinates `x1`, `x2` with a grid composed of `N1`× `N2` collocation points.
3. Next implement the `getLossTerms(x1, x2, u, N1, N2, C, force)` function. The residual is given as Eq. (4.2) and relies on the displacement prediction `u`, the coordinates `x1`, `x2`, the number of grid points `N1`, `N2`, and the material matrix `C`. For the boundary conditions, consult Eq. (4.22) in which the traction `force` is given as t. We assume that no body force is acting on the plate, i.e., $p(x_1, x_2) = 0$.
4. Run the training loop in order to learn the solution using `exercise17.ipynb`. Try changing the boundary conditions in order to simulate case 2 from Fig. 4.5b. Tune the hyperparameters correspondingly. Does the optimization become more challenging, and if it does, why?

References

1. Psichogios, Dimitris C., and Lyle H., Ungar. 1992. A hybrid neural network-first principles approach to process modeling. *AIChE Journal* 38(10):1499–1511. ISSN: 0001-1541, 1547-5905. https://doi.org/10.1002/aic.690381003.
2. Lagaris, I. E., A. Likas, and D. I. Fotiadis. 1998. Artificial neural networks for solving ordinary and partial differential equations. *IEEE Transactions on Neural Networks* 9(5):987–1000. ISSN: 10459227. https://doi.org/10.1109/72.712178.
3. Hirn, Matthew, Stéphane, Mallat, and Nicolas, Poilvert. 2017. Wavelet scattering regression of quantum chemical energies. *Multiscale Modeling & Simulation* 15(2):827–863. tex.ids= Hirn2017a arXiv: 1605.04654. ISSN: 1540-3459, 1540-3467. https://doi.org/10.1137/16M1075454.
4. Mallat, Stéphane. 2016. Understanding deep convolutional networks. *Philosophical Transactions of the Royal Society A: Mathematical, Physical and Engineering Sciences* 374(2065):20150203. tex.ids= Mallat2016a, ISSN: 1364-503X, 1471-2962. https://doi.org/10.1098/rsta.2015.0203.

5. Kollmannsberger, Stefan, et al. 2021. *Deep learning in computational mechanics: an introductory course*. eng. Studies in computational intelligence, vol. 977. Cham: Springer. ISBN: 978-3-030-76587-3 978-3-030-76586-6.
6. Abadi, Martín, et al. 2015. *TensorFlow: large-scale machine learning on heterogeneous systems*.
7. Raissi, Maziar. 2020. *maziarraissi/PINNs*. tex.ids= Raissi2020c, originalDate: 2018-01-21T04:04:32Z.
8. Samaniego, Esteban, et al. 2019. An energy approach to the solution of partial differential equations in computational mechanics via machine learning: concepts, implementation and applications. arXiv:1908.10407 *[cs, math, stat]* arXiv: 1908.10407.
9. Roache, Patrick J. 2002. Code verification by the method of manufactured solutions. *Journal of Fluids Engineering* 124(1):4–10. Publisher: American Society of Mechanical Engineers Digital Collection, ISSN: 0098-2202. https://doi.org/10.1115/1.1436090.
10. Liu, Dong C., and Jorge, Nocedal. 1989. On the limited memory BFGS method for large scale optimization. *Mathematical Programming* 45(1–3):503–528. ISSN: 0025-5610, 1436-4646. https://doi.org/10.1007/BF01589116.
11. Sundström, Bengt, ed. 2010. *Handbook of solid mechanics*. eng. OCLC: 938957431. Stockholm: Department of Solid Mechanics. ISBN: 978-91-972860-4-6.
12. Berg, Jens, and Kaj, Nyström. 2018. A unified deep artificial neural network approach to partial differential equations in complex geometries. *Neurocomputing* 317:28–41. arXiv: 1711.06464. ISSN: 09252312. https://doi.org/10.1016/j.neucom.2018.06.056.
13. Henkes, Alexander, Henning, Wessels, and Rolf, Mahnken. 2022. Physics informed neural networks for continuum micromechanics. *Computer Methods in Applied Mechanics and Engineering* 393:114790. ISSN: 0045-7825. https://doi.org/10.1016/j.cma.2022.114790.
14. Hughes, Thomas J. R. 2000. *The finite element method: Linear static and dynamic finite element analysis*. Mineola, NY: Dover Publications. ISBN: 978-0-486-41181-1.
15. Kollmannsberger, S., et al. 2018. A hierarchical computational model for moving thermal loads and phase changes with applications to selective laser melting. Computers & Mathematics with Applications 75(5):1483–1497. ISSN: 0898-1221. https://doi.org/10.1016/j.camwa.2017.11.014.
16. Paolini, Alexander, Stefan, Kollmannsberger, and Ernst, Rank. 2019. Additive manufacturing in construction: A review on processes, applications, and digital planning methods. *Additive Manufacturing* 30:100894. ISSN: 2214-8604. https://doi.org/10.1016/j.addma.2019.100894.
17. Kollmannsberger, Stefan, et al. 2019. Accurate prediction of melt pool shapes in laser powder bed fusion by the non-linear temperature equation including phase changes: Model validity: Isotropic versus anisotropic conductivity to capture AM Benchmark Test AMB2018-02. *Integrating Materials and Manufacturing Innovation* 8(2):167–177. ISSN: 2193-9764, 2193-9772. https://doi.org/10.1007/s40192-019-00132-9.
18. Stein, Michael. 1987. Large sample properties of simulations using Latin hypercube sampling. *Technometrics* 29(2):143–151. ISSN: 0040-1706, 1537-2723. https://doi.org/10.1080/00401706.1987.10488205.
19. Nabian, Mohammad Amin, and Hadi, Meidani. 2019. A deep neural network surrogate for high-dimensional random partial differential equations. *Probabilistic Engineering Mechanics* 57:14–25. arXiv: 1806.02957. ISSN: 02668920. https://doi.org/10.1016/j.probengmech.2019.05.001.
20. Raissi, M., P., Perdikaris, and G. E., Karniadakis. 2019. Physics-informed neural networks: A deep learning framework for solving forward and inverse problems involving nonlinear partial differential equations. *Journal of Computational Physics* 378:686–707. ISSN: 00219991. https://doi.org/10.1016/j.jcp.2018.10.045.

21. Tartakovsky, Alexandre M., et al. 2018. Learning parameters and constitutive relationships with physics informed deep neural networks. tex.ids= Tartakovsky2018a.
22. Kondor, Risi. 2018. N-body networks: A covariant hierarchical neural network architecture for learning atomic potentials. arXiv:1803.01588 *[cs]*. tex.ids= Kondor2018a arXiv: 1803.01588.

Chapter 5
Advanced Physics-Informed Neural Networks

Abstract Physics-informed neural networks (PINNs) solve differential equations by approximating the solution with a neural network and minimizing the corresponding residual. Due to the popularity of PINNs, a multitude of extensions and variations has been proposed. This chapter explores advanced methodologies and extensions of PINNs, such as the deep energy method, variational approaches, and improved boundary condition handling. These techniques are demonstrated through basic one-dimensional and two-dimensional examples using static bars and plates. Furthermore, the connection to the finite element method is established, showcasing the similarities and limitations of physics-informed neural networks.

As seen in Chap. 4, physics-informed neural networks are not competitive with classical approaches, such as the finite element method. Therefore, this chapter explores the most prominent techniques to improve their performance.[1, 2]

A major contribution to the cost of physics-informed neural networks is automatic differentiation. The computational time increases in a power-law relationship with respect to the degree of differentiation, as depicted in Fig. 5.1a (obtained with the implementation in Listing 4.2). Thus, lowering the differentiation order within the cost function is highly beneficial. This is one of the main motivations for variations of physics-informed neural networks based on the weak form of the differential equation instead of the strong form, as was used in Chap. 4. Most prominent are the deep energy method (also called deep Ritz method), variational physics-informed neural networks, and weak adversarial networks. All of these will be treated in the upcoming Sect. 5.1, where a focus will be laid on the deep energy method, as it performs the best—with the least amount of hyperparameter tuning. It is, however, limited to conservative systems, i.e., systems that conserve the potential energy, which are often encountered in mechanics. For non-conservative systems (and inverse problems[3]), more general approaches, such as variational physics-informed neural networks or

[1] Note that these techniques are still not sufficient to make physics-informed neural networks and its variations competitive with classical discretization techniques.

[2] All runtimes within this chapter were obtained with a laptop equipped with a 11th Gen Intel Core I7-1165G7 CPU and 32 GBs of RAM.

[3] The deep energy method is not generally applicable to inverse problems.

L. Herrmann et al., *Deep Learning in Computational Mechanics*,
https://doi.org/10.1007/978-3-031-89529-6_5

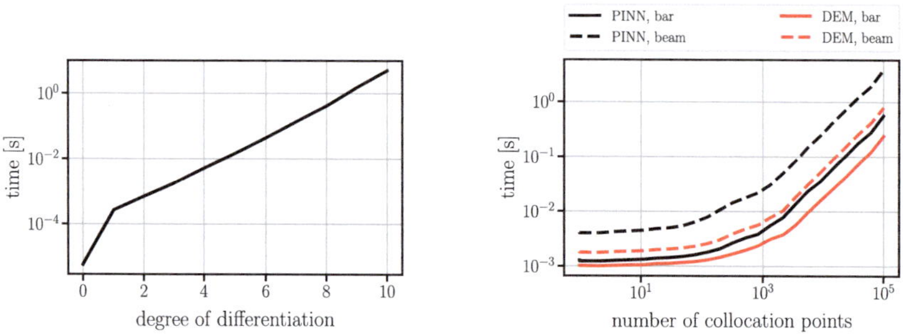

(a) Computational time to compute higher-order derivatives using automatic differentiation(using Listing 4.2), which includes the time for forward propagation (provided as the zero degree of differentiation)

(b) Computational time of a single epoch of a physics-informed neural network (PINN) and the deep energy method (DEM) for correspondingly the bar and beam equation

Fig. 5.1 Computational time associated with automatic differentiation

weak adversarial networks, must be employed. The impact of lowering the differentiation order becomes apparent in Fig. 5.1b, where a physics-informed neural network is compared to the deep energy method for correspondingly the bar and beam equation. The bar equation reduces from second to first order, while the beam equation reduces from fourth to second order. The difference in computation time becomes even more pronounced with more collocation points, which also follows a power-law relation.

The added benefit of considering the residual in its weak form, i.e., fulfilling the partial differential equation in an integral sense, is that weak singularities, such as the stress singularity in the plate in membrane action example from Sect. 4.2.2, can be considered appropriately. In continuum mechanics, the weak formulation can exploit that the integral of the stress is well-defined despite a stress singularity. Thus, stress singularities can be handled, which, in contrast, was not the case for the strong form, as seen in the results from Sect. 4.2.2.

Besides the physics-informed neural network variations, many improvements have been made without changing the residual formulation. Examples include the different methods of handling boundary conditions, optimization algorithms, or distribution of collocation points, as already touched on in Chap. 4. Section 5.2 systematically summarizes the main ideas. As the advantages of the different approaches depend highly on the problem at hand, they are provided, such that they can easily be used. Thus, the mathematical and implementational aspects are discussed without delving into results. Topics include optimizers, sampling, boundary conditions, differentiation, the ansatz space, and transfer learning. These can be employed independently of the equation and physics-informed neural network type (including the variations from Sect. 5.1).

5.1 Variations

In the following, we cover the deep energy method [4, 5] (also called deep Ritz method), variational physics-informed neural networks [12, 13], and weak adversarial networks [1]. Worthy of mentioning are also the deep least-squares method [2] and the deep Galerkin method [3], which are, however, not covered in this chapter.

5.1.1 Deep Energy Method

An alternative method to the data-driven inference with physics-informed neural networks for physical systems is the deep energy method. This approach was introduced by [5] as the deep energy method and by [4] as the deep Ritz method. Instead of considering partial differential equations and minimizing the corresponding residual, the system's potential energy is minimized. However, this is only valid for physical systems that fulfill the principle of minimum potential energy, such as, for example, structural mechanical problems under conservative loads.[4]

The basic idea of using a neural network to approximate the solution $u(x)$ of a partial differential equation is the same as for the physics-informed neural network. The boundary conditions can again either be enforced strongly or through penalization in the loss function. The Neumann boundary conditions are, however, often already part of the potential energy. Furthermore, a strong enforcement of the boundary conditions leads to an unconstrained optimization problem instead of a constrained optimization problem, as solely the potential energy has to be minimized. This makes the minimization easier and thereby improves the solution estimate $u(x)$.

The potential energy Π_{tot} can be expressed by the sum of the internal Π_{i} and the external energy Π_{e}

$$\Pi_{\text{tot}} = \Pi_{\text{i}} + \Pi_{\text{e}}. \tag{5.1}$$

The internal energy is computed by an integral of the internal energy density over the domain, which in structural mechanics is the product of the strains and stresses. The external energy is given by the negative work performed by the load. The external energy is defined as negative, as bringing the system back to its original configuration, including the applied load, requires positive work. Both energy components require the computation of an integral that depends on the solution $u(x)$. As the solution is only known at specific collocation points, a numerical integration scheme has to be applied. One of the simplest choices is the trapezoidal rule, described in textbooks like [7], where the domain is split into trapezoids, which makes it possible to approximate

[4] A non-conservative load refers to loads, whose work is path-dependent and thereby cannot be recovered during load reversal. Contributing factors to a system's non-conservative nature include, but are not limited to, frictional forces, aerodynamic drag, damping, plastic deformation, viscoelastic behavior, and path-following loads.

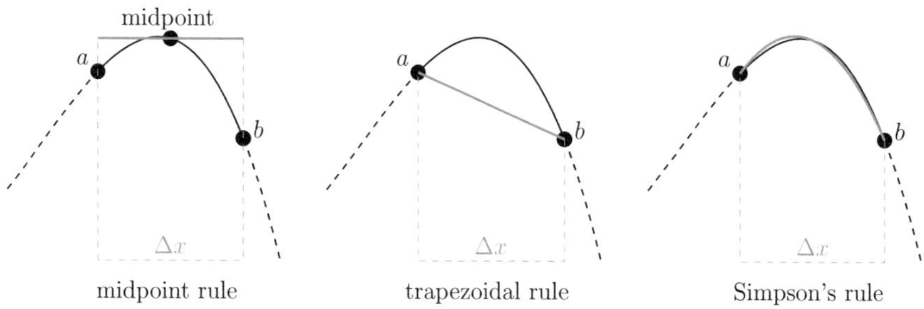

Fig. 5.2 Numerical integration schemes

the area and thereby the integral, as shown in Fig. 5.2. The integration scheme can be expressed as

$$\int_a^b f(x)dx \approx \frac{\Delta x}{2}[f(x_0) + 2f(x_1) + 2f(x_2) + \cdots + 2f(x_{n-1}) + f(x_n)], \tag{5.2}$$

where $\Delta x = (b - a)/n$ is the distance between the integration points $x_i = a + i\Delta x$. An alternative but very similar numerical integration scheme is the Simpson's rule, also described in [7], which computes the integral by splitting the domain into subdomains and approximates the integrand with quadratic polynomials, as seen in Fig. 5.2. The scheme is given as

$$\int_a^b f(x)dx \approx \frac{\Delta x}{3}[f(x_0) + 4f(x_1) + 2f(x_2) + 4f(x_3) + \cdots + 4f(x_{n-1}) + f(x_n)], \tag{5.3}$$

where Δx and x_i are calculated in the same way as for the trapezoidal rule. Simpson's rule is more accurate but requires three instead of two integration points. A third alternative from the same family is the midpoint rule, where the domain of the integral is split up into rectangles, illustrated in Fig. 5.2. Therefore, the integration scheme is the simplest and can be expressed as

$$\int_a^b f(x)dx \approx \sum_{i=0}^{n} f\left(\frac{x_{i+1} - x_i}{2}\right)\Delta x, \tag{5.4}$$

where Δx and x_i are computed as before. This rule is the most inaccurate but very useful, as the function is only evaluated at the midpoints, not the boundaries. This is useful if an integrand has a singularity at the boundary, which then can be avoided in the evaluation of the integral. The three integration schemes are summarized in Table 5.1. Another famous integration scheme that could be used is Gaussian quadrature, as typically employed in the finite element method [8].

The evaluation of the integral slightly increases the complexity of the computations and makes the optimization problem more challenging. It is therefore recom-

Table 5.1 Comparison of three numerical integration schemes. The distance between the integration points is given as $\Delta x = (b-a)/n$ and the integration points are given as $x_i = a + i\Delta x$

	Order of integration	Integration scheme
Midpoint rule	0	$\sum_{i=0}^{n} f(\frac{x_{i+1}-x_i}{2})\Delta x$
Trapezoidal rule	1	$\frac{\Delta x}{2}[f(x_0)+2f(x_1)+2f(x_2)+\cdots+2f(x_{n-1})+f(x_n)]$
Simpson's rule	2	$\frac{\Delta x}{3}[f(x_0)+4f(x_1)+2f(x_2)+4f(x_3)+\cdots+f(x_n)]$

mended by [5] to use a strong enforcement of the boundary conditions so that the optimization problem is unconstrained.

Another drawback of the deep energy method is that there is no inbuilt regularization, as was empirically observed and is demonstrated in Sect. 5.1.2 for the physics-informed neural network in [6]. In conventional physics-informed neural networks, overfitting is not a problem. However, in the deep energy method, overfitting can occur, causing unphysical oscillations of the solution $u(x)$. These oscillations lead to an excessive and unphysical external energy Π_e, which may grow toward $-\infty$, while not increasing the internal energy Π_i accordingly. An explanation for this phenomenon will follow below. In practice, the computed internal energy is not increasing, as it is computed by the same set of integration points. If a different set of integration points is used, an enormous internal energy would be observed, and one would discover that the potential energy was not successfully minimized. To avoid overfitting with this method, one remedy is early stopping regularization, as it directly counteracts this issue. Here, a different set of integration points is used to test if the potential energy also decreases for this set of points or only for the training set. If overfitting is encountered, the training is stopped early, and the best solution $u(x)$ for the testing set is used.

The deep energy method also has advantages. For example, it handles cases with weak singularities, such as in stresses or even the distributed load $p(x)$ better. This is because the singularity never has to be explicitly evaluated, thanks to the numerical integration. In the integral sense, the quantities are well-defined. For the physics-informed neural network, these cases pose a problem where the singularity either has to be approximated or avoided, which can decrease the quality of the solution significantly. Moreover, the deep energy method naturally avoids the evaluation of higher-order derivatives of the network with respect to the input variable and instead computes the integral of the solution. The evaluation of that integral turns out to be computationally much less demanding than the computation of the spared derivative evaluations. In essence, the integral is computationally not very different than the summation performed over the mean squared error of each collocation point in, e.g., Eq. (4.13). The integral is also composed over a summation with specialized coefficients, as seen in Table 5.1. Thus, the integral is, at most, slightly more expensive than the summation in the physics-informed neural network.

5.1.2 One-dimensional Static Model

The one-dimensional static bar from Sect. 4.2.1 is revisited to illustrate how the deep energy method works. The approach can thereby also be easily compared to the one-dimensional static bar implemented with the physics-informed neural network, illustrated in Fig. 4.2. The simplest version of the static bar uses a linear elastic formulation. The goal is to estimate the longitudinal displacement $u(x)$ under a certain distributed load $p(x)$ and concentrated loads F_i. The internal energy is given as

$$\Pi_\mathrm{i} = \int_\Omega \frac{1}{2} EA \left(\frac{du}{dx}\right)^2 dx, \tag{5.5}$$

where E is the Young's modulus and A is the cross-sectional area of the bar. The external energy is defined as

$$\Pi_\mathrm{e} = -\int_\Omega p(x)u(x)dx - \sum_i F_i u(x_i). \tag{5.6}$$

The potential energy is then computed as the sum of the internal and external energy in Equation (5.1) that has to be minimal for the correct displacement $u(x)$. To compute the internal and external energy, either the trapezoidal, Simpson's, or midpoint rule can be used. For the static bar, a strong enforcement of the boundary conditions is very easy. Therefore, the cost function is calculated directly with the potential energy

$$C = \Pi_\mathrm{tot} = \Pi_\mathrm{i} + \Pi_\mathrm{e}, \tag{5.7}$$

where the Neumann boundary conditions, in form of the concentrated forces, are already part of the external energy. The Dirichlet boundary conditions are applied to the prediction $z(x)$ produced by the neural network, i.e., via strong enforcement. This is achieved using a multiplication with a function $a(x)$ and a summation with a function $b(x)$

$$u(x) = a(x)z(x) + b(x). \tag{5.8}$$

The functions $a(x)$ and $b(x)$ are chosen such that the Dirichlet boundary conditions are fulfilled. For homogeneous boundary conditions, $b(x) = 0$, while $a(x)$ is chosen such that $a(x)$ is zero at the boundaries and non-zero inside the domain. For a clamped bar on both sides and a length of L, the Dirichlet boundary conditions are given as

$$u(0) = 0, \tag{5.9}$$

$$u(L) = 0. \tag{5.10}$$

A possible choice for the strong application of the boundary conditions is $a(x) = (x - L)x$ and $b(x) = 0$. The choice is, however, not unique.

To further illustrate the deep energy method, a specific example is now presented. The same manufactured solution as for the physics-informed static bar example in Sect. 4.2.1 is chosen as

$$u(x) = \sin(2\pi x), \tag{5.11}$$

for $x \in \Omega = [0, 1]$ and $L = 1$. The corresponding distributed load is found by insertion into the differential equation

$$p(x) = 4\pi^2 \sin(2\pi x) \tag{5.12}$$

with the cross-sectional properties being $EA = 1$. Using the boundary conditions Eqs. (5.9) and (5.10) lead to a displacement that can be expressed via the prediction $z(x)$ of the neural network.

$$u(x) = (x - L)xz(x). \tag{5.13}$$

The deep energy method uses the same `buildModel` and `getDerivative` functions as described in Sect. 4.2.1 to create the neural network and calculate the derivatives. The displacements are found with the modified neural network output.

```
def getDisplacement(x):
    z = model(x)
    u = (x - 1) * x * z
    return u
```

Furthermore, a numerical integration scheme has to be implemented. Here, the trapezoidal rule is chosen, where it is assumed that the points are distributed equidistantly.

```
def trapezoidalIntegration(y, x):
    dx = x[1] - x[0]
    result = torch.sum(y)
    result = result - (y[0] + y[-1]) / 2
    return result * dx
```

Finally, the internal energy Π_i and external energy Π_e can be computed with the following two functions.

```
def getInternalEnergy(u, x, EA):
    strain = getDerivative(u, x)
    internalEnergyDensity = 0.5 * EA(x) * strain ** 2
    internalEnergy = trapezoidalIntegration(internalEnergyDensity, x)
    return internalEnergy

def getExternalEnergy(u, x, p):
    externalEnergyDensity = -p(x) * u
    externalEnergy = trapezoidalIntegration(externalEnergyDensity, x)
    return externalEnergy
```

Everything can then be combined to calculate the potential energy Π_{tot} for the presented example.

```
model = buildModel(1, 10, 1)
x = torch.linspace(0, 1, 10, requires_grad=True).view(-1, 1)
EA = lambda x: 1 + 0 * x
p = lambda x: 4 * torch.pi ** 2 * torch.sin(2 * torch.pi * x)

u = getDisplacement(x)
internalEnergy = getInternalEnergy(u, x, EA)
externalEnergy = getExternalEnergy(u, x, p)
potentialEnergy = internalEnergy + externalEnergy
```

Now it is possible to minimize the cost function C, which is the potential energy Π_{tot}, via a stochastic gradient descent or more sophisticated schemes like the L-BFGS optimizer. After the minimization it is possible to make an accurate prediction of the displacement $u(x)$. Algorithm 11 summarizes how the deep energy method applied to a static bar works. When comparing it to Algorithm 9 for the physics-informed neural network, the only difference is the cost function C.

Algorithm 11 Training the deep energy method for the static solution of the problem described in Equation (4.4)

Require: training data for boundary condition $\{\tilde{x}^i_{\mathcal{B}}, \tilde{u}^i_{\mathcal{B}}\}_{i=1}^{m_{\mathcal{B}}}$
generate $m_{\mathcal{R}(\mathcal{R})}$ integration points with a uniform distribution $\{x^i_{\mathcal{R}}\}_{i=1}^{m_{\mathcal{R}}}$
define network architecture (input, output, hidden layers, hidden neurons)
initialize network parameters $\boldsymbol{\Theta}$: weights $\{\boldsymbol{W}^l\}_{l=1}^{L}$ and biases $\{\boldsymbol{b}^l\}_{l=1}^{L}$ for all layers L
set hyperparameters for optimizer (epochs, learning rate $\alpha, \dots$)

for all epochs **do**
  $\frac{\partial \hat{\boldsymbol{u}}_b}{\partial x} \leftarrow \frac{\partial u_{NN}}{\partial x}(\tilde{\boldsymbol{x}}_{\mathcal{B}}; \boldsymbol{\Theta})$
  $\hat{\boldsymbol{u}} \leftarrow u_{NN}(\boldsymbol{x}; \boldsymbol{\Theta})$
  compute Π_{i}, Π_{e} using numerical integration ▷ cf. Equations (5.5) and (5.6)
  compute cost function: $C \leftarrow \Pi_{\text{tot}} = \Pi_{\text{i}} + \Pi_{\text{e}}$
  update parameters: $\boldsymbol{\Theta} \leftarrow \boldsymbol{\Theta} - \alpha \frac{\partial C}{\partial \boldsymbol{\Theta}}$ ▷ Adam or L-BFGS
end for

The computed displacements and the analytical solution for the discussed example are compared in Fig. 5.3, where, again, L-BFGS was used during training. One observes that the estimation of the solution $u(x)$ is very accurate, as no noticeable discrepancies are observable. Additionally, Fig. 5.3 shows the training history. Here, one can see the different energy terms over the number of training epochs. Conver-

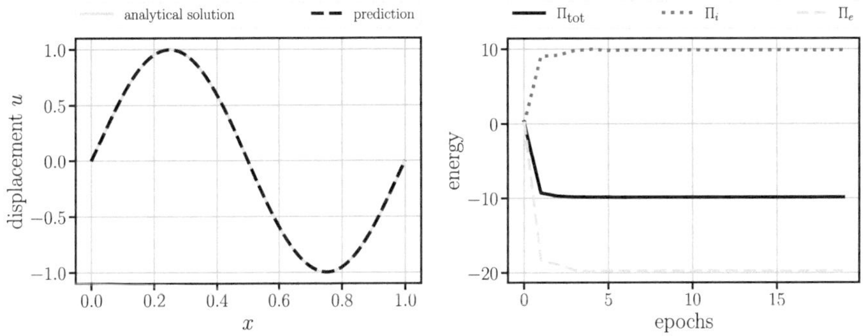

(a) Displacements $u(x)$ (b) Training history

Fig. 5.3 Linear elastic static bar example. Left: the displacements $u(x)$ computed with the deep energy method and the analytic solution are compared. Right: the training history is illustrated by showing the cost function, i.e., the potential energy for each epoch

gence is reached within 40 epochs, which is faster than what was achieved with the physics-informed neural network taking about 100 epochs, as seen in Fig. 4.4b.

Interestingly, the found potential energy is negative. This is due to the way the external energy was defined. Clapeyron's theorem for linear elasticity [9] states that the potential energy is equal to half of the external work. This is similarly reflected in Fig. 5.4.

The overfitting issue with the deep energy method can be illustrated with the same example by increasing the number of training epochs without introducing any regularization to the neural network. This is shown in Fig. 5.4. It is easy to spot the point at which overfitting occurs. The energy values start diverging from there on, where the external and potential energy go toward $-\infty$. The overfitting can also be seen by looking at the bar's displacements, which now inhibit large errors.

As stated previously, overfitting can be avoided by introducing early stopping. For this, a second (finer) integration grid is employed as a validation grid. By evaluating the energy contributions Eqs. (5.5) and (5.6) on a training grid and validation grid, a training and validation error are established. Only the integrals evaluated on the training grid are used for the optimization. These separate grids enable us to determine the point of overfitting. Figure 5.5 shows the evolution of the energy contributions on the training grid and a validation grid, which has twice as many integration points as the training grid. Although the potential energy decreases on the training grid, the potential energy on the validation grid increases, enabling us to choose the best possible solution at the minimum of the potential energy obtained on the validation grid.

Interestingly, the difference only occurs between the internal energies Π_i and not the external energies Π_e. The integral of the internal energy from Eq. (5.5) relies on the first derivative of the displacement du/dx, while the integral of the external energy from Eq. (5.6) is computed with the displacement u directly. As only the internal energies Π_i differ on the two grids, the error must originate from the derived

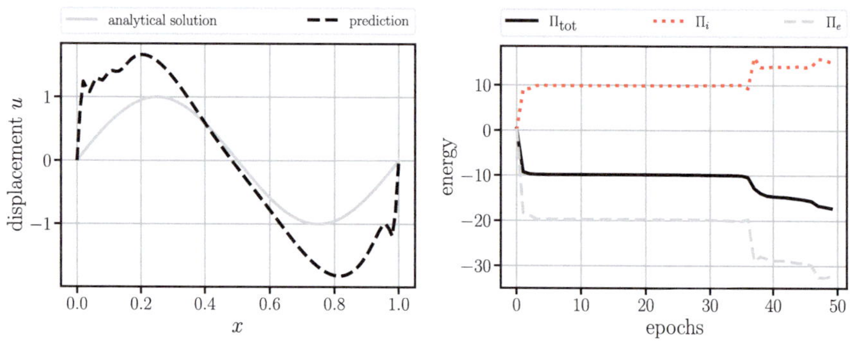

(a) Displacements $u(x)$ (b) Training history

Fig. 5.4 Linear elastic static bar example. Left: the displacements $u(x)$ computed with the deep energy method and the analytic solution are compared. Right: the training history is illustrated by showing the cost function, i.e., the potential energy for each epoch. Overfitting occurs after 22 epochs

Fig. 5.5 Training history from Fig. 5.4 showing evolution of energy contributions on both a training and validation grid

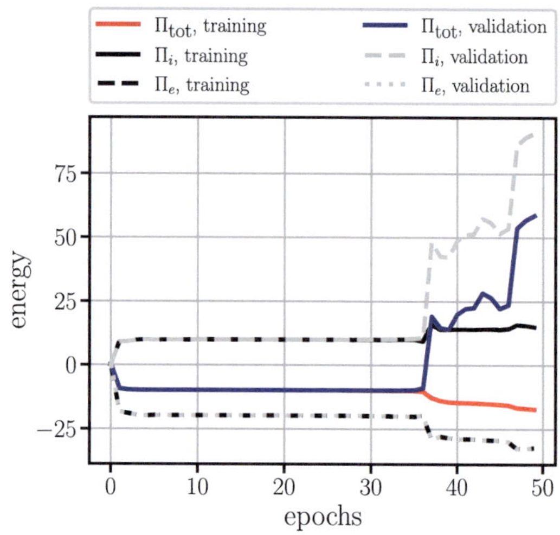

displacement du/dx, i.e., the strain. From the training history in Fig. 5.5, it can be concluded that a case exists in which the displacement's magnitude increases (decreasing the external energy Π_e) without increasing the strain. Such a situation can be constructed. Consider the toy example illustrated in Fig. 5.6. The training grid points (red) are located on sections with zero slope, i.e., zero strain. Thus, the internal energy Π_i is zero. Simultaneously, an infinite maximum displacement can be reached by increasing u_{max} to an arbitrary degree. The external energy Π_e would be $-\infty$. The energy values are thus zero and $-\infty$ on the training grid. By placing additional validation grid points (blue), which experience large slopes, i.e., strains, this erroneous case is easy to detect. This is also what was achieved in Fig. 5.5 by considering a validation grid. Thus, the overfitted displacements from Fig. 5.4a

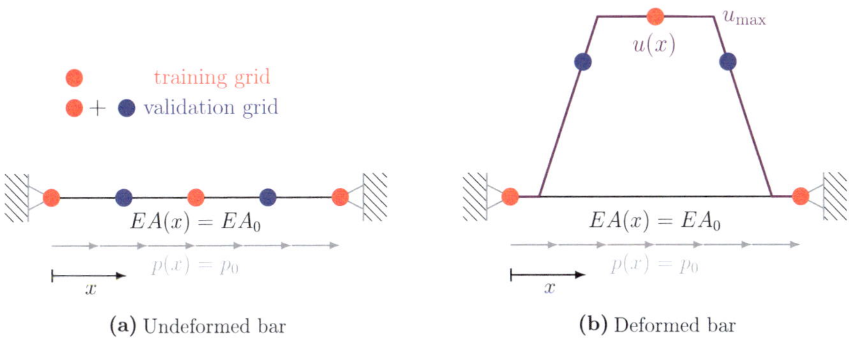

Fig. 5.6 Toy example explaining overfitting phenomenon in the deep energy method

occur due to spurious (unseen) oscillations (similar to those resulting from Gibbs phenomena) between the grid points, falsifying the derivatives from the automatic differentiation.

A similar explanation was provided in [10], where remedies besides early stopping are suggested. Probabilistic integration methods, such as Monte Carlo integration [11, 30], are one option. Another possibility is to adaptively refine the training grid based on the validation error. Yet another way is to use numerical differentiation instead of automatic differentiation, thus avoiding these spurious effects entirely.

5.1.3 Two-dimensional Static Model

The plate in membrane action from Sect. 4.2.2 is also revisited. Remember that the physics-informed neural network struggled with case 2 from Fig. 4.5 with the stress singularity at the right bottom corner. Case 2 will, therefore, be considered with the deep energy method. The first step is to formulate the potential energy Π_{tot} to be used as cost function C. The internal energy Π_{i} is defined as

$$\Pi_{\text{i}} = \frac{1}{2}\int_{\Omega} \varepsilon \cdot \boldsymbol{\sigma} d\Omega, \tag{5.14}$$

where the stresses $\boldsymbol{\sigma}$ and strains ε are computed with Eqs. (4.15) and (4.17) from the displacement prediction $\hat{\boldsymbol{u}} = u_{NN}(x_1, x_2; \boldsymbol{\Theta})$. The external energy Π_{e} is composed of the contribution of the distributed load $\boldsymbol{p}$ and inhomogeneous Neumann boundary conditions, defined in terms of traction $\boldsymbol{t}$.

$$\Pi_{\text{e}} = -\int_{\Omega} \boldsymbol{p} \cdot \boldsymbol{u} d\Omega - \int_{\Gamma_N} \boldsymbol{t} \cdot \boldsymbol{u} d\Gamma_N \tag{5.15}$$

The boundary conditions of case 2 are summarized in Fig. 4.5. The Neumann boundary conditions are incorporated via the external energy Π_e from Eq. (5.15). The Dirichlet boundary conditions will be handled in the same way as in Sect. 4.2.2, namely with a strong enforcement. Reusing the neural network output modification from Eq. (4.22)

$$\boldsymbol{u} = \begin{pmatrix} x_1 x_2 z_1 \\ x_1 x_2 z_2 \end{pmatrix}, \tag{5.16}$$

guarantees that the bottom edge and left edge are fixed in correspondingly the x_2 and x_1 direction. Thus, the cost function to be minimized can be simply written as

$$C = \Pi_{\text{tot}} = \Pi_\text{i} + \Pi_\text{e}. \tag{5.17}$$

An implementation is provided in the context of Exercise E.19.

The resulting displacement field after 400 epochs with Adam is depicted in Fig. 5.7. A decent solution is obtained within 100 epochs, as seen by the training history in Fig. 5.7a, which is considerably faster than the 4000 epochs it took for the physics-informed neural network to converge (cf. Fig. 4.9a). In addition, each epoch is cheaper due to the lowered order of differentiation. The deep energy computation time per epoch on 20×20 integration points is $\sim 4.0 \cdot 10^{-3}$ s, while for the physics-informed neural network, the computation time is $\sim 2.3 \cdot 10^{-2}$ s per epoch on the same grid (employed as collocation points). However, compared to a finite element simulation with 15×15 elements with integrated Legendre polynomials of up to degree $p = 2$ corresponding to 1922 degrees of freedom, which takes $\sim 9.1 \cdot 10^{-2}$ s, the deep energy method would have to reach a better prediction than the finite element result after 21 epochs. This is not the case. The strain fields obtained by deriving the displacement predictions with Eq. (4.17) are depicted in Fig. 5.8. Although the deep energy method comes much closer to the finite element results than the physics-informed neural network, the results are still distinguishable

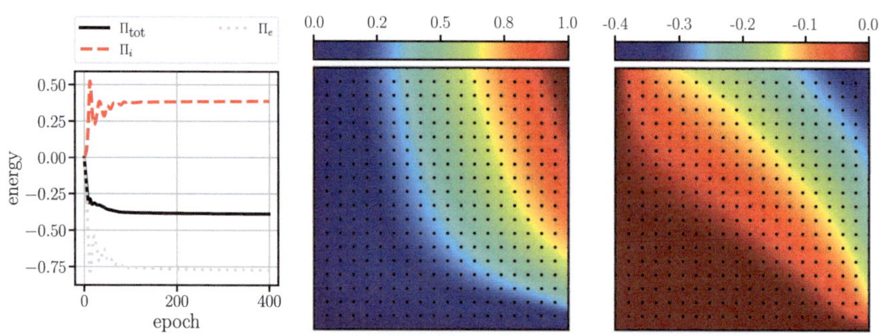

(a) Training history (b) u_1 (c) u_2

Fig. 5.7 Displacement prediction of case 2 from Fig. 4.5 using the deep energy method. The 20×20 integration points are indicated by the black circles

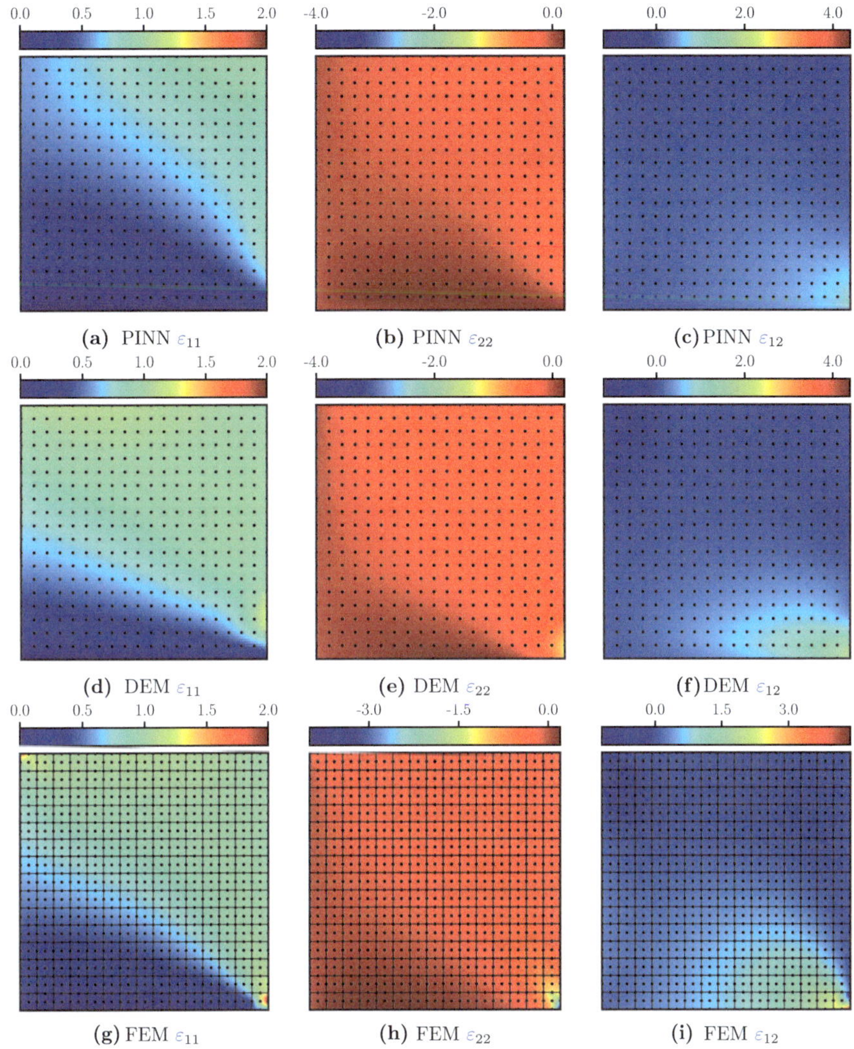

Fig. 5.8 Comparison of strains obtained from a physics-informed neural network (PINN), the deep energy method (DEM), and the finite element method (FEM)

by visual inspection and insufficient for engineering applications. However, take note of how the stress singularity is captured much better with the deep energy method than with the physics-informed neural network.

5.1.4 Variational Physics-Informed Neural Networks

Unfortunately, the deep energy method can by construction only be applied to conservative systems, i.e., systems that can be defined by solely using a potential. Variational physics-informed neural networks [12, 13] provide a more generalized alternative, which relies on variational formulations to lower the order of differentiation. First, the strong form of the partial differential equation is multiplied with test functions $v_i(x)$ and integrated over the domain Ω. Next, the derivative order is lowered through integration by parts.

To demonstrate the procedure, consider the strong form of the bar equation defined in Eq. (4.4). Multiplication with test functions $v_i(x)$ and integration over the domain Ω, yields

$$\int_\Omega \frac{d}{dx}\left(EA\frac{du}{dx}\right) v_i d\Omega + \int_\Omega p v_i d\Omega = 0, \forall v_i, \tag{5.18}$$

which, with integration by parts, can be rewritten as

$$-\int_\Omega EA\frac{du}{dx}\frac{dv_i}{dx} d\Omega + \int_\Gamma EA\frac{du}{dx} v_i d\Gamma + \int_\Omega p v_i d\Omega = 0, \forall v_i. \tag{5.19}$$

Note that the integral over the boundary Γ evaluates to zero for homogeneous Neumann boundary conditions. At the same time, the test functions v_i are selected such that they vanish at Dirichlet boundaries, which are enforced either via a penalty term (Eq. (4.12)) or through a strong enforcement (Eq. 5.8). Thus, the boundary integral is only non-zero for inhomogeneous Neumann boundary conditions and can be rewritten with the Neumann boundary conditions, i.e., forces F, from Eq. (4.5)

$$\int_\Gamma EA\frac{du}{dx} v_i d\Gamma = \int_{\Gamma_N} F v_i d\Gamma_N, \tag{5.20}$$

which for the one-dimensional bar can be reduced to a summation over the boundaries with applied forces F.

The variational formulation enables us to construct a variational loss $\mathcal{L}_\mathcal{V}$ from the prediction $\hat{u} = u_{NN}(x; \boldsymbol{\Theta})$. For a single test function v_i, the loss $\mathcal{L}_{\mathcal{V}_i}$ can be formulated as

$$\mathcal{L}_{\mathcal{V}_i} = \left(-\int_\Omega EA\frac{d\hat{u}}{dx}\frac{dv_i}{dx} d\Omega + \int_{\Gamma_N} F v_i d\Gamma_N + \int_\Omega p v_i d\Omega\right)^2. \tag{5.21}$$

As Eq. (5.19) has to be zero for each test function v_i, the expression is squared in Eq. (5.21). The contributions are summed up over the number of test functions $m_\mathcal{V}$, yielding the variational loss

$$\mathcal{L}_\mathcal{V} = \frac{1}{m_\mathcal{V}} \sum_{i=1}^{m_\mathcal{V}} \mathcal{L}_{\mathcal{V}_i}. \tag{5.22}$$

As multiple test functions are required in order to converge to a proper solution, variational physics-informed neural networks are more expensive than the deep energy method, where the solution u essentially acts as a single test function (cf. Eqs. (5.5) and (5.6)).

Dirichlet boundary conditions can be incorporated either via strong enforcement or an additional loss term $\mathcal{L}_{\mathcal{B}}$. Here, we use strong enforcement, leading to the cost function

$$C = \mathcal{L}_{\mathcal{V}}. \tag{5.23}$$

The strong enforcement of the Dirichlet boundary conditions can directly be reused to modify the test functions. Consider homogeneous Dirichlet boundary conditions at both ends, i.e., $x = 0, x = L$, as defined in Eqs. (5.9) and (5.10). Any valid test function can be constructed by multiplication with $(L - x)x$. Note that the test functions also have to be zero at inhomogeneous Dirichlet boundaries. Alternatively, the test functions can directly be constructed, so that Dirichlet boundaries are zero. An option would be

$$v_i(x) = \sin\left(\frac{i\pi}{L}x\right). \tag{5.24}$$

With the cost and test functions established, the cost function can be implemented with the following code snippet.

```
def getCostFunctionVPINN(x, u, v, EA, distLoad, vB, F):
    # derivative computation of all test functions
    dvdx = torch.zeros(v.shape)
    for i in range(v.shape[1]):
        dvdx[:, i:(i + 1)] = getDerivative(v[:, i:(i + 1)], x, 1)

    integrand = EA(x) * getDerivative(u, x, 1) * dvdx
    lossPDE = - trapezoidalIntegration(integrand, x)

    integrand = distLoad(x) * v
    lossPDE += trapezoidalIntegration(integrand, x)

    for i in range(len(uB)):  # loop over Neumann boundary conditions
        lossPDE += F[i][0] * vB[i]

    lossPDE = torch.sum(lossPDE ** 2)
    return lossPDE
```

The additional inputs are the test functions `v`, the test function evaluation at the boundaries `vB`, and the Neumann boundary conditions `F`. The complete code can be found in Exercise E.18.

5.1.5 Weak Adversarial Networks

Variational physics-informed neural networks introduce yet another user-defined quantity, namely the test functions. Their performance is highly dependent on the choice of test functions. Weak adversarial networks [1] replace the user-defined test functions with neural networks, hence making the test functions learnable. Thus the variational loss from Eq. (5.21) is simply enhanced with the learnable test functions $\hat{\boldsymbol{v}} = v_{NN}(x; \boldsymbol{\Theta}^v)$:

$$\mathcal{L}_{\mathcal{V}_i} = \left(-\int_\Omega EA \frac{d\hat{u}}{dx} \frac{d\hat{v}_i}{dx} d\Omega + \int_{\Gamma_N} F \hat{v}_i d\Gamma_N + \int_\Omega p \hat{v}_i d\Omega \right)^2. \tag{5.25}$$

Again, the test functions have to be zero at all Dirichlet boundary conditions. This is best achieved via strong enforcement applied on the neural network output. The test functions are learned within a minimax optimization

$$\min_{\boldsymbol{\Theta}^u} \max_{\boldsymbol{\Theta}^v} C, \tag{5.26}$$

in which the solution prediction $\hat{u} = u_{NN}(x; \boldsymbol{\Theta}^u)$ is continuously challenged.

This minimax optimization is easily incorporated in the prior variational physics-informed neural network implementation. By defining two neural networks, one for the displacement u and one for the test functions v,

```
modelu = buildModel(1, 10, 1)
modelv = buildModel(1, 10, numberOfTestFunctions)
```

two predictions can be made, namely $\hat{u}$ and $\hat{\boldsymbol{v}}$.

```
uPred = modelu(x)
vPred = modelv(x)
```

Both models are incorporated in the optimizer construction, such that their parameters are adapted during training.

```
optimizer = torch.optim.Adam((list(modelu.parameters())
                             + list(modelv.parameters())), lr)
```

To ensure that the parameters of the test function neural network v_{NN} are adjusted such that the cost function C is maximized, the sign of the gradients with respect to $\boldsymbol{\Theta}^v$ is inverted after the backpropagation `cost.backward()` (but before the optimization step `optimizer.step()`).

```
for i in range(len(list(modelv.parameters()))):
    list(modelv.parameters())[i].grad = \
  - list(modelv.parameters())[i].grad
```

This extension is placed within the training loop over all epochs. Again, the full code is presented in Exercise E.18.

5.1.6 *Test Functions in Variational Physics-Informed Neural Networks*

Finally, a last remark has to be made on the connection between physics-informed neural networks, variational physics-informed neural networks, and weak adversarial networks. Essentially, the only distinction between them is the test function. As discussed, weak adversarial networks are simply variational physics-informed neural networks with neural networks as test functions. Also, physics-informed neural networks can be derived from variational physics-informed neural networks with Dirac delta functions $v_i(x) = \delta(x - x^i)$ as test functions, where x^i is the considered integration point. Starting from Eq. (5.21) and inserting the Dirac delta function as test function results in

$$\mathcal{L}_{\mathcal{V}_i} = \left(- \int_\Omega EA \frac{d\hat{u}}{dx} \frac{d\delta(x - x^i)}{dx} d\Omega + \int_{\Gamma_N} F\delta(x - x^i) d\Gamma_N + \int_\Omega p\delta(x - x^i) d\Omega \right)^2 . \quad (5.27)$$

We start with the last integral, as it is the easiest. Using the sifting (or sampling) property of the Dirac delta function, the integral can be evaluated as

$$\int_\Omega p(x)\delta(x - x^i) d\Omega = p(x^i). \quad (5.28)$$

The same principle applies to the second integral, which reduces to zero at all points but the inhomogeneous Neumann boundary.

$$\int_{\Gamma_N} F(x)\delta(x - x^i) d\Gamma_N = \begin{cases} F(x^i) & \text{if } x^i \text{ on } \Gamma_N \\ 0 & \text{else} \end{cases} \quad (5.29)$$

The last integral can be rewritten with the generalized derivative of the Dirac delta (assuming that the integration point x^i lies within the domain Ω), followed again by exploiting the sifting property.

$$\begin{aligned} - \int_\Omega E(x)A(x) \frac{d\hat{u}(x)}{dx} \frac{d\delta(x - x^i)}{dx} d\Omega &= \int_\Omega \frac{d}{dx} \left(E(x)A(x) \frac{d\hat{u}(x)}{dx} \right) \delta(x - x^i) d\Omega \\ &= \frac{d}{dx} \left(E(x^i)A(x^i) \frac{d\hat{u}(x^i)}{dx} \right) \end{aligned} \quad (5.30)$$

Combining the three integral evaluations as in Equation (5.27) at integration points not located on inhomogeneous Neumann boundary conditions yields

$$\mathcal{L}_{\mathcal{V}_i} = \left(\frac{d}{dx} \left(E(x^i)A(x^i) \frac{d\hat{u}(x^i)}{dx} \right) + p(x^i) \right)^2 . \quad (5.31)$$

After summation over all test functions, i.e., integration points x^i, as defined in Equation (5.22), the expression

$$\mathcal{L}_{\mathcal{V}} = \frac{1}{m_{\mathcal{V}}} \sum_{i=1}^{m_{\mathcal{V}}} \left(\frac{d}{dx} \left(E(x^i) A(x^i) \frac{d\hat{u}(x^i)}{dx} \right) + p(x^i) \right)^2 = \frac{1}{m_{\mathcal{V}}} \sum_{i=1}^{m_{\mathcal{V}}} \left(\hat{r}(x^i) \right)^2 \tag{5.32}$$

becomes indistinguishable from the residual loss $\mathcal{L}_{\mathcal{R}}$ introduced in Eq. (4.13).

5.2 Extensions

The following provides a brief overview of popular extensions to physics-informed neural networks, which may improve accuracy or reduce the computational effort. The focus lies mostly on the mathematical and implementational aspects. Unless stated otherwise, the implementations can be further explored in Exercise E.20. Only the section on the ansatz space (Sect. 5.2.5) delves into results, which highlights the connection between the finite element method and physics-informed neural networks.

5.2.1 Optimizer

The optimizer plays a crucial role during the training process of any neural network. In deep learning, Adam (cf. Sect. 2.8.2) is the default choice. However, as seen in Sects. 4.2.1 and 5.1.2, the L-BFGS (cf. Sect. 2.8.3) can reduce the number of required epochs for physics-informed neural networks. A common choice is to start the optimization procedure with Adam in order to approach an optimum and follow it with L-BFGS to quickly converge towards the optimum, as was done in Sect. 4.2.3. Note, however, that the L-BFGS is a second-order optimization algorithm in which each epoch becomes more expensive. In PyTorch, the optimizer is simply selected with

```
optimizer = torch.optim.Adam(model.parameters(), lr)
```

for Adam, or

```
optimizer = torch.optim.LBFGS(model.parameters(), lr)
```

for the L-BFGS. Many other optimizers exist in PyTorch, such as the in Sect. 2.8 discussed stochastic gradient descent `torch.optim.SGD`, Adagrad `torch.optim.Adagrad`, or RMSprop `torch.optim.RMSprop`. For first-order optimizers, the optimizer can simply be called after the backward propagation call `cost.backward()`:

```
optimizer.step()
```

The L-BFGS implementation, however, requires a closure function, which computes the cost function and performs the backward propagation. This is necessary, as the cost function and gradient computation have to be performed multiple times within a single epoch. A simple supervised learning example is provided below:

```
def closure():
    optimizer.zero_grad()  # reset gradients
    yPred = model(xLabel)
    cost = getCostFunction(yPred, yLabel)
    cost.backward()
    return cost
```

which is called with the same optimization step as before.

```
optimizer.step(closure)
```

Another popular technique to improve the optimization performance is to adjust the learning rate, which is achieved with learning rate schedulers. Often, the learning rate is decreased over time [14] in, e.g., a step-like manner or with an exponential decay. The intention is to allow for more fine-grained adjustments as the minimum is approached, thus avoiding unnecessary oscillations (cf. Fig. 2.14b). Simultaneously, this allows for a larger initial learning rate, which speeds up convergence (cf. Fig. 2.14c). However, also increases in learning rate are employed. Cosine annealing [15] adjusts the learning rate based on the cosine curve in the hope of escaping local minima. In PyTorch, the learning rate scheduler `scheduler` is called after the optimizer update.

```
optimizer.step()
scheduler.step()
```

Many different types are provided in PyTorch. Polynomial decay has, for example, been shown to be beneficial for physics-informed neural networks. The corresponding update rule is defined as

$$\alpha = (b \cdot \text{epoch} + 1)^a, \tag{5.33}$$

where a and b are hyperparameters. This can be defined with a custom learning rate scheduler relying on lambda functions.

```
lr_lambda = lambda epoch: (b * epoch + 1) ** a
scheduler = torch.optim.lr_scheduler.LambdaLR(optimizer, lr_lambda)
```

A good default choice for the two hyperparameters is $a = -0.5$, $b = 0.2$.

5.2.2 Sampling

Another important aspect is the distribution of the collocation points. The simplest choice are equidistantly spaced points, which, however, can lead to suboptimal learning.

```
x = np.linspace(0, L, N)
```

A better choice are randomly distributed points from a probability distribution. These do, however, not necessarily fill out the entire domain of interest, as seen in Fig. 5.9 with the uniform distribution.

```
x = np.random.uniform(0, L, N)
```

Latin hypercube sampling resolves this issue by ensuring that samples are more evenly distributed across the entire range of each dimension.

```
sampler = scipy.stats.qmc.LatinHypercube(d=1)  # dimensionality 1
x = sampler.random(N - 2) * L
```

Another deterministic option is the use of integration points, such as the Gauss-Legendre points.

```
x = (0.5 * np.polynomial.legendre.leggauss(N - 2)[0] + 0.5) * L
```

All four options are compared in Fig. 5.9.

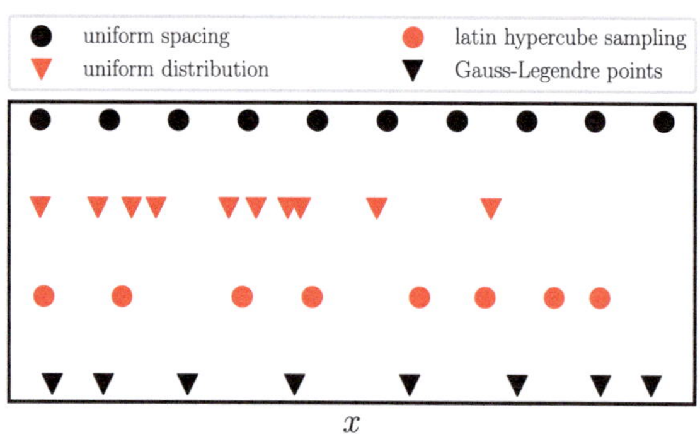

Fig. 5.9 Four sampling strategies, where the two black ones are deterministic and the two red ones are probabilistic

5.2.3 Boundary and Initial Conditions

The imposition of boundary (and initial[5]) conditions is challenging for physics-informed neural networks. The straightforward approach of enforcing boundary conditions via a penalty term (weak enforcement) $\mathcal{L}_{\mathcal{B}}$, as in Eq. (4.12) leads to an imbalance in the cost function C. Imagine, for example, a neural network first learning the physics of the problem by minimizing the differential equation residual $\mathcal{L}_{\mathcal{R}}$, which can be achieved non-uniquely without fulfilling the boundary conditions. To improve the prediction, the network now has to minimize the boundary loss $\mathcal{L}_{\mathcal{B}}$. This might not necessarily be possible without temporarily violating the differential equation, which would lead to a larger residual $\mathcal{L}_{\mathcal{R}}$ and cost function C. In such a situation, the network is stuck in a local optimum. Although only a thought experiment, it illustrates how an unfavourable competition between the loss terms can and may lead to unsatisfactory optimization results.

The best remedy is to employ strong enforcement for the boundary conditions, such that only one loss term remains, i.e., only $\mathcal{L}_{\mathcal{R}}$, thus yielding an unconstrained minimization problem. For Dirichlet boundary conditions, this is achieved with

$$u(x) = a(x)z(x) + b(x). \tag{5.34}$$

Here, $a(x)$ is zero at the boundaries and non-zero within the domain. At the same time $b(x)$ matches the Dirichlet boundary conditions, such that $u(x)$ satisfies the boundary conditions independent of the neural network output $z(x)$. This procedure poses challenges in complex domains, as finding suitable functions $a(x)$, $b(x)$ becomes more difficult. Furthermore, it is only applicable to Dirichlet boundary conditions.

A more generic procedure relies on a weighting of the loss terms.

$$C = \kappa_{\mathcal{R}}\mathcal{L}_{\mathcal{R}} + \kappa_{\mathcal{B}}\mathcal{L}_{\mathcal{B}} \tag{5.35}$$

These weights $\boldsymbol{\kappa} = \{\kappa_{\mathcal{R}}, \kappa_{\mathcal{B}}\}$ can be considered hyperparameters and tuned manually. Alternatively, an automatic weighting procedure can be employed, which maximizes the cost function C with respect to the weights. The intention is to focus on the more challenging loss term. This is achieved with a minimax optimization[6]

$$\min_{\boldsymbol{\Theta}} \max_{\boldsymbol{\kappa}} C. \tag{5.36}$$

In PyTorch, the weights `weights` are simply added to the optimizer `optimizer`.

```
weights = torch.tensor([1., 1.], requires_grad=True)
optimizer.add_param_group({'params': weights})
```

[5] The following concepts apply to initial conditions as well, although they are not mentioned further to declutter the notation.

[6] In practice, the weighting factors $\boldsymbol{\kappa}$ are thresholded at an upper limit (or penalized) to prevent a growth towards infinity.

```
optimizer.param_groups[-1]['lr'] = lr  # different
    learning rate can be specified
```

The weights are thereby considered during the backpropagation. In the training loop, the sign of the gradients with respect to the weights is switched to ensure a maximization.

```
# within the training loop
cost = weights[0] * lossPDE + weights[1] * lossBC
cost.backward()
weights.grad *= -1
optimizer.step()
```

Taking this automatic weighting idea one step further, each individual collocation point can be considered a single loss term (or equality constraint). Inspired by the augmented Lagrangian method [31], a modified residual loss term can be formulated

$$\mathcal{L}_{\mathcal{R}} = \frac{1}{m_{\mathcal{R}}} \sum_{i=1}^{m_{\mathcal{R}}} \kappa_{\mathcal{R}_i} (\hat{r}(x_{\mathcal{R}}^i))^2 \tag{5.37}$$

with individual weights $\kappa_{\mathcal{R}_i}$ for each collocation point $x_{\mathcal{R}}^i$. Again, a minimax optimization is performed with respect to the weights $\boldsymbol{\kappa} = \{\kappa_{\mathcal{R}_i}\}_{i=1}^{m_{\mathcal{R}}}$. This introduces an adaptive attention on individual collocation points instead of only the two loss terms $\mathcal{L}_{\mathcal{R}}, \mathcal{L}_{\mathcal{B}}$. The disadvantage is that the weights are fixed to the collocation points, which prevents collocation points from being resampled. Competitive physics-informed neural networks resolve this issue by sampling the weights from a second neural network $\hat{\kappa}_i = K_{NN}(x_{\mathcal{R}}^i; \boldsymbol{\Theta}^\kappa)$. The parameters $\boldsymbol{\Theta}^\kappa$ are updated by maximizing the cost function C. This approach is closely related to the weak adversarial networks from Sect. 5.1.5 in which we considered the weak instead of the strong form of the differential equation.

5.2.4 *Differentiation*

Although automatic differentiation is the usual choice for physics-informed neural networks, it is accompanied by several disadvantages compared to numerical differentiation. As seen in connection with the deep energy method in Sect. 5.1.2, automatic differentiation is not always reliable, as spurious oscillations between points may arise, leading to incorrect derivatives. Furthermore, automatic differentiation requires an arithmetic connection between coordinates and network outputs (if they are to be differentiated with respect to the coordinates), which is not the case for convolutional neural networks. Another detriment is the computational overhead. Numerical differentiation offers a much faster (for low dimensions, as seen in Fig. 5.1a for one dimension) and more reliable (not susceptible to unexpected derivatives due to small spurious oscillations, as, e.g., encountered in Sect. 5.1.2) alternative.

The only requirement is a structured grid. A prominent choice is the finite difference method. The first-order stencils with three adjacent grid points x_{i-1}, x_i, x_{i+1} is given below.

$$\text{forward difference: } \frac{dy(x_i)}{dx} \approx \frac{y(x_{i+1}) - y(x_i)}{x_{i+1} - x_i} \tag{5.38}$$

$$\text{central difference: } \frac{dy(x_i)}{dx} \approx \frac{y(x_{i+1}) - y(x_{i-1})}{x_{i+1} - x_{i-1}} \tag{5.39}$$

$$\text{backward difference: } \frac{dy(x_i)}{dx} \approx \frac{y(x_i) - y(x_{i-1})}{x_i - xi - 1} \tag{5.40}$$

These stencils can be implemented in the same recursive manner as in the automatic differentiation wrapper from Listing 4.2, yielding a `getDerivative` function usable in physics-informed neural networks with structured grids.

```
def getDerivativeFiniteDifference(y, x, n):
    if n == 0:
        return y
    else:
        dydx = y * 0
        # central difference
        dydx[1:-1] = (y[:-2] - y[2:]) / (x[:-2] - x[2:])
        # forward difference
        dydx[0] = (y[1] - y[0]) / (x[1] - x[0])
        # backward difference
        dydx[-1] = (y[-1] - y[-2]) / (x[-1] - x[-2])
        return getDerivativeFiniteDifference(dydx, x, n - 1)
```

5.2.5 *Ansatz*

Finally, the ansatz space will be discussed. This involves the neural network architecture but also classical finite element spaces. In the latter case, the neural network space can be projected onto the finite element space (finite element interpolated neural networks), or alternatively, the neural network space is restricted such that it conforms to the finite element space (hierarchical deep learning neural networks).

5.2.5.1 Neural Networks

A simple extension to conventional fully connected neural networks are adaptive activation functions [16]. The idea is to introduce a learnable parameter within the activation itself. An example is the parametric ReLU (PReLU) in which the negative slope (see Fig. 3.11) becomes a trainable parameter. This extension enlarges the neural network space and can help with convergence, as explored in [17]. In PyTorch, the PReLU activation function can simply be selected in the following manner.

```
activationFunction = torch.nn.PReLU(init=0.2)
```

A more invasive modification, is the use of a different neural network architecture—tailored to the problem at hand. For temporal problems, this might be recurrent neural networks, while spatial domains are better suited to convolutional neural networks. We will focus on the latter. As discussed in Sect. 3.9.1, convolutional neural networks are helpful when working with images. Physical domains and solution fields can often be viewed as images; see, e.g., the displacement and strain fields of the plate in membrane action from Sect. 4.2.2. These fields can be learned with much fewer neural network parameters when using convolutional neural networks. The difference to fully connected neural networks is that the prediction is not performed from coordinates but from images. These images can contain relevant information, such as the geometry, but can in principle be selected arbitrarily. When only a single solution is learned, as is the case in physics-informed neural networks, the best performance is reached with random noise as input. The change in input also means that the neural network output is no longer arithmetically connected to the coordinates. Thus, numerical differentiation has to be employed to obtain the prediction derivatives, as explained in Sect. 5.2.4. This arithmetic disconnection similarly prevails for the temporal derivatives when recurrent neural networks are employed.

5.2.5.2 Finite Element Space

In the finite element method (see [8, 18, 19] for introductory textbooks), the domain of interest is partitioned into subdomains, called elements. Each element is interpolated by n_S shape functions $N_i^e(x)$ with coefficients u_i^e (degrees of freedom), which yields the element solution field on the element domain Ω_e[7]

$$u^e(x) = \sum_{i=1}^{n_S} N_i^e(x) u_i^e, \quad x \in \Omega_e. \tag{5.41}$$

Importantly, degrees of freedom associated with the boundaries of the element[8] are shared between elements neighboring at the corresponding boundary. Thus, a coupling of the global solution field $u(x)$ across the elements is established.

A possible choice are linear shape functions, which in one dimension are defined as

[7] The somewhat unconventional notation relying on global coordinates x instead of local coordinates ξ is motivated by the upcoming paragraphs, where global coordinates simplify the notation.

[8] A degree of freedom is associated with the boundary if its corresponding shape function can be non-zero at the boundary.

$$
\begin{aligned}
N_1^e &= \frac{x - x_2^e}{x_1^e - x_2^e}, \\
N_2^e &= \frac{x - x_1^e}{x_2^e - x_1^e}.
\end{aligned}
\tag{5.42}
$$

Here x_1^e, x_2^e are the coordinates of the element boundaries. Thus, the element solution field $u^e(x_1^e)$ takes the value u_1^e at x_1^e, and $u^e(x_2^e)$ takes the value u_2^e at x_2^e (cf. Eq. (5.41)). This interpolation relies on polynomials of order $p = 1$ with $n_S = 2$ shape functions. More expressive finite element spaces can be constructed with higher order polynomials [20].

The global solution field is retrieved by an assembly over all element solution fields. Practically, this can be achieved with a summation over all n_E elements

$$
u(x) = \sum_{e=1}^{n_E} \alpha^e(x) u^e(x) = \sum_{e=1}^{N_E} \alpha^e(x) \sum_{i=1}^{n_S} N_i^e(x) u_i^e \tag{5.43}
$$

using an element weighting function $\alpha^e(x)$

$$
\alpha^e(x) = \begin{cases} 1 & \text{if } x \in \Omega_e, \\ 0 & \text{else.} \end{cases} \tag{5.44}
$$

In order to declutter the notation of the upcoming paragraphs, a global shape function $N_i(x)$ is introduced through a composition of the element weighting function and element shape function.

$$
N_i(x) = \alpha^e(x) N_j^e(x) \tag{5.45}
$$

Each of the $n = n_S \cdot n_E$ global shape functions $N_i(x)$ is then associated to a degree of freedom u_i. Note that the degrees of freedom can be shared among multiple shape functions, i.e., some degrees of freedom are associated to multiple shape functions (of neighboring elements).

The constructed finite element space bears multiple advantages over the neural network space employed in physics-informed neural networks. Firstly, spatial derivatives of the solution field are obtained by deriving the shape functions directly

$$
\frac{du(x)}{dx} = \sum_{i=1}^{n} \frac{dN_i(x)}{dx} u_i, \tag{5.46}
$$

as the degrees of freedom u_i do not depend on x. This is cheaper than automatic differentiation but still exact in contrast to numerical differentiation. Secondly, the finite element space is linear with respect to the degrees of freedom u_i—unlike the neural network space with respect to its parameters $\boldsymbol{\Theta}$. This enables us to directly solve for the optimal degrees of freedom u_i when facing linear problems, such as in linear elasticity. However, also for non-linear problems, extremely efficient solution procedures have been established. The clear disadvantage of the finite element space, when compared to the neural network space, is that it is much smaller. However, despite the neural networks' superior expressivity, no efficient solution procedures—that match the finite element method's speed—have yet been established (see the comparisons in Sect. 5.2.5.5). To this end, researchers have attempted to combine the two methodologies in order to exploit the advantages of each.

5.2.5.3 Finite Element Interpolated Neural Networks

Finite element interpolated neural networks [21] employ a neural network to predict the degrees of freedom $\hat{u}_i = u_{NN}(x_i; \boldsymbol{\Theta})$ from their corresponding global coordinates x_i. With the predicted degrees of freedom, the global displacement field $\hat{u}(x)$ can be evaluated with Equation (5.43). By evaluating the internal energy

$$\Pi_{\mathrm{i}} = \frac{1}{2}\int_{\Omega} EA\left(\frac{du}{dx}\right)^2 d\Omega = \frac{1}{2}\int_{\Omega} EA\left(\sum_{i=1}^{n}\frac{dN_i}{dx}u_i\right)^2 d\Omega \tag{5.47}$$

and external energy (assuming homogeneous Neumann boundary conditions)

$$\Pi_{\mathrm{e}} = -\int_{\Omega} pu d\Omega = -\int_{\Omega} p\sum_{i=1}^{n} N_i(x)u_i d\Omega \tag{5.48}$$

the total potential energy $\Pi_{\mathrm{tot}} = \Pi_{\mathrm{i}} + \Pi_{\mathrm{e}}$ is established and minimized. Dirichlet boundary conditions are directly imposed on the finite element space by setting the corresponding degrees of freedom, i.e., not predicting them. This is equivalent to the strong enforcement.

Thus, finite element interpolated neural networks project/interpolate a neural network space onto a finite element space. This, however, reduces the learnable space while simultaneously also decreasing efficiency, as gradient descent optimization embedded within an automatic differentiation framework is required—instead of directly solving for the degrees of freedom. In other words, the method produces, at best, the finite element solution with a less efficient solution algorithm. The method is, however, by design able to learn mappings of multiple solutions in contrast to the subsequent method.

5.2.5.4 Hierarchical Deep Learning Neural Networks

Hierarchical deep learning neural networks (HiDeNNs) [22–26] also reduce the space to the finite element space (with enforced Dirichlet boundary conditions) and learn the solution through gradient descent optimization. The difference is, however, that custom networks are constructed such that they only span the finite element space (see

Sect. 10.3.1 in Chap. 10 for more details). However, according to the authors, these networks are inefficient. Instead, they recommend coding the shape functions (e.g., Eq. (5.42)) directly and embedding them in an automatic differentiation framework in order to obtain the displacement and energy derivatives. This yields the same results but with less computational overhead. Thus, despite the method's name, no neural networks (or deep learning) are employed.

Nevertheless, we can consider the idea of solving the finite element method with gradient descent optimization. However, one additional caveat has to be mentioned. As the required gradients are available analytically, the automatic differentiation can additionally be removed. The displacement gradients are obtainable from the shape function derivatives, as seen in Eq. (5.46). The internal energy from Eq. (5.47) is differentiated with respect to the degrees of freedom u_j using

$$\frac{\partial \Pi_{\mathrm{i}}}{\partial u_j} = \int_\Omega EA \sum_{\substack{\forall k \text{ whose} \\ N_k \text{are} \\ \text{associated} \\ \text{with } u_j}} \left(\left(\sum_{i=1}^{n} \frac{dN_i}{dx} u_i \right) \frac{dN_k}{dx} \right) d\Omega \tag{5.49}$$

$$= \int_\Omega EA \sum_{\substack{\forall k \text{ whose} \\ N_k \text{are} \\ \text{associated} \\ \text{with } u_j}} \left(\frac{dN_k^2}{dx} u_k \right) d\Omega. \tag{5.50}$$

Analogously, the derived external energy (cf. Equation (5.48)) can be obtained with (again assuming homogeneous Neumann boundary conditions)

$$\frac{\partial \Pi_{\mathrm{e}}}{\partial u_j} = \int_\Omega p \sum_{\substack{\forall k \text{ whose} \\ N_k \text{are} \\ \text{associated} \\ \text{with } u_j}} N_k d\Omega. \tag{5.51}$$

Thus, the total potential energy gradient can be computed as

$$\frac{\partial \Pi_{\mathrm{tot}}}{\partial u_j} = \frac{\partial \Pi_{\mathrm{i}}}{\partial u_j} + \frac{\partial \Pi_{\mathrm{e}}}{\partial u_j}. \tag{5.52}$$

Stripping the hierarchical deep learning neural network from automatic differentiation (simultaneously reducing the computational effort) leaves us with the finite element method solved with gradient descent optimization, which is clearly less efficient than solving the system directly. Again, at best, the solution of the finite element method can be obtained.

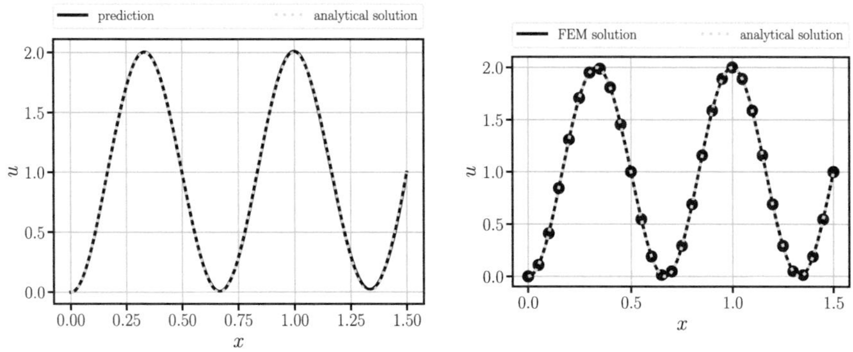

(a) Deep energy method (b) Finite element method, $p = 2$

Fig. 5.10 Comparison of deep energy method and solution obtained with the finite element method

5.2.5.5 Comparison to the Finite Element Method

To underline the postulations of the prior paragraphs, a numerical example is considered. Specifically, we consider the static bar equation from Eq. (4.4) with the manufactured solution

$$u(x) = 1 - \cos(3\pi x). \tag{5.53}$$

The Dirichlet boundary conditions $u(0) = 0$ and $u(L) = 1$, where the length of the bar is $L = 3/2$. For the cross-sectional properties of $EA(x) = x^2 + 1$, the distributed load $p(x)$ can be derived as

$$p(x) = -6x\pi \sin(3\pi x) - 9\pi^2(x^2 + 1)\cos(3\pi x) \tag{5.54}$$

This allows a gradual transition from the physics-informed neural network to the finite element method. The intermediate steps are finite element-interpolated neural networks, hierachic deep learning neural networks with and without automatic differentiation. As seen in Sect. 5.1.2, the deep energy method with strong enforcement is the most performant physics-informed neural network for the problem at hand. The predictions of the two extremes, i.e., the deep energy method and the finite element method, are depicted in Fig. 5.10. Both approaches capture the solution well. More interestingly is the computational effort behind these results. Table 5.2 provides an overview of the computational times[9] needed to reach an accuracy (measured in terms of the relative error in potential energy). A point-wise error is provided in Fig. 5.12.

[9] Runtimes were obtained with a laptop equipped with an 11th Gen Intel Core I7-1165G7 CPU and 32 GB of RAM. For the investigated case, a GPU does not speed up computations, as the problem is not large enough. For more complex problems (with more collocation or integration points), a GPU would lead to acceleration. This would, however, not improve the performance relative to the finite element method, as the number of required epochs would increase multiple orders of magnitude [27, 28], making the finite element method even more favorable.

Table 5.2 Comparison of computation time and solution quality for different neural network-based approaches (deep energy method, finite element interpolated neural networks and HiDeNN) and the finite element method

Method	Relative error in potential energy	Elapsed time per epoch	Number of epochs	Total elapsed time
Deep energy method	0.31%	$\sim 8.1 \cdot 10^{-4}$ s	$\sim$ 1000	$\sim$ 0.81 s
FE-interpolated NNs, $p = 1$	4.6%	$\sim 1.3 \cdot 10^{-2}$ s	$\sim$ 400	$\sim$ 5.3 s
HiDeNN, $p = 1$	3.9%	$\sim 1.3 \cdot 10^{-2}$ s	$\sim$ 100	$\sim$ 1.3 s
HiDeNN without AD, $p = 1$	3.9%	$\sim \mathbf{1.7 \cdot 10^{-3}}$ s	$\sim$ 100	$\sim$ 0.17 s
FEM, $p = 1$	**3.8**%	$\sim 4.6 \cdot 10^{-2}$ s	**1**	$\sim$ **0.046** s
FEM, $p = 2$	0.24%	$\sim 5.2 \cdot 10^{-2}$ s	1	$\sim$ 0.052 s

In Table 5.2, the best possible results with a finite element space with polynomials of degree $p = 1$ are indicated in bold. The lowest error is achieved by the finite element method, which is also the optimal result within the given ansatz space. The fastest time per epoch is reached by the hierarchical deep learning neural network without automatic differentiation, where automatic differentiation is responsible for approximately one order of magnitude. However, as the finite element method does not rely on iterations, it requires the least amount of total computation time—being one to two orders of magnitudes faster than the other approaches. In the bigger picture, the deep energy method with the neural network space yields a better result than the finite element space with $p = 1$. However, simply increasing the polynomial order to $p = 2$ makes the finite element method outperform the deep energy method in terms of both accuracy and computational effort. Figure 5.11 provides a similar picture as a graphical representation of the training history, in which the rate of improvement during training is visible. The implementations used to investigate the methods are studied further in Exercise E.21.

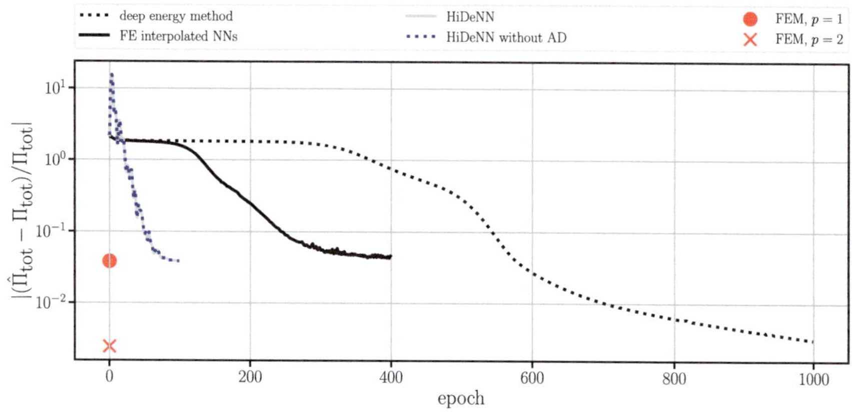

Fig. 5.11 Comparison of training histories for different neural network-based approaches (deep energy method, finite element interpolated neural networks and HiDeNN) and the finite element method

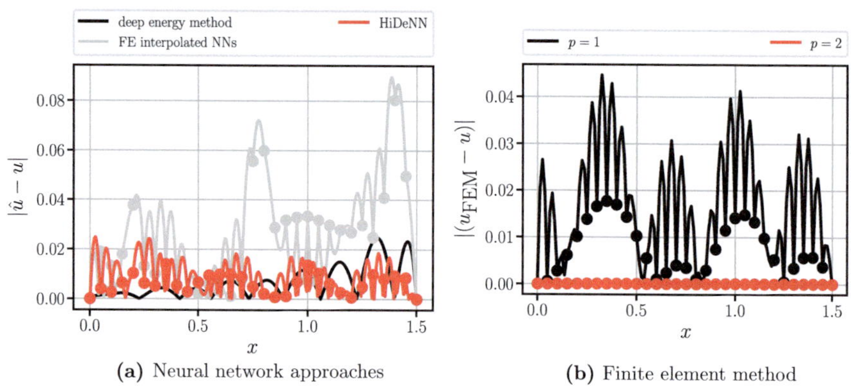

Fig. 5.12 Comparison of error in prediction of neural network-based methods (deep energy method, finite element interpolated neural networks and HiDeNN) with the finite element method

Exercises

E.18 Variations of Physics-Informed Neural Networks (C)

The physics-informed neural network variations discussed in Sect. 5.1 are to be compared. To this end, we first consider a bar in the domain $x \in [0, 1]$ with the cross-sectional properties $EA = 2$. The displacement and distributed load are given as

$$u(x) = 1 - \cos(3\pi x),$$
$$p(x) = -18\pi^2 \cos(3\pi x).$$

A homogeneous Dirichlet boundary condition at $x = 0$ and a homogeneous Neumann boundary condition at $x = 1$ are considered.

1. An implementation of the variations is provided in `exercise 18 .ipynb`. The different methods can be selected by assigning the cost function type with `getCostFunction`. These include physics-informed neural networks `getCostFunctionPINN`, the deep Galerkin method `getCostFunctionDGM`, the deep energy method `getCostFunctionDEM`, variational physics-informed neural networks `getCostFunctionVPINN`, and weak adversarial networks. To enable weak adversarial networks, use the variational physics-informed neural network cost function and select a neural network as test functions provided as `modelW`. Study the differences in implementation and compare the methods for the method at hand. After tuning each method, which method seems the most robust?
2. An additional problem is considered. A bar is given in the domain $x \in [0, 1]$ with the cross-sectional properties $EA = 1$ and the displacement
$$u(x) = -x^{0.65} + 0.65x.$$
Derive the distributed load $p(x)$ that has to be applied for the given displacement. Additionally, identify from the displacement $u(x)$ the Dirichlet boundary condition at $x = 0$ and the Neumann boundary condition at $x = 1$.
3. Exchange the new problem setup in `exercise 18 .ipynb` and compare the methods again. Which problem do you encounter?
Hint: Try changing the integration technique with `getIntegral`. Why does the integration technique alleviate the encountered issue?

E.19 Deep Energy Method for a Plate in Membrane Action (C)

Revisiting the physics-informed neural network implementation for a plate in membrane action from Exercise E.17, we consider case 1 in Fig. 4.5a. In summary, a plate of size 1×1 under plane stress with material properties $E = 1$, $\nu = 0.3$ and a traction of $t = 1$ on the right edge is studied. The boundary conditions are listed in Table 4.1. The code from Exercise E.17 is modified to accommodate the deep energy method in `exercise 19 .ipynb`. The modifications are indicated in red.

1. Implement the loss term computation in `exercise 19 .ipynb` as `getLossTerms (x1, x2, u, N1, N2, L1, L2, C, force, alpha)`, in which the internal and external energy are to be computed. The internal energy is obtained using the displacement prediction `u` at `x1, x2`, and the material stiffness matrix `C` in Voigt notation. The external energy relies on the applied traction `force` and the predicted displacement `u` (at the right edge). To perform the integration, the

domain size is provided as `L1`× `L2` and the number of collocation points `N1`× `N2`. Again, no body load acts on the plate. Ignore the last input `alpha` for now.
2. Solve the problem with `exercise 19 .ipynb` and compare the computational effort with the physics-informed neural network implementation from Exercise E.17.
3. We now want to consider more complex geometries. To this end, we introduce a spatially dependent indicator function $\alpha(x_1, x_2)$, which is zero if no material is present at x_1, x_2 and one if there is. This is similar to the finite cell method [20, 29], a fictitious domain method. Thus, a geometry can be defined through α. One constraint is imposed on α, namely that it is one at the boundary to simplify the boundary treatment. With this constraint in mind, the indicator function is incorporated in the internal energy as

$$\Pi_i = \frac{1}{2}\int_\Omega \alpha(x_1, x_2)\boldsymbol{\sigma} \cdot \boldsymbol{\varepsilon} d\Omega.$$

Modify your implementation of `getLossTerms(x1, x2, u, N1, N2, L1, L2, C, force, alpha)` to take the indicator function `alpha` into account.
4. Modify the indicator function in `getIndicator(x1, x2)` (currently one everywhere) to be zero in a centered circular hole. The radius is set to 0.2.
5. Tune the deep energy method to learn the correct displacement field of a plate with a hole. A reference solution obtained with the finite element method is provided in Fig. 5.13.
6. Consider the physics-informed neural network from Exercise E.17. Would it be sufficient to incorporate an indicator function in order to perform simulations on complex geometries? Can you think of a simpler alternative to the indicator function for physics-informed neural networks?

E.20 Extensions for Physics-Informed Neural Networks (C)

A physics-informed neural network for the bar equation is considered. Implementations of the extensions discussed in Sect. 5.2 are provided in `exercise 20 .ipynb`. These include the optimizer, sampling, loss term weighting (manual and automatic), strong enforcement of boundary conditions, weighting per collocation point, learning rate scheduling, activation functions, numerical differentiation, convolutional neural networks, and even a feature encoding layer. This exercise's intended purpose is to provide a physics-informed neural network with the most important extensions, which can be used for your own projects. However, to familiarize yourself, first take a closer look at each implementation, and next try activating and deactivating each of the different extensions. Which approaches have the greatest impact on the learning outcome and computational effort?

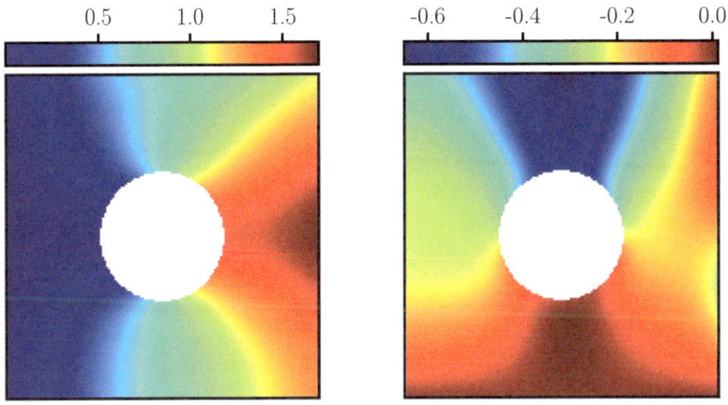

(a) u_1 (b) u_2

Fig. 5.13 Displacement prediction of a plate with a hole under uni-axial tension solved with the finite element method

E.21 Variations of the Ansatz Space (C)

The implementations of the different approaches from Sect. 5.2.5 relying on different ansatz spaces are provided in `exercise21.ipynb`. To this end, try to reproduce the results from Table 5.2. Do you arrive at a similar conclusion? Compare the timings and errors with the finite element solver provided as `FESolver.py`, where you can increase the polynomial degree p or number of elements n to reduce the discretization error at the price of a higher computational cost.

Method	MSE of solution u	Elapsed time per epoch	Number of epochs	Total elapsed time
Deep energy method				
FE-interpolated NNs, $p = 1$				
HiDeNN, $p = 1$				
HiDeNN without AD, $p = 1$				

References

1. Yaohua, Zang et al. 2020. Weak adversarial networks for high-dimensional partial differential equations. en. *Journal of Computational Physics 411*: 109409. issn: 0021-9991. https://doi.org/10.1016/j.jcp.2020.109409.
2. Zhiqiang, Cai et al. 2020. Deep least-squares methods: An unsupervised learning-based numerical method for solving elliptic PDEs. en. *Journal of Computational Physics 420*: 109707. issn: 0021-9991. https://doi.org/10.1016/j.jcp.2020.109707.

3. Justin, Sirignano, and K. Spiliopoulos. 2018. DGM: A deep learning algorithm for solving partial differential equations. en. *Journal of Computational Physics 375*: 1339–1364. arXiv: 1708.07469. issn: 00219991. https://doi.org/10.1016/j.jcp.2018.08.029.
4. Weinan, E., and B. Yu. (2018). The Deep Ritz Method: A deep learning-based numerical algorithm for solving variational problems. en. *Communications in Mathematics and Statistics 6.1*: 1–12. issn: 2194-6701, 2194-671X. https://doi.org/10.1007/s40304-018-0127-z.
5. Vien, Minh Nguyen-Thanh, X. Zhuang, and T. Rabczuk. 2019. A deep energy method for finite deformation hyperelasticity. en. *European Journal of Mechanics—A/Solids*, 103874. issn: 09977538. https://doi.org/10.1016/j.euromechsol.2019.103874.
6. M. Raissi, P. Perdikaris, and G.E. Karniadakis. 2019. Physics-informed neural networks: A deep learning framework for solving forward and inverse problems involving nonlinear partial differential equations". en. *Journal of Computational Physics 378*: 686–707. issn: 00219991. https://doi.org/10.1016/j.jcp.2018.10.045.
7. Ward, E. Cheney, and D.R. Kincaid. 2012. *Numerical Mathematics and Computing*. English. 7th. tex.ids= Cheney2012a. Cengage Learning. isbn: 978-1-133-10371-4.
8. Thomas, J. R. Hughes. 2000. *The finite element method: Linear static and dynamic finite element analysis*. Mineola, NY: Dover Publications. isbn: 978-0-486-41181-1.
9. Fosdick, Roger, and L. Truskinovsky. 2003. About Clapeyron's theorem in linear elasticity. *Journal of Elasticity 72*: 145–172.
10. Jon, A. Rivera et al. 2022. On quadrature rules for solving Partial Differential Equations using Neural Networks. en. *Computer Methods in Applied Mechanics and Engineering 393*: 114710. issn: 0045-7825. https://doi.org/10.1016/j.cma.2022.114710.
11. 14 - Monte carlo integration I: Basic concepts. en. In: *Physically based rendering*, ed. by Matt Pharr and Greg Humphreys, 631–660, Jan 2004. Burlington: Morgan Kaufmann. isbn: 978-0-12-553180-1. https://doi.org/10.1016/B978-012553180-1/50016-8.
12. E. Kharazmi, Z. Zhang, and G. E. Karniadakis. 2019. Variational Physics-Informed Neural Networks For Solving Partial Differential Equations, Nov. 2019. en. arXiv:1912.00873 [physics, stat]. tex.ids= Kharazmi2019a.
13. Ehsan, Kharazmi, Z. Zhang, and G. E. M. Karniadakis. 2021. hp-VPINNs: Variational physics-informed neural networks with domain decomposition. en. *Computer Methods in Applied Mechanics and Engineering 374*: 113547. issn: 0045-7825. https://doi.org/10.1016/j.cma.2020.113547.
14. Kaichao, You et al. 2019. *How Does Learning Rate Decay Help Modern Neural Networks?*, Sept 2019. arXiv:1908.01878 [cs, stat]. https://doi.org/10.48550/arXiv.1908.01878.
15. Ilya, Loshchilov and F. Hutter. 2017. *SGDR: Stochastic Gradient Descent with Warm Restarts*, May 2017. arXiv:1608.03983 [cs, math]. https://doi.org/10.48550/arXiv.1608.03983.
16. Kaiming, He et al. 2015. *Delving Deep into Rectifiers: Surpassing Human-Level Performance on ImageNet Classification*, Feb 2015. arXiv:1502.01852 [cs] version: 1. https://doi.org/10.48550/arXiv.1502.01852.
17. Ameya D. Jagtap and G. Em Karniadakis. 2020. Adaptive activation functions accelerate convergence in deep and physics-informed neural networks. *Journal of Computational Physics 404*: 109136. arXiv: 1906.01170. issn: 00219991. https://doi.org/10.1016/j.jcp.2019.109136.
18. Robert, Davis Cook and R. D. Cook, eds. 2001. *Concepts and applications of finite element analysis*, 4th ed. New York, NY: Wiley. isbn: 978-0-471-35605-9.
19. Hans, Petter Langtangen and K.-A. Mardal. 2019. *Introduction to numerical methods for variational problems*. eng. Texts in computational science and engineering 21. Cham: Springer. isbn: 978-3-030-23790-5 978-3-030-23787-5.
20. Alexander, Düster, E. Rank and B. Szabó. 2017. The *p*-Version of the finite element and finite cell methods. en. In: *Encyclopedia of Computational Mechanics Second Edition*, ed. by Erwin, Stein, René Borst, and Thomas J. R. Hughes, 1st ed., 1–35, Dec. 2017. Wiley. isbn: 978-1-119-00379-3 978-1-119-17681-7. https://doi.org/10.1002/9781119176817.ecm2003g.
21. Santiago, Badia, W. Li, and A. F. Martín. 2024. Finite element interpolated neural networks for solving forward and inverse problems. *Computer Methods in Applied Mechanics and Engineering 418*: 116505. issn: 0045-7825. https://doi.org/10.1016/j.cma.2023.116505.

22. Lei, Zhang et al. 2021. Hierarchical deep-learning neural networks: Finite elements and beyond. en. *Computational Mechanics 67.1*: 207–230. issn: 1432-0924. https://doi.org/10.1007/s00466-020-01928-9.
23. Lei, Zhang et al. 2022. HiDeNN-TD: Reduced-order hierarchical deep learning neural networks. en. *Computer Methods in Applied Mechanics and Engineering 389*: 114414. issn: 0045-7825. https://doi.org/10.1016/j.cma.2021.114414.
24. Ye, Lu et al. 2023. Convolution Hierarchical Deep-learning Neural Networks (C-HiDeNN): Finite elements, isogeometric analysis, tensor decomposition, and beyond. en. *Computational Mechanics 72.2*: 333–362. issn: 1432-0924. https://doi.org/10.1007/s00466-023-02336-5.
25. Chanwook, Park et al. 2023. Convolution hierarchical deep-learning neural network (C-HiDeNN) with graphics processing unit (GPU) acceleration. en. *Computational Mechanics 72.2*: 383–409. issn: 1432-0924. https://doi.org/10.1007/s00466-023-02329-4.
26. Hengyang, Li et al. 2023. Convolution hierarchical deep-learning neural network tensor decomposition (C-HiDeNN-TD) for high-resolution topology optimization. en. *Computational Mechanics 72.2*: 363–382. issn: 1432-0924. https://doi.org/10.1007/s00466-023-02333-8.
27. Khemraj, Shukla et al. 2020. Physics-Informed neural network for ultrasound nondestructive quantification of surface breaking cracks. en. *Journal of Nondestructive Evaluation 39.3*: 61. issn: 1573-4862. https://doi.org/10.1007/s10921-020-00705-1.
28. Majid, Rasht-Behesht et al. 2022. Physics-Informed Neural Networks (PINNs) for wave propagation and full waveform inversions. en. *Journal of Geophysical Research: Solid Earth 127.5* issn: 2169-9313, 2169-9356. https://doi.org/10.1029/2021JB023120.
29. Jamshid, Parvizian, A. Düster, and E. Rank. 2007. Finite cell method. en. *Computational Mechanics 41.1*: 121–133. issn: 1432-0924. https://doi.org/10.1007/s00466-007-0173-y.
30. William, J. Morokoff and Russel E. Caflisch. 1995. Quasi-Monte carlo integration. en. *Journal of Computational Physics 122.2*: 218–230. issn: 0021-9991. https://doi.org/10.1006/jcph.1995.1209.
31. Lu, Lu et al. 2021. Physics-Informed neural networks with hard constraints for inverse design. en. *SIAM Journal on Scientific Computing 43.6*: B1105–B1132. issn: 1064-8275, 1095-7197. https://doi.org/10.1137/21M1397908. DOI: https://doi.org/10.1137/21M1397908.

Chapter 6
Machine Learning in Computational Mechanics

Abstract Deep learning is only a subfield of machine learning. Other techniques in machine learning include decision trees, reinforcement learning, clustering, dimensionality reduction, bayesian inference, or even evolutionary optimization. To provide an initial idea of machine learning more generally, this chapter discusses the application of machine learning methods beyond neural networks, emphasizing their use in dimensionality reduction, reduced order models, and model identification. Techniques such as singular value decomposition, principal components analysis, reduced order modeling, sparse identification of non-linear dynamical systems (SINDy), clustering, and support vector machines are covered. The focus is on demonstrating how these methods complement neural networks and simulation.

Before delving further into deep learning, this chapter provides a broader perspective on machine learning with instructive examples from computational mechanics. By highlighting selected, prominent methods, some of which are fundamental for the upcoming chapters, we aim to provide an impression of machine learning and insights into some of its foundational techniques. These include dimensionality reduction techniques (Sect. 6.1), reduced order models (Sect. 6.2), and sparse identification of non-linear dynamical systems (Sect. 6.3). Subsequently, an outlook to clustering algorithms (Sect. 6.4) and classification without neural networks (Sect. 6.5) is provided. While these techniques stand isolated within this work, a glimpse of machine learning beyond deep learning is offered. For a thorough introduction to machine learning methods in engineering, the reader is referred to [1]—a resource that has significantly influenced this chapter. Further recommended reading for machine learning in general includes [2–4].

6.1 Dimensionality Reduction

This section introduces dimensionality reduction, a cornerstone technique in machine learning that reduces complex data structures, enabling more efficient and improved data analysis, but also faster simulations. Two linear reduction techniques will be explored, namely singular value decomposition and principal components analysis.

L. Herrmann et al., *Deep Learning in Computational Mechanics*,
https://doi.org/10.1007/978-3-031-89529-6_6

A prominent non-linear approach is the autoencoder, briefly introduced in Sect. 3.9.3 and to be discussed further in Chap. 8. Other methods not discussed in this work include feature embedding, factor analysis, linear discriminant analysis (for linear techniques), as well as locally linear embedding, isometric mapping and t-distributed stochastic neighbor embedding (for non-linear techniques).

6.1.1 Singular Value Decomposition

Singular value decomposition is, first and foremost, a unique matrix factorization [1, 5, 6]. A matrix $A \in \mathbb{C}^{n\times m}$ can be decomposed into

$$\boldsymbol{A} = \boldsymbol{U}\boldsymbol{\Sigma}\boldsymbol{V}^*, \tag{6.1}$$

with $\boldsymbol{U} \in \mathbb{C}^{n\times n}$, $\boldsymbol{\Sigma} \in \mathbb{R}_0^{+^{n\times m}}$ and $\boldsymbol{V}^* \in \mathbb{C}^{m\times m}$. The decomposition is visualized in Fig. 6.1. The three matrices are found from eigenvectors and eigenvalues, where the columns of $\boldsymbol{U}$ consist of the eigenvectors from $\boldsymbol{A}\,\boldsymbol{A}^*$ and the columns of $\boldsymbol{V}$ consist of the eigenvectors from $\boldsymbol{A}^*\boldsymbol{A}$. As $\boldsymbol{A}\boldsymbol{A}^*$, and $\boldsymbol{A}^*\boldsymbol{A}$, are symmetric, both $\boldsymbol{U}$ and $\boldsymbol{V}$ are unitary matrices with the desirable property of orthonormality.

$$\boldsymbol{U}^*\boldsymbol{U} = \boldsymbol{I} \tag{6.2}$$

$$\boldsymbol{V}^*\boldsymbol{V} = \boldsymbol{I} \tag{6.3}$$

The singular value matrix $\boldsymbol{\Sigma}$ is composed of the square root of the corresponding eigenvalues in descending order on the diagonal entries. The non-diagonal entries are zero. Note that non-zero eigenvalues for $\boldsymbol{U}$ and $\boldsymbol{V}$ are identical. Noteworthy is also the consistency of the sign of the eigenvectors in $\boldsymbol{U}$ and $\boldsymbol{V}$ to one another. Inconsistencies will lead to an incorrect reconstruction of $\boldsymbol{A}$. These inconsistencies can be checked by considering a reformulation of Equation (6.1).

$$\boldsymbol{\Sigma} = \boldsymbol{U}^*\boldsymbol{A}\boldsymbol{V} \tag{6.4}$$

If a diagonal entry of the singular value matrix $\boldsymbol{\Sigma}$ has an opposite sign, there is an inconsistency, which can be resolved by changing the sign of the corresponding eigenvector entries in either $\boldsymbol{U}$ or $\boldsymbol{V}$.

A basic implementation of the singular value decomposition in Python is provided below, in which we assume a real matrix $\boldsymbol{A}$ and, therefore, use transposed matrices instead of complex conjugate transposed matrices. First, the eigenvalues and eigenvectors are extracted from $\boldsymbol{A}\boldsymbol{A}^\intercal$, which are sorted to obtain $\boldsymbol{U}$. This is analogously performed with $\boldsymbol{A}^\intercal\boldsymbol{A}$ to compute $\boldsymbol{V}$. The square roots of the non-zero eigenvalues (identical for `eigenvaluesU` and `eigenvaluesV`) are placed in the diagonal entries of $\boldsymbol{\Sigma}$. Lastly, the check from Eq. (6.4) is used in order to make the decomposition consistent.

```
eigenvaluesU, U = np.linalg.eig(A @ np.transpose(A))
eigenvaluesU[np.abs(eigenvaluesU) < 1e-10] = 0.
index = np.argsort(eigenvaluesU)[::-1]
eigenvaluesU = eigenvaluesU[index]  # sort eigenvalues
U = U[:, index]  # sort eigenvectors

eigenvaluesV, V = np.linalg.eig(np.transpose(A) @ A)
eigenvaluesV[np.abs(eigenvaluesV) < 1e-10] = 0.
index = np.argsort(eigenvaluesV)[::-1]
eigenvaluesV = eigenvaluesV[index]  # sort eigenvalues
V = V[:, index]  # sort eigenvectors

S = np.zeros((len(U), len(V)))
for i in range(min(len(U), len(V))):
    S[i, i] = np.sqrt(eigenvaluesU[i])

# correction to make U and V consistent
SCorrected = np.transpose(U) @ A @ V
for i in range(min(len(U), len(V))):
    V[:, i] = V[:, i] * np.sign(SCorrected[i, i])
```

The original matrix A can be reconstructed with Eq. (6.1).

```
AReconstructed = U @ S @ np.transpose(V)
```

Alternatively, the built-in implementation in the NumPy package can be used.

```
U, S, VTransposed = np.linalg.svd(A)
AReconstruct = U @ np.diag(S) @ VTransposed  # valid for square A
```

Until now, the singular value decomposition has purely been presented as a matrix decomposition technique. However, as seen in Fig. 6.1, for non-square matrices, i.e., $m \neq n$, the matrix A can be reconstructed exactly with a smaller reduced set of matrices U, Σ, V. The reason herefore, are the zero eigenvalues stored in Σ in the additional dimensions of either AA^* (when $n > m$) or A^*A (when $n > m$). Thus the matrices U, Σ can be truncated when $n > m$, and the matrices V, Σ can be truncated when $n < m$, while still producing an accurate reconstruction of A (see Fig. 6.1). This reduction is referred to as the economy singular value decomposition and is formulated as

$$
\begin{aligned}
\text{for } n > m \quad & A = \tilde{U}\tilde{\Sigma}V^*, \\
\text{for } n < m \quad & A = U\tilde{\Sigma}\tilde{V}^*.
\end{aligned}
\tag{6.5}
$$

Building on this concept, the truncated singular value decomposition truncates the set of matrices $\bar{U}$, $\bar{\Sigma}$, $\bar{V}$ beyond the mismatch in dimensionality, enabling an approximate reconstruction of A (with a reduced basis). Specifically, the truncated singular

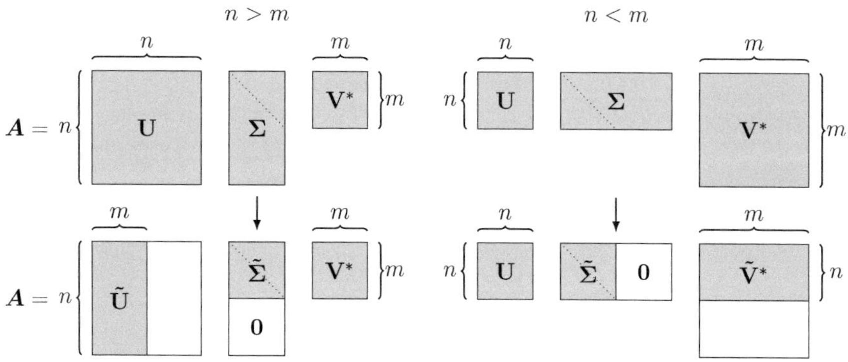

Fig. 6.1 Singular value decomposition of matrix $\boldsymbol{A}$. The top illustrates the full singular value decomposition for either $n > m$ or $n < m$. At the same time, the bottom showcases the economy singular value decomposition, in which no information is lost, i.e., $\boldsymbol{A}$ is still decomposed exactly. Inspired by [1]

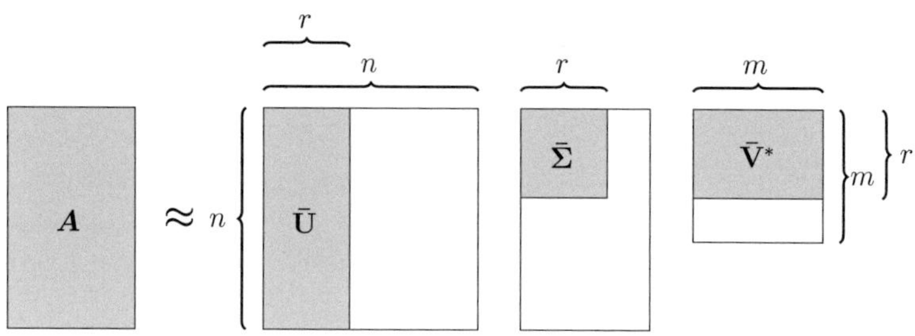

Fig. 6.2 Truncated singular value decomposition with truncation level r, which reconstructs $\boldsymbol{A}$ in an approximate sense. Inspired by [1]

value decomposition is an optimal low-rank approximation in a least squares sense. A truncation up to the rth singular value, is given by

$$\boldsymbol{A} \approx \bar{\boldsymbol{U}} \bar{\boldsymbol{\Sigma}} \bar{\boldsymbol{V}}^*, \tag{6.6}$$

where $\bar{\boldsymbol{U}} \in \mathbb{C}^{n\times r}$, $\bar{\boldsymbol{\Sigma}} \in \mathbb{R}_0^{+^{r\times r}}$ and $\bar{\boldsymbol{V}} \in \mathbb{C}^{m\times r}$. A visual interpretation is provided in Fig. 6.2, where the truncation of non-zero singular values in $\boldsymbol{\Sigma}$, i.e., the loss of information, becomes clear.

The truncation is straightforward to implement using the previous code performing the singular value decomposition.

```
r = 10  # level of truncation
AReconstructed = U[:, :r] @ S[:r, :r] @ np.transpose(V[:, :r])
```

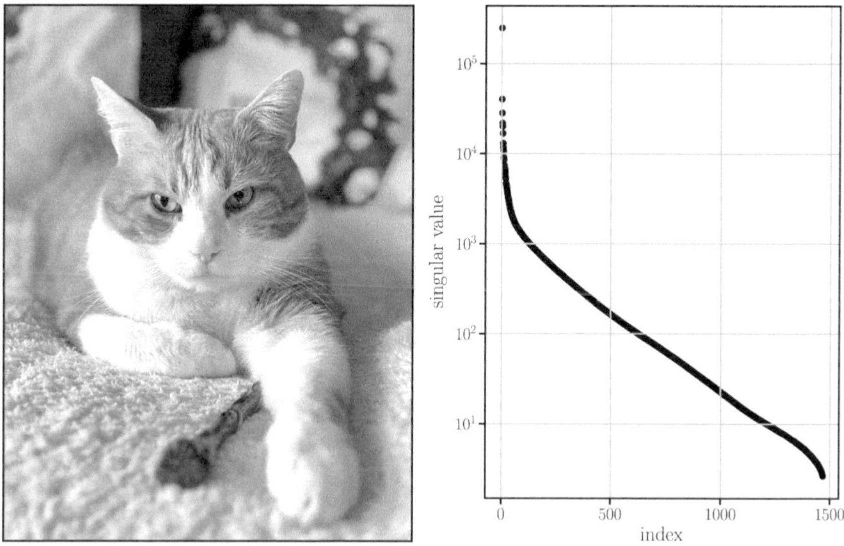

(a) Original image (of Keks) (b) Singular values

Fig. 6.3 Image compression with the singular value decomposition. Example inspired by Mordecai in [1]

6.1.1.1 Image Compression

Before proceeding to applications within mechanics, the prominent application of image compression is discussed. A gray-scale image can be interpreted as a matrix with pixel entries in the x_1 and x_2 direction. Consider the image provided in Fig. 6.3a. The singular value decomposition identifies the most prominent modes in x_1 and x_2 and correspondingly stores them in $\boldsymbol{V}$ and $\boldsymbol{U}$. The importance of each individual mode is seen in terms of the singular values, as depicted in Fig. 6.3b. Due to the sharp fall in magnitude, a drastic truncation should not affect an image reconstruction significantly. The original image size of Fig. 6.3a is 1868×1468, yielding 1868 modes of size 1868 in the x_2 direction (stored in $\boldsymbol{U}$) and 1468 modes of size 1468 in the x_1 direction (stored in $\boldsymbol{V}$). Without loss of information, the number of modes can be reduced to 1468 for both axes with the economy singular value decomposition. Reducing the number of modes further in both dimensions, i.e., employing the truncated singular value decomposition, enables an approximative compression. Figure 6.4 illustrates the reconstruction for truncation degrees $r = 1, 2, 10, 20, 100, 200$, corresponding to a compression ratio of 0.116%, 0.243%, 1.22%, 2.43%, 12.5%, 25.8%. The compression ratio is computed by dividing the number of entries in the original matrix $\boldsymbol{A}$, i.e., nm and the number of entries of $\bar{\boldsymbol{U}}$, $\bar{\boldsymbol{\Sigma}}$, $\bar{\boldsymbol{V}}$, i.e., $nr + rr + mr$.[1] The quality of approximation can be assessed prior to compression by consulting the singular

[1] Greater compression ratios can be achieved by performing the singular value decomposition over multiple images, as seen in [1]. For example, with a library of face images, a basis can be identified

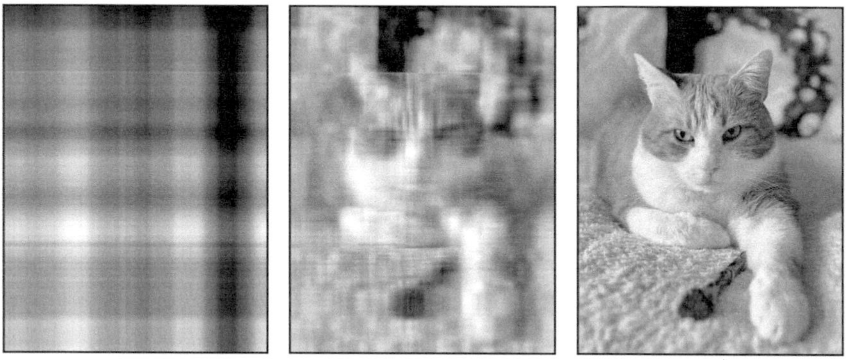

(a) $r = 1$ (b) $r = 10$ (c) $r = 100$

(d) $r = 2$ (e) $r = 20$ (f) $r = 200$

Fig. 6.4 Image compression (using Keks from Fig. 6.3a) with the singular value decomposition illustrating the effect of different levels of truncation r

values stored in $\mathbf{\Sigma}$, depicted in Fig. 6.3b. A lot of information is contained in the first $\sim$100 modes, visible by the sharp drop in the magnitude of the singular values. This aligns with the resulting reconstruction in Fig. 6.4, where the reconstructed image with a truncation level of $r = 200$ is barely distinguishable from the original image in Fig. 6.3a.

6.1.1.2 Identification of a Reduced Basis

In the image compression example, we have essentially identified a reduced basis from which we can reconstruct an approximation with the correct singular values. This can similarly be applied to data from simulations. Consider, the following spatio-temporal function

---

with which yet unseen faces can be reconstructed efficiently; see eigenfaces [7, 8] used in facial recognition [9, 10].

$$y = f(x,t) = \sin(2\pi x) + 2\sin(4\pi x)xt^2 + x^2t + 5. \tag{6.7}$$

Assuming we extract $f(x,t)$ from measurements or simulations, the results can be stored in an $m \times n$ snapshot matrix $\boldsymbol{X}$

$$\boldsymbol{X} = \begin{pmatrix} f(x_1,t_1) & f(x_2,t_1) & \cdots & f(x_n,t_1) \\ f(x_1,t_2) & f(x_2,t_2) & \cdots & f(x_n,t_2) \\ \vdots & \vdots & \vdots & \vdots \\ f(x_1,t_m) & f(x_2,t_m) & \cdots & f(x_n,t_m) \end{pmatrix} \tag{6.8}$$

consisting of m temporal snapshots of spatial measurements of size n. Specifically, we consider spatial measurements at distinct coordinates $x_1, x_2, \ldots, x_n$, the features, and the snapshots in time as the samples. The snapshot matrix $\boldsymbol{X}$ can be decomposed with singular value decomposition, yielding

$$\boldsymbol{X} = \boldsymbol{U}\boldsymbol{\Sigma}\boldsymbol{V}^*. \tag{6.9}$$

The temporal modes can be extracted from the columns in $\boldsymbol{U}$. By contrast, the spatial modes can be identified from the columns in $\boldsymbol{V}$.[2] The first four identified modes are depicted in Fig. 6.5 (based on $n = 100$, $m = 40$).[3] Only the first three modes seem reasonable. This is confirmed by considering the singular values with Fig. 6.6, where only the first three are non-zero. Thus, mode four and onwards are simply numerical noise and thereby useless. This aligns with the chosen function $f(x,t)$ from Eq. (6.7), which consists of only three linearly independent spatial modes. Thus,

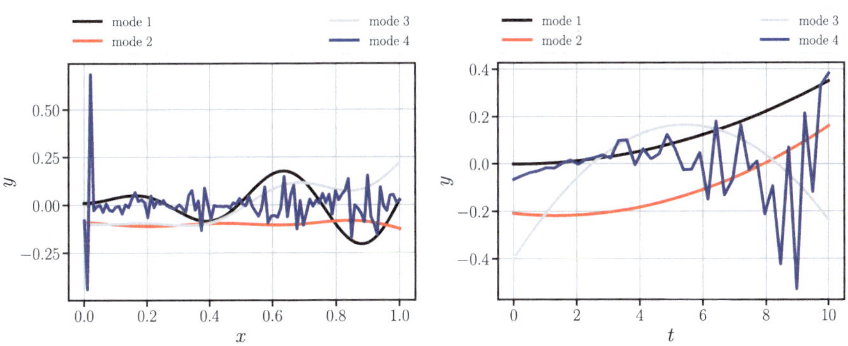

(a) Spatial modes (b) Temporal modes

Fig. 6.5 Identification of the first four spatial and temporal modes of the snapshot matrix $\boldsymbol{X}$

[2] This behavior would be interchanged if we considered the transpose $\boldsymbol{X}^\intercal$ instead of $\boldsymbol{X}$.

[3] Note that the identified spatial modes are invariant to the defined time-interval as soon as a sufficiently large time-interval is selected. Thus, in the case of $f(x,t)$ being a solution to a differential equation, the identified spatial modes can be used as a reduced basis. Hereby, the spatial discretization is reduced from n to 3.

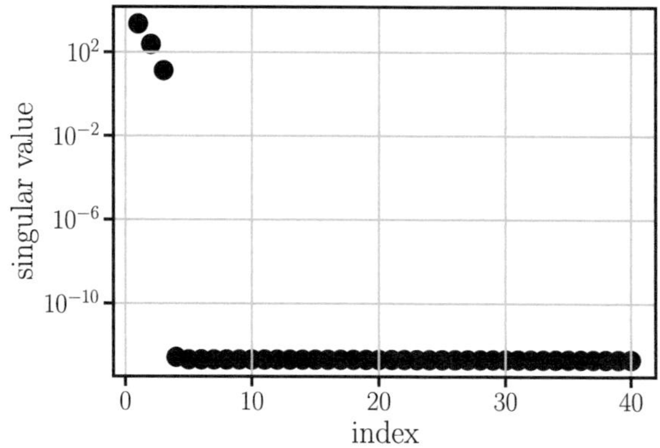

Fig. 6.6 Modeshape coefficients of the snapshot matrix $\boldsymbol{X}$

the singular value decomposition enabled the identification of three distinct modes, which accurately capture the behavior of $f(x, y)$.

In other words, the snapshot can be fully recreated given only the first three mode shapes, enabling a truncated decomposition (with $r = 3$) of the form

$$\boldsymbol{X} = \bar{\boldsymbol{U}}\bar{\boldsymbol{\Sigma}}\bar{\boldsymbol{V}}^{*}. \tag{6.10}$$

Yet another way of viewing this is that $\bar{\boldsymbol{V}} \in \mathbb{R}^{n \times r}$ enables a projection onto a lower dimensional space governed by the time-dependent coefficients $\boldsymbol{c}(t) \in \mathbb{R}^{m \times r}$ (containing all m time steps):

$$\boldsymbol{c}(t) = \boldsymbol{X}\bar{\boldsymbol{V}} \tag{6.11}$$

An approximate reconstruction is enabled by

$$\boldsymbol{X} \approx \boldsymbol{c}(t)\bar{\boldsymbol{V}}^{\intercal}, \tag{6.12}$$

obtained by exploiting the orthonormal property of $\boldsymbol{V}$ and thereby $\bar{\boldsymbol{V}}$.

To exemplify the usefulness of this projection, assume that $\boldsymbol{f}$ is the right-hand-side of a system of equations, which you are attempting to solve for

$$\boldsymbol{K}\boldsymbol{y} = \boldsymbol{f}. \tag{6.13}$$

Thus, the goal is to identify $\boldsymbol{y}$. Prior knowledge of similar systems is stored in a matrix $\boldsymbol{X}$, which in each column stores a solution to a similar system. Using a truncated $\bar{\boldsymbol{V}}$ of $\boldsymbol{X}$ obtained through the singular value decomposition, the system is projected into a lower dimensional system

$$\boldsymbol{K}\boldsymbol{y}\bar{\boldsymbol{V}} = \boldsymbol{f}\bar{\boldsymbol{V}}, \tag{6.14}$$

$$\boldsymbol{K}\boldsymbol{c} = \boldsymbol{f}\bar{\boldsymbol{V}}. \tag{6.15}$$

This reduces the number of unknowns to the level of truncation used for $\bar{V}$. More details on these projections will follow in Sect. 6.2, where the projection will be applied to a differential equation.

6.1.2 Principal Components Analysis

Related to the singular value decomposition is the principal components analysis, which similarly harnesses orthogonal transformations, enabling reduced representations [11, 12] (see [13] for a thorough textbook and [14] for insightful applications). The procedure begins with calculating the mean of each column $\boldsymbol{A} \in \mathbb{R}^{n \times m}$ (for each feature across all the samples)

$$\overline{a}_j = \frac{1}{n} \sum_{i=1}^{n} A_{ij}, \tag{6.16}$$

which is used to assemble a mean matrix

$$\overline{\boldsymbol{A}} = \begin{pmatrix} 1 \\ \vdots \\ 1 \end{pmatrix} \overline{\boldsymbol{a}}. \tag{6.17}$$

Subtracting the mean matrix $\overline{\boldsymbol{A}}$ from the original data $\boldsymbol{A}$, yields the mean-subtracted data, i.e., the centered data for each feature

$$\boldsymbol{B} = \boldsymbol{A} - \overline{\boldsymbol{A}}. \tag{6.18}$$

By computing the eigenvectors of the matrix-matrix product $\boldsymbol{B}^* \boldsymbol{B}$, the principal components $\boldsymbol{w}_j$ are retrieved. The first principal component $\boldsymbol{w}_1$ is the one associated with the largest eigenvalue (the eigenvalues of $\boldsymbol{B}^* \boldsymbol{B}$ are referred to as the principal values of $\boldsymbol{A}$). As the matrix-matrix product $\boldsymbol{B}^* \boldsymbol{B}$ is proportional to the covariance matrix $\frac{1}{n-1} \boldsymbol{B}^* \boldsymbol{B}$, the principal components capture the directions in which the data varies the most, i.e., the modes maximizing the variation.

The principal components are stored in the columns of $\boldsymbol{W}$ which is orthonormal

$$\boldsymbol{W}^* \boldsymbol{W} = \boldsymbol{I}. \tag{6.19}$$

This is similar to $\boldsymbol{U}$ and $\boldsymbol{V}$ in the singular value decomposition. Note that the principal components are equivalent to the singular modes in $\boldsymbol{V}$ from the singular value decomposition if $\boldsymbol{A}$ is centered around the mean, i.e., $\overline{\boldsymbol{A}} = \boldsymbol{0}$ or $\boldsymbol{B} = \boldsymbol{A}$. Singular value decomposition also distinguishes itself, as it is more than a reduction technique. It is a general matrix decomposition method. For a more detailed comparison; see [15].

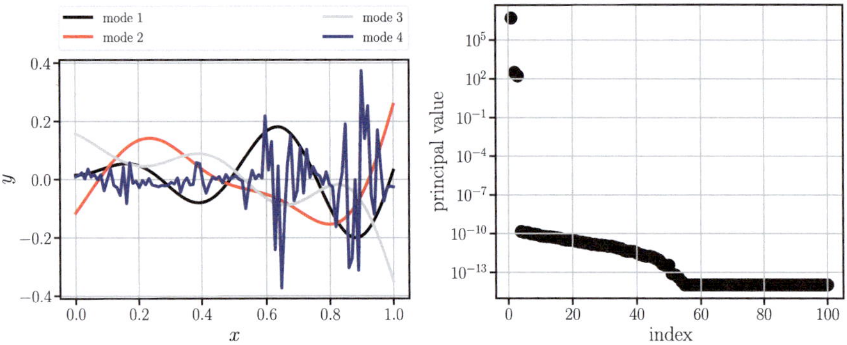

(a) Principal components (spatial modes) (b) Principal values

Fig. 6.7 Principal components and values of the snapshot matrix $\boldsymbol{X}$

Principal components analysis can be performed with the following code snippet in Python. First, the data is centered around the mean and followed by the eigendecomposition.

```
# mean of each feature across all samples
Xbar = np.mean(X, axis=1, keepdims=True)
B = X - Xbar
principalValues, W = np.linalg.eig(np.transpose(B) @ B)
index = np.argsort(principalValues)[::-1]
principalValues = principalValues[index]  # sort eigenvalues
W = W[:, index]  # sort eigenvectors
```

6.1.2.1 Identification of a Reduced Basis

Revisiting the snapshot matrix $\boldsymbol{X}$ from Eq. (6.8) in Sect. 6.1.1.2, we perform a principal components analysis on $\boldsymbol{X}$. This yields a matrix $\boldsymbol{W}$ containing the principal components $\boldsymbol{w}_j$ in its columns. These principal components correspond to the mean-centered spatial modes, i.e., the modes maximizing the variation. Similar as with the singular values, only the first three principal values are non-zero, as seen in Fig. 6.7a with the corresponding spatial modes depicted in Fig. 6.7b. Again, the fourth mode onward is simply numerical noise.

Similar as with the singular value decomposition, the principal components can be used as a means to a projection to a lower dimensional space

$$\boldsymbol{c}(t) = (\boldsymbol{X} - \overline{\boldsymbol{X}})\bar{\boldsymbol{W}}. \tag{6.20}$$

When compared to the singular value decomposition projection from Eq. (6.11), the only difference is the subtraction of the mean matrix $\overline{\boldsymbol{X}}$. The approximate reconstruction is enabled by

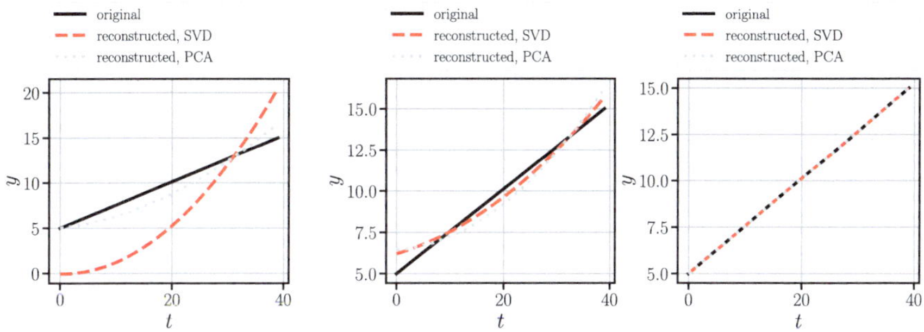

(a) Using one mode (b) Using two modes (c) Using three modes

Fig. 6.8 Reconstruction of the snapshot matrix $\boldsymbol{X}$ with singular value decomposition (SVD) and principal components analysis (PCA). Visualized at the right-most coordinate x_n

$$\boldsymbol{X} \approx \boldsymbol{c}(t)\bar{\boldsymbol{V}}^{\mathsf{T}} + \overline{\boldsymbol{X}}. \tag{6.21}$$

The mean-centering leads to an additional operation with the mean matrix $\overline{\boldsymbol{X}}$. Both projection approaches are demonstrated in Fig. 6.8 where the reconstructions with three different truncation levels are shown.

In summary, the only difference in the projection between singular value decomposition and principal components is the mean centering. In a broader sense, singular value decomposition is a general matrix decomposition technique, while principal components analysis is only used for dimensionality reduction. Both techniques are linear transformations and will yield insufficient decompositions for non-linearly related data. An alternative for non-linear dimensionality reduction are autoencoders, discussed in Sect. 3.9.3 and Chap. 8.

6.2 Reduced Order Models

As discussed in Sects. 6.1.1.2 and 6.1.2.1, singular value decomposition and principal components analysis can be used to identify a reduced basis. To demonstrate how a reduced basis is incorporated in a simulation [16], we consider the finite element method.

6.2.1 Finite Element Method for Dynamics

Before introducing the reduced basis, we summarize the finite element method for dynamical systems. Consider the spatially discretized finite element equations, i.e., the system of ordinary differential equations

$$\boldsymbol{M}\ddot{\boldsymbol{u}}(t) + \boldsymbol{K}\boldsymbol{u}(t) = \boldsymbol{F}(t). \tag{6.22}$$

Here $\ddot{\boldsymbol{u}}$ is the double temporal derivative of the degrees of freedom $\boldsymbol{u}$, $\boldsymbol{M}$ is the global mass matrix, $\boldsymbol{K}$ is the global stiffness matrix, and $\boldsymbol{F}$ is the global force vector. The system of ordinary differential equations can be solved by various integration schemes, like basic finite difference schemes [17–19], Runge-Kutta methods [20, 21], and schemes from the Newmark family [22]. For example using a central difference approximation of $\ddot{\boldsymbol{u}}$ with a time step size of Δt

$$\dot{\boldsymbol{u}}(t) \approx \frac{\boldsymbol{u}(t+\Delta t) - \boldsymbol{u}(t+\Delta t)}{2\Delta t} \tag{6.23}$$

$$\ddot{\boldsymbol{u}}(t) \approx \frac{\boldsymbol{u}(t+\Delta t) - 2\boldsymbol{u}(t) + \boldsymbol{u}(t-\Delta t)}{\Delta t^2} \tag{6.24}$$

allows Eq. (6.22) to be rewritten. At an equilibrium at time t, this yields

$$\boldsymbol{M}\left(\frac{\boldsymbol{u}(t+\Delta t) - 2\boldsymbol{u}(t) + \boldsymbol{u}(t-\Delta t)}{\Delta t^2}\right) + \boldsymbol{K}\boldsymbol{u}(t) \approx \boldsymbol{F}(t), \tag{6.25}$$

$$\frac{1}{\Delta t^2}\boldsymbol{u}(t+\Delta t)\boldsymbol{M} \approx \boldsymbol{F}(t) - \left(\boldsymbol{K} - \frac{2}{\Delta t^2}\boldsymbol{M}\right)\boldsymbol{u}(t) - \frac{1}{\Delta t^2}\boldsymbol{M}\boldsymbol{u}(t-\Delta t). \tag{6.26}$$

The resulting system of algebraic equation can be solved by inverting the mass matrix $\boldsymbol{M}$, thus estimating the solution $\boldsymbol{u}(t+\Delta t)$ at the next time step. As the next time step depends on the previous two time steps, the scheme is a two-step scheme and requires a special treatment of the first timestep. By considering the initial conditions, i.e., the initial displacement $\boldsymbol{u}(0)$ and initial velocity $\dot{\boldsymbol{u}}(0)$ (and Eq. (6.23)), an expression for $\boldsymbol{u}(-\Delta t)$ is obtained

$$\dot{\boldsymbol{u}}(0) \approx \frac{\boldsymbol{u}(\Delta t) - \boldsymbol{u}(-\Delta t)}{2\Delta t}, \tag{6.27}$$

$$\boldsymbol{u}(-\Delta t) \approx \boldsymbol{u}(\Delta t) - 2\Delta t\dot{\boldsymbol{u}}(0). \tag{6.28}$$

Inserting Eq. (6.28) into Eq. (6.26) for the equilibrium at the first time step $t = \Delta t$ gives us the adapted scheme for the first time step

$$\frac{1}{\Delta t^2}\boldsymbol{M}\boldsymbol{u}(\Delta t) = \boldsymbol{F}(0) - \left(\boldsymbol{K} - \frac{2}{\Delta t^2}\boldsymbol{M}\right)\boldsymbol{u}(0) - \frac{1}{\Delta t^2}\boldsymbol{M}\boldsymbol{u}(-\Delta t), \tag{6.29}$$

$$\frac{2}{\Delta t^2}\boldsymbol{M}\boldsymbol{u}(\Delta t) = \boldsymbol{F}(0) - \left(\boldsymbol{K} - \frac{2}{\Delta t^2}\boldsymbol{M}\right)\boldsymbol{u}(0) - \frac{2}{\Delta t}\boldsymbol{M}\dot{\boldsymbol{u}}(0). \tag{6.30}$$

The scheme can be implemented in Python relying on the sparse solver `spsolve` from the SciPy library to efficiently invert the mass matrix $\boldsymbol{M}$. Furthermore we rely on the fictive functions `computeGlobalSystem`, `applyDirichletBoundaryConditions`, `applyNeumannBoundaryConditions` for the computation of the global system

```python
from scipy.sparse.linalg import spsolve

dofs  # number of degrees of freedom

# computation of system matrices
K, M, F = computeGlobalSystem()  # (dofs, dofs), (dofs,
    dofs), (dofs)
K, F = applyDirichletBoundaryConditions()
F = applyNeumannBoundaryConditions(t=0)  # time-dependent

# homogeneous initial conditions
u0 = np.zeros(dofs)
du0dt = np.zeros(dofs)

# first time step
rhs = F - (F - (K - 2. / dt ** 2 * M) @ u0 - 2. / dt * M
    @ du0dt)
lhs = 2. / dt ** 2 * M
u1 = spsolve(lhs, rhs)

# time-stepping scheme
for i in range(1, timesteps):
    F = applyNeumannBoundaryConditions(t=i * dt)
    rhs = F - (K - 2. / dt ** 2 * M) @ u1 - 1. / dt ** 2
    * M @ u0
    lhs = 1. / dt ** 2 * M
    u2 = spsolve(lhs, rhs)
    u0 = u1
    u1 = u2
```

6.2.2 Reduced Order Modeling with Finite Elements

Now, that the dynamic finite element procedure has been clarified, we enhance it with a reduced order model. We assume that we have obtained a reduced basis $\boldsymbol{\Psi} = \bar{\boldsymbol{V}} \in \mathbb{R}^{n \times r}$ with the truncated singular value decomposition from a snapshot matrix $\boldsymbol{X} = (\boldsymbol{u}(t_1), \boldsymbol{u}(t_2), \dots, \boldsymbol{u}(t_m))^\intercal \in \mathbb{R}^{n \times m}$, where $\boldsymbol{u}(t_i)$ are the global finite element degrees of freedom at time t_i. Relying on the projection $\boldsymbol{\Psi} \in \mathbb{R}^{n \times r}$, we can project the vector of degrees of freedom $\boldsymbol{u}(t) \in \mathbb{R}^n$ onto a reduced basis $\boldsymbol{c}(t) \in \mathbb{R}^r$—also referred to as Galerkin expansion

$$\boldsymbol{u}(t) \approx \boldsymbol{\Psi} \boldsymbol{c}(t). \tag{6.31}$$

Due to the truncation r, this is an approximation. The approximation may then be inserted into the semi-discrete finite element equations from Eq. (6.22), yielding

$$\boldsymbol{M}\boldsymbol{\Psi}\ddot{\boldsymbol{c}}(t) + \boldsymbol{K}\boldsymbol{\Psi}\boldsymbol{c}(t) = \boldsymbol{F}(t), \tag{6.32}$$

which now are n equations with r degrees of freedom. By pre-multiplying with $\boldsymbol{\Psi}^{\mathsf{T}}$, the system is projected onto the reduced space defined by $\boldsymbol{\Psi}$, effectively reducing the number of equations to r.

$$\boldsymbol{\Psi}^{\mathsf{T}}\boldsymbol{M}\boldsymbol{\Psi}\ddot{\boldsymbol{c}}(t) + \boldsymbol{\Psi}^{\mathsf{T}}\boldsymbol{K}\boldsymbol{\Psi}\boldsymbol{c}(t) = \boldsymbol{\Psi}^{\mathsf{T}}\boldsymbol{F}(t) \tag{6.33}$$

Thus instead of solving for $\boldsymbol{u}(t) \in \mathbb{R}^n$, we solve for $\boldsymbol{c}(t) \in \mathbb{R}^r$, where $r \ll n$. The full solution vector can be recovered with

$$\boldsymbol{u}(t) \approx \boldsymbol{c}(t)\boldsymbol{\Psi}^{\mathsf{T}}. \tag{6.34}$$

The procedure is also referred to as proper orthogonal decomposition,[4] as the solution is projected onto a lower-dimensional subspace, relying on a set of orthogonal basis functions, i.e., the modes identified by singular value decomposition (or principal components analysis).

The introduced projection leads to only minor modifications of the implementation presented above. Subsequent to the initial system matrix computations, the projection is applied to the system matrices and vectors.

```
u0 = u0 @ psi
du0dt = du0dt @ psi
F = F @ phi
K = np.transpose(psi) @ K @ psi)
M = np.transpose(psi) @ M @ psi
```

During the time-stepping, the code remains almost the same, apart from the projections back to the global system and projecting the force vector to the reduced system at each time step. The three changes to the original system are highlighted in orange.

```
# first time step
rhs = F - (F - (K - 2. / dt ** 2 * M) @ u0 - 2. / dt * M @ du0dt)
lhs = 2. / dt ** 2 * M
u1 = spsolve(lhs, rhs)
u1Full = u1 @ np.transpose(psi)  # projection to full system

# time-stepping scheme
```

[4] An alternative procedure for model order reduction is the proper generalized decomposition, which instead of relying on previous data, reduces the complexity a priori through the decomposition of high-dimensional problems into a sum of separable, lower-dimensional functions; see [23] for a recent review. It is particularly suitable for high-dimensional problems.

```
for i in range(1, timesteps):
    F = applyNeumannBoundaryConditions(t=i * dt)
    F = F @ psi
    rhs = F - (K - 2. / dt ** 2 * M) @ u1 - 1. / dt ** 2 * M @ u0
    lhs = 1. / dt ** 2 * M
    u2 = spsolve(lhs, rhs)
    u2Full = u2 @ np.transpose(psi)  # projection to full system
    u0 = u1
    u1 = u2
```

6.2.3 *Cantilever Beam*

To illustrate the impact of a reduced basis on a simulation, we consider a cantilever beam with a time-independent point load in the x_2 direction, depicted schematically in Fig. 6.9 and with simulated deflections in Fig. 6.10. The slim cantilever beam is to be simulated with two-dimensional finite elements under small deformations; see Fig. 6.10. Thus a reduction from two dimensions to one dimension should be possible, as known from Euler-Bernoulli beam theory. Specifically, a 40×4 grid of uniform bilinear elements is employed to represent the beam. The resulting displacement field in the x_1 direction is depicted in Fig. 6.9 at four distinct snapshots. Performing a singular value decomposition on the snapshot matrix $\boldsymbol{X} = (\boldsymbol{u}(t_1), \boldsymbol{u}(t_2), \dots, \boldsymbol{u}(t_m))^\intercal$ consisting of $m = 5000$ timesteps with a step size of $\Delta t = 5 \cdot 10^{-6}$ (see Fig. 6.13a), yields the projection $\boldsymbol{\Psi}$. The first four spatial modes contained in $\boldsymbol{\Psi}$ are shown in Fig. 6.11. These modes align with the mode shapes $m_n(x)$ from the Euller-Bernoulli beam theory. The mode shapes are given as [24]

$$m_n(x_1) = \cosh(\beta_n x_1) - \cos(\beta_n x_1) - \sigma_n(\sinh(\beta_n x_1) - \sin(\beta_n x_1)), \tag{6.35}$$

where

$$\sigma_n = \frac{\sinh(\beta_n L) - \sin(\beta_n L)}{\cosh(\beta_n L) + \cos(\beta_n L)}. \tag{6.36}$$

The characteristic equation is used to determine β_n

$$\cos(\beta_n L)\cosh(\beta_n L) = -1 \tag{6.37}$$

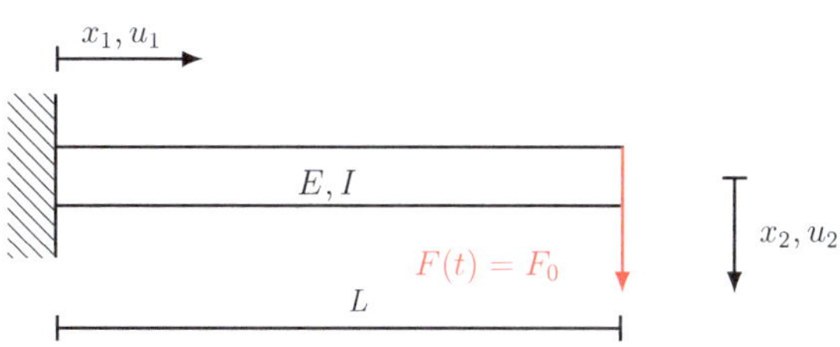

Fig. 6.9 Cantilever beam

Table 6.1 Modeshape parameters upto $n = 4$

	$n = 1$	$n = 2$	$n = 3$	$n = 4$
$\beta_n L$	1.8751	4.6941	7.8548	10.9955
σ_n	0.7341	1.0185	0.9992	1.0000

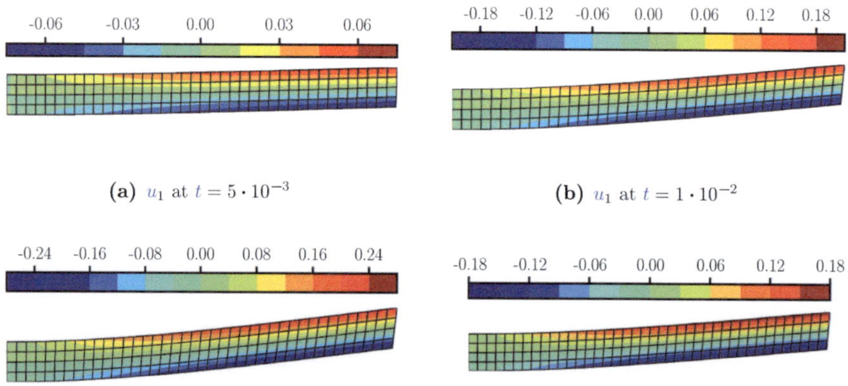

(a) u_1 at $t = 5 \cdot 10^{-3}$ **(b)** u_1 at $t = 1 \cdot 10^{-2}$

(c) u_1 at $t = 5 \cdot 1.5^{-2}$ **(d)** u_1 at $t = 2 \cdot 10^{-2}$

Fig. 6.10 Two-dimensional dynamic finite element analysis of a cantilever beam. The contour plot shows the displacement u_1 in the x_1 direction

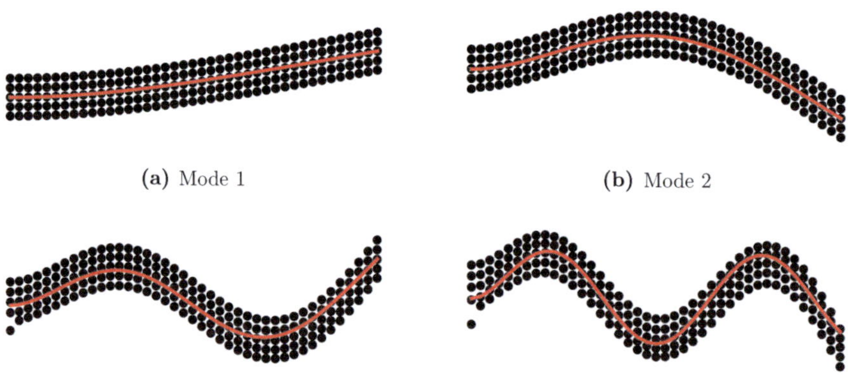

(a) Mode 1 **(b)** Mode 2

(c) Mode 3 **(d)** Mode 4

Fig. 6.11 First four identified mode shapes

and yields the mode shape parameters presented in Table 6.1 and depicted in Fig. 6.11.

The singular values, indicating the importance of each mode are provided in Fig. 6.12, where a stagnation is seen after the fifth singular value. Therefore we employ a truncation level of $r = 5$, thus reducing the system from 410 degrees of

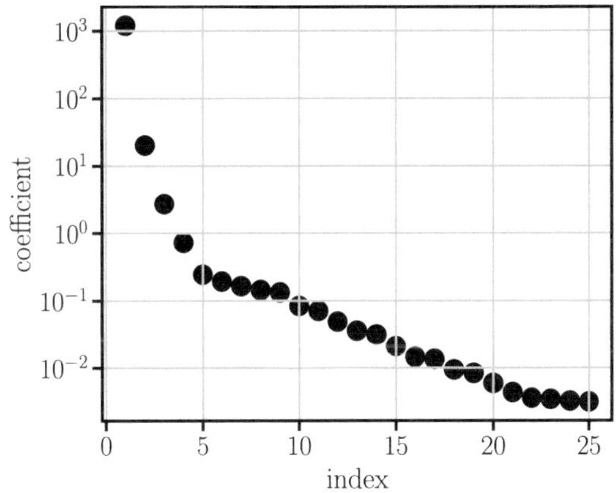

Fig. 6.12 Singular values

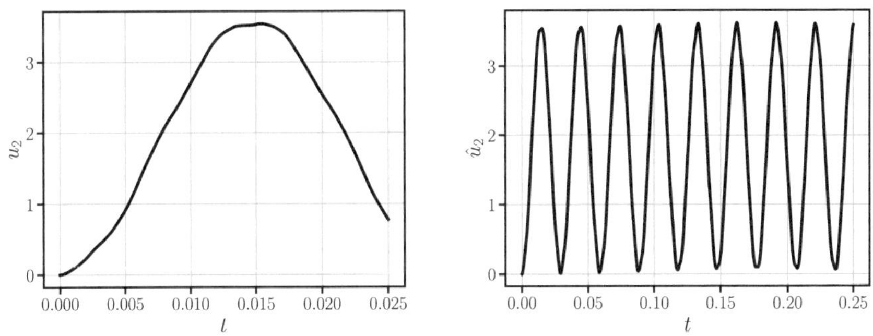

(a) Measurement until $t = 0.025$ **(b)** Simulation with reduced model until $t = 0.25$

Fig. 6.13 Measurement of full system and simulation with reduced system

freedom to five. Now the reduced model is used to solve the model until $t = 0.25$ using measurements until $t = 0.025$, i.e., 50,000 timesteps, seen in Fig. 6.13. The full system required $2.87 \cdot 10^{-3}$ s per time step, while the reduced system only takes $1.65 \cdot 10^{-4}$ s per time step. Thus a speed-up of about one order of magnitude is achieved.

6.3 Sparse Identification of Non-linear Dynamical Systems SINDy

Sparse identification of non-linear dynamical systems [25], abbreviated by SINDy, deals with the discovery of dynamic systems of the form

$$\frac{d}{dt}\boldsymbol{x}(t) = \boldsymbol{f}(\boldsymbol{x}(t)). \tag{6.38}$$

In other words it is a procedure to solve inverse problems composed of a system of ordinary differential equations (in an interpretable manner).

The basic procedure of SINDy can be summarized by the following four steps.

1. Sample temporal snapshots of $\boldsymbol{x}(t_i) = (x_1(t_i)x_2(t_i)\cdots x_m(t_i))^{\mathsf{T}}$ and its first time derivative $\dot{\boldsymbol{x}}(t_i) = (\dot{x}_1(t_i)\dot{x}_2(t_i)\cdots \dot{x}_m(t_i))^{\mathsf{T}}$ in corresponding snapshot matrices $\boldsymbol{X}$ and $\dot{\boldsymbol{X}}$.

$$\boldsymbol{X} = \begin{pmatrix} - \ \boldsymbol{x}(t_1) \ - \\ - \ \boldsymbol{x}(t_2) \ - \\ \vdots \\ - \ \boldsymbol{x}(t_n) \ - \end{pmatrix}, \dot{\boldsymbol{X}} = \begin{pmatrix} - \ \dot{\boldsymbol{x}}(t_1) \ - \\ - \ \dot{\boldsymbol{x}}(t_2) \ - \\ \vdots \\ - \ \dot{\boldsymbol{x}}(t_n) \ - \end{pmatrix} \tag{6.39}$$

   The time derivative $\dot{\boldsymbol{x}}(t_i)$ can either be found through separate measurements or numerically from $\boldsymbol{x}(t_i)$ with for example finite differences.

2. Select a library $\boldsymbol{\Theta}(\boldsymbol{X})$ of candidate functions, e.g., polynomials, for the sparse regression.

$$\boldsymbol{\Theta}(\boldsymbol{X}) = \begin{pmatrix} 1 & x_1(t_1) & x_2(t_1) & \cdots & x_1(t_1)^2 & x_2(t_1)^2 & x_1(t_1)x_2(t_1) & \cdots \\ 1 & x_1(t_2) & x_2(t_2) & \cdots & x_1(t_2)^2 & x_2(t_2)^2 & x_1(t_2)x_2(t_2) & \cdots \\ \vdots & \vdots & \vdots & \vdots & \vdots & \vdots & \vdots & \vdots \\ 1 & x_1(t_n) & x_2(t_n) & \cdots & x_1(t_n)^2 & x_2(t_n)^2 & x_1(t_n)x_2(t_n) & \cdots \end{pmatrix} \tag{6.40}$$

3. The sparse regression problem can now be formulated as

$$\dot{\boldsymbol{X}} = \boldsymbol{\Theta}(\boldsymbol{X})\boldsymbol{\Xi}, \tag{6.41}$$

   where the goal is to identify sparse vectors of coefficients $\boldsymbol{\Xi}$ corresponding to the candidate functions.

4. Compute the sparse vector of coefficients using sequential thresholded least squares; see Algorithm 12, or the least absolute shrinkage and selection operator (LASSO).

The sequential thresholded least squares algorithm iteratively solves the least squares regression, as discussed in Chap. 2, while simultaneously thresholding small coefficients to zero. This yields a sparse set of coefficients $\boldsymbol{\Xi}$. With the candidate library $\boldsymbol{\Theta}$ and the identified sparse vector of coefficients $\boldsymbol{\Xi}$, an interpretable system of ordinary equations is identified.

The sequential least squares algorithm can be implemented in Python with the following code snippet. The function `sequentialThresholdedLeastSquares` will be employed in the example in Sect. 6.3.1.

```
def sequentialThresholdedLeastSquares(theta, dXdt, k=10, tol=1e-1):
    Xi = np.linalg.lstsq(theta, dXdt, rcond=None)[0]
    for i in range(k):
        smallindices = abs(Xi) < tol
```

```python
        Xi[smallindices] = 0
        for j in range(len(dXdt[1])):
            bigindices = ~smallindices[:, j]
            Xi[bigindices, j] = np.linalg.lstsq(theta[:, bigindices],
                                                 dXdt[:, j],
                                                 rcond=None)[0]
    return Xi
```

Algorithm 12 Sequential thresholded least squares algorithm

Require: snapshot matrix of first time derivatives $\dot{X}$, library evaluated with snapshots $\mathbf{\Theta}(X)$, number of iteration k, sparse tolerance t

Initial least squares regression
$\Xi = (\mathbf{\Theta}^\top \mathbf{\Theta})^{-1} \mathbf{\Theta}^\top \dot{X}$

for all k **do**
  Identify indices i of sparse terms
  $i_{\text{sparse}} = |\Xi| < t$
  $\Xi i = 0$

  Perform the least squares regression sequentially for the number of degrees of freedom m on the non-sparse terms.
  $j = 1$
  for all m **do**
    $i_{\text{nonsparse}} = \neg i_{\text{sparse}}[:, j]$
    $\Xi[i_{\text{nonsparse}}, j] = (\mathbf{\Theta}[i_{\text{nonsparse}}, :]^\top \mathbf{\Theta}[i_{\text{nonsparse}}, :])^{-1} \mathbf{\Theta}[i_{\text{nonsparse}}, :]^\top \dot{X}[:, j]$
    $j = j + 1$
  end for

end for

6.3.1 Learning a Differential Equation

To illustrate SINDy, consider the following system of differential equations

$$\begin{aligned} \dot{x}_1 &= -x_1 + 2x_2, \\ \dot{x}_2 &= -x_1 + x_2. \end{aligned} \tag{6.42}$$

Using the initial conditions

$$x_1(0) = 1, \tag{6.43}$$
$$x_2(0) = 1, \tag{6.44}$$

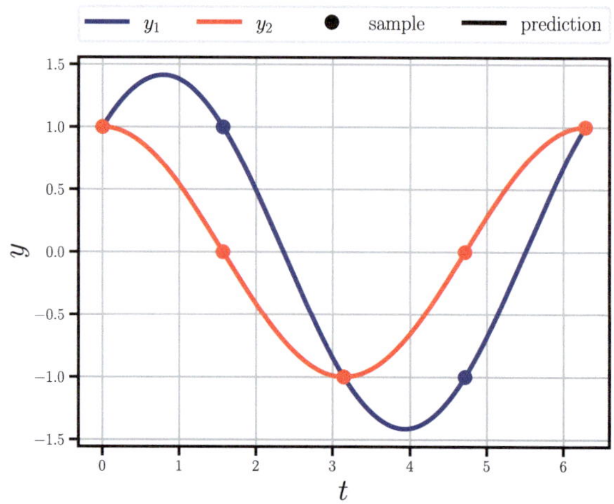

Fig. 6.14 Prediction based on system identified with SINDy

results in the solution

$$x_1(t) = \sin(t) + \cos(t), \tag{6.45}$$
$$x_2(t) = \cos(t). \tag{6.46}$$

By sampling the solution and its time derivative in form of snapshot matrices $\boldsymbol{X}$, $\dot{\boldsymbol{X}}$ at five timesteps (see Fig. 6.14), a SINDy system can be constructed. Using a candidate library composed of $1, x_1, x_2, x_1^2, x_2^2, x_1x_2$ and performing the sparse regression from Eq. (6.41), yields the coefficients $\boldsymbol{\Xi}$ listed below.

	$\dot{x}_1 =$	$\dot{x}_2 =$
1	0	0
x_1	-1	-1
x_2	2	1
x_1^2	0	0
x_2^2	0	0
x_1x_2	0	0

In this case, SINDy is able to retrieve the original system from Eq. (6.42). The identified system can be solved in order to predict solutions beyond the measurements in the snapshot matrix, as illustrated in Fig. 6.14.

Implementation-wise, the identification was achieved with the code presented below. After creating the snapshot matrices $\boldsymbol{X}$, $\dot{\boldsymbol{X}}$, a library candidate function is established using a lambda function. Lastly, the sparse set of coefficients is computed with the previously implemented `sequentialThresholdedLeastSquares` function.

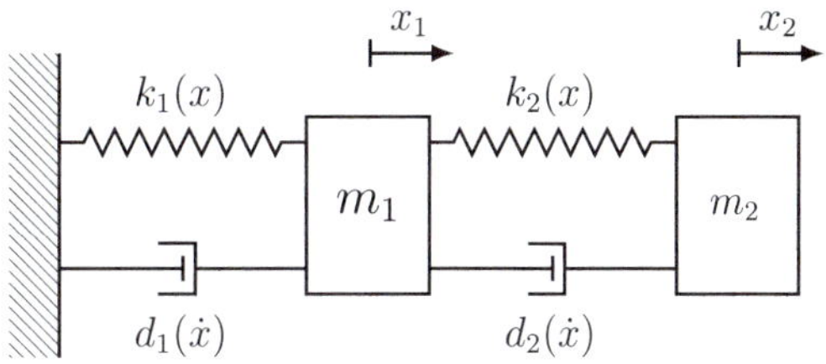

Fig. 6.15 Two-dimensional mass-spring-damper model from structural dynamics

```
# hyperparameters
tol = 1e-2
k = 10

# snapshot sampling
t = np.linspace(0, 2 * np.pi, 5)
t = np.expand_dims(t, axis=1)
x1 = np.sin(t) + np.cos(t)
x2 = np.cos(t)
dx1dt = np.cos(t) - np.sin(t)
dx2dt = -np.sin(t)

# time history collection
X = np.concatenate((x1, x2), axis=1)
dXdt = np.concatenate((dx1dt, dx2dt), axis=1)

# library of candidate functions
theta = lambda X: np.transpose(np.array([X[0] * 0 + 1, X[0], X[1],
                                         X[0] ** 2, X[1] ** 2,
                                         X[0] * X[1]]))

# solving the SINDy system
Xi = sequentialThresholdedLeastSquares(theta(np.transpose(X)), dXdt, k,
     tol)
```

6.3.2 *Structural Dynamics Model*

SINDy is now applied to fundamental multi-degree of freedom systems from structural dynamics. Consider the two-degree of freedom system depicted in Fig. 6.15. The goal is to identify the stiffness terms $k_1(x)$, $k_2(x)$, the damping $d_1(\dot{x})$, $d_2(\dot{x})$, and the masses m_1, m_2 from measurements of the deflections x_1, x_2. Note that the stiffness, and damping may follow any arbitrary relation and not necessarily a linear relation.

The governing equations of the system are derived from an equilibrium state, yielding

$$\begin{aligned} m_1\ddot{x}_1 + k_1(x_1) + d_1(\dot{x}_1) + k_2(x_1 - x_2) + d_2(\dot{x}_1 - \dot{x}_2) = 0, \\ m_2\ddot{x}_2 + k_2(x_2 - x_1) + d_2(\dot{x}_2 - \dot{x}_1) = 0. \end{aligned} \tag{6.47}$$

For the system of second order equations, two initial conditions per ordinary differential equation must be formulated to make the solution x_1, x_2 well-posed.

$$x_1(0) = x_{1_0} \tag{6.48}$$

$$\dot{x}_1(0) = \dot{x}_{1_0} \tag{6.49}$$

$$x_2(0) = x_{2_0} \tag{6.50}$$

$$\dot{x}_2(0) = \dot{x}_{2_0} \tag{6.51}$$

However, as SINDy is only able to handle systems of first-order differential equations (Eq. (6.38)), the system must be rewritten as system of first-order differential equations.[5] By introducing a substitution for the first-order derivatives $y_1 = x_1, y_2 = \dot{x}_1, y_3 = x_2, y_4 = \dot{x}_2$, the two equations from Eq. (6.47) can be rewritten as

$$\dot{y}_1 = y_2, \tag{6.52}$$

$$\dot{y}_2 = \frac{-k_1(y_1) - d_1(y_2) - k_2(y_1 - y_3) - d_2(y_2 - y_4)}{m_1}, \tag{6.53}$$

$$\dot{y}_3 = y_4, \tag{6.54}$$

$$\dot{y}_4 = \frac{-k_2(y_3 - y_1) - d_2(y_4 - y_2)}{m_2}. \tag{6.55}$$

This yields four first-order differential equations with the four unknowns y_1, y_2, y_3, y_4.

6.3.2.1 Linear Example

First consider the case, where the springs and dampers behave linearly, i.e.,

$$k_i(x) = K_i x, \tag{6.56}$$

$$d_i(\dot{x}) = D_i \dot{x}. \tag{6.57}$$

[5] This is possible for any system of higher order ordinary differential equations. Thus SINDy is applicable to all systems of ordinary differential equations, but not to partial differential equations in its original formulation. Extensions to partial differential equations do however exist [26, 27].

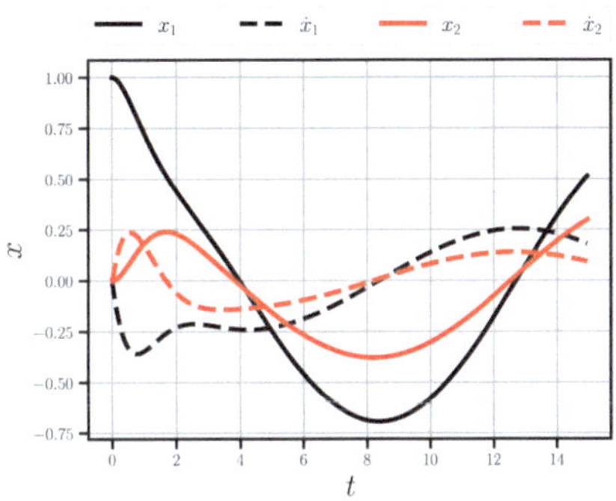

(a) Measurement

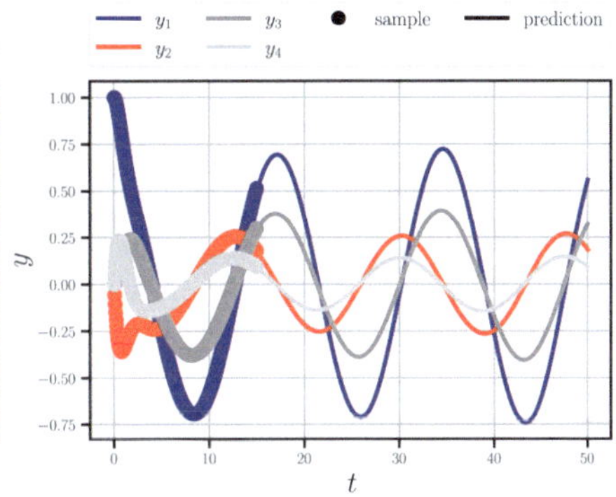

(b) Prediction based on SINDy

Fig. 6.16 Linear multi-degree of freedom system example

This allows us to express the system of equations from Eq. (6.52) in matrix form

$$\dot{\mathbf{y}} = \begin{pmatrix} 0 & 1 & 0 & 0 \\ -\frac{K_1+K_2}{m_1} & -\frac{D_1+D_2}{m_1} & \frac{K_2}{m_1} & \frac{D_2}{m_1} \\ 0 & 0 & 0 & 1 \\ \frac{K_1}{m_2} & \frac{D_1}{m_2} & -\frac{K_2}{m_2} & -\frac{D_2}{m_2} \end{pmatrix} \begin{pmatrix} y_1 \\ y_2 \\ y_3 \\ y_4 \end{pmatrix}. \tag{6.58}$$

The system parameters are provided in Table 6.2. The corresponding measurements of the system's response are conducted for 15 s with 300 snapshots, as depicted in Fig. 6.16a. A candidate library is constructed with polynomials up to degree $p = 2$. The identified system is presented below and compared to the true system (see Fig. 6.16b), which can directly be computed from Eq. (6.58).

Table 6.2 Parameters of linear multi-degree of freedom system

Parameters	K_1	K_2	D_1	D_2	M_1	M_2
Values	2 N/m	4 N/m	2 Ns/m	3 Ns/m	5 kg	2 kg

	$\dot{y}_1 =$	$\dot{y}_2 =$	$\dot{y}_3 =$	$\dot{y}_4 =$
1	0	0	0	0
y_1	0	−1.1340	0	0.9341
y_2	0.9999	−0.9677	0	0.9628
y_3	0	1.8778	0	−1.8779
y_4	0	1.4562	0.9997	−1.4564
y_1^2	0	0	0	0
y_2^2	0	0	0	0
y_3^2	0	0	0	0
y_4^2	0	0	0	0
$y_1 y_2$	0	0	0	0
$y_1 y_3$	0	0	0	0
$y_1 y_4$	0	0	0	0
$y_2 y_3$	0	0	0	0
$y_2 y_4$	0	0	0	0
$y_3 y_4$	0	0	0	0

(a) SINDy estimation

	$\dot{y}_1 =$	$\dot{y}_2 =$	$\dot{y}_3 =$	$\dot{y}_4 =$
1	0	0	0	0
y_1	0	−1.2	0	1
y_2	1	−1	0	1
y_3	0	2	0	−2
y_4	0	1.5	1	−1.5
y_1^2	0	0	0	0
y_2^2	0	0	0	0
y_3^2	0	0	0	0
y_4^2	0	0	0	0
$y_1 y_2$	0	0	0	0
$y_1 y_3$	0	0	0	0
$y_1 y_4$	0	0	0	0
$y_2 y_3$	0	0	0	0
$y_2 y_4$	0	0	0	0
$y_3 y_4$	0	0	0	0

(b) Original system from which the data was sampled from

Thus the original system is recovered approximately and is used to compute an extended time history, shown in Sect. 6.3.2.1 and compared to the measurements used to identify the system.

6.3.2.2 Non-linear Example

Until now, the true power of SINDy has not been exploited, as only linear systems were considered. To this end, we now employ non-linear spring and damper relations, summarized in Table 6.3. Again measurements are conducted. Due to the non-linearity, more measurements are required. Specifically, 25 s with 500 snapshots are recorded; see Fig. 6.17a.

Table 6.3 Functions of non-linear multi-degree of freedom system

Parameters	$k_1(x)$	$k_2(x)$	$d_1(\dot{x})$	$d_2(\dot{x})$	M_1	M_2
Functions	x	$2x^3$	$\dot{x}$	1	5 kg	2 kg

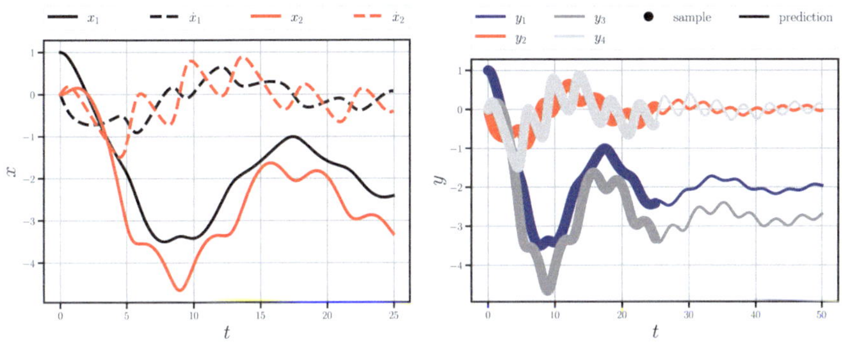

(a) Measurements

(b) Prediction based on system identified with SINDy

Fig. 6.17 Non-linear multi-degree of freedom system example

The resulting system in comparison to the original system is summarized below.

	$\dot{y}_1 =$	$\dot{y}_2 =$	$\dot{y}_3 =$	$\dot{y}_4 =$
1	0	−0.2029	0	−0.4986
y_1	0	−0.2006	0	0
y_2	0.9998	−0.2119	0	0
y_3	0	0	0	0
y_4	0	0	0.9998	0
y_1^2	0	0	0	0
y_3^2	0	0	0	0
$y_1 y_3$	0	0	0	0
y_1^3	0	−0.3844	0	0.9756
y_3^3	0	0.3935	0	−0.9915
$y_1^2 y_3$	0	1.1652	0	−2.9479
$y_1 y_3^2$	0	−1.1741	0	2.9634

(a) SINDy estimation

	$\dot{y}_1 =$	$\dot{y}_2 =$	$\dot{y}_3 =$	$\dot{y}_4 =$
1	0	−0.2	0	−0.5
y_1	0	−0.2	0	0
y_2	1	−0.2	0	0
y_3	0	0	0	0
y_4	0	0	1	0
y_1^2	0	0	0	0
y_3^2	0	0	0	0
$y_1 y_3$	0	0	0	0
y_1^3	0	−0.4	0	1
y_3^3	0	0.4	0	−1
$y_1^2 y_3$	0	1.2	0	−3
$y_1 y_3^2$	0	−1.2	0	3

(b) Original system from which the data was sampled from

Again only an approximative reconstruction is possible, which however after a forward simulation until 50 s fits the measurement data well, as seen in Fig. 6.17b.

6.4 Clustering

As touched upon in Sect. 2.4, clustering is an unsupervised task, in which a dataset is partitioned in a number of clusters. One way of viewing this is k-means clustering, where the dataset is subdivided into k clusters. Each datapoint belongs to the cluster with the closest mean. This can be formulated as a minimization problem

$$\arg\min_{\boldsymbol{\mu}_j} \sum_{j=1}^{k} \sum_{\boldsymbol{x}_i | y(\boldsymbol{x}_i)=y_j} ||\boldsymbol{x}_i - \boldsymbol{\mu}_j||_2^2, \tag{6.59}$$

where $\boldsymbol{\mu}_j$ is the mean of the jth cluster (or the cluster's center) computed from the datapoints associated with the cluster, i.e., $y_j = y(\boldsymbol{x}_i)$. Thereby the distances of the cluster's mean and of the samples associated with the cluster are to be minimized. This is, however, a NP-hard problem, whose global optimum is not attainable, due to the required computational resources for larger problems. Heuristic algorithms without guarantee of convergence do however exist. An example is Lloyds algorithm [28] summarized in Algorithm 13, which essentially is comprised of two parts:

1. computation of distances d_{ij}, used to identify closest cluster y_i by taking the minimum distance $\min_j d_{ij}$
2. recomputation of cluster centers $\boldsymbol{\mu}_j$ with the mean of the associated data points $\boldsymbol{x}_i \in D_j$

These are iterated upon until convergence.

Algorithm 13 Lloyds algorithm for k-means clustering [28]

Require: data $\boldsymbol{x}_i$, number of iterations, number of clusters
  Initialize cluster centers $\boldsymbol{\mu}_j$
  for all iterations **do**
    for all samples $\boldsymbol{x}_i$ **do**
      for all cluster centers $\boldsymbol{\mu}_j$ **do**
        Compute Euclidean distances $d_{ij} = ||\boldsymbol{\mu}_j - \boldsymbol{x}_i||_2^2$
      end for
      Identify associated cluster $y_i = \min_j d_{ij}$
    end for
    for all cluster centers $\boldsymbol{\mu}_j$ **do**
      Update cluster centers $\boldsymbol{\mu}_j = \frac{1}{m}\sum_{i=1}^{m} \boldsymbol{x}_i$
    end for
  end for

The algorithm can be implemented in Python with the following code snippet.

```
k = 3  # number of clusters
# initial guess
centers = [np.random.uniform(np.min(data, axis=0),
                             np.max(data, axis=0)) for i in range(k)]
classification = np.zeros(len(data), dtype=np.int64)
for iteration in range(numberOfIterations):
    for j in range(len(data)):
        distances = [np.linalg.norm(centers[i] - data[j], 2) for i in range(k)]
        classification[j] = np.argmin(distances)

```

```
    centers = [np.mean(data[classification == i], axis=0) for i in range(
     k)]
```

An alterntive is to directly use the implementation provided by the Scikit-Learn (`sklearn`) package.

```
from sklearn.cluster import KMeans
kmeans = KMeans(n_clusters=k, random_state=0).fit(data)
classification = kmeans.labels_
centers = kmeans.cluster_centers_
```

When performing clustering in high-dimensional spaces a challenge is identifying relevant features beforehand. To this end, the dimensionality reduction techniques from Sects. 6.1.1 and 6.1.2 are beneficial. These can be used to reduce high-dimensional data to extract the most relevant features which subsequently can be employed in clustering algorithms. Still clustering remains a useful technique in exploring design spaces, particularly due to its ability to effectively distinguish between different design options.

6.4.1 *Cross-Section Clustering*

As a didactical example from structural engineering, we consider the cross-section of an I-beam; see Fig. 6.18. The area moments of inertia (expressing the geometrical stiffness against bending in beams), and the area (expressing the longitudinal stiffness against tension/compression in bars) are given as

$$I_1 = \frac{bh^3}{12} + \frac{B}{12}(H^3 - h^3), \tag{6.60}$$

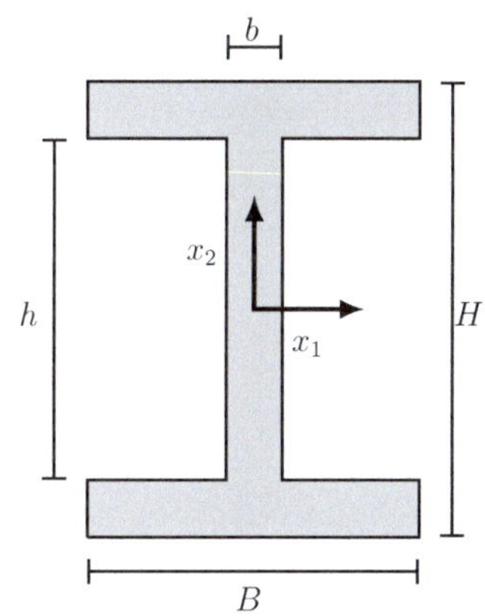

Fig. 6.18 I-beam cross-section

Table 6.4 Distributions used to generate three distinct cross-sections

	H	h	B	b
Cross-section 1	$\mathcal{U}(1, 2)$	$\mathcal{U}(0.5, H)$	$\mathcal{U}(2, 4)$	$\mathcal{U}(1, B)$
Cross-section 2	$\mathcal{U}(4, 5)$	$\mathcal{U}(0.5, H)$	$\mathcal{U}(2, 3)$	$\mathcal{U}(0.5, b)$
Cross-section 3	$\mathcal{U}(3.5, 3.5)$	$\mathcal{U}(3, H)$	$\mathcal{U}(3.5, 4.5)$	$\mathcal{U}(3, B)$

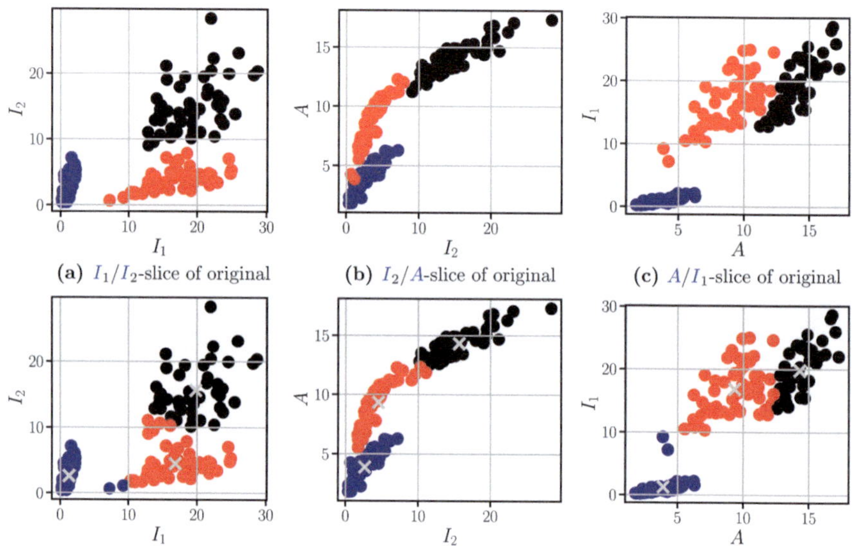

(a) I_1/I_2-slice of original (b) I_2/A-slice of original (c) A/I_1-slice of original

(d) I_1/I_2-slice of identified (e) I_2/A-slice of identified (f) A/I_2-slice of identified

Fig. 6.19 Original and identified clusters of I-beam cross-sections. The cluster centers are indicated by silver crosses, and the data's association to each cluster is indicated by either blue, red, or black

$$I_2 = \frac{b^3 h}{12} + \frac{B^3}{12}(H - h), \tag{6.61}$$

$$A = HB - h(B - b). \tag{6.62}$$

Three distinct distributions of I-beams were generated with uniform distributions $\mathcal{U}$ as summarized in Table 6.4. The resulting clusters (from 50 samples per I-beam distribution) are illustrated in Fig. 6.19a to c as slices of the three-dimensional space composed of I_1, I_2, A. Using Lloyds algorithm from Algorithm 13, similar clusters are identified (without labeled data); see Fig. 6.19. The resulting accuracy is 95%, quantified in terms of misclassified samples relative to the total number of samples. To showcase how the algorithm operates, the first four iterations are depicted in Fig. 6.20. This illustrates how the centers are moved (recomputed) after each iteration and how the samples are potentially assigned to the new clusters.

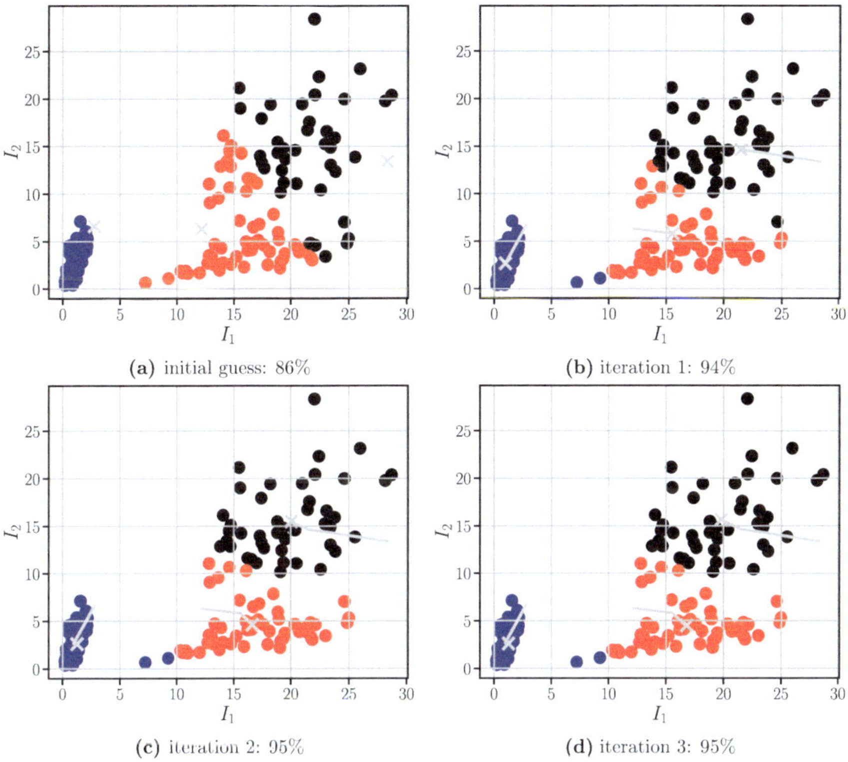

(a) initial guess: 86%

(b) iteration 1: 94%

(c) iteration 2: 95%

(d) iteration 3: 95%

Fig. 6.20 History of k-means clustering algorithm. Convergence is reached at the third iteration

6.5 Support Vector Machines

Lastly, we will touch on a classification algorithm: support vector machines [29, 30], that can be considered the predecessor of neural networks. Yet due to its simplicity and interpretability, support vector machines are still a useful tool. In contrast to the clustering approaches, support vector machines are a supervised learning algorithm relying on labeled datasets. More specifically, they are a binary classification algorithm, i.e., distinguishing only between two labels (-1 and 1).[6]

The simplest form are linear support vector machines, in which a hyperplane is constructed in order to separate between two categories. The hyperplane is defined implicitly with the weights $\boldsymbol{w}$ and bias b

---

[6] Binary classification algorithms are easily extended to multi-classification tasks. In the one-vs-one approach, N classes are handled by $N(N-1)/2$ classifiers, where each classifier separates one class from another. This is only good for small numbers of classes. For large number of classes, one-vs-rest classifiers should be employed. Here N classifiers are trained for N classes, where each separates ono class from all the others. Direct multiclass support vector machines offer yet another possibility [31, 32].

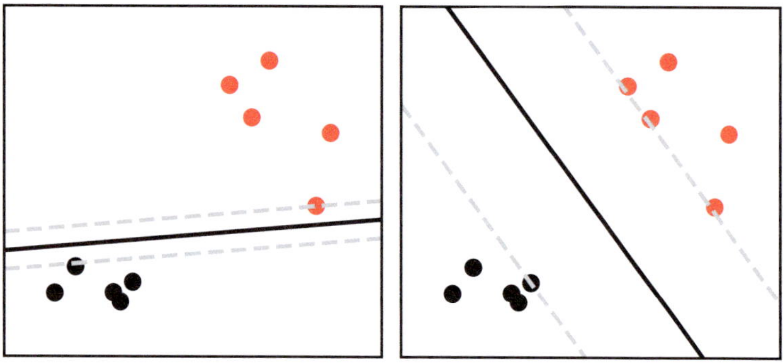

(a) Minimal margin (b) Maximal margin

Fig. 6.21 Concept of support vector machines. The hyperplane (in two dimensions a line) separates the two categories with the largest possible margin, wherein the margin is defined as the distance between the hyperplane and the closest points, indicated by the gray dashed lines. The closest points are the support vectors

$$\boldsymbol{w} \cdot \boldsymbol{x} + b = 0 \tag{6.63}$$

The hyperplane is defined to separate the data in the best possible manner, which can be framed as maximizing the margin between the closest points to the hyperplane, as illustrated in Fig. 6.21.

With the hyperplane from Eq. (6.63), a prediction $\hat{y}$ can be made, depending on which side of the plane the datapoint lies

$$\hat{y} = \text{sign}(\boldsymbol{w} \cdot \boldsymbol{x} + b). \tag{6.64}$$

Thus the prediction is either -1 or 1 depending on the position $\boldsymbol{x}$. In order to determine the hyperplane, an optimization is formulated. Firstly the data should be classified correctly and secondly the margin should be maximized. The following loss achieves the latter:

$$\mathcal{L}(\tilde{y}_i, \hat{y}(\tilde{\boldsymbol{x}}_i)) = \begin{cases} 0 \text{ if } \tilde{y}_i = \text{sign}(\boldsymbol{w} \cdot \tilde{\boldsymbol{x}}_i + b) \\ 1 \text{ if } \tilde{y}_i \neq \text{sign}(\boldsymbol{w} \cdot \tilde{\boldsymbol{x}}_i + b) \end{cases} \tag{6.65}$$

relying on labeled data $\tilde{\boldsymbol{x}}$, $\tilde{y}$. If a datapoint is mislabeled, the cost increases by one. In addition a constraint of

$$\min_i |\tilde{\boldsymbol{x}}_i \cdot \boldsymbol{w}| = 1 \tag{6.66}$$

ensures that the closest data points (the support vectors) are at a standardized distance (one) from the decision boundary, facilitating the maximization of the margin. The loss $\mathcal{L}$ from Eq. (6.65) is combined with a regularization term. The regularization term ensures a maximization of the margin, as the margin is inversely proportional to the weights' magnitude.[7]

$$C = \sum_{i=1}^{m} \mathcal{L}(\tilde{y}_i, \hat{y}(\tilde{\boldsymbol{x}}_i)) + \frac{1}{2}||\boldsymbol{w}||^2 \tag{6.67}$$

The cost C is however not differentiable due to $\mathcal{L}$, which prevents the use of gradient-based optimization. Instead a popular alternative is the piece-wise differentiable Hinge loss function

$$\mathcal{L}_{\mathcal{H}}(\tilde{y}_i, \hat{y}(\tilde{\boldsymbol{x}}_i)) = \max(0, 1 - \tilde{y}_i \hat{y}_i). \tag{6.68}$$

Inserting $\mathcal{L}_{\mathcal{H}}$ in place of $\mathcal{L}$ in Eq. (6.67) provides a differentiable cost function C. In addition, the Hinge loss automatically enforces the constraint from Eq. (6.66). In order to determine the optimal hyperplane, the cost function is minimized with a gradient-based optimization, as discussed in Chapter 2. The gradients are obtained analytically, as

$$\frac{\partial \mathcal{L}_{\mathcal{H}}}{\partial \boldsymbol{w}} = \begin{cases} -\tilde{y}_i \tilde{\boldsymbol{x}}_i \text{ if } 1 - \tilde{y}_i \hat{y}_i > 0 \\ 0 \text{ else }, \end{cases} \tag{6.69}$$

$$\frac{\partial \mathcal{L}_{\mathcal{H}}}{\partial b} = \begin{cases} -\tilde{y}_i \text{ if } 1 - \tilde{y}_i \hat{y}_i > 0 \\ 0 \text{ else }. \end{cases} \tag{6.70}$$

A corresponding Python implementation is given as follows. The cost function with the Hinge loss $\mathcal{L}_{\mathcal{H}}$ is defined first.

```
def getCostFunction(x, y, w, b):
    regularizationTerm = 0.5 * np.sum(w*w)
    hingeLoss = 0
    for i in range(len(x)):
        hingeLoss += max(0, 1 - y[i] * ((np.dot(w, x[i])) + b))
    cost = hingeLoss + regularizationTerm
    return cost
```

Next the gradient computation can be hard-coded using the analytical derivatives.

```
def getGradients(x, y, w, b):
    dcostdw = w
    dcostdb = 0
    for i in range(len(x)):
        if (1 - y[i] * ((np.dot(w, x[i])) + b)) > 0:
```

[7] Consider the distance of the hyperplane to the position $\boldsymbol{x}$, which is $\frac{|\boldsymbol{w} \cdot \boldsymbol{x} + b|}{||\boldsymbol{w}||}$. For the closest point, it holds that $\boldsymbol{w} \cdot \boldsymbol{x} = 1$ (or -1), which leads to the distance between hyperplane and closest point, i.e., margin becomes $\frac{1}{||\boldsymbol{w}||}$.

```
            dcostdw -= y[i] * x[i]
            dcostdb -= y[i]
    return dcostdw, dcostdb
```

Finally, gradient descent is implemented relying on the cost function and gradient computation.

```
for epoch in range(epochs):
    cost = getCostFunction(x, y, w, b)
    dcostdw, dcostdb = getGradients(x, y, w, b)

    w -= lr * dcostdw
    b -= lr * dcostdb
```

Revisiting the cross-section example from Sect. 6.4.1, we now only consider the two area moments of inertia I_1 and I_2 for two distinct clusters; see Fig. 6.22a. The support vector machine finds the hyperplane with the maximum margin. By rewriting the implicit formula from Equation (6.63), an explicit expression of the hyperplane can be formulated as

$$I_2 = \frac{-w_1 I_1 - b}{w_2}, \tag{6.71}$$

which is used in Fig. 6.22a. The clear downside of the linear support vector machine, is that only linear decision boundaries can be created. Non-linear support vector machines overcome this limitation through an encoding $e(x)$ of the input x. Typically, this entails a projection onto a higher-dimensional feature space in which a separation by a linear, but high-dimensional hyperplane is possible. The prediction from Eq. (6.64) is only affected by a minor modification.

$$\hat{y} = \text{sign}(\boldsymbol{w} \cdot e(\boldsymbol{x}) + b) \tag{6.72}$$

The remaining methodology is the same. One downside of non-linear support vector machines, is the curse of dimensionality invoked by very high-dimensional feature space encodings. This issue can be overcome by kernel support vector machines exploiting the kernel trick; see [1] for details.

To illustrate the power of non-linear support vector machines, consider the example presented in Fig. 6.22b. Here a sine-like separation boundary is enabled through projection into a higher-dimensional space using the encoding

$$e(x_1, x_2) = \begin{pmatrix} x_1 \\ x_2 \\ \sin(\pi x_1) \\ \sin(2\pi x_1) \end{pmatrix}. \tag{6.73}$$

In the expanded space, the hyperplane is still only linear. Only after projection back to the space x_1, x_2 does the hyperplane seem non-linear, as is the case in Fig. 6.22b. Again, as in Eq. (6.71) an explicit formula can be formulated for the hyperlane, by solving for the variable x_2, which however is not generally possible.

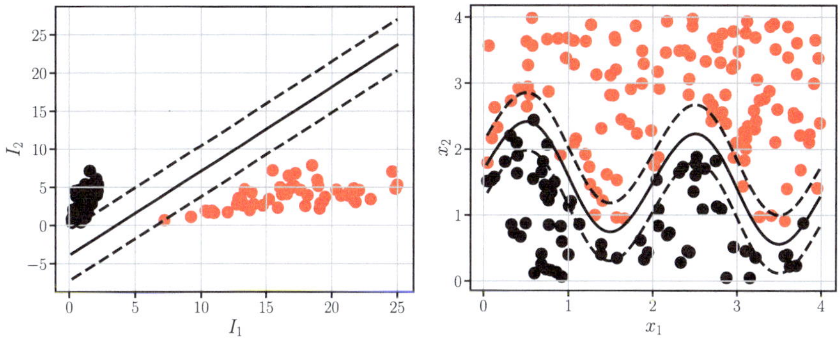

(a) Linear support vector machine with the cross-section example

(b) Non-linear support vector machine with a sine-like separation

Fig. 6.22 Support vector machine examples

Furthermore note that in Fig. 6.22b the support vector machine is not able to minimize the Hinge loss entirely towards zero, as the margins are not free of data points.

Important to note is also that the quality of separation depends entirely on the chosen encoding, which is a challenging task. An automatic way of finding a viable encoding e is through the linear dimensionality reduction techniques from Sects. 6.1.1 and 6.1.2 or even non-linear transformations, such as autoencoders from Chapter 8.

Exercises

E.22 Singular Value Decomposition (P & C)

1. Given the matrix $\boldsymbol{A}$ below, perform a singular value decomposition of $\boldsymbol{A}$.

$$\boldsymbol{A} = \begin{pmatrix} 3 & 1 \\ 0 & 4 \end{pmatrix}$$

2. Implement a function that performs the singular value decomposition on an arbitrary matrix $\boldsymbol{A} \in \mathbb{R}^{n \times m}$. The output should consist of the three matrices $\boldsymbol{U} \in \mathbb{R}^{n \times n}$, $\boldsymbol{\Sigma} \in \mathbb{R}^{n \times m}$, $\boldsymbol{V} \in \mathbb{R}^{m \times m}$, such that $\boldsymbol{U}\boldsymbol{\Sigma}\boldsymbol{V}^\top = \boldsymbol{A}$.
3. Consider the following matrix

$$\boldsymbol{A} = \begin{pmatrix} 1 & 0 & 0 & 2 & 0 \\ 0 & 0 & 0 & 0 & 1 \\ 3 & 4 & 1 & 1 & 2 \\ 2 & 0 & 0 & 4 & 0 \\ 0 & 0 & 0 & 0 & 2 \end{pmatrix}$$

(a) Original image (of Günther)

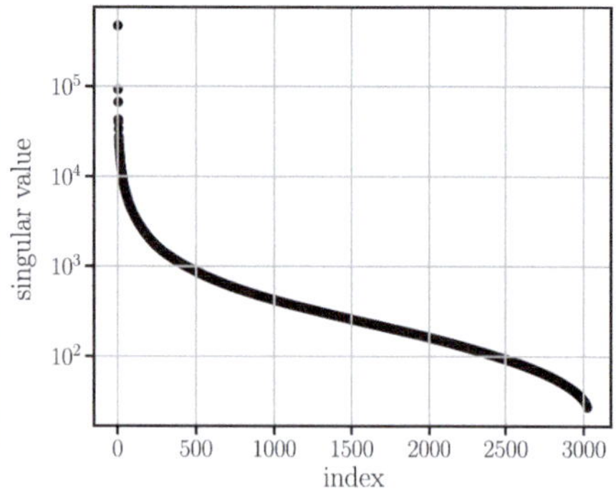

(b) Singular values

Fig. 6.23 Image compression to be performed in this exercise using singular value decomposition

and perform a singular value decomposition with the implemented function. Then, truncate the decomposition up to the rth singular value. What is the lowest possible truncation that decomposes $\boldsymbol{A}$ exactly and why?

4. Now perform a truncated singular value decomposition on an image (e.g., provided in Fig. 6.23) in order to see the compression capabilities discussed in Sect. 6.1.1.1. The image can be loaded with the Pillow library using `from PIL import Image`. Before operating on the image, the image should be converted to a grayscale image to reduce the data to a two-dimensional matrix. The following code snippet showcases how to transform the image `GuentherSmall.jpg`.

```
img = Image.open('GuentherSmall.jpg')
data = np.asarray(img)

# conversion to grayscale image
rgbWeights = [0.2989, 0.5870, 0.1140]
data = np.dot(data[..., :3], rgbWeights)  # shape is
    (1125, 1500)
```

5. What is the lowest possible truncation level r (for the image in Fig. 6.23 with pixel dimensions 3024×4032) without inducing a loss of information? What is the difference to the image considered in Figs. 6.3 and 6.4? What are the corresponding compression ratios for $r = 1, 2, 10, 20, 100, 200$ and the previously identified truncation level?

E.23 Introduction to Reduced Order Models (C)

A function $f(x, t)$ is defined in the spatial domain $x \in [0, 1]$ and the temporal domain $t \geq 0$

$$f(x, t) = 3\cos(2\pi x)t^2 + \sin(6\pi x)t + 4\cos(12\pi x).$$

1. Sample the function in the given spatial domain with 100 spatial point from $t = 0$ until $t = 10$ with a timestep size of $\Delta t = 0.5$. Collect the results in a snapshot matrix $\boldsymbol{X}$ as defined in Sect. 6.2, i.e., each column is one snapshot of the spatial domain for a given time.
2. Perform a singular value decomposition of the snapshot matrix $\boldsymbol{X}$
   Hint: Use the function `np.linalg.svd` for the singular value decomposition.
3. Plot the spatial modes found in the matrix $\boldsymbol{V}$. What do you observe and why?
   Hint: Take a closer look at the singular value matrix $\boldsymbol{\Sigma}$.
4. Plot the temporal modes found in the matrix $\boldsymbol{U}$. The singular value decomposition is often employed in the context of reduced order models. In computational mechanics, typically, only the spatial modes are of interest. Give a possible explanation for this.

E.24 Reduced Order Models with Finite Elements (C)

A cantilever beam is considered as shown in Fig. 6.24 with an external load $F(t) = f \cos(\omega t)$ on the free edge and a Young's modulus of $E = 2, 10, 000$ and mass density of $7.8 \cdot 10^{-6}$. Initially, the amplitude of the external load is defined as $f = 100$ and the frequency as $\omega = 0$, thus removing its time dependency. It is modeled with the finite element method, using bilinear quadrilateral elements. The discretization consists of $N_1 = 40$ elements in the x_1-direction and $N_2 = 4$ elements in the x_2-direction. Out of simplicity, all elements have the dimension 2×2, leading them to being identical to their definition in the parametric space.

1. The finite element code to simulate the time-dependent behavior is provided in `exercise 24 .ipynb` accompanied by visualization tools. The first step is to define the model and simulation parameters as defined above, which then can be used to construct an instance of the `DynamicCantileverBeam` class. This is achieved by calling the corresponding constructor `DynamicCantileverBeam (n1, n2, E, nu, rho, f, omega)`. After this, solve the model using the member function `Solve(n, dt)`, where `n` is the number of timesteps and `dt` is the step size. Run the simulation for 5000 time steps and a step size of $5 \cdot 10^{-6}$. The results can be visualized with `plotField(x0, x1, y, title)` and `plotDeformedField(x0, x1, u1, u2, y, title, s=1)` at specific

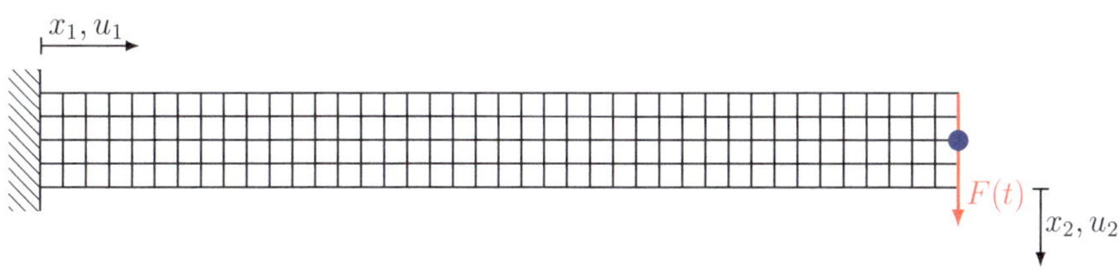

Fig. 6.24 Finite element mesh of the cantilever beam from Fig. 6.9

time instances, and with `plotTimeHistoryAtNode(x1, x2, y, x10, x20, dt, ylabel)` for the temporal history at a specific node at location `x10, x20`. Plot the time history at the right-most middle node marked in blue in Fig. 6.24.

2. The goal is now to perform additional simulation steps with a reduced order model. Thus, a reduced basis has to be identified. For this purpose, collect the previous simulation results in a snapshot matrix $(n \times m)$, available as the member `U`, where n is the number of degrees of freedom. Then, perform a singular value decomposition on the matrix `U`. For this task, define the function `ROM(U, r)`, where `r` is the truncated number of freedoms. The function should return the reduced basis with dimensions $(n \times r)$ and the full singular value matrix.
3. Plot the modeshape coefficients, i.e., the singular values, using the utility function `plotModeshapeCoefficients(s, r)`, where `s` is the singular value matrix. Determine an appropriate truncation for the reduced basis.
4. Plot the first four Euler-Bernoulli modeshapes from Eq. (6.35) together with the estimated modes found via the singular value decomposition. Use the utility function `plotModeshapes(phi, x, j, s, dim, modeshape` for this purpose, where `j` selects the jth mode shape, `s` is the deformation scaling factor, `dim` the dimension (0 for x_1, 1 for x_2), and `modeshape` is the analytical mode shape specified as lambda function.
5. Take a closer look at the member function `Solve(n, dt, phi=None)` from the `DynamicCantileverbeam` class. The possibility of including a reduced basis `phi` exists. The code is, however, incomplete. Add the modifications to allow solving the problem on a reduced basis.
6. Finally, solve the finite element model with the reduced basis for $10n$ timesteps and compare the computation times with the previous simulation.

E.25 Introduction to SINDy (P & C)

Consider the following two functions

$$x_1(t) = \sin(2t),$$
$$x_2(t) = 2\cos(2t) - \sin(2t).$$

1. Create a snapshot matrix $\boldsymbol{X}$ consisting of x_1, x_2 and its temporal derivative $\dot{\boldsymbol{X}}$ with three timesteps, starting at $t = 0$ and a stepsize of $\Delta t = \frac{\pi}{4}$.
2. Perform a SINDy using the snapshot matrix and the candidate functions $1, x_1(t), x_2(t)$ to recover the underlying system of differential equations.
3. A framework for SINDy is provided in `exercise25.ipynb`. Before SINDy can be performed, the sequential thresholded least squares algorithm has to be implemented as the function `sequentialThresholdedLeastSquares(theta, dXdt, k, tol`, where `theta` contains the library of candidate functions, `dXdt`

is the snapshot of the derivatives, `k` the number of sequential iterations, and `tol` the cut-off tolerance for the sparsity enforcement.
4. Modify the driver file to the example from above to verify your results and implementation.

E.26 SINDy for Structural Dynamics (P & C)

Consider the following system (depicted in Fig. 6.25) with two degrees of freedom with the following governing system of second-order differential equations.

$$m_1\ddot{x}_1 + K_1x_1 + D_1\dot{x}_1 + K_2(x_1 - x_2) = 0$$
$$m_2\ddot{x}_2 + k_3(x_2) + K_2(x_2 - x_1) = 0$$

All springs and dampers are linear, except $k_3 = K_3x^3$. The system parameters are $K_1 = 1$, $K_2 = 2$, $K_3 = 1$, $D_1 = 1$, $m_1 = 5$, $m_2 = 2$.

1. Rewrite the system of second-order differential equations into a system of first-order differential equations.
2. A modified driver file from Exercise E.25, namely `exercise26.ipynb` is to be used to identify the system above. To this end, the driver file contains a different multi-degree of freedom system. Modify the system in order to discover the system above. Are the system parameters identified correctly? Try adjusting the parameters, such as the number and range of data points, the candidate functions, and the sequential thresholded least squares parameters. What do you observe?

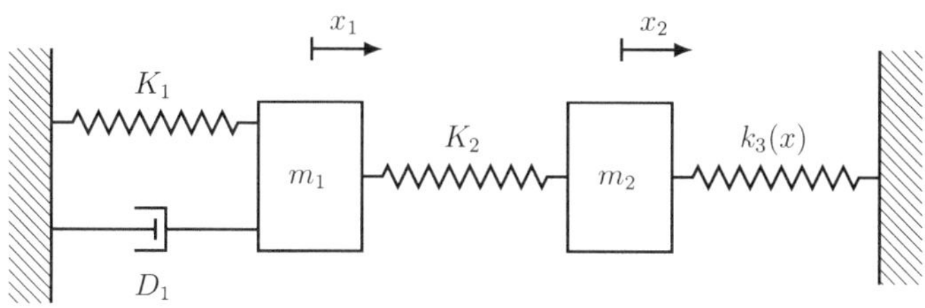

Fig. 6.25 Two-dimensional mass-spring-damper model from structural dynamics

References

1. Brunton, Steven L., and Jose, Nathan, Kutz. 2022. *Data-driven science and engineering: Machine learning, dynamical systems, and control*. Cambridge, United Kingdom, New York, NY: Cambridge University Press. ISBN: 978-1-00-909848-9.
2. Bishop, Christopher M. 2006. *Pattern recognition and machine learning*. eng. Softcover reprint of the original 1st edn (corrected at 8th printing 2009). Information science and statistics. New York, NY: Springer New York. ISBN: 978-1-4939-3843-8.
3. Alpaydin, Ethem. 2020. *Introduction to machine learning*. 4th edn. Adaptive computation and machine learning series. Cambridge, Massachusetts: The MIT Press. ISBN: 978-0-262-04379-3.
4. Russell, Stuart J., and Peter, Norvig. 2022. *Artificial intelligence: A modern approach*. eng. 4th edn., global edition. Pearson series in artificial intelligence. Harlow: Pearson. ISBN: 978-1-292-40113-3.
5. Stewart, G. W. 1993. On the early history of the singular value decomposition. *SIAM Review* 35(4): 551–566. ISSN: 0036-1445.
6. Strang, Gilbert. *Introduction to linear algebra*. eng. 5th edn. Wellesley: Cambridge Press. ISBN: 978-0-9802327-7-6.
7. Sirovich, L., and M. Kirby. 1987. Low-dimensional procedure for the characterization of human faces. *JOSA A* 4(3): 519–524. ISSN: 1520-8532. https://doi.org/10.1364/JOSAA.4.000519.
8. Kirby, M., and L. Sirovich. 1990. Application of the Karhunen-Loeve procedure for the characterization of human faces. *IEEE Transactions on Pattern Analysis and Machine Intelligence* 12(1): 103–108. ISSN: 1939-3539. https://doi.org/10.1109/34.41390.
9. Swets, D.L., and J.J. Weng. 1996. Using discriminant eigenfeatures for image retrieval. *IEEE Transactions on Pattern Analysis and Machine Intelligence* 18(8): 831–836. ISSN: 1939-3539. https://doi.org/10.1109/34.531802.
10. Belhumeur, P.N., J.P. Hespanha, and D.J. Kriegman. 1997. Eigenfaces versus Fisherfaces: Recognition using class specific linear projection. *IEEE Transactions on Pattern Analysis and Machine Intelligence* 19(7): 711–720. ISSN: 1939-3539. https://doi.org/10.1109/34.598228.
11. Pearson, Karl. 1901. LIII. On lines and planes of closest fit to systems of points in space. *The London, Edinburgh, and Dublin Philosophical Magazine and Journal of Science* 2(11): 559–572. ISSN: 1941-5982, 1941-5990. https://doi.org/10.1080/14786440109462720.
12. Hotelling, H. 1933. Analysis of a complex of statistical variables into principal components. *Journal of Educational Psychology* 24(6): 417–441. ISSN: 1939-2176. https://doi.org/10.1037/h0071325.
13. Jolliffe, I.T. 2002. *Principal component analysis*. 2nd ed. Springer series in statistics. New York: Springer. ISBN: 978-0-387-95442-4.
14. Hastie, Trevor, Robert, Tibshirani, and Jerome, Friedman. 2009. *The elements of statistical learning: Data mining, inference, and prediction*, 2nd edn. Springer series in statistics. tex.ids= Hastie2009a, Hastie2009b. New York: Springer. ISBN: 978-0-387-84857-0. https://doi.org/10.1007/978-0-387-84858-7.
15. Gerbrands, Jan J. 1981. On the relationships between SVD, KLT and PCA. *Pattern recognition. 1980 Conference on Pattern Recognition* 14(1): 375–381. ISSN: 0031-3203. https://doi.org/10.1016/0031-3203(81)90082-0.
16. Benner, Peter et al., eds. 2021. *Model order reduction*. OCLC: on1111771838. Berlin; Boston: De Gruyter. ISBN: 978-3-11-050043-1 978-3-11-067140-7 978-3-11-050044-8.
17. Hughes, Thomas J.R. 2000. *The finite element method: Linear static and dynamic finite element analysis*. Mineola, NY: Dover Publications. ISBN: 978-0-486-41181-1.
18. Langtangen, Hans Petter, Svein, Linge, and Springer International Publishing AG. 2018. *Finite difference computing with PDEs a modern software approach*. English. OCLC: 1049590683. ISBN: 978-3-319-85666-7.
19. Langtangen, Hans Petter, and Kent-Andre, Mardal. 2019. *Introduction to numerical methods for variational problems*. eng. Texts in computational science and engineering, vol. 21. Cham: Springer. ISBN: 978-3-030-23790-5 978-3-030-23787-5.

20. Butcher, J.C. 1964. Implicit Runge-Kutta processes. *Mathematics of Computation* 18(85): 50–64. ISSN: 0025-5718. https://doi.org/10.2307/2003405.
21. Dormand, J.R., and P.J. Prince. 1980. A family of embedded Runge-Kutta formulae. *Journal of Computational and Applied Mathematics* 6(1): 19–26. ISSN: 0377-0427. https://doi.org/10.1016/0771-050X(80)90013-3.
22. Newmark, Nathan M. 1959. A method of computation for structural dynamics. *Journal of the Engineering Mechanics Division* 85(3): 67–94. ISSN: 0044-7951, 2690-2427. https://doi.org/10.1061/JMCEA3.0000098.
23. Chinesta, Francisco, Pierre, Ladeveze, and Elías, Cueto. 2011. A short review on model order reduction based on proper generalized decomposition. *Archives of Computational Methods in Engineering* 18(4): 395–404. ISSN: 1886-1784. https://doi.org/10.1007/s11831-011-9064-7.
24. Inman, D.J. 2014. *Engineering vibration*. 4th edn. Boston: Pearson. ISBN: 978-0-13-287169-3.
25. Brunton, Steven L., Joshua L. Proctor, and J. Nathan Kutz. 2016. Discovering governing equations from data by sparse identification of nonlinear dynamical systems. *Proceedings of the National Academy of Sciences* 113(15): 3932–3937. ISSN: 0027-8424, 1091-6490. https://doi.org/10.1073/pnas.1517384113.
26. Rudy, Samuel H., et al. 2017. Data-driven discovery of partial differential equations. *Science Advances* 3(4): e1602614. ISSN: 2375-2548. https://doi.org/10.1126/sciadv.1602614.
27. Schaeffer, Hayden. 2017. Learning partial differential equations via data discovery and sparse optimization. *Proceedings of the Royal Society A: Mathematical, Physical and Engineering Sciences* 473(2197): 20160446. ISSN: 1364-5021, 1471-2946. https://doi.org/10.1098/rspa.2016.0446.
28. Lloyd, S. 1982. Least squares quantization in PCM. *IEEE Transactions on Information Theory* 28(2): 129–137. ISSN: 1557-9654. https://doi.org/10.1109/TIT.1982.1056489.
29. Boser, Bernhard E., Isabelle M. Guyon, and Vladimir N. Vapnik. 1992. A training algorithm for optimal margin classifiers. In: *Proceedings of the fifth annual workshop on Computational learning theory*. COLT '92. New York, NY, USA: Association for Computing Machinery, 144–152. ISBN: 978-0-89791-497-0. https://doi.org/10.1145/130385.130401.
30. Cortes, Corinna, and Vladimir, Vapnik. 1995. Support-vector networks. *Machine Learning* 20(3): 273–297. ISSN: 1573-0565. https://doi.org/10.1007/BF00994018.
31. Weston, J., and C. Watkins. 1999. Support vector machines for multi-class pattern recognition. In: *ESANN'1999 proceedings - European symposium on artificial neural networks*.
32. Crammer, Koby, and Yoram, Singer. 2001. On the algorithmic implementation of multiclass kernel-based vector machines. *Journal of Machine Learning Research* 2: 265–292. ISSN: ISSN 1533-7928.

Chapter 7
Material Modeling with Neural Networks

Abstract The formulation of material models is both an art and a science, as it requires extensive knowledge and skillful tuning by the practitioner for every new material model. Machine learning techniques offer an automatized pipeline to this potentially tedious process but must be designed appropriately to deliver reliable results. This chapter considers both data-driven and physics-aware/physics-augmented approaches. It starts with elastic materials in one dimension, explains how neural networks can model linear and hyperelastic behavior, and how incorporating physical laws improves predictions drastically. Extensions to viscoelastic materials explore architectures like recurrent neural networks and thermodynamic consistency constraints. For three-dimensional solids, advanced techniques such as invariant-based and deformation gradient-based physics-augmented neural networks are introduced. The chapter provides practical examples to convey these methods didactically and highlights their potential to transform current material modeling by integrating learning (from data) techniques with established physical principles.

In continuum field theories such as mechanics, thermal or chemical diffusion, electromagnetics, or coupled, multi-physical problems, the governing partial differential equations require constitutive models that describe the material behavior. For instance, in linear elasticity the Cauchy stress $\boldsymbol{\sigma}$ is related to the infinitesimal strain tensor ε by Hooke's law for isotropic materials as $\boldsymbol{\sigma} = \lambda\,\mathrm{tr}(\varepsilon)\,\boldsymbol{I} + 2\mu\,\varepsilon$, where λ and μ are the Lamé parameters, i.e., the elastic coefficients. Similarly, in magnetostatics, the magnetic flux density $\boldsymbol{b}$ can be related to the magnetic field intensity $\boldsymbol{h}$ by $\boldsymbol{b} = \mu_b\,\boldsymbol{h}$, where μ_b is the magnetic permeability. However, such simple linear material laws often have limited applicability. In finite strain continuum mechanics, which must be used to model polymers, rubbers, or biomaterials that can undergo large deformations, the stress-strain relation is non-linear, i.e., the first Piola-Kirchhoff stress tensor $\boldsymbol{P} = \boldsymbol{P}(\boldsymbol{F})$ depends nonlinearly on the deformation gradient $\boldsymbol{F}$. Likewise, in magnetostatics, due to the saturation observed in most magnetic materials, also the relation $\boldsymbol{b} = \boldsymbol{b}(\boldsymbol{h})$ is, in fact, non-linear.

The formulation of material models is both an art and a science. On the one hand, constitutive models must fulfill certain physical and mathematical requirements. In solid mechanics, they should be independent of the choice of coordinate system, consider that a material cannot be compressed or expanded infinitely, respect the laws

L. Herrmann et al., *Deep Learning in Computational Mechanics*,
https://doi.org/10.1007/978-3-031-89529-6_7

of thermodynamics, and ensure stability of numerical schemes. On the other hand, the most suitable functional description of a non-linear constitutive model depends highly on the material at hand. Thus, many different forms of non-linear elastic material models have been developed over the past decades, each usually applicable to certain (classes of) materials. However, due to the nonlinearities, anisotropies, and multiphysically coupled behaviors of modern materials, such as reinforced composites, multilayered laminates, microstructured tissues, and biomaterials, or additively manufactured metamaterials (see, e.g., the complex microstructures in Fig. 8.11), defining or identifying suitable constitutive models becomes increasingly difficult. Even more so when the material behavior is not energy conserving, i.e., when dissipative effects such as viscous damping or plasticity (or hysteresis for magnetics) must be considered.

Thus, the general approximation capabilities of artificial neural networks and also other techniques, such as Gaussian process or sparse regression, have motivated the use of machine learning for formulating more universal material models, which has been a rapidly expanding field of research in recent years. Machine learning methods are generally able to represent highly non-linear functional relationships. However, as ingredients to numerical simulations that must deliver accurate, reliable, and trustworthy outcomes, machine learning-based constitutive models should fulfill physical and mathematical conditions. Classical, analytically formulated material models typically adhere to (most of) these requirements and thus this knowledge should not be disregarded when applying machine learning for material modeling. Furthermore, analytical models can usually be characterized based on a minimal amount of experimental data. Nevertheless, they generalize well and behave reasonably also for unseen load cases. Such behavior is also highly desirable for machine learning-based material models, as experimental characterization data is usually scarce and expensive to obtain.

The aim of this Chapter is to outline how machine learning can be employed in material modeling. In particular, the focus is to convey how neural networks can be *informed* or *augmented* with constitutive conditions. This generally not only improves the accuracy of neural network-based material models but also greatly enhances their generalization capabilities, which means that they always deliver reliable and reasonable predictions. Furthermore, it greatly reduces the required training data, essential for its practical applicability. The implementations of the discussed models can be found at www.deeplearningincomputationalmechanics.com.

7.1 Elastic Material Modeling in One Dimension

In this section, the essential ideas and approaches for the formulation of *physics-augmented* neural networks for constitutive modeling are introduced with applications to one-dimensional mechanics. The deformation of a one-dimensional bar is described by its displacement $u(x)$, which must fulfill the governing differential equation $-\sigma' = p$ (as previously encountered in Sect. 4.2.1), where σ is the axial stress and p the axially applied load. The stress σ is related to the strain $\varepsilon = u'$ by a material law $\sigma = \sigma(\varepsilon)$.

7.1.1 Linear Elasticity

Usually, the material behavior is assumed as linear elastic and described by Hooke's law as $\sigma = E\,\varepsilon = E\,u'$, where E is the Young's modulus of the material. This would be sufficient to describe the behavior of a metal bar at small strains. However, for instance, a rubbery bar could undergo large tensile strains of 100% or more, and the stress-strain relation $\sigma(\varepsilon)$ could be highly non-linear in this scenario. Figure 7.1a shows data for such a non-linear stress-strain relation, which could stem from conducting a uni-axial tension test experimentally. As can also be seen in Fig. 7.1, a linear elastic material model that exactly fits the elastic modulus at $\varepsilon = 0$ with $E = 3.2$ describes this relation insufficiently for larger strains. A suitable material model should be able to capture the non-linear stress-strain behavior and also be able to provide reasonable stress values in compression for $-1 < \varepsilon < 0$ and in tension for $\varepsilon > 2$.

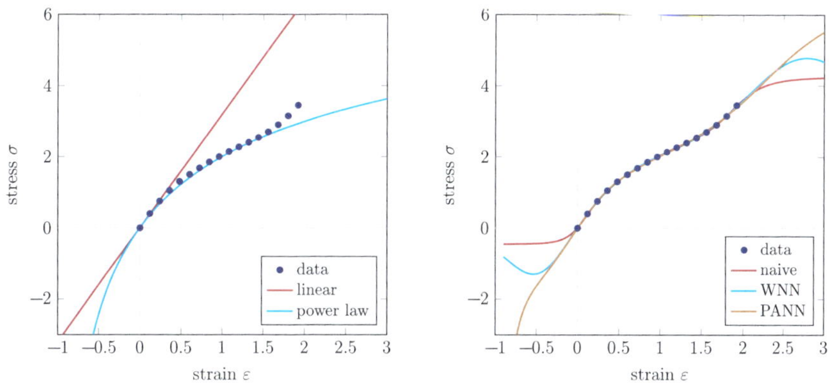

(a) Non-linear stress-strain data and evaluation of linear elastic and power law (Equation (7.5)) analytical models

(b) Evaluation of machine learning models σ_{NN} (naive), W_{NN} (standard neural network that approximates stress-strain law indirectly through the strain energy potential W) and W_{PANN} (physics-augmented neural network)

Fig. 7.1 Stress-strain data and material models for one-dimensional elasticity. Details on the models are discussed in Sect. 7.1.3

7.1.2 Naive Neural Networks

Now, to apply machine learning for non-linear material modeling, a straightforward idea is to formulate a fully connected neural network as a surrogate material model: $\sigma_{NN}(\varepsilon; \boldsymbol{\Theta})$. Figure 7.1b shows the predictions of such a neural network model (using two hidden layers with eight neurons each and Sigmoid activations, and a linear output layer) when trained on the dataset $\{(\tilde{\varepsilon}_k, \tilde{\sigma}_k)\}_{k=1}^{m_\mathcal{D}}$ with $m_\mathcal{D} = 19$ data points using the stress mean squared error. While an excellent fit can be observed within the range of the training data, the model delivers fairly unreasonable predictions outside the training range as the stresses plateau for compressive strains with $\varepsilon < 0$ as well as for larger tensile strains with $\varepsilon > 2$.

7.1.3 Hyperelasticity

Rubbers can still behave *elastically* even at large strains, i.e., deformations should be fully reversible, and the stress should solely depend on the current strain. This means that no energy is dissipated during deformation, and the material model should respect energy conservation. Thus, the mechanical power $\dot{W}$, i.e., the rate of change of the internal energy density W of the material, should be equal to the product of stress and strain rate [4]: $\dot{W} = \sigma\,\dot{\varepsilon}$. Now, a *hyperelastic* material model defines the internal (or *strain*) energy density solely as a function of the strain,

$$W = W(\varepsilon) \quad \Rightarrow \quad \dot{W}(\varepsilon) = \frac{\partial W}{\partial \varepsilon}\,\dot{\varepsilon} \quad \Rightarrow \quad \sigma(\varepsilon) = \frac{\partial W}{\partial \varepsilon}(\varepsilon), \tag{7.1}$$

from which the stress is obtained by differentiation. In this way, the material model not only fulfills energy conservation by construction but also allows the formulation of additional conditions for physically reasonable material behavior, such as [1]:

Energy and stress normalization: $$W|_{\varepsilon=0} = 0, \quad \sigma|_{\varepsilon=0} = 0, \tag{7.2}$$

Growth conditions: $$W \to \infty \text{ as } \varepsilon \to -1 \wedge \varepsilon \to \infty, \tag{7.3}$$

Convexity: $$\frac{\partial^2 W}{\partial \varepsilon^2} \geq 0 \quad \forall \varepsilon \in \,]-1, \infty[. \tag{7.4}$$

The conditions from Eq. (7.2) seem intuitively reasonable since stresses should be zero when the material is undeformed, and thus, no potential energy is stored in it. Likewise, it is intuitive that Eq. (7.3) must hold since an infinite amount of energy should be required to compress material to zero length/volume (i.e., as $\varepsilon \to -1$) and also to stretch it infinitely ($\varepsilon \to \infty$). The latter condition, Eq. (7.4), means that a material must always have non-negative "*stiffness*" and increasing stress for increasing strain. It ensures that the solution of the boundary value problem is well-defined.

7.1.3.1 Analytical Models

In fact, Hooke's law of linear elasticity $\sigma(\varepsilon) = E\,\varepsilon$ can be derived from the hyperelastic strain energy potential $W(\varepsilon) = \frac{1}{2}E\,\varepsilon^2$. However, this model does not fulfill the growth condition (Eq. 7.3) for $\varepsilon \to -1$, and it leads to unphysical behavior as the material is severely compressed; see Fig. 7.1a.

An alternative could be a "*power law*" strain energy potential with $p > 0,\ p \neq 1$:

$$W(\varepsilon) = \frac{E}{p(p-1)}\big((1+\varepsilon)^p - p\varepsilon - 1\big) \quad \Rightarrow \quad \sigma(\varepsilon) = \frac{E}{p-1}\big((1+\varepsilon)^{p-1} - 1\big), \tag{7.5}$$

which fulfills all constitutive conditions. Choosing $E = 3.2$ to match the elastic modulus at $\varepsilon = 0$ and $p = 0.7$, as shown in Fig. 7.1a, this model fits the highly non-linear stress-strain data fairly well and behaves reasonably also in compression, but cannot capture the stiffening observed in the data for $\varepsilon > 1.3$.

7.1.3.2 Hyperelastic Neural Networks

To obtain a hyperelastic model using machine learning, we can express the strain energy density by a fully connected neural network as $W_{NN}(\varepsilon; \boldsymbol{\Theta})$. Then, the stress prediction for this hyperelastic neural network can be obtained by differentiating the neural network with respect to its input using backpropagation as $\sigma(\varepsilon) = \frac{\partial W_{NN}}{\partial \varepsilon}$. As the strain energy W is typically not observable in experiments, also this neural network model can only be trained on stress-strain data using the mean squared error of the stresses, i.e., the derivatives of W_{NN} with respect to its inputs, similar to *Sobolev training*, seen in Sect. 3.8. Similar to physics-informed neural networks from Chaps. 4 and 5, the input-derivative of the neural network is evaluated in the loss function, which entails that also the parameter-derivatives of this derivative term need to be computed for the overall gradient of the loss function. Thus, both the evaluation and the training of W_{NN} are computationally more expensive and time-consuming compared to the naive σ_{NN}. As can be seen in Fig. 7.1b, expressing the strain energy potential as a fully connected neural network W_{NN} with just one hidden layer with eight neurons, Sigmoid activations, and a linear output layer, this machine learning model is highly accurate within the training range but also does not generalize well and shows highly unreasonable extrapolation in compression and tension.

7.1.3.3 Physics-Augmented Neural Networks

However, the formulation in terms of W allows to *augment* the hyperelastic neural network with the further constitutive requirements. To ensure convexity; see Eq. (7.4), the fully connected neural network can be made convex by employing an *input-convex*[1] neural network (icNN) architecture $W_{icNN}(\varepsilon; \boldsymbol{\Theta})$, which uses a convex and non-decreasing activation function, such as Softplus, and only non-negative weights in all hidden layers except the first one [20]. Now, the normalization conditions from Eq. (7.2) can be enforced by adding two terms to W_{icNN}, which depend on its evaluation for $\varepsilon = 0$. The growth conditions in Eq. (7.3) are ensured by adding two analytically defined terms $W^{gr}_{-1}(\varepsilon) = a\big((1+\varepsilon)^{-1} + \varepsilon - 1\big)$ and $W^{gr}_{\infty}(\varepsilon) = b\,\varepsilon^2$, which exhibit the desired growth behavior (and must also be normalized and convex) and can be scaled by either pre-defined or trainable weight factors $a, b > 0$. Overall, the *physics-augmented neural network* (PANN) one-dimensional hyperelastic material model is formulated as

$$W_{PANN}(\varepsilon) = W_{icNN}(\varepsilon) - W_{icNN}\big|_{\varepsilon=0} - \frac{\partial W_{icNN}}{\partial \varepsilon}\bigg|_{\varepsilon=0} \varepsilon + a\Big(\frac{1}{1+\varepsilon} + \varepsilon - 1)\Big) + b\,\varepsilon^2. \quad (7.6)$$

Again, the training of this model is performed only on stress-strain data using the mean squared error. Figure 7.1b also shows the results for this machine learning model, where W_{icNN} is using just one hidden layer with eight neurons, Softplus activations, and arbitrary weights, a linear output layer with non-negative weights, and trainable a, b. For this physics-augmented neural network model, not only the fit in the training regime is very good, but the model also delivers reasonable and thus robust predictions in extrapolation with the stresses tending to $-\infty$ as $\varepsilon \to 0$ and increasing stresses for $\varepsilon > 2$.

7.1.4 Discussion

The much-improved generalization of the physics-augmented neural network model compared to an only hyper/-elastic and a naive stress-strain model underpin the importance of considering prior knowledge, here in the form of the constitutive conditions (Eqs. (7.2), (7.3) and (7.4)), as inductive biases in the formulation of machine learning models. In contrast to *physics-informed* neural networks, which fulfill the governing equations of a problem only weakly, the *physics-augmented* neural network architecture for material modeling enforces all conditions strongly,

---

[1] *Convex function*: A function $f(x)$ is convex if, for any two points x_1 and x_2, the line segment connecting $f(x_1)$ and $f(x_2)$ lies above or on the graph of the function. Mathematically, $f(\lambda x_1 + (1-\lambda)x_2) \leq \lambda f(x_1) + (1-\lambda) f(x_2)$ for $\lambda \in [0, 1]$.

(Parametrized) input-convex function: A function $f(x; \boldsymbol{\Theta})$ is input-convex if it is convex in the input x for fixed parameters $\boldsymbol{\Theta}$. This is useful in optimization problems where $\boldsymbol{\Theta}$ is learned, and convexity in x ensures tractable solutions.

as the model architecture is designed in a way that ensures that all these requirements are always fulfilled by construction.

7.2 Viscoelastic Material Modeling in One Dimension

In fact, most materials do not exhibit perfectly elastic, energy-conserving behavior but dissipate energy during deformation through various *inelastic* effects such as viscoelasticity, plasticity, softening, damage, or fracture [1]. In these cases, the stress depends not only on the current strain but also on the history of the deformation in terms of past strain or stress states, rates, maximum values, or similar. Thus, (hyper-) elastic material models, in which the stress only depends on the current strain, cannot be used to model these material behaviors.

For instance, soft, rubbery materials are typically viscoelastic, i.e., they show a dependence of the stress-strain behavior on the strain rate $\dot{\varepsilon}$ and stress-relaxation over time; see Fig. 7.2. To fit viscoelastic material models, one typically uses stress-strain data over time, i.e., a dataset of the form $\{(t_k, \varepsilon_k, \sigma_k)\}_{k=1}^{m_\mathcal{D}}$. This dataset should include stress-strain curves ("*load paths*") for more than one applied strain rate $\dot{\varepsilon}$ and ideally also unloading or relaxation tests.

7.2.1 *Naive Fully Connected Neural Networks*

Several data-driven machine learning approaches appear to be viable for modeling viscoelastic material behavior. As the stress in an inelastic material depends on the history of stress, strain, and/or their rates, a straightforward approach would be stress

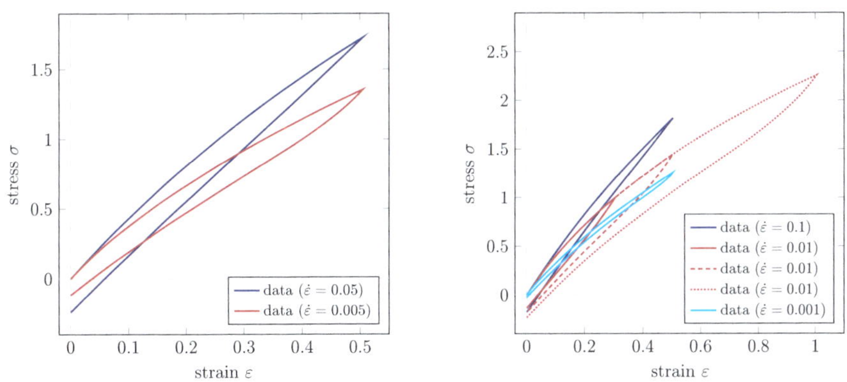

(a) Loading and unloading training data (b) Loading and unloading test data

Fig. 7.2 Data used for training and evaluation of viscoelastic one-dimensional models prediction using a fully connected neural network that takes the history in terms of a certain number k of past states also as input

$$\sigma_i = \sigma_{NN}(\varepsilon_i,\ \Delta t_i,\ \ \varepsilon_{i-1},\ \Delta t_{i-1},\ \sigma_{i-1},\ \ldots,\ \varepsilon_{i-k},\ \Delta t_{i-k},\ \sigma_{i-k}). \tag{7.7}$$

Such models can be expected to perform well when trained on a sufficient amount of data. However, similar to the elastic σ_{NN}, they extrapolate poorly when exceeding the training ranges of strains, stresses, or rates/time step sizes; see [21] for examples. Furthermore, the number of past states to be included must be determined heuristically (as few as possible and as many as necessary). Additionally, neural network sizes in terms of layers and number of neurons per layer must be fairly large, as the number of neural network inputs is relatively large.

This naive, fully connected neural network model is demonstrated in Fig. 7.3. The model architecture, which is illustrated in Fig. 7.3a, consists of a fully connected neural network with five hidden layers with 40 nodes in each layer, using Softplus

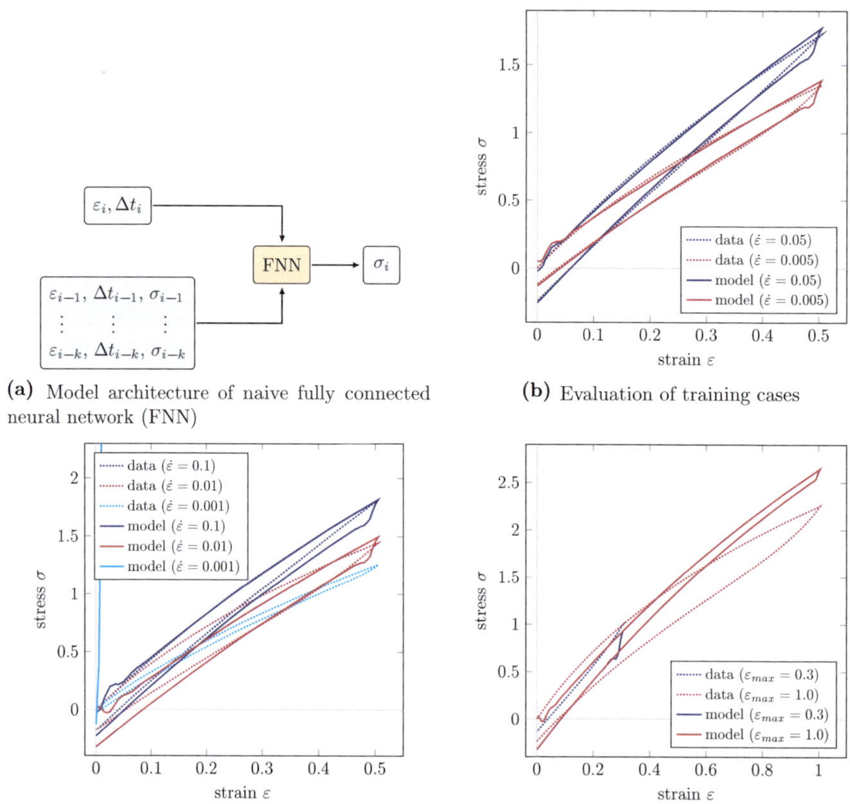

(a) Model architecture of naive fully connected neural network (FNN)

(b) Evaluation of training cases

(c) Test cases with other strain rates

(d) Test cases with other max. strains

Fig. 7.3 Naive fully connected neural network model for one-dimensional viscoelasticity activations and a linear output layer. As inputs, $k = 3$ states from previous time steps are used in addition to the current strain and time step size.

The model is trained on two load curves with loading up to a strain of $\varepsilon_{\max} = 0.5$ and subsequent unloading, which is carried out at the strain rates $\dot{\varepsilon} = 0.05$ and $\dot{\varepsilon} = 0.005$. The data and the predictions of the trained model (after 16,000 epochs using Adam with a learning rate of $\alpha = 0.001$) are shown in Fig. 7.3b. While the agreement with the training data, as well as with test cases at strain rates $\dot{\varepsilon} = 0.1$ and $\dot{\varepsilon} = 0.01$ shown in Fig. 7.3c is satisfactory, the evaluation for $\dot{\varepsilon} = 0.001$ completely fails. Also the predictions for using $\dot{\varepsilon} = 0.01$ but unloading at $\varepsilon_{\max} = 0.3$ and $\varepsilon_{\max} = 1.0$ shown in Fig. 7.3d are recognizably worse.

7.2.2 *Naive Recurrent Neural Networks*

Alternatively, the time series character of stress prediction, of course, also motivates the use of recurrent neural networks, such as LSTMs, that were introduced in Sect. 3.9.6. There, the idea is to obtain the stress prediction from a hidden state $\boldsymbol{h}$ that is updated from time step to time step by the LSTM

$$(\boldsymbol{h}_i, \boldsymbol{c}_i) = \boldsymbol{h}_{\mathrm{LSTM}}(\varepsilon_i, \Delta t_i, \ \boldsymbol{h}_{i-1}, \boldsymbol{c}_{i-1}) \qquad \rightarrow \qquad \sigma_i = \sigma_{NN}(\boldsymbol{h}_i). \tag{7.8}$$

Such a recurrent neural network architecture with an LSTM and a fully connected neural network is generally more concise than the above-mentioned fully connected neural network since the involved neural networks take much fewer inputs. When training on a sufficient amount of data, it is usually able to generalize, i.e., "*learn the physics*", better and thus performs better in extrapolation. Nevertheless, such models may still give unphysical, unreasonable predictions; see again [21] for examples.

Also, this naive recurrent neural network model is demonstrated here; see Fig. 7.4. The model architecture, which is illustrated in Fig. 7.4a, consists of an LSTM cell with 16 units. Its hidden state is input to a fully connected with three hidden layers with 40 nodes in each layer, using Softplus activations and a linear output layer to predict the stress. The training of this model (16,000 epochs using Adam with a learning rate of $\alpha = 0.001$) converges significantly better than the previous naive fully connected neural network, which can be recognized from the good fit in Fig. 7.4b. Nevertheless, the predictions for $\dot{\varepsilon} = 0.001$ shown in Fig. 7.4c and using $\dot{\varepsilon} = 0.01$ but unloading at $\varepsilon_{\max} = 0.3$ and $\varepsilon_{\max} = 1.0$ shown in Fig. 7.4d are completely off and unphysical. Thus, this model exhibits strong overfitting and is unable to generalize.

7.2.3 *Analytic Models*

As data-driven, completely "*physics-unaware*" models do not provide the desired generalization capabilities, one should again take inspiration from classical, analytical

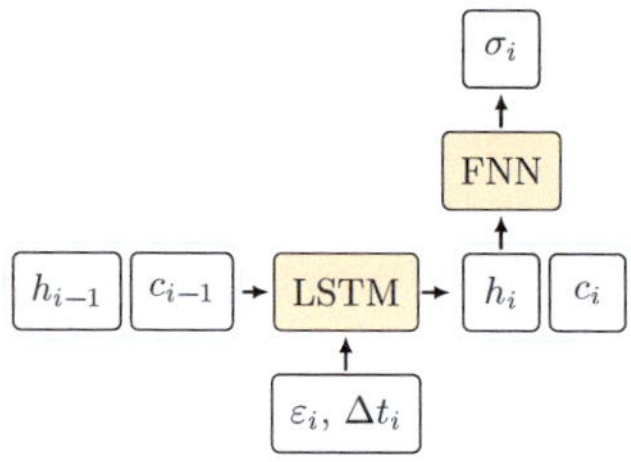

(a) Model architecture of naive recurrent neural network (RNN) (using a fully connected neural network (FNN))

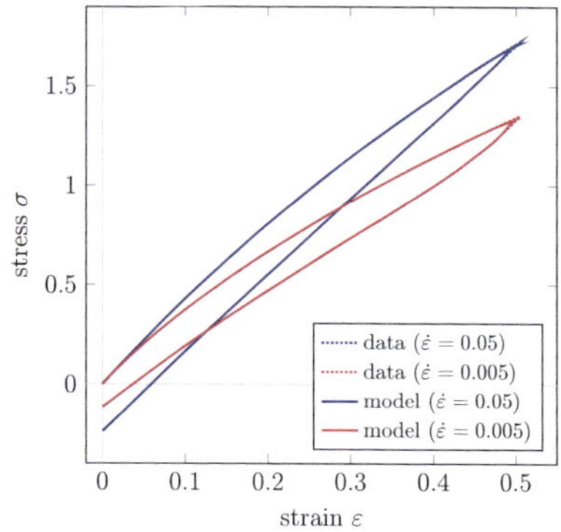

(b) Evaluation of training cases

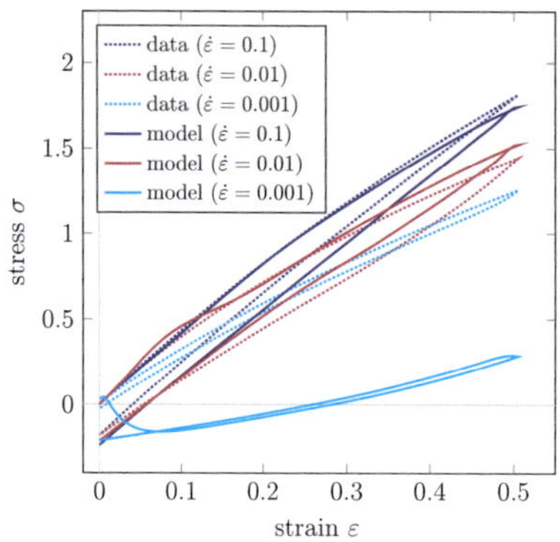

(c) Test cases with other strain rates

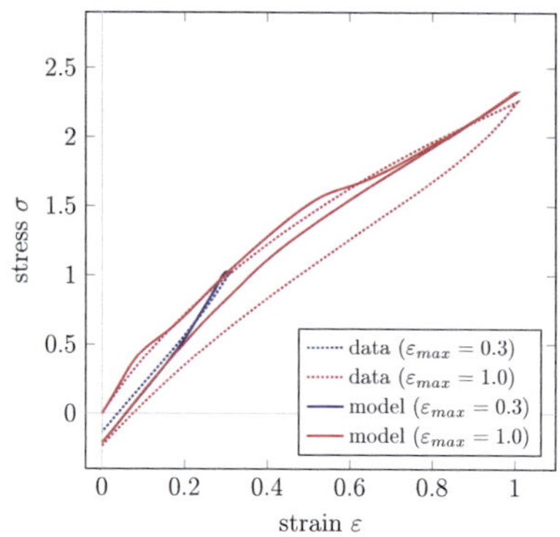

(d) Test cases with other max. strains

Fig. 7.4 Naive recurrent neural network model for one-dimensional viscoelasticity

modeling. In mechanics, analogies to rheological, spring-damper models are typically used to model viscoelastic material behavior analytically [1].

A typical, simple choice is the "*standard linear solid*" or "*Zener*" model, which consists of two parallel branches, an elastic spring and, in parallel to that, a "*Maxwell element*" containing an elastic spring and a viscous dashpot in series; see Fig. 7.5a. In this case, the strain on the spring in the elastic branch is equal to the total (applied) strain ε, while the strain on the spring in the Maxwell element is equal to the total strain minus the (unknown) strain on the dashpot γ. The latter is a so-called *internal variable*, as its state is usually not observable in practice [22]. The total internal

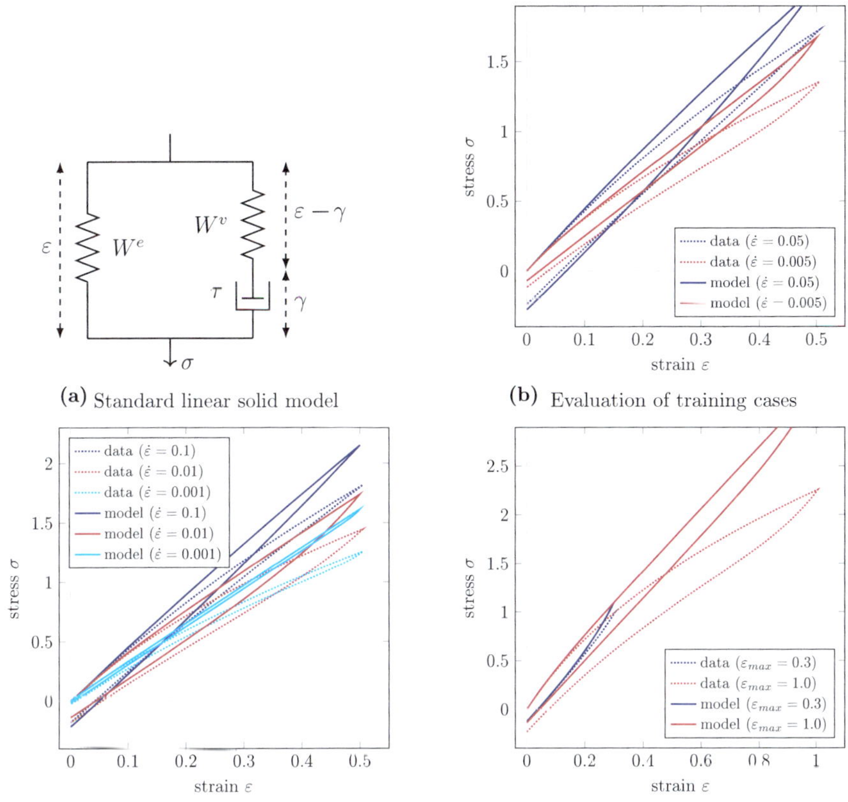

(a) Standard linear solid model

(b) Evaluation of training cases

(c) Test cases with other strain rates

(d) Test cases with other max. strains

Fig. 7.5 Analytical standard linear solid model for one-dimensional viscoelasticity

energy stored in the material, which consists of the strain energies W^e and W^v of both springs, reads as

$$W(\varepsilon, \gamma) = W^e(\varepsilon) + W^v(\varepsilon - \gamma). \tag{7.9}$$

Thus, the total stress is then obtained as the sum of the stresses on both springs

$$\sigma(\varepsilon, \gamma) = \frac{\partial W}{\partial \varepsilon}(\varepsilon, \gamma) = \frac{\partial W^e}{\partial \varepsilon}(\varepsilon) + \frac{\partial W^v}{\partial \varepsilon}(\varepsilon - \gamma). \tag{7.10}$$

In the standard linear solid model, the potentials take the form

$$W^e(\varepsilon) = \tfrac{1}{2} E\, \varepsilon^2, \qquad W^v(\varepsilon - \gamma) = \tfrac{1}{2} E^v\, (\varepsilon - \gamma)^2, \tag{7.11}$$

where E and E^v are the Young's moduli of the springs in the elastic and Maxwell branches, respectively. Thus, the stress follows from Eq. (7.10) as

$$\sigma(\varepsilon, \gamma) = E\,\varepsilon + E^v\,(\varepsilon - \gamma). \tag{7.12}$$

To complete the model and determine the internal variable γ, an evolution equation for $\dot{\gamma}$ is required, which is given for the linear solid model by the linear ordinary differential equation

$$\dot{\gamma} = \frac{1}{\tau}(\varepsilon - \gamma), \tag{7.13}$$

where $\tau > 0$ is the relaxation time of the dashpot. This evolution equation must be integrated over time, e.g., using a numerical technique such as explicit (forward) or implicit Euler, to determine the viscous strain $\gamma(t)$ for a given strain $\varepsilon(t)$ in order to evaluate the stress $\sigma(t)$ using Eq. (7.10).

Figure 7.5 shows the evaluation of the viscoelastic standard linear solid model with parameters $E = 3.2$, $E^v = 1.4$, and $\tau = 10$, which are manually chosen to match the slopes of the training data at $\varepsilon = 0$ in Fig. 7.5b (not by a least-squares fit). The model generalizes well—which is the key advantage of analytical models based on physical principles and insights—as the predictions in Fig. 7.5c, d match well for small strains. However, the strong nonlinearity of the material cannot be captured, and deviations for larger strains are unacceptable.

In fact, the data used in this section was generated by a more general analytical model, which can employ non-linear potentials; see Eq. (7.5), and several parallel spring-damper branches.

7.2.4 *Thermodynamic Consistency*

Importantly, viscoelastic (and any inelastic) model formulations should fulfill the second law of thermodynamics, i.e., the Clausius-Duhem inequality, which states that the dissipation D must always be non-negative

$$D = \sigma\,\dot{\varepsilon} - \dot{W} \geq 0 \quad \forall t. \tag{7.14}$$

This is fulfilled by the linear viscoelastic material model, defined as

$$\begin{aligned} D &= \left(\frac{\partial W^e}{\partial \varepsilon} + \frac{\partial W^v}{\partial \varepsilon}\right)\dot{\varepsilon} - \left(\frac{\partial W^e}{\partial \varepsilon}\dot{\varepsilon} + \frac{\partial W^v}{\partial \varepsilon}\dot{\varepsilon} + \frac{\partial W^v}{\partial \gamma}\dot{\gamma}\right) \\ &= -\frac{\partial W^v}{\partial \gamma}\dot{\gamma} = \frac{E^v}{\tau}(\varepsilon - \gamma)^2 \geq 0 \quad \forall \varepsilon, \gamma. \end{aligned} \tag{7.15}$$

Generally, for a model formulated in terms of $W(\varepsilon, \gamma)$ to be thermodynamically consistent, the evolution equation can be chosen according to Onsager's principle [23] as

$$\dot{\gamma} = -f(\varepsilon, \gamma) \frac{\partial W}{\partial \gamma} \quad \Rightarrow \quad D = -\frac{\partial W}{\partial \gamma} \dot{\gamma} = f(\varepsilon, \gamma) \left(\frac{\partial W}{\partial \gamma} \right)^2 \geq 0, \tag{7.16}$$

where $f(\varepsilon, \gamma) \geq 0$ is required.

7.2.5 Viscoelastic Physics-Augmented Neural Networks

Thus, a *viscoelastic physics-augmented neural network*, i.e., a thermodynamically consistent, *physics-aware* neural network model for viscoelasticity (in [23] called "*variational Onsager neural network*"), can be formulated by parameterising the internal energy $W = W^e_{PANN} + W^v_{PANN}$ with two hyperelastic physics-augmented neural networks $W^e_{PANN}(\varepsilon)$ and $W^v_{PANN}(\varepsilon - \gamma)$ analogous to Eq. (7.6) and expressing the evolution equation using Eq. (7.16) as

$$\dot{\gamma} = -f_{pNN}(\varepsilon, \gamma) \frac{\partial W}{\partial \gamma} = -f_{pNN}(\varepsilon, \gamma) \frac{\partial W^v_{PANN}}{\partial \gamma} = f_{pNN}(\varepsilon, \gamma) \frac{\partial W^v_{PANN}}{\partial \varepsilon}, \tag{7.17}$$

where $f_{pNN}(\varepsilon, \gamma) \geq 0$ is also parameterized by a "*positive*" fully connected neural network (pNN), which must be constrained to deliver only non-negative outputs, e.g., by applying Softplus as output activation and non-negative biases in the output layer.

For training the viscoelastic physics augmented neural network on a dataset with stress-strain load paths over time, the mean squared error as loss function can be used. However, the time dependence of the problem necessitates that for every load path, the stress output is generated sequentially by (numerical) integration of γ and subsequent stress evaluation. Consequently, the stress values at time step t_i depend on the values of γ in all previous time steps, e.g., when using a forward Euler with time step size Δt_i as

$$\begin{aligned} \gamma_i &= \gamma_{i-1} + \Delta t_i \, f_{pNN}(\varepsilon_{i-1}, \gamma_{i-1}) \frac{\partial W^v_{PANN}}{\partial \varepsilon}(\varepsilon_{i-1} - \gamma_{i-1}), \\ \sigma_i &= \frac{\partial W^e_{PANN}}{\partial \varepsilon}(\varepsilon_i) + \frac{\partial W^v_{PANN}}{\partial \varepsilon}(\varepsilon_i - \gamma_i). \end{aligned} \tag{7.18}$$

Thus, a viscoelastic physics-augmented neural network has, in fact, a recurrent neural network-like structure, as its state from one time step serves as input for the next one. In this context, γ, which is not directly observable or measurable in practice, can be seen as the latent, hidden variable of the recurrent neural network. For the training of the viscoelastic physics-augmented neural network, i.e., the simultaneous determination of the parameters of the three fully connected neural networks W^e_{PANN}, W^v_{PANN}, and f_{pNN}, this means that "*backpropagation through time*" is required to determine the derivatives of the loss function, i.e., the mean squared error with respect to the parameters of W^v_{PANN} and f_{pNN}.

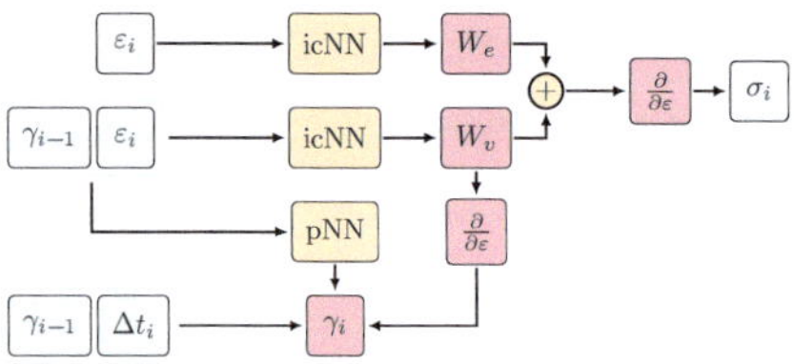

(a) Model architecture of viscoelastic physics-augmented neural network (using input-convex neural networks (icNNs) and a positive fully connected neural network (pNN))

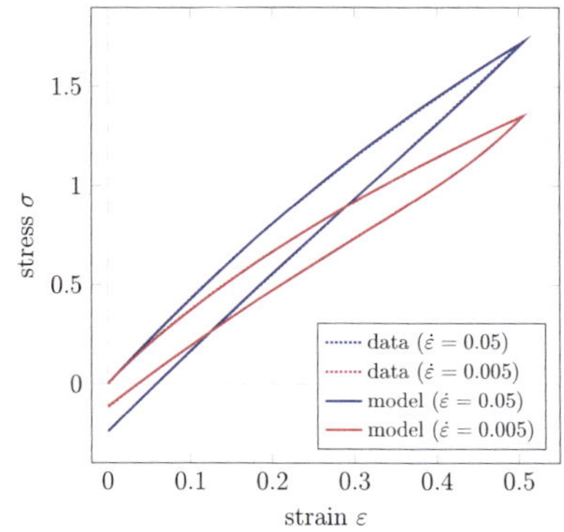

(b) Evaluation of training cases

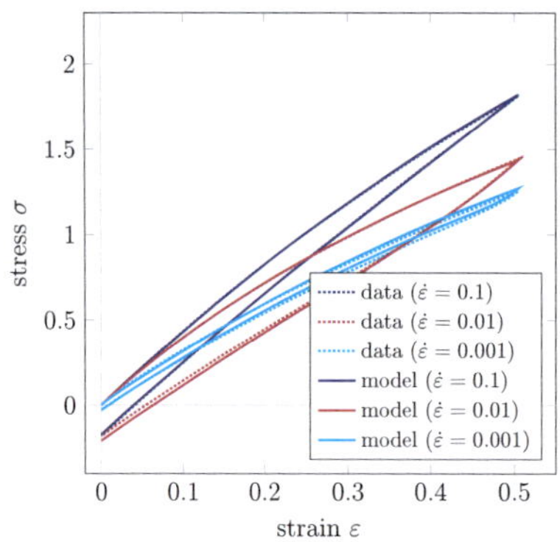

(c) Test cases with other strain rates

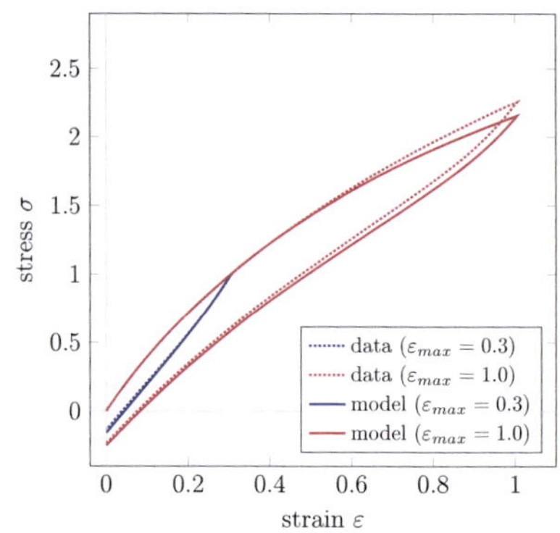

(d) Test cases with other max. strains

Fig. 7.6 Viscoelastic physics-augmented neural network model for one-dimensional viscoelasticity

The application of the viscoelastic physics-augmented neural network is demonstrated in Fig. 7.6. For the potentials, W^e and W^v, input-convex neural networks with one hidden layer with 16 nodes, Softplus activations, and a linear output layer with non-negative weights are used. The positive fully connected neural network f_{pNN} for obtaining $\dot{\gamma}$ also uses one hidden layer with 16 nodes and Softplus activations and a linear output layer. The training using Adam with a learning rate of $\alpha = 0.001$ converged well after 12,000 epochs, as can be seen from the perfect agreement in Fig. 7.6b. Due to its physics-augmented formulation that resembles the analytical models for viscoelasticity in combination with the non-linear approximation capabilities of neural networks, this model also delivers excellent predictions for the extrapolation cases in Fig. 7.6c, d. This clearly demonstrates the advantages of physics-augmented neural network models.

7.2.6 Weakly Thermodynamics-Based Neural Networks

While the structure of the hyperelastic potentials W^e_{PANN} and W^v_{PANN} in the viscoelastic physics-augmented neural network model is only constrained by the physical requirements of hyperelasticity, in Eq. (7.16) a somewhat specific choice for the structure of the evolution equation for the internal variable γ was made, which ensures thermodynamic consistency.

A more general ansatz was presented in [24], where the evolution of γ is simply represented in terms of an ordinary differential equation with a non-linear right-hand side, which is parameterized by a fully connected neural network, i.e., as a *neural ordinary differential equation*:

$$\dot{\gamma} = f_{NN}(\varepsilon, \gamma). \tag{7.19}$$

While this approach is very flexible and justified by Truesdell's equipresence axiom [25], it cannot a priori ensure thermodynamic consistency. Thus, for the training of such a "*thermodynamics-based neural network*" model, which consists of W_{PANN} (or W^e_{PANN} and W^v_{PANN} instead) and f_{NN}, a regularizing loss term must be added that aims to avoid violating the dissipation inequality $D \geq 0$:

$$\mathcal{L}_D = \text{ReLU}(-D) = \text{ReLU}\left(\frac{\partial W_{PANN}}{\partial \gamma} f_{NN}(\varepsilon, \gamma)\right). \tag{7.20}$$

Though this approach only weakly ensures thermodynamic consistency, it appears to learn to respect the dissipation inequality very well and delivers excellent results in [24]. Furthermore, in [24], it is also discussed how such a model can be trained efficiently, since similar to the viscoelastic physics-augmented neural network and Eq. (7.18), also here the internal variables for each load path must be determined by numerical integration of Eq. (7.19).

Further variants of this idea include the direct prediction of the internal variable γ or its increment $\Delta\gamma$ by a fully connected neural network [26] and the prediction of γ as state variable in a recurrent neural network [27], where the weak fulfillment of the dissipation inequality is always ensured within the loss function via Eq. (7.20). For a detailed comparison of these approaches, see [21].

7.2.7 Generalized Standard Materials

Another alternative to Onsager's principle from Eqs. (7.16) and (7.17), which also ensures thermodynamic consistency by construction, is the generalization of the definition of the evolution equation by the *generalized standard materials* (GSM) concept [2, 21, 28]. Here, besides the internal energy density $W(\varepsilon, \gamma)$, a *dissipation potential* $\phi(\varepsilon, \gamma, \dot{\gamma})$ must be defined. The evolution of the internal variable γ is the implicitly deduced Biot equation (using a compatibility condition)

$$\frac{\partial W}{\partial \gamma}(\varepsilon, \gamma) + \frac{\partial \phi}{\partial \dot{\gamma}}(\varepsilon, \gamma, \dot{\gamma}) = 0 \quad \forall t. \tag{7.21}$$

To ensure thermodynamic consistency, the dissipation potential $\phi(\varepsilon, \gamma, \dot{\gamma})$ must also fulfill certain physical requirements [21]: (i) convexity in $\dot{\gamma}$, $\frac{\partial^2 \phi}{\partial \dot{\gamma}^2} \geq 0$, (ii) normalization in $\dot{\gamma}$, $\phi(\varepsilon, \gamma, 0) = 0$, and (iii) $\dot{\gamma} = 0$ being a minimum, $\frac{\partial \phi}{\partial \dot{\gamma}}(\varepsilon, \gamma, 0) = 0$.

In fact, the evolution equation (Eq. (7.16)) can also be derived from the generalized standard materials concept with:

$$\phi(\varepsilon, \gamma, \dot{\gamma}) = \frac{1}{2f(\varepsilon, \gamma)} \dot{\gamma}^2 \tag{7.22}$$

$$\Rightarrow \quad \frac{\partial W}{\partial \gamma} + \frac{\partial \phi}{\partial \dot{\gamma}} = -\frac{\partial W^v}{\partial \varepsilon} + \frac{1}{f(\varepsilon, \gamma)} \dot{\gamma} \stackrel{!}{=} 0 \quad \Leftrightarrow \quad \dot{\gamma} = f(\varepsilon, \gamma) \frac{\partial W^v}{\partial \varepsilon}. \tag{7.23}$$

Here, a relatively simple form of ϕ was assumed, but the generalized standard materials concept would allow for much more complex non-linear interrelations of the variables $\varepsilon, \gamma, \dot{\gamma}$. Thus, a challenge for formulating generalized standard materials models is the concrete formulation of the dissipation potential, which should fulfill the above-mentioned requirements.

7.2.8 Generalized Standard Material-Neural Networks

Now, in its most general form, a *generalized standard material-neural network* (GSM-NN) model can be defined by parameterizing $W_{NN}(\varepsilon, \gamma; \boldsymbol{\Theta}^W)$ and $\phi_{NN}(\varepsilon, \gamma, \dot{\gamma}; \boldsymbol{\Theta}^\phi)$ using neural networks. To consider all physical conditions, one could parameterize W_{NN} using a physics-augmented neural network with two inputs and that is convex in both ε, γ and fulfills normalization conditions with respect to both arguments—or as before by two hyperelastic physics-augmented neural networks as $W_{NN} = W^e_{PANN}(\varepsilon) + W^v_{PANN}(\varepsilon - \gamma)$. The dissipation potential $\phi_{NN}(\varepsilon, \gamma, \dot{\gamma}; \boldsymbol{\Theta}^\phi)$ should be a partially input-convex fully connected neural network [20], as it requires convexity in $\dot{\gamma}$, but not in ε and γ, as well as positive in ε and γ; see [21] for further details on suitable architectures.

For a generalized standard materials model, evaluation of the stress (for prediction and training) is fairly involved compared to the viscoelastic physics-augmented neural network, cf. Eq. (7.18). In each time step t_i, the non-linear compatibility condition (Eq. (7.21)) has to be solved for γ_i by introducing a numerical discretization of $\dot{\gamma}_i$, e.g., using backward Euler, and using a Newton-Raphson method, before the stress can be evaluated:

$$\begin{aligned} &\text{solve} \quad \frac{\partial W_{NN}}{\partial \gamma}(\varepsilon_i, \gamma_i) + \frac{\partial \phi_{NN}}{\partial \dot{\gamma}}(\varepsilon_i, \gamma_i, \dot{\gamma}_i) = 0 \quad \text{for } \dot{\gamma}_i \quad \text{with} \quad \gamma_i = \gamma_{i-1} + \Delta t_i \dot{\gamma}_i, \\ &\sigma_i = \frac{\partial W_{NN}}{\partial \varepsilon}(\varepsilon_i, \gamma_i). \end{aligned} \tag{7.24}$$

As internal variables are not observable in practice, data of γ is usually not available for training. Thus, the initial training of the evolution of γ is highly challenging (for GSM-NNs as well as viscoelastic physics-augmented neural networks) as errors accumulate over time. As an expedient, [29] proposed to train and use auxiliary fully connected neural networks or recurrent neural networks for each load path during training that simply provide $\gamma_i = \gamma_{NN}(t_i)$. W_{NN} and ϕ_{NN} can then be evaluated using these predictions for γ_i and the backward finite difference $\dot{\gamma}_i \approx (\gamma_i - \gamma_{i-1})/\Delta t_{i-1}$, which greatly eases their training. However, to ensure consistency, an additional loss term is required that minimizes the error of the Biot equation (Eq. (7.21)). This helps to correctly train the GSM-NN, which can then be used for prediction without the additional fully connected neural networks according to Eq. (7.24). The model architectures used for training and prediction are illustrated in Fig. 7.7. For further details about the training of GSM-NN models, variants for the model formulations, and their evaluation (for one-dimensional as well as three-dimensional visco-hyperelastic models), refer to [3, 21, 29].

Also, the application of the GSM-NN model is demonstrated here; see Fig. 7.8. The model consists of four fully connected neural networks and an LSTM cell, which is only used during training to predict the internal variable γ together with a standard fully connected neural network (three hidden layers with 16 nodes each and Softplus activations, and a linear output layer). The internal energy density $W_{NN}(\varepsilon, \gamma; \boldsymbol{\Theta}^W)$ is modeled as a physics-augmented neural network with an input-convex neural network with three hidden layers with 16 nodes, Softplus activations, and a linear output layer. To fulfill the physical requirements, the dissipation potential is represented as $\phi_{NN} = \phi_{NN}^{\text{conv}}(\dot{\gamma}; \boldsymbol{\Theta}^{\phi_{\text{conv}}}) \cdot \phi_{NN}^{+}(\varepsilon, \gamma; \boldsymbol{\Theta}^{\phi_+})$. Here, ϕ_{NN}^{+} is a positive fully connected neural network with three hidden layers with 16 nodes each, Softplus activations, and a linear output layer, while ϕ_{NN}^{conv} is a convex input-convex neural network with the same architecture. The training using the Adam optimizer with a learning rate of $\alpha = 0.001$ converges as well as for the viscoelastic physics-augmented neural network and is stopped after 16,000 epochs, delivering a perfect agreement of data and predictions; see Fig. 7.8a. Also, for this model, the test cases in Fig. 7.8b, c show excellent predictions and extrapolation behavior. Thus, this more general physics-augmented architecture was also able to learn the rate dependence and nonlinearity of the material from a fairly small amount of training data.

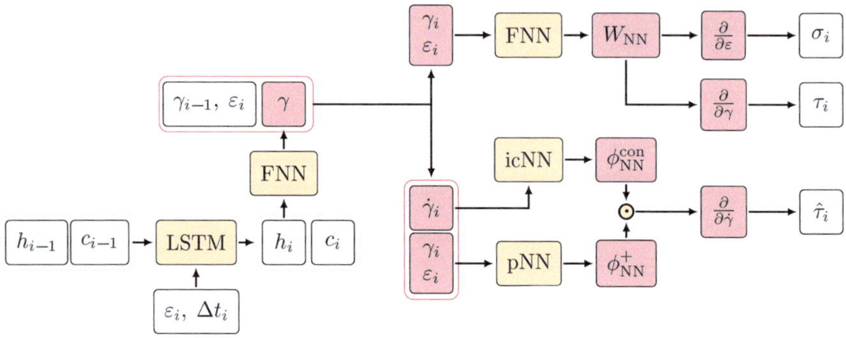

(a) Architecture for training with additional LSTM (on the left side)

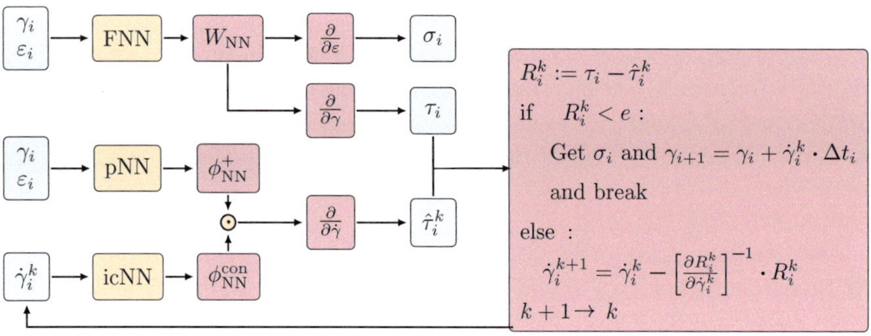

(b) Architecture for prediction with solution of Biot equation (on the right side)

Fig. 7.7 GSM-NN model architectures used for training and prediction (relying on fully connected neural networks (FNNs), a positive fully connected neural network (pNN), and an input-convex neural network (icNN))

7.2.9 *Extensions*

Importantly, the above-mentioned concepts of modeling dissipative behaviors such as viscous damping, plastic, damage, or other effects and combinations thereof can be extended to include more than one internal variable.

When formulating neural network-based constitutive models in analogy to rheological models, several parallel branches and elements in serial can be employed, each with their associated internal variable. By using two or more Maxwell elements, more complex viscoelastic behavior could be captured, or by including a friction element (visco-), elastoplastic behavior could be modeled. Depending on the element type, e.g., elastic, viscous, or plastic, certain structures for the internal energy potentials and evolution equations can be concluded; see the viscoelastic physics-augmented neural network introduced here. These model structures, if implemented correctly, will ensure thermodynamic consistency and further physical requirements and still be (to some extent) physically interpretable. Often, the training of its neural network

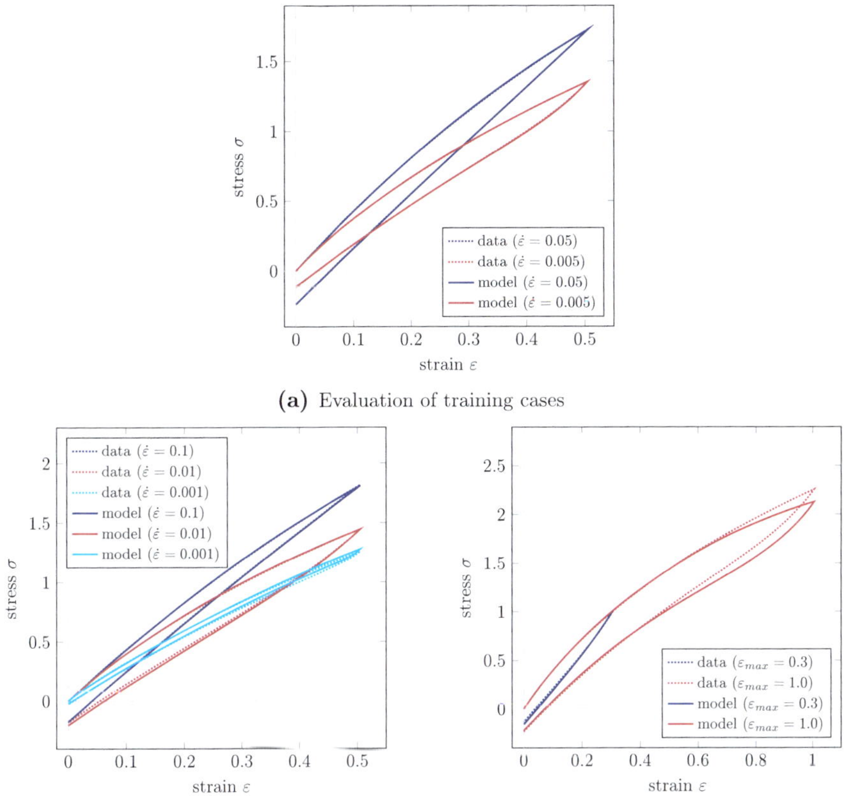

(a) Evaluation of training cases

(b) Test cases with other strain rates

(c) Test cases with other max. strains

Fig. 7.8 GSM-NN model for one-dimensional viscoelasticity

building blocks can also be eased by separating physical effects from the training data. For instance, first data from quasi-static, low strain rate experiments can be used to identify a hyperelastic potential W^e_{PANN}. Subsequently, to obtain a viscoelastic physics-augmented neural network model, the viscoelastic potential W^v_{PANN} and evolution equation in terms of f_{pNN} can be trained using stress-strain data for higher strain rates or from relaxation tests. For this, W^e_{PANN} could be used to initialize W^v_{PANN} and f_{pNN} could be pre-trained on a constant relaxation time, which could potentially ease the training.

On the other hand, the generalized standard materials approach can also be used with several internal variables. However, here, the interpretability of the internal variables gets lost, as they cannot be easily classified as viscous or plastic strains anymore. Furthermore, the training effort increases significantly with the number of internal variables. Further approaches, results, and discussions on the formulation of GSM-NNs for one-dimensional elasto-(visco-) plasticity can be found in [21].

7.3 Physics-Augmented Neural Networks for Three-Dimensional Solids

In finite deformation continuum mechanics, a material model is required to close the partial differential equation that governs the deformation field $\boldsymbol{u} : \Omega \to \mathbb{R}^3$, which is the balance of linear momentum

$$-\operatorname{div} \boldsymbol{P}(\boldsymbol{u}) = \boldsymbol{p}. \tag{7.25}$$

For hyperelastic, i.e., non-dissipative, reversible material behavior, the first Piola-Kirchhoff stress tensor depends solely on the deformation gradient tensor $\boldsymbol{F} = \boldsymbol{I} + \nabla \boldsymbol{u}$. To model this relationship, the above-mentioned concepts for machine learning-based material modeling can also be transferred to three-dimensional, solid mechanical constitutive models. In particular, as we have already seen above, *physics-augmented neural network* models can be formulated, which combine theoretical rigor with the flexibility of machine learning. Here, we outline such approaches for three-dimensional hyperelasticity, i.e., for reversible, energy-conserving material behavior.

7.3.1 Hyperelastic Material Theory

Similar to Sect. 7.1, the relation $\boldsymbol{P}(\boldsymbol{F})$ has to fulfill a number of conditions in order to describe a physically reasonable and mathematically well-posed material behavior [4]:

- **Thermodynamic consistency**: The material model must adhere to the laws of thermodynamics. Energy conservation of an elastic material can be ensured by formulating the constitutive model in terms of potentials such as the internal, Gibbs, or Helmholtz free energy densities. In hyperelastic material theory, the stress tensor $\boldsymbol{P}$ is derived as the gradient of the internal (strain) energy density $W(\boldsymbol{F})$ with respect to the deformation gradient $\boldsymbol{F}$:

$$\boldsymbol{P}(\boldsymbol{F}) = \frac{\partial W}{\partial \boldsymbol{F}} \qquad \Rightarrow \qquad \dot{W} - \boldsymbol{P} : \dot{\boldsymbol{F}} = 0. \tag{7.26}$$

- **Balance of angular momentum**: The material model must also ensure that the momentum balance equation is fulfilled, which can be expressed as

$$\boldsymbol{P} \cdot \boldsymbol{F}^{\mathsf{T}} = \boldsymbol{F} \cdot \boldsymbol{P}^{\mathsf{T}} \quad \forall \boldsymbol{F}, \boldsymbol{P}. \tag{7.27}$$

  This implies symmetry of the Cauchy stress tensor $\boldsymbol{\sigma} = J^{-1}\, \boldsymbol{P} \cdot \boldsymbol{F}^{\mathsf{T}}$ and the second Piola-Kirchhoff stress tensor $\boldsymbol{S} = \boldsymbol{F}^{-1} \cdot \boldsymbol{P}$.

- **Objectivity**: The material response must be independent of a change of observer. The objectivity condition formalizes this.

$$W(\boldsymbol{Q} \cdot \boldsymbol{F}) = W(\boldsymbol{F}) \quad \forall \boldsymbol{Q} \in SO(3) \tag{7.28}$$

- **Material symmetry**: Furthermore, the material response must reflect underlying symmetries of the material's microstructure, i.e., its (an-) isotropy:

$$W(\boldsymbol{F} \cdot \boldsymbol{Q}) = W(\boldsymbol{F}) \quad \forall \boldsymbol{Q} \in G. \tag{7.29}$$

  Here, $G \subseteq O(3)$ is the symmetry group of the material, e.g., $G = SO(3)$ for an isotropic material with direction-independent response.
- **Normalization**: When a body is in its undeformed configuration $\boldsymbol{F} = \boldsymbol{I}$, it should be stress-free, and no internal energy should be stored in it:

$$W(\boldsymbol{F} = \boldsymbol{I}) = 0 \quad \text{and} \quad \boldsymbol{P}(\boldsymbol{F} = \boldsymbol{I}) = \boldsymbol{0}. \tag{7.30}$$

- **Coercivity**: The compression of a body towards zero volume and to infinite volume should both be associated with an infinite amount of internal energy stored in it, which is expressed by the growth conditions:

$$W(\boldsymbol{F}) \to \infty \quad \text{as} \quad J \to 0^+ \quad \text{and} \quad J \to \infty, \tag{7.31}$$

  where $J = \det \boldsymbol{F}$ is a measure of volumetric deformation.
- **Polyconvexity**: Ellipticity is a mathematical requirement on the strain energy density, which ensures material stability and is thus favorable for convergence of non-linear finite element simulations. This condition is ensured by the polyconvexity of the strain energy density, i.e., when W can be formulated as a convex function in terms of the deformation gradient $\boldsymbol{F}$, its cofactor $\operatorname{cof} \boldsymbol{F} = J\,\boldsymbol{F}^{-\mathsf{T}}$, and its determinant J:

$$W(\boldsymbol{F}) = \mathcal{P}(\boldsymbol{F}, \operatorname{cof} \boldsymbol{F}, J). \tag{7.32}$$

  When the function $\mathcal{P} : GL^+(3) \times GL^+(3) \times \mathbb{R} \to \mathbb{R}$ is twice continuously differentiable, which is required for finite element implementations, polyconvexity is equivalent to the Hessian of the strain energy being positive semi-definite.

Analytically formulated material models usually aim at ensuring all of the above-mentioned constitutive requirements [5]. To ensure objectivity and material symmetry of isotropic materials, they are typically expressed in terms of three invariants of the objective right Cauchy-Green strain tensor $\boldsymbol{C} = \boldsymbol{F}^{\mathsf{T}}\boldsymbol{F}$:

$$I_1 = \operatorname{tr}(\boldsymbol{C}), \qquad I_2 = \operatorname{tr}(\operatorname{cof} \boldsymbol{C}), \qquad I_3 = \det(\boldsymbol{C}). \tag{7.33}$$

It can be easily verified that these three scalars fulfill Eqs. (7.28) and (7.29) and are polyconvex. Then, the further mathematical structure of $W(I_1, I_2, I_3)$ is chosen to fulfill the normalization, coercivity, and polyconvexity conditions. For other types of material symmetries, usually, further invariants are added for the specific anisotropy class [4]. In fact, not all commonly used material models are polyconvex since (i) it is difficult to formulate very general polyconvex functions, e.g., mixed polynomials such as $(I_1\, I_2)^2$ are not polyconvex, and (ii) anisotropic invariants may not be polyconvex.

7.3.2 Invariant-Based Physics-Augmented Neural Networks

As already motivated and demonstrated above in Sect. 7.1, machine learning and, in particular, neural networks can be used to formulate hyperelastic material models that are both theoretically grounded and highly flexible. In particular, in [6], invariant-based, polyconvex hyperelastic material models were formulated using input-convex neural networks and extended with the normalization requirements in [7]. Similar to the model in Eq. (7.6), these *physics-augmented* neural network models can fulfill all above-mentioned requirements from Eqs. (7.26) and (7.32) by construction of a well thought-out model architecture:

$$
\begin{aligned}
W^{I}_{PANN}(\boldsymbol{F};\boldsymbol{\Theta}) &= W_{icNN}(I_1, I_2, I_3, I_3^*) + W_{en} + W_{st}(J) + W_{gr}(J) \\
\text{with}\quad W_{en} &= -\,W_{icNN}\big|_{\boldsymbol{F}=\boldsymbol{I}}, \\
W_{st}(J) &= -\,2\left(\frac{\partial W_{icNN}}{\partial I_1} + 2\frac{\partial W_{icNN}}{\partial I_2} + \frac{\partial W_{icNN}}{\partial I_3} + \frac{\partial W_{icNN}}{\partial I_3^*}\frac{\partial I_3^*}{\partial I_3}\right)\Bigg|_{\boldsymbol{F}=\boldsymbol{I}}(J-1), \\
W_{gr}(J) &= a\big(J + J^{-1} - 2\big)^2.
\end{aligned}
\tag{7.34}
$$

Using the invariants I_1, I_2, I_3, and $I_3^* = -2J$ as inputs to the input-convex neural network potential W_{icNN}, the model is thermodynamically consistent, momentum-conserving, objective, isotropic, and polyconvex. The two terms W_{en} and W_{st}, which depend on the evaluation of W_{icNN} and its derivatives for $\boldsymbol{F} = \boldsymbol{I}$, guarantee the energy and stress normalization, respectively. The growth term W_{gr} with (fixed or learnable) parameter $a > 0$ ensures coercivity. Importantly, it should be noted that all terms are chosen so that they do not violate any condition, e.g., the normalization terms must also be objective, isotropic, and polyconvex. This model architecture is also visualized in Fig. 7.9a.

Similar to Sect. 7.1, W^{I}_{PANN} can be trained given only stress-strain datasets of the form $\{\boldsymbol{F}_k, \boldsymbol{P}_k\}_{k=1}^{m_{\mathcal{D}}}$, where $\boldsymbol{F}_k$ are deformation gradients and $\boldsymbol{P}_k$ the corresponding first Piola-Kirchhoff stresses along one additional load path. The loss function, i.e., the mean squared error is then simply comparing the actual stress $\boldsymbol{P}_k$ from the dataset with the predicted ones $\boldsymbol{P}(\boldsymbol{F}_k) = \frac{\partial W^{I}_{PANN}}{\partial \boldsymbol{F}}$.

As shown in [7], for isotropic materials, such a physics-augmented neural network may even be trained only on a single experimental load case such as uni-axial tension or compression and nevertheless may still provide reasonable predictions, i.e., generalize well also for other load cases. This is illustrated in Fig. 7.9, where Fig. 7.9b

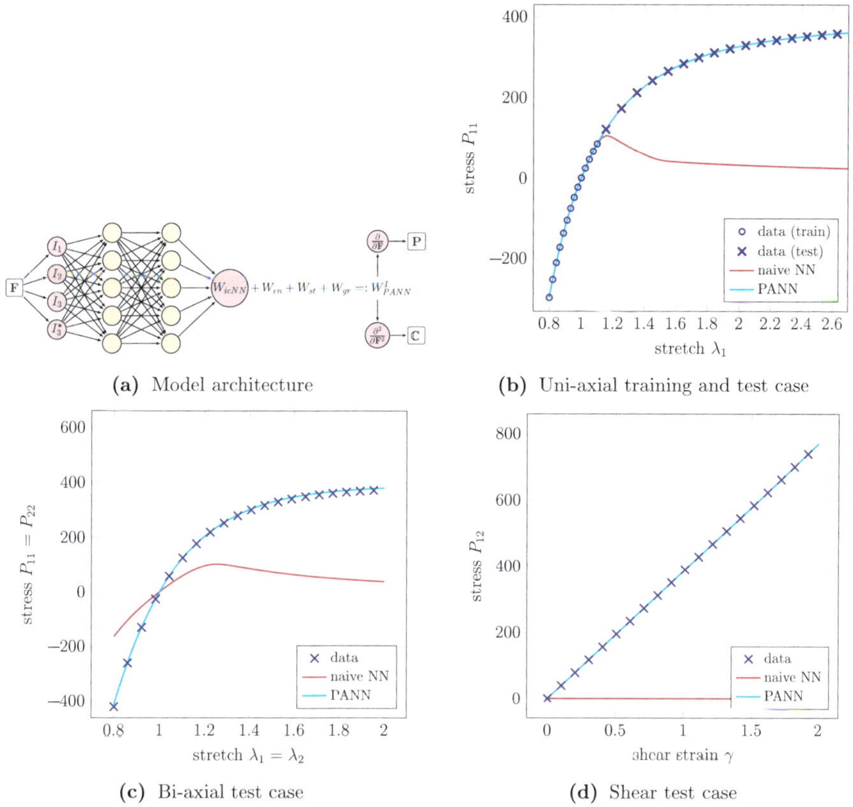

Fig. 7.9 Physics-augmented neural network model for three-dimensional hyperelasticity

shows the small amount of data points generated from a simple, analytical isotropic Neo-Hooke model for uni-axial compression and tension (with $0.8 \leq \lambda_1 \leq 1.1$) used for model training. The test data includes extrapolation behavior in uni-axial tension; see also Fig. 7.9b, as well as bi-axial compression and tension; see Fig. 7.9c, and simple shear; see Fig. 7.9d. While a naive neural network that directly predicts $\boldsymbol{P}(\boldsymbol{F})$ can fit the training data well but does not generalize at all with very poor predictions for the test cases, the physics-augmented neural network model W^I_{PANN} shows excellent agreement with the data also for the extrapolation cases.

Note, however, that assessing the generalization in practice can be difficult since experimental data may often only be available for very few different load cases. On the other hand, when constitutive models are to be fitted to synthetically generated data, e.g., from the numerical homogenization of microstructured materials, cf. [6, 8], it is advisable to ensure a suitable sampling of the space of deformation gradients $\boldsymbol{F} \in GL(3)^+$, which can for instance be achieved by concentric sampling techniques on manifolds, cf. [9].

7.3.3 Deformation Gradient-Based Physics-Augmented Neural Networks

The use of polyconvex invariants and an input-convex neural network architecture ensure polyconvexity and thus ellipticity of W^I_{PANN}, but also severely restrict the flexibility of physics-augmented neural network material models; see [6] for an example of a highly non-linear cubic metamaterial. Thus, [6] also proposed a hyperelastic model architecture that directly uses $\boldsymbol{F}$, cof $\boldsymbol{F}$, and J as inputs:

$$W^F_{PANN}(\boldsymbol{F};\boldsymbol{\Theta}) = \frac{1}{n_G}\sum_{\boldsymbol{Q}\in G}\left\{W_{icNN}(\boldsymbol{F}\cdot\boldsymbol{Q}, \operatorname{cof}\boldsymbol{F}\cdot\boldsymbol{Q}, J) - W_{icNN}\big|_{\boldsymbol{F}=\boldsymbol{Q}}\right\} + a\big(J + J^{-1} - 2\big)^2. \tag{7.35}$$

Here, the input-convex neural network potential has 19 input values (nine components each for $\boldsymbol{F}$ and cof $\boldsymbol{F}$, plus the scalar J). The material symmetry is ensured by a *group symmetrization*, i.e., the normalized summation of the evaluation of the input-convex neural network for all n_G elements of the material symmetry group G, which can also consider the energy normalization.

While this approach is more flexible, it does not ensure the balance of angular momentum, objectivity, and stress normalization by construction. However, such conditions can also be enforced *weakly* by *informing* the models about them, either by enhancing the training data or the loss function to reflect these requirements. For instance, in [6] a *data augmentation* approach was proposed for informing W^F_{PANN} about objectivity; see Eq. (7.28), which entails the stress condition $\boldsymbol{P}(\boldsymbol{F}) = \boldsymbol{Q}\cdot\boldsymbol{P}(\boldsymbol{Q}\cdot\boldsymbol{F})$. Assuming a dataset consisting of certain pairs of deformation gradients and stresses $\{\boldsymbol{F}_k, \boldsymbol{P}_k\}_{k=1}^{m_\mathcal{D}}$, the augmented dataset includes rotated pairs $\{\boldsymbol{Q}_j\cdot\boldsymbol{F}_k, \boldsymbol{Q}_j\cdot\boldsymbol{P}_k\}_{k=1}^{m_\mathcal{D}}$, where $\boldsymbol{Q}_j \in SO(3)$ are three-dimensional rotation matrices that should be somehow uniformly sampled from $SO(3)$. In this way, during training, the model is evaluated at additional rotated deformations, and the stress mean squared error is computed also for the rotated stresses. In this way, the model is informed about the objectivity condition, as it will learn this rotation invariance from the data. Similarly, instead of augmenting the data, the loss function could also be modified.

7.3.4 Other Variants of Physics-Augmented Machine Learning Models

Besides input-convex neural networks, other neural network-based approaches such as neural differential equations [30] can be used to ensure polyconvexity of invariant-based the strain energy densities W. As already mentioned above, polyconvexity may also be an overly restrictive condition, which may be violated, especially by microstructured materials with highly non-linear behaviors. Thus, many authors have also proposed models that employ simple, fully connected neural networks instead of input-convex neural networks, either using invariants or directly $\boldsymbol{C}$ or the Green-Lagrange strain $\boldsymbol{E} = \frac{1}{2}(\boldsymbol{C} - \boldsymbol{I})$ as inputs. For models that use $\boldsymbol{C}$ or $\boldsymbol{E}$ as input, also input-convex neural networks have been employed and shown to improve generalization [29, 31]. However, the convexity in these tensors is not physically meaningful in terms of ellipticity. Furthermore, weak enforcement of some conditions through constraints was also employed, e.g., by [30, 32].

Besides neural networks, similar concepts for hyperelastic material modeling have also been developed using other machine learning approaches. For instance, *constitutive artificial neural networks* [33] employ a simplified, neural network-like architecture that also predicts the strain energy in terms of invariants and can also incorporate all constitutive requirements. Similarly, large and versatile material model libraries composed of simple functions such as polynomials of the invariants can be pre-defined, and sparse regression techniques can be used to identify material models that are then composed just of a few terms [34]. Alternatively, also kriging/-Gaussian process regression can be employed instead of neural networks [10, 11]. Such approaches possess advantages over neural networks, particularly when training data is scarce. Here, it has again been shown that invariant-based models $W(I_1, \ldots)$ that predict the strain energy in terms of invariants yield much better generalization compared to naive models $\boldsymbol{P}(\boldsymbol{F})$ or $\boldsymbol{S}(\boldsymbol{C})$ that predict stresses directly in terms of strains, as they inherently consider constitutive conditions such as thermodynamic consistency, objectivity, or material symmetry.

7.3.5 Parameterized Models

Neural network-based material models also allow for the straightforward inclusion of parameter dependencies as additional model inputs. This is particularly important and relevant when microstructured materials (see, e.g., Fig. 8.11) are to be modeled, for instance, to consider volume fractions or orientations of fibers in reinforced composites, process parameters in molding or additive manufacturing, or spatial variations in biological materials and tissues. For neural network-based strain energies, a parameter vector $\boldsymbol{\theta}$ can simply be considered as an additional input, see, e.g., the $W(\boldsymbol{C}, \boldsymbol{\theta})$ model in [35], or the $W(I_1, \ldots, \boldsymbol{\theta})$ model in [12], where the neural network architecture was carefully adjusted to preserve the fulfillment of all constitutive conditions, in particular the convexity. In [31], a more involved but very versatile and intriguing route was pursued by parameterizing all weights and biases of their $W_{icNN}(\boldsymbol{C})$ model through a separate neural network. This enabled a straightforward insurance of the convexity of the $W_{icNN}(\boldsymbol{C})$ model through non-negative weights and enabled the incorporation of the complex microstructural dependence on the parameters of the investigated three-dimensional lattice metamaterials.

7.3.6 Beyond Hyperelasticity

Furthermore, in recent years, researchers have started to develop physics-augmented neural network models for inelastic material behaviors in three dimensions, i.e., extending the approaches presented for one-dimensional visco-elasticity in Sect. 7.2. Efforts on this end include models for hyperelasticity combined with plasticity, viscous dissipation, damage, or growth; see, e.g., [3, 13–15, 24, 36], to name a few. Likewise, researchers have started to extend the physics-augmented neural network concepts to multiphysically coupled problems such as electro-elasticity [16], magneto-elasticity [17], or thermo-elasticity [18] at finite deformations. For a comprehensive review of these developments, see [19].

References

1. Simo, J. C. and Thomas J. R. Hughes. 1998. *Computational inelasticity*. en. Interdisciplinary applied mathematics, vol 7. New York: Springer. isbn: 978-0-387-97520-7.
2. David, González, Francisco Chinesta, and Elías Cueto. 2019. Thermodynamically consistent data-driven computational mechanics. *Continuum Mechanics and Thermodynamics*.
3. Max, Rosenkranz et al. 2024. Viscoelasticty with physics-augmented neural networks: Model formulation and training methods without prescribed internal variables. *Computational Mechanics*. issn: 0178-7675, 1432-0924. https://doi.org/10.1007/s00466-024-02477-1.
4. Šilhavý, M. 1997. *The mechanics and thermodynamics of continuous media*. Berlin, Heidelberg: Springer.
5. Gerhard, A. Holzapfel. 2000. *Nonlinear solid mechanics: A continuum approach for engineering*. Chichester, New York: Wiley. isbn: 978-0-471-82304-9.
6. Dominik, K. Klein et al. 2021. Polyconvex anisotropic hyperelasticity with neural networks. *Journal of the Mechanics and Physics of Solids*.
7. Lennart, Linden et al. 2023. Neural networks meet hyperelasticity: A guide to enforcing physics. *Journal of the Mechanics and Physics of Solids*.
8. Til, Gärtner, Mauricio Fernández, and Oliver Weeger. 2021. Nonlinear multiscale simulation of elastic beam lattices with anisotropic homogenized constitutive models based on artificial neural networks. *Computational Mechanics 68.5*: 1111–1130. issn: 1432-0924. https://doi.org/10.1007/s00466-021-02061-x.
9. Oliver, Kunc, and Felix Fritzen. Generation of energy-minimizing point sets on spheres and their application in mesh-free interpolation and differentiation. *Advances in Computational Mathematics 45.5-6*: 3021–3056. issn: 1019-7168, 1572-9044. https://doi.org/10.1007/s10444-019-09726-5.
10. Jan, Niklas Fuhg, Michele Marino, and Nikolaos Bouklas. 2022. Local approximate gaussian process regression for data-driven constitutive laws: Development and comparison with neural networks. *Computer Methods in Applied Mechanics and Engineering*.
11. Nathan, Ellmer et al. 2024. Gradient enhanced gaussian process regression for constitutive modelling in finite strain hyperelasticity. In: *Computer Methods in Applied Mechanics and Engineering*.
12. Dominik, K. Klein et al. 2023. Parametrized polyconvex hyperelasticity with physics-augmented neural networks. *Data-Centric Engineering*.
13. Jan, Niklas Fuhg et al. 2023. Modular machine learning-based elastoplasticity: Generalization in the context of limited data. *Computer Methods in Applied Mechanics and Engineering*.

14. Hagen, Holthusen et al. 2024. Theory and implementation of inelastic constitutive artificial neural networks. *Computer Methods in Applied Mechanics and Engineering 428*: 117063.issn: 00457825. https://doi.org/10.1016/j.cma.2024.117063.
15. Kian, P. Abdolazizi, Kevin Linka, and Christian J. Cyron. 2024. Viscoelastic constitutive artificial neural networks (vCANNs)–A framework for data-driven anisotropic nonlinear finite viscoelasticity. *Journal of Computational Physics*.
16. Dominik, K. Klein et al. 2022. Finite electro-elasticity with physics-augmented neural networks. *Computer Methods in Applied Mechanics and Engineering*.
17. Karl, A. Kalina et al. 2024. Neural network-based multiscale modeling of finite strain magneto-elasticity with relaxed convexity criteria. *Computer Methods in Applied Mechanics and Engineering*.
18. Jan, N. Fuhg et al. 2024. Polyconvex neural network models of thermoelasticity. *Journal of the Mechanics and Physics of Solids 192*: 105837. issn: 0022-5096. https://doi.org/10.1016/j.jmps.2024.105837.
19. Jan, N. Fuhg et al. 2024. A review on data-driven constitutive laws for solids. *Archives of Computational Methods in Engineering*. https://doi.org/10.1007/s11831-024-10196-2.
20. Brandon, Amos, Lei Xu, and J. Zico Kolter. 2017. Input convex neural networks. en. In: *Proceedings of the 34th International Conference on Machine Learning*, 146–155, July 2017. PMLR.
21. Max, Rosenkranz et al. 2023. A comparative study on different neural network architectures to model inelasticity. en. *International Journal for Numerical Methods in Engineering*, nme.7319. issn: 0029-5981, 1097-0207. https://doi.org/10.1002/nme.7319.
22. Bernard, D. Coleman, and Morton E. Gurtin. 1967. Thermodynamics with internal state variables. *The Journal of Chemical Physics 47.2*: 597–613. issn: 0021-9606. https://doi.org/10.1063/1.1711937.
23. Shenglin, Huang et al. 2022. Variational onsager neural networks (VONNs): A thermodynamics-based variational learning strategy for non-equilibrium PDEs. *Journal of the Mechanics and Physics of Solids 163*: 104856. issn: 0022-5096. https://doi.org/10.1016/j.jmps.2022.104856.
24. Filippo, Masi, and Ioannis Stefanou. Evolution TANN and the identification of internal variables and evolution equations in solid mechanics. *Journal of the Mechanics and Physics of Solids*: 105245. issn: 0022-5096. https://doi.org/10.1016/j.jmps.2023.105245.
25. Truesdell, C. 1960. Modern theories of materials. *Transactions of The Society of Rheology 4*: 9–22. https://doi.org/10.1122/1.548873.
26. Filippo, Masi et al. 2021. Thermodynamics-based artificial neural networks for constitutive modeling. en. *Journal of the Mechanics and Physics of Solids 147*: 104277. issn: 0022-5096. https://doi.org/10.1016/j.jmps.2020.104277.
27. Thermodynamically Consistent Machine-Learned Internal State Variable Approach for Data-Driven Modeling of Path-Dependent Materials. *Computer Methods in Applied Mechanics and Engineering 402*: 115348. issn: 0045-7825. https://doi.org/10.1016/j.cma.2022.115348.
28. Alexander, Mielke. 2006. A mathematical framework for generalized standard materials in the rate-independent case. In: *Multifield Problems in Solid and Fluid Mechanics*, Ed. by Rainer, Helmig, Alexander Mielke, and Barbara I. Wohlmuth. Lecture Notes in Applied and Computational Mechanics. Berlin, Heidelberg: Springer. https://doi.org/10.1007/978-3-540-34961-7_12.
29. Faisal, As'ad, and Charbel Farhat. A Mechanics-Informed Neural Network Framework for Data-Driven Nonlinear Viscoelasticity. en. In: *AIAA SCITECH 2023 Forum*. National Harbor, MD & Online: American Institute of Aeronautics and Astronautics, Jan 2023. isbn: 978-1-62410-699-6. https://doi.org/10.2514/6.2023-0949.
30. Vahidullah, Tac, Francisco Sahli Costabal, and Adrian B. Tepole. 2022. Data-driven tissue mechanics with polyconvex neural ordinary differential equations. en. *Computer Methods in Applied Mechanics and Engineering 398*: 115248. issn: 0045-7825. https://doi.org/10.1016/j.cma.2022.115248.

31. Li, Zheng, Dennis M. Kochmann, and Siddhant Kumar. 2024. HyperCAN: Hypernetwork-driven deep parameterized constitutive models for metamaterials. *Extreme Mechanics Letters 72*: 102243. issn: 2352-4316. https://doi.org/10.1016/j.eml.2024.102243.
32. Patrick, Weber, Jeremy Geiger, and Werner Wagner. 2021. Constrained neural network training and its application to hyperelastic material modeling. en. *Computational Mechanics 68.5*: 1179–1204. issn: 1432-0924. https://doi.org/10.1007/s00466-021-02064-8.
33. Kevin, Linka et al. 2021. Constitutive artificial neural networks: A fast and general approach to predictive data-driven constitutive modeling by deep learning. en. *Journal of Computational Physics 429*: 110010. issn: 0021-9991. https://doi.org/10.1016/j.jcp.2020.110010.
34. Moritz, Flaschel, Siddhant Kumar, and Laura De Lorenzis. 2021. Unsupervised discovery of interpretable hyperelastic constitutive laws. en. *Computer Methods in Applied Mechanics and Engineering 381*: 113852. issn: 0045-7825. https://doi.org/10.1016/j.cma.2021.113852.
35. Mauricio, Fernández, Felix Fritzen, and Oliver Weeger. 2022. Material modeling for parametric, anisotropic finite strain hyperelasticity based on machine learning with application in optimization of metamaterials. en. *International Journal for Numerical Methods in Engineering 123.2*: 577–609. issn: 0029-5981, 1097-0207. https://doi.org/10.1002/nme.6869.
36. Nikolaos, N. Vlassis, and WaiChing Sun. 2021. Sobolev training of thermodynamic-informed neural networks for interpretable elasto-plasticity models with level set hardening. en. *Computer Methods in Applied Mechanics and Engineering 377*: 113695. issn: 0045-7825. https://doi.org/10.1016/j.cma.2021.113695.

Chapter 8
Generative Artificial Intelligence

Abstract Generative artificial intelligence aims at learning data distributions from samples, enabling it to generate new samples that are distinct yet resemble the original data distributions. Prominent methodologies include variational autoencoders, generative adversarial networks, diffusion models, and transformers, all of which are covered in detail in this chapter. Furthermore, their usage is explored in the context of computational mechanics, focusing on applications such as dimensionality reduction, data synthesis for microstructures, and anomaly detection. By providing practical examples, the chapter demonstrates how these methods can transform traditional workflows in engineering and simulation.

In recent years, generative models have revolutionized the field of deep learning, offering powerful tools for understanding and synthesizing complex data. Unlike discriminative models that focus on predicting labels given input data, generative models aim to capture the underlying distribution of the data itself. This capability opens up a plethora of applications, from creating realistic images and designing new materials to optimizing complex systems in computational mechanics. We refer to [1] for an in-depth review in the field of design, and to [2] for a hands-on textbook.

To date, the most impactful deep learning generative models (see Fig. 8.1 for conceptual illustrations) that have shaped the field are

- variational autoencoders (VAEs),
- generative adversarial networks (GANs),
- diffusion models,
- flow-based models,
- transformer models.

This chapter focuses on variational autoencoders, generative adversarial networks, and diffusion models. We will provide basic examples that explain their architectures and functions. Furthermore, we will explore how these models can be applied to computational mechanics. For a deep dive into the promising field of normalizing flows, we refer the reader to [3].

Despite the unprecedented success of transformer-based generative models (generative pretrained transformers, i.e., GPTs, bidirectional encoder representations from transformers, i.e., BERT, etc.), they are not treated in-depth in this chapter.

L. Herrmann et al., *Deep Learning in Computational Mechanics*,
https://doi.org/10.1007/978-3-031-89529-6_8

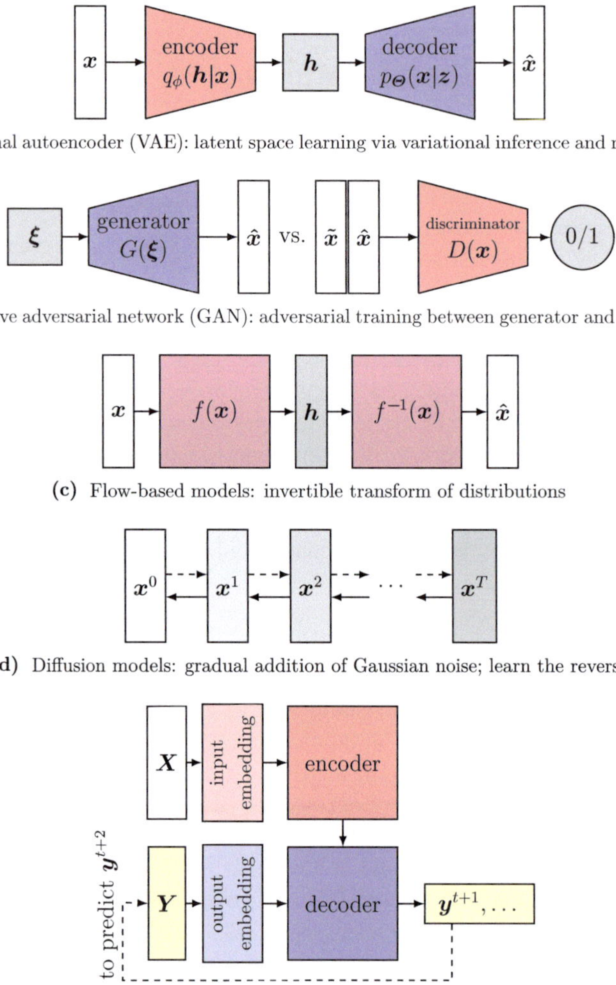

(a) Variational autoencoder (VAE): latent space learning via variational inference and reconstruction

(b) Generative adversarial network (GAN): adversarial training between generator and discriminator

(c) Flow-based models: invertible transform of distributions

(d) Diffusion models: gradual addition of Gaussian noise; learn the reverse

(e) Transformer: self-attention mechanism for sequence modeling

Fig. 8.1 Overview of generative models. Adapted from [4]

Originally developed for sequence modeling in natural language processing, they use self-attention mechanisms to capture long-range dependencies in data. They extend beyond text to images, signals, and other data types, perform well with large datasets, can model complex relationships, and have even been extended to computer vision (vision transformers) [5]. At the point of writing, their application to other domains is a highly dynamic area of research [6–8]. In computational mechanics, they could be used to model sequential data like load histories or time-dependent material behaviors. An example is the application of transformers to the thermal analysis in additive manufacturing [9] or replacing expensive simulations of particle collisions at the Large Hadron Collider [10].

Currently, generative models already have multiple existing applications in computational mechanics but the fast-paced developments also offer potential to advance the field in the future. By generating a diverse set of design candidates, engineers can explore innovative solutions that meet specific performance criteria. Enhancing datasets through data augmentation (cf. Sect. 3.7.5) allows for training more robust predictive models, especially when real-world data is scarce or expensive to obtain. In anomaly detection, usually a classic pattern recognition task, generative approaches have emerged as a promising alternative, potentially outperforming classical outlier detection methods.

8.1 Autoencoders

Autoencoders [11–15] are neural network architectures designed to learn a compact representation of the input data and then reconstruct it as closely as possible. They consist of two main components:

- Encoder $e : X \rightarrow H$ projects data from the high-dimensional input space X to a lower-dimensional representation, also called the latent space H,
- Decoder $d : H \rightarrow X$ attempts to reconstruct the original input from the compressed representation.

The encoder $e(\boldsymbol{x})$, maps the high-dimensional input $\boldsymbol{x}$ to a low-dimensional latent vector $\boldsymbol{h}$. From the latent vector, the decoder $d(\boldsymbol{h})$ tries to reconstruct the high-dimensional input vector $\hat{\boldsymbol{x}}$. The parameters of the encoder and the decoder are learned by minimizing the error between $\boldsymbol{x}$ and $d(e(\boldsymbol{x}))$:

$$C = \|d(e(\boldsymbol{x})) - \boldsymbol{x}\|^2 . \tag{8.1}$$

Figure 8.2 illustrates the concept of a basic autoencoder using a fully connected neural network with five input and output neurons and two neurons in the latent space (i.e., the single hidden layer). The quantity of interest is the compressed representation in terms of the learned latent space $\boldsymbol{h}$. As this is achieved without labeled data, the learning in an autoencoder is considered unsupervised. While basic autoencoders are powerful tools for dimensionality reduction (complementing the linear techniques discussed in Sect. 6.1) and feature extraction, they are not inherently generative models. The reason is that the latent space into which they convert their inputs and where their encoded vectors lie may not be continuous or allow easy interpolation between encodings; see Fig. 8.3a. The difficulty lies in how to sample the latent space such that it matches the distribution of the encoder's output $e(\boldsymbol{x})$. Generally, autoencoders serve as a foundation for more advanced architectures that can perform generative tasks by producing new outputs from randomly sampled latent spaces with the decoder.

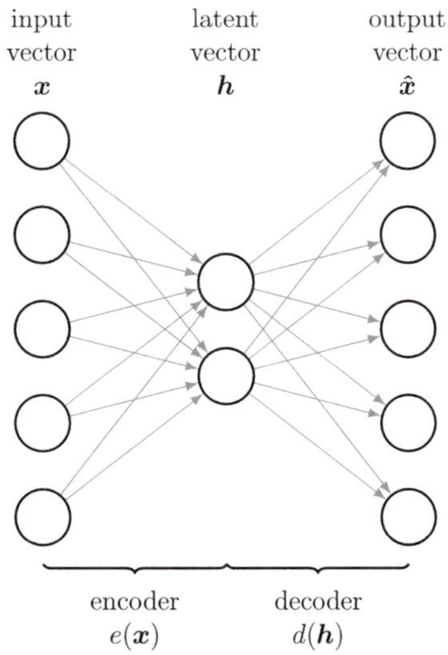

Fig. 8.2 Fully connected autoencoder: Input $\boldsymbol{x}$ is mapped to low-dimensional latent vector $\boldsymbol{h}$ via the encoder $e(\boldsymbol{x})$ and back to the output $\hat{\boldsymbol{x}}$ via the decoder $d(\boldsymbol{h})$. The autoencoder is trained, such that the output approximates the input $e(d(\boldsymbol{x})) \approx \boldsymbol{x}$

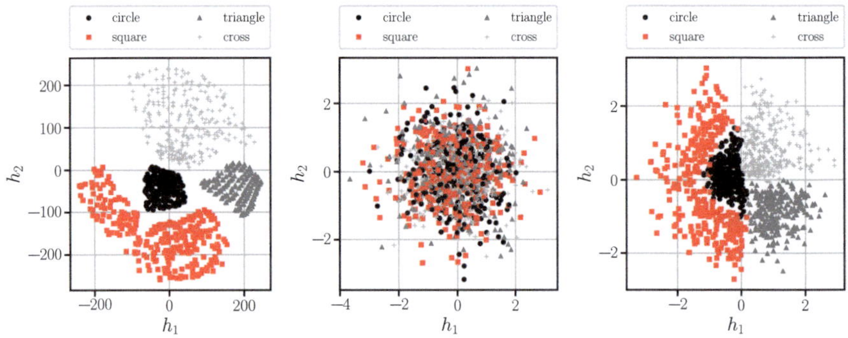

(a) Optimizing purely for reconstruction loss leads to a structured but discontinuous distribution of encodings in the latent space

(b) Optimizing using pure Kullback-Leibler divergence loss structures the encodings randomly around zero, which makes it impossible for the decoder to produce anything meaningful from it

(c) Optimizing using both reconstruction loss and Kullback-Leibler divergence loss creates a meaningful clusters centered around zero, which makes it easy to sample and to interpolate

Fig. 8.3 The fundamental problem with autoencoders, for generation, is that the latent space they convert their inputs to and where their encoded vectors lie may not be continuous, thereby preventing easy interpolation

Implementing a simple fully connected autoencoder in PyTorch using `torch.nn.Module` is straightforward. The encoder $e(\boldsymbol{x})$ is defined as follows.

Listing 8.1 Fully connected encoder to be used in an autoencoder

```
class Encoder(torch.nn.Module):
    def __init__(self, latent_dims):
        super(Encoder, self).__init__()
        self.linear1 = torch.nn.Linear(784, 512)
        self.linear2 = torch.nn.Linear(512, latent_dims)

    def forward(self, x):
        x = torch.flatten(x, start_dim=1)
        x = F.relu(self.linear1(x))
        return self.linear2(x)
```

The decoder $d(\boldsymbol{h})$ inverts the structure of the encoder.

```
class Decoder(torch.nn.Module):
    def __init__(self, latent_dims):
        super(Decoder, self).__init__()
        self.linear1 = torch.nn.Linear(latent_dims, 512)
        self.linear2 = torch.nn.Linear(512, 784)

    def forward(self, h):
        h = F.relu(self.linear1(h))
        h = torch.sigmoid(self.linear2(h))
        return h.reshape((-1, 1, 28, 28))
```

Combining the encoder and decoder yields a basic autoencoder.

```
class Autoencoder(torch.nn.Module):
    def __init__(self, latent_dims):
        super(Autoencoder, self).__init__()
        self.encoder = Encoder(latent_dims)
        self.decoder = Decoder(latent_dims)

    def forward(self, x):
        h = self.encoder(x)
        return self.decoder(h)
```

The autoencoder is trained in an end-to-end fashion, optimizing the weights of the encoder $\boldsymbol{\Theta}_e$ and decoder $\boldsymbol{\Theta}_d$ simultaneously by minimizing the error between the encoder input $\boldsymbol{x}$ and decoder output $\hat{\boldsymbol{y}}$.

```
def train(autoencoder, data, epochs=20):
    opt = torch.optim.Adam(autoencoder.parameters())
    for epoch in range(epochs):
        for x, *_ in data: # any labels in data are not
    needed
```

```
            x = x.to(device) # GPU
            opt.zero_grad()
            x_hat = autoencoder(x)
            loss = ((x - x_hat)**2).sum()
            loss.backward()
            opt.step()
    return autoencoder
```

Following early descriptions of autoencoders [12–14], a classic example of the use for dimensionality reduction is the work of Hinton and Salakhutdinov, who showed that deep autoencoders outperform principal components-based approaches on image and text data [11]. A very popular autoencoder architecture for feature extraction is the U-net (to be discussed in Sect. 8.4), which was primarily designed for image segmentation in the medical domain [16]. Currently, the U-net forms the backbone of most diffusion models (see Sect. 8.5), the state-of-the-art image generation architectures [17, 18]. For a more detailed overview of autoencoders, see [15].

8.2 Variational Autoencoder

In traditional autoencoders, the only constraint on the latent vector representation is that latent vectors should be easily decodable back into the original input. Consequently, the latent space $\boldsymbol{h}$ can become disjointed and discontinuous (Fig. 8.3a). Variational autoencoders (VAEs) [19, 20] address this issue by introducing a probabilistic framework. Instead of mapping inputs deterministically to latent vectors as $\boldsymbol{h} = e(\boldsymbol{x})$, variational autoencoders map inputs to a probability distribution over latent vectors, from which a latent vector is sampled. This probabilistic mapping enhances the decoder's robustness in reconstructing latent vectors. Specifically, rather than mapping the input $\boldsymbol{x}$ to a latent vector $\boldsymbol{h} = e(\boldsymbol{x})$, we map it to a mean vector $\boldsymbol{\mu}(\boldsymbol{x})$ and a standard deviation vector $\boldsymbol{\sigma}(\boldsymbol{x})$. These vectors parameterize a diagonal Gaussian distribution $\mathcal{N}(\boldsymbol{\mu}, \boldsymbol{\sigma}^2)$, from which we sample a latent vector $\boldsymbol{h} \sim \mathcal{N}(\boldsymbol{\mu}, \boldsymbol{\sigma}^2)$. This approach is typically implemented by replacing the last layer of a traditional encoder with two layers that output $\boldsymbol{\mu}(\boldsymbol{x})$ and $\boldsymbol{\sigma}(\boldsymbol{x})$, respectively. To ensure positivity, an exponential activation function is often applied to $\boldsymbol{\sigma}(\boldsymbol{x})$. However, this modification alone does not entirely resolve the issue of discontinuities in the latent space, as the outputted means may differ significantly, and the standard deviations may be small. To mitigate this, we introduce an auxiliary loss that penalizes the distribution $p(\boldsymbol{h} \mid \boldsymbol{x})$ for deviating from the standard normal distribution $\mathcal{N}(\boldsymbol{0}, \boldsymbol{I})$. This penalty term is the Kullback-Leibler (KL) divergence [21] between $p(\boldsymbol{h} \mid \boldsymbol{x})$ and $\mathcal{N}(\boldsymbol{0}, \boldsymbol{I})$, given by

$$\mathrm{KL}\left(\mathcal{N}(\boldsymbol{\mu}, \boldsymbol{\sigma}^2) \parallel \mathcal{N}(\boldsymbol{0}, \boldsymbol{I})\right) = \frac{1}{2} \sum_{i=1}^{d} \left(\sigma_i^2 + \mu_i^2 - \ln \sigma_i^2 - 1\right), \tag{8.2}$$

where d is the dimensionality of the latent space, and μ_i and σ_i are the ith components of $\boldsymbol{\mu}$ and $\boldsymbol{\sigma}$, respectively. This expression applies to multivariate Gaussian distributions with diagonal covariance matrices. Extending it to our case involves summing the Kullback-Leibler divergence over each dimension. Normally, sampling operations are non-differentiable and, therefore, do not allow the error to be backpropagated through the network. Here, a reparameterization trick comes into play. Assuming $\boldsymbol{h}$ is a random variable following a Gaussian distribution, it can be expressed as

$$\boldsymbol{h} = \boldsymbol{\mu}(\boldsymbol{x}) + \boldsymbol{\sigma}(\boldsymbol{x})\epsilon, \quad \epsilon \sim \mathcal{N}(\boldsymbol{0}, \boldsymbol{I}). \tag{8.3}$$

This auxiliary loss is beneficial for two reasons. First, this allows the encoder network to be trained using gradient descent despite the non-differentiable sampling operation. Second, by penalizing the Kullback-Leibler divergence, we encourage the latent vectors to form a more centralized and continuous distribution. Essentially, we constrain the encoder to produce latent vectors that approximate a standard Gaussian distribution, which the decoder can then effectively reconstruct. To train the variational autoencoder, we simply combine the reconstruction loss $\|d(e(\boldsymbol{x})) - \boldsymbol{x}\|$ and the Kullback-Leibler divergence loss (Eq. 8.2):

$$C = \|\boldsymbol{x} - \hat{\boldsymbol{x}}\|^2 + \mathrm{KL}\left(\mathcal{N}\left(\boldsymbol{\mu}(\boldsymbol{x}), \boldsymbol{\sigma}\right), \mathcal{N}(\boldsymbol{0}, \boldsymbol{I})\right). \tag{8.4}$$

Figure 8.4 illustrates the variational autoencoder architecture schematically. To implement the `VariationalEncoder` class, we adapt the `Encoder` class (Listing 8.1) to produce $\boldsymbol{\mu}(\boldsymbol{x})$ and $\boldsymbol{\sigma}(\boldsymbol{x})$, and use these to sample a latent vector $\boldsymbol{h}$ via the reparameterization trick. Additionally, this class keeps track of the Kullback-Leibler divergence loss term during training.

```
class VariationalEncoder(torch.nn.Module):
    def __init__(self, latent_dims):
        super(VariationalEncoder, self).__init__()
        self.linear1 = torch.nn.Linear(784, 512)
        self.linear2 = torch.nn.Linear(512, latent_dims)
        self.linear3 = torch.nn.Linear(512, latent_dims)

        self.N = torch.distributions.Normal(0, 1)
        self.N.loc = self.N.loc.cuda()
        self.N.scale = self.N.scale.cuda()
        self.kl = 0

    def forward(self, x):
        x = torch.flatten(x, start_dim=1)
        x = F.relu(self.linear1(x))
        mu =  self.linear2(x)
```

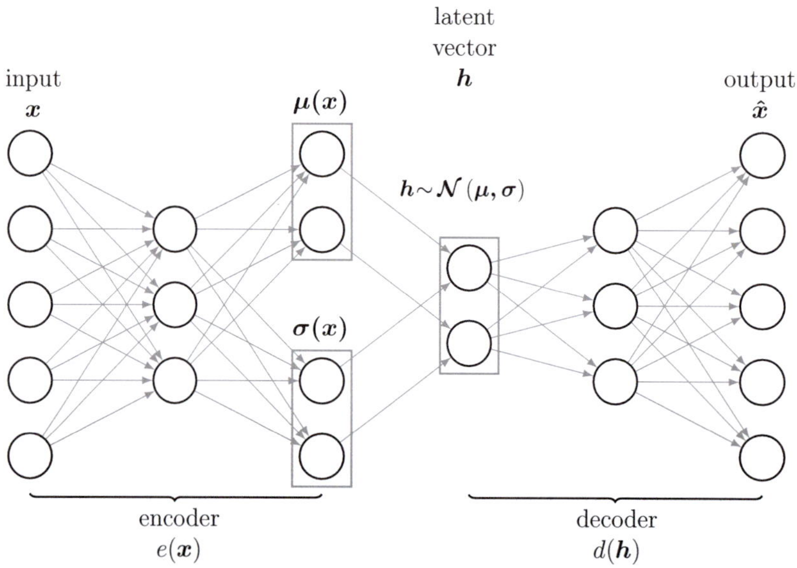

Fig. 8.4 Variational autoencoder: The input $\boldsymbol{x}$ is mapped to the parameters of a latent distribution—mean $\boldsymbol{\mu}$ and standard deviation $\boldsymbol{\sigma}$—via the encoder $e(\boldsymbol{x})$. A latent vector $\boldsymbol{h}$ is sampled from this distribution using the reparameterization trick: $\boldsymbol{h} = \boldsymbol{\mu} + \boldsymbol{\sigma}\epsilon$, where $\epsilon \sim \mathcal{N}(\mathbf{0}, \boldsymbol{I})$. The decoder $d(\boldsymbol{h})$ maps $\boldsymbol{h}$ back to the reconstructed output $\hat{\boldsymbol{x}}$. The variational autoencoder is trained to minimize the reconstruction error and the Kullback-Leibler divergence between the approximate posterior $q(\boldsymbol{z}|\boldsymbol{x})$ and the prior $p(\boldsymbol{h})$

```
        sigma = torch.exp(self.linear3(x))
        z = mu + sigma*self.N.sample(mu.shape)
        self.kl = (sigma**2 + mu**2 - torch.log(sigma) - 1/2).sum()
        return z
```

To obtain the `VariationalAutoencoder`, the encoder is replaced with `VariationalEncoder`.

```
class VariationalAutoencoder(torch.nn.Module):
    def __init__(self, latent_dims):
        super(VariationalAutoencoder, self).__init__()
        self.encoder = VariationalEncoder(latent_dims)
        self.decoder = Decoder(latent_dims)

    def forward(self, x):
        z = self.encoder(x)
        return self.decoder(z)
```

The cost used in the training loop now includes the Kullback-Leibler divergence loss.

```
def train(autoencoder, data, epochs=20):
    opt = torch.optim.Adam(autoencoder.parameters())
    for epoch in range(epochs):
        for x, y in data:
            x = x.to(device)
            opt.zero_grad()
            x_hat = autoencoder(x)
            loss = ((x - x_hat)**2).sum() + autoencoder.encoder.kl
            loss.backward()
            opt.step()
    return autoencoder
```

Examples of the generative capabilities of variational autoencoders include synthetic face generation [22] or music manipulation [23]. For further reading on the fundamentals of variational autoencoders, see [26].

8.3 Generative Adversarial Networks

Generative adversarial networks [27] emulate data distributions by setting up a two-player adversarial game between two neural networks (see Fig. 8.5):

- the generator G_{NN},
- the discriminator D_{NN}.

The generator creates predictions $\hat{\boldsymbol{x}} = G_{NN}(\boldsymbol{\xi}; \boldsymbol{\Theta}_G)$ from random noise $\boldsymbol{\xi}$, while the discriminator attempts to distinguish between these generated predictions $\hat{\boldsymbol{x}}$ from real data $\tilde{\boldsymbol{x}}$. The discriminator assigns a probability score $\hat{p} = D_{NN}(\boldsymbol{x}; \boldsymbol{\Theta}_D)$ which evaluates the probability of a data point $\boldsymbol{x}$ being real or generated. The quality of both the generator and the discriminator can be expressed via the following cost function:

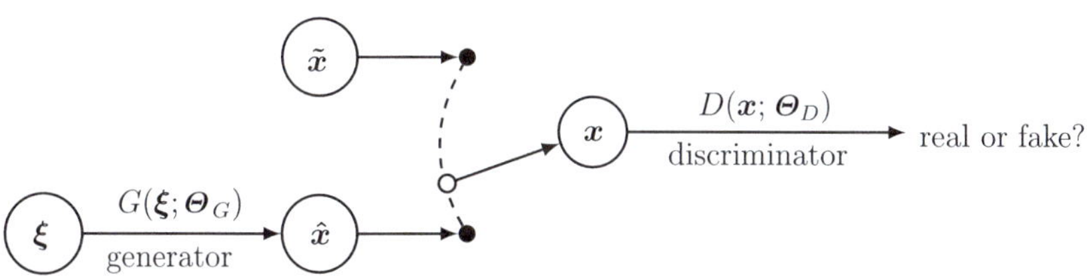

Fig. 8.5 Generative adversarial network: The generator G creates fake samples $\hat{\boldsymbol{x}}$ from random noise $\boldsymbol{\xi}$, while the discriminator D attempts to distinguish between real samples $\tilde{\boldsymbol{x}}$ and fake samples $\hat{\boldsymbol{x}}$. The adversarial training of the generator and discriminator is a two-player game where the generator tries to fool the discriminator, and the discriminator tries to distinguish between real and fake samples. Adapted from [24]

$$C = \frac{1}{m_D}\sum_{i=1}^{m_D}\log\Big[D_{NN}(\tilde{\boldsymbol{x}}_i;\boldsymbol{\Theta}_D)\Big] + \frac{1}{m_G}\sum_{i=1}^{m_G}\log\Big[1 - D_{NN}\big(G_{NN}(\boldsymbol{\xi}_i;\boldsymbol{\Theta}_G);\boldsymbol{\Theta}_D\big)\Big]. \tag{8.5}$$

Here, m_D real samples and m_G generated samples are used for training. The goal for the generator is to minimize the cost function, implying that the discriminator fails to distinguish between real and generated samples. However, the discriminator strives to maximize the cost. Therefore, this is formulated as a minimax optimization problem

$$\min_{\boldsymbol{\Theta}_G}\max_{\boldsymbol{\Theta}_D} C. \tag{8.6}$$

Convergence is ideally reached at the Nash equilibrium [25] (from game theory), where the discriminator always outputs a probability of 1/2, signifying its inability to distinguish between real and generated samples.

The following code snippet presents the implementation of a deep convolutional-generative adversarial network (DC-GAN) architecture [28] where `nc` is the number of channels, `nz` is the size of the latent vector, `ngf` is the size of feature maps in the generator, and `ndf` is the size of feature maps in discriminator.

```
class Generator(torch.nn.Module):
    def __init__(self):
        super(Generator, self).__init__()
        self.main = torch.nn.Sequential(
            torch.nn.ConvTranspose2d( nz, ngf * 8, 4, 1, 0, bias=False),
            torch.nn.BatchNorm2d(ngf * 8),
            torch.nn.LeakyReLU(0.2,True),
            torch.nn.ConvTranspose2d(ngf * 8, ngf * 4, 4, 2, 1, bias=False),
            torch.nn.BatchNorm2d(ngf * 4),
            torch.nn.LeakyReLU(0.2, True),
            torch.nn.ConvTranspose2d( ngf * 4, ngf * 2, 4, 2, 1, bias=False),
            torch.nn.BatchNorm2d(ngf * 2),
            torch.nn.LeakyReLU(0.2, True),
            torch.nn.ConvTranspose2d( ngf * 2, ngf, 4, 2, 1, bias=False),
            torch.nn.BatchNorm2d(ngf),
            torch.nn.LeakyReLU(0.2, True),
            torch.nn.ConvTranspose2d( ngf, nc, 4, 2, 1, bias=False),
            torch.nn.Tanh()
        )

    def forward(self, input):
```

```
        return self.main(input)
```

```
class Discriminator(torch.nn.Module):
    def __init__(self):
        super(Discriminator, self).__init__()
        self.main = torch.nn.Sequential(
            torch.nn.Conv2d(nc, ndf, 4, 2, 1, bias=False),
            torch.nn.LeakyReLU(0.2, inplace=True),
            torch.nn.Conv2d(ndf, ndf * 2, 4, 2, 1, bias=False),
            torch.nn.BatchNorm2d(ndf * 2),
            torch.nn.LeakyReLU(0.2, inplace=True),
            torch.nn.Conv2d(ndf * 2, ndf * 4, 4, 2, 1, bias=False),
            torch.nn.BatchNorm2d(ndf * 4),
            torch.nn.LeakyReLU(0.2, inplace=True),
            torch.nn.Conv2d(ndf * 4, ndf * 8, 4, 2, 1, bias=False),
            torch.nn.BatchNorm2d(ndf * 8),
            torch.nn.LeakyReLU(0.2, inplace=True),
            torch.nn.Conv2d(ndf * 8, 1, 4, 1, 0, bias=False),
            torch.nn.Sigmoid()
        )

    def forward(self, input):
        return self.main(input)
```

In each iteration of the training loop, the discriminator network `netD` is first updated by calculating its loss on a batch of real images and then on a batch of fake images generated by the generator `netG`; these losses are combined, gradients are computed, and the weights of `netD` are updated accordingly. Next, the generator network `netG` is updated by generating new fake images and calculating the loss with labels set as real (to deceive the discriminator); gradients are computed, and `netG` is updated.

```
iters = 0
for epoch in range(num_epochs):
    # For each batch in the dataloader
    for i, data in enumerate(dataloader, 0):

        netD.zero_grad()
        real_cpu = data.to(device)
        b_size = real_cpu.size(0)
        label = torch.full((b_size,), real_label,
        dtype=torch.float, device=device)
```

```
        output = netD(real_cpu).view(-1)
        errD_real = criterion(output, label)
        errD_real.backward()
        D_x = output.mean().item()

        noise = torch.randn(b_size, nz, 1, 1,
        device=device)
        fake = netG(noise)
        label.fill_(fake_label)

        output = netD(fake.detach()).view(-1)

        errD_fake = criterion(output, label)
        errD_fake.backward()

        D_G_z1 = output.mean().item()
        errD = errD_real + errD_fake
        optimizerD.step()

        netG.zero_grad()
        label.fill_(real_label)
        output = netD(fake).view(-1)

        errG = criterion(output, label)

        errG.backward()
        D_G_z2 = output.mean().item()

        optimizerG.step()
        iters += 1
```

To generate a batch of fake data, we apply the generator to random Gaussian noise.

```
noise = torch.randn(batch_size, nz, 1, 1, device=device)
with torch.no_grad():
    fake = netG(fixed_noise).detach().cpu()
```

Unfortunately, generative adversarial networks can be challenging to train. Problems like mode collapse [29] can arise. Here, the generator learns only a few modes from the training data. In the extreme case, only a single sample from the training data is learned, yielding a low discriminator score yet an undesirable outcome. To combat mode collapse, design diversity can either be promoted in the learning algorithm or the cost [29, 30]. Another challenge lies in balancing the training of the two neural networks. If the discriminator learns too quickly and manages to distinguish all generated samples, the gradient of the cost function (Eq. (8.5)) with respect to

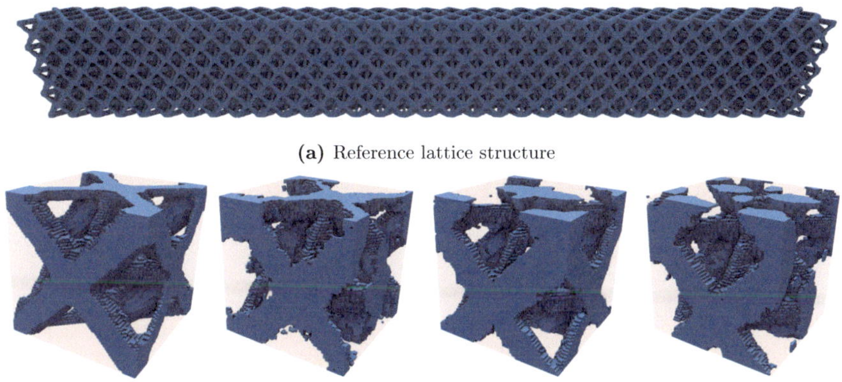

(a) Reference lattice structure

(b) Reference cell (c) Generated cell (d) Generated cell (e) Generated cell

Fig. 8.6 Lattice structures generated by a generative adversarial network serving as an example of generative adversarial network's capabilities. By exploiting the periodicity of the full lattice structure (**a**), only individual subcells are generated. Note that the generated cells (**c**–**e**) are not centered (as in (**b**)) because the generative adversarial network learned to reproduce any periodic cell within the full lattice structure (**a**). For details on the specific generative adversarial network, see [35], which has been applied to the depicted lattice structure in [36]

the weights becomes zero, halting further progress. A possible remedy is to use the Wasserstein distance in the cost function [31], as was, e.g., used to generate the lattice structures in Fig. 8.6.[1]

Additionally, generative adversarial networks can be modified to include inputs that control the generated data. This can be achieved in a supervised manner with conditional generative adversarial networks [37]. The conditional generative adversarial network does not just receive random noise but also an additional input. This supplementary input is considered by the discriminator, which assesses whether the input-output pairs are real or generated. An unsupervised alternative is InfoGANs [38], which disentangle the input information, i.e., the random input $\boldsymbol{\xi}$, defining the generated data. This is achieved by introducing an additional parameter c, a latent code to the generator $G_{NN}(\boldsymbol{\xi}, c; \boldsymbol{\Theta}_G)$. To ensure that the parameter is used by the neural network, the cost (Eq. (8.5)) is extended by a mutual information term [39] $\mathcal{L}_I(c, G_{NN}(\boldsymbol{x}, c; \boldsymbol{\Theta}_G))$ ensuring that the generated data varies meaningfully based on the input latent code c.

In comparison to variational autoencoders, generative adversarial networks typically generate higher-quality data. However, the advantage of autoencoders lies in their ability to construct a well-structured latent space, where proper sampling leads to smooth interpolations in the generated space. In other words, small changes in the latent space correspond to small changes in the generated space—a characteristic

[1] Using a single (computed tomography-scanned) lattice structure from [32–34] (Fig. 8.6a) and sampling periodic unit cells (such as Fig. 8.6b), a Wasserstein generative adversarial network was trained to predict representative lattice structures (Fig. 8.6c, e).

not inherent to generative adversarial networks. To achieve smooth interpolations, autoencoders can be combined with generative adversarial networks [40], where the autoencoder acts as the generator in the generative adversarial network framework, employing both an autoencoder loss and a generative adversarial network loss. Fittingly, this technique is referred to as variational autoencoder generative adversarial network.

8.4 Autoencoders in Computer Vision: U-Net

Autoencoders have also given rise to a variety of network architectures that are considered state-of-the-art in computer vision. Simply reusing the network architecture in a supervised sense, i.e., learning labeled data $\hat{\boldsymbol{y}}$ from inputs $\boldsymbol{x}$, is beneficial, as it learns to predict $\hat{\boldsymbol{y}}$ from a compressed space. Here, it identifies only the relevant features, which make less prone to overfitting. Furthermore, working on a lower-dimensional space in the bulk of the network is computationally more efficient due to fewer neural connections and parameters. The most prominent architecture is the U-net [16], named after the resemblance of its architecture with an uppercase U; see Fig. 8.7. A decoder network downsamples an input with convolutional and pooling layers, while an encoder upsamples the latent space with convolutional and deconvolutional layers. Additionally, skip connections connect the encoded and decoded layers prior to downsampling in order to allow information to travel more directly to the output without compression in between. This enables a feature extraction with the encoder-decoder structure without loss of information through the skip connections. An example is provided in Fig. 8.7. The network was developed in the context of biomedical image segmentation. Many extensions are currently deemed state-of-the-art in computer vision [41, 42]. In addition, U-nets form the backbone of the most powerful image generation models, which we will explore in the following Sect. 8.5.

8.5 Diffusion Models

Diffusion models [43, 44, 46, 47] are a relatively recent addition to the class of generative models. They achieve exceptional performance in image [17, 18] and audio generation [45, 48, 49] and are more stable to train compared to adversarial training in generative adversarial networks. Diffusion models generate data by reversing a gradual noising procedure, which is inspired by non-equilibrium thermodynamics and comprises the following two processes:

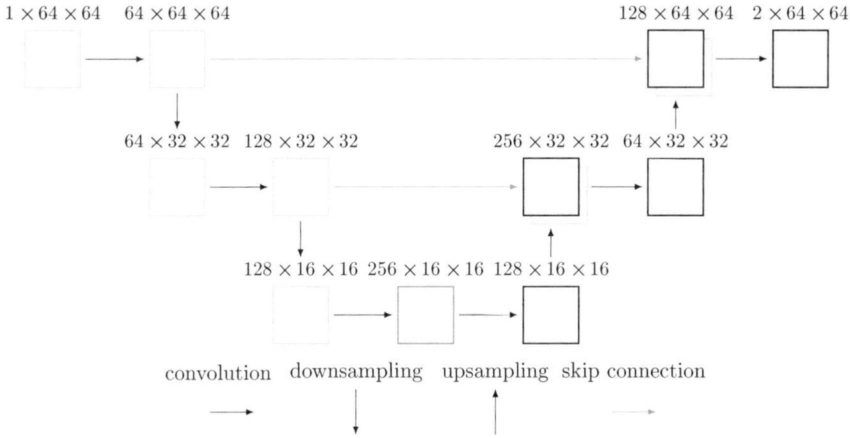

Fig. 8.7 A two-layer deep U-net: The input image of size 1 channel $\times$ 64 $\times$ 64 is decoded to an image of size 128 channels $\times$ 16 $\times$ 16 from which it is upsampled again, yielding an output of size 2 channels $\times$ 64 $\times$ 64. In between skip connections ensure the direct flow of information between the encoder and decoder. Adapted from [16]

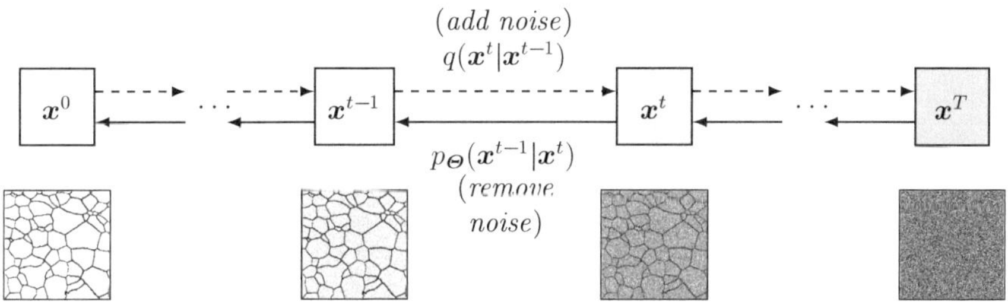

Fig. 8.8 Forward and reverse diffusion processes in a diffusion model. Starting from a sample $\boldsymbol{x}^0$, noise is gradually added through the known forward diffusion process $q(\boldsymbol{x}^t|\boldsymbol{x}^{t-1})$ to reach pure Gaussian noise $\boldsymbol{x}^T$ (dashed arrows). The network is trained to reverse the diffusion process $p_{\boldsymbol{\Theta}}(\boldsymbol{x}^{t-1}|\boldsymbol{x}^t)$ by sequentially removing noise until the original input is reconstructed (solid arrow). Adapted from [46]

- A forward diffusion process q: a Markov chain of diffusion steps that slowly adds random noise to data $\boldsymbol{x}$,
- A reverse denoising diffusion process $p_{\boldsymbol{\Theta}}$: a neural network that learns to reverse the diffusion process to construct desired data samples from the noise.

Note the similarity to autoencoders. Instead of learning a mapping to a hidden random state $\boldsymbol{h}_i$, the encoding is prescribed as the iterative application of Gaussian noise [2]. The forward diffusion process, as illustrated in Fig. 8.8, is a step-wise transformation of a sample $\boldsymbol{x}^0$ to a noisy Gaussian random variable $\boldsymbol{x}^T \sim \mathcal{N}(\boldsymbol{0}, \boldsymbol{I})$. Let $\boldsymbol{x}^0$ be a sample from a real data distribution $q(\boldsymbol{x}^0)$ (e.g., images). Starting from $\boldsymbol{x}^0$ we add noise at each time step $t = 1, \ldots, T$, such that the mean of $\boldsymbol{x}^t$ is gradually moved towards zero and its variance gradually increases towards unity. In other words, the goal is to gradually transform the complex data distribution $q(\boldsymbol{x}^0)$ into a known Gaussian prior $q(\boldsymbol{x}^T)$ through a forward diffusion process defined as

$$q(\boldsymbol{x}^t \mid \boldsymbol{x}^{t-1}) = \mathcal{N}\Big(\boldsymbol{x}^t;\ \sqrt{1-\beta^t}\,\boldsymbol{x}^{t-1},\ \beta^t \boldsymbol{I}\Big). \tag{8.7}$$

Here, $\mathcal{N}$ is a normal distribution with mean $\sqrt{1-\beta^t}\,\boldsymbol{x}^{t-1}$ and variance $\beta^t \boldsymbol{I}$. Note that $\boldsymbol{x}^t$ is the output of the forward diffusion process at step t, while $\boldsymbol{x}^{t-1}$ is an input.[2] The parameter β follows a predefined variance schedule $\{\beta^t\}_{t=1}^T$, where $0 < \beta^1 < \beta^2 < \cdots < \beta^T < 1$. The schedule $\{\beta^t\}$ itself can be designed in various ways (e.g., linear, quadratic, or cosine-based). As $T \to \infty$, the final sample $\boldsymbol{x}^T$ approaches an isotropic Gaussian distribution.

To generate a sample from the distribution we apply the reparameterization trick (see Eq. 8.3) to obtain

$$\boldsymbol{x}^t = \sqrt{1-\beta^t}\,\boldsymbol{x}^{t-1} + \sqrt{\beta^t}\,\epsilon, \tag{8.8}$$

where $\epsilon \sim \mathcal{N}(\boldsymbol{0}, \boldsymbol{I})$.

Since this process can be interpreted as a Markov chain, the joint probability of $\boldsymbol{x}^1, \ldots, \boldsymbol{x}^T$ given the initial sample $\boldsymbol{x}^0$ factorizes as

$$q(\boldsymbol{x}^{1:T} \mid \boldsymbol{x}^0) = \prod_{t=1}^{T} q(\boldsymbol{x}^t \mid \boldsymbol{x}^{t-1}). \tag{8.9}$$

Because sums of Gaussian variables remain Gaussian, we can obtain a sample $\boldsymbol{x}^t$ at any noise level t given $\boldsymbol{x}^0$ without iterating through all previous steps. Specifically,

$$q(\boldsymbol{x}^t \mid \boldsymbol{x}^0) = \mathcal{N}\Big(\boldsymbol{x}^t;\ \sqrt{\bar{\alpha}^t}\,\boldsymbol{x}^0,\ \big(1-\bar{\alpha}^t\big)\boldsymbol{I}\Big), \tag{8.10}$$

where $\alpha^t = 1 - \beta^t$ and $\bar{\alpha}^t = \prod_{s=1}^t \alpha^s$. The computable form can be written as

$$\boldsymbol{x}^t = \sqrt{\bar{\alpha}^t}\boldsymbol{x}^0 + \sqrt{1-\bar{\alpha}^t}\epsilon, \quad \epsilon \sim \mathcal{N}(\boldsymbol{0}, \boldsymbol{I}). \tag{8.11}$$

If the conditional probability $q(\boldsymbol{x}^{t-1}|\boldsymbol{x}^t)$ was known, the process could be reversed by starting from random Gaussian noise $\boldsymbol{x}^T$ and gradually removing noise to obtain a sample $\boldsymbol{x}^0$ from the real data distribution. However, assuming the reverse diffusion process is also Gaussian we can employ a neural network with $\boldsymbol{\Theta}$ to estimate the mean and variance of $q(\boldsymbol{x}^{t-1}|\boldsymbol{x}^t)$. Then, the inverse diffusion process can be parameterized as

$$p_{\boldsymbol{\Theta}}(\boldsymbol{x}^{t-1}|\boldsymbol{x}^t) = \mathcal{N}\left(\boldsymbol{x}^{t-1};\ \mu_{\boldsymbol{\Theta}}(\boldsymbol{x}^t, t),\ \Sigma_{\boldsymbol{\Theta}}(\boldsymbol{x}^t, t)\right), \tag{8.12}$$

[2] The vertical bar in $q(\boldsymbol{x}^t|\boldsymbol{x}^{t-1})$ denotes a conditional probability where $\boldsymbol{x}^{t-1}$ is a random variable from the previous time step. In $\mathcal{N}\left(\boldsymbol{x}^t; \sqrt{1-\beta^t}\,\boldsymbol{x}^{t-1}, \beta^t\,\boldsymbol{I}\right)$ the semicolon seperates the variable (or point of evaluation) $\boldsymbol{x}^t$ from the parameters μ^t and σ^{2^t}.

where the mean μ_Θ and variance Σ_Θ are learned and conditioned on the current sample $\boldsymbol{x}^t$ and the time step t. Like the forward process, the reverse process can be seen as a Markov chain, and the joint probability of a sequence of samples can be written as a product of conditionals and the marginal probability of $\boldsymbol{x}^T$:

$$p_\Theta(\boldsymbol{x}^{0:T}) = p(\boldsymbol{x}^T) \prod_{t=1}^{T} p_\Theta(\boldsymbol{x}^{t-1}|\boldsymbol{x}^t). \tag{8.13}$$

Here, the distribution $p(\boldsymbol{x}^T) = q(\boldsymbol{x}^T)$, which corresponds to the known distribution $\mathcal{N}(\mathbf{0}, \boldsymbol{I})$. Thus, the generation process starts with Gaussian noise and then samples from the learned individual steps of the reverse process.

In practice, the neural network is designed to model the mean μ_Θ while keeping the variance Σ_Θ fixed as proposed by [46]. This simplifies the learning task and focuses the neural network's capacity on modeling the mean of the reverse conditional distribution.

To derive an objective function for learning the mean of the backward diffusion process, the combination of the forward process q and the reverse process p_Θ can be interpreted within a variational autoencoder framework. This allows the use of the variational lower bound (for details, see [20]) to minimize the negative log-likelihood with respect to the ground truth data sample $\boldsymbol{x}^0$. The variational lower bound decomposes into a sum of losses at each time step t:

$$L = L^0 + L^1 + \cdots + L^T. \tag{8.14}$$

Due to the Gaussian nature of the processes q and p_Θ, each term (except L^0) in the loss function is the Kullback-Leibler divergence between two Gaussian distributions, which can be explicitly expressed as an L^2-loss with respect to their means.

Ho et al. introduced another improvement to the reverse diffusion process. By reparameterizing the mean as

$$\mu_\Theta(\boldsymbol{x}^t, t) = \frac{1}{\sqrt{\alpha^t}} \left(\boldsymbol{x}^t - \frac{\beta^t}{\sqrt{1 - \bar{\alpha}^t}} \epsilon_\Theta(\boldsymbol{x}^t, t) \right), \tag{8.15}$$

we can train the neural network to predict the added noise $\epsilon_\Theta(\boldsymbol{x}^t, t)$ at a given noise level t instead of predicting the mean. This simplifies the final objective function L^t to minimizing the difference between the true noise ϵ and the predicted noise $\epsilon_\Theta(\boldsymbol{x}^t, t)$:

$$\left\| \epsilon - \epsilon_\Theta(\boldsymbol{x}^t, t) \right\|^2 = \left\| \epsilon - \epsilon_\Theta \left(\sqrt{\bar{\alpha}^t}\, \boldsymbol{x}^0 + \sqrt{1 - \bar{\alpha}^t}\, \epsilon, t \right) \right\|^2, \tag{8.16}$$

where $\epsilon \sim \mathcal{N}(\mathbf{0}, \boldsymbol{I})$. Here, $\boldsymbol{x}^0$ is the original data sample, $\boldsymbol{x}^t$ is obtained directly from the forward process at noise level t, and $\epsilon_\Theta(\boldsymbol{x}^t, t)$ is the neural network's prediction.

The implementation of a diffusion model is demonstrated on a small toy example. A very naive way to facilitate the forward diffusion process is to corrupt the input with some noise.

```
def corrupt(x, amount):
    """Corrupt the input x by mixing it with noise according to amount"""
    noise = torch.rand_like(x)
    amount = amount.view(-1, 1, 1, 1)  # Sort shape so broadcasting works
    return x * (1 - amount) + noise * amount
```

In contrast to autoencoders or generative adversarial networks, the dimensionality of the input variable does not change throughout the diffusion and the reverse diffusion process. For structured data like images, the primary choice is a U-net architecture (see Sect. 8.4) as it preserves the dimensionality of the input. In this simplified example, the U-net tries to predict $\boldsymbol{x}^0$ given a noisy input $\boldsymbol{x}^{\{1 \ldots T\}}$.

```
class BasicUNet(torch.nn.Module):
    """A minimal UNet implementation."""

    def __init__(self, in_channels=1, out_channels=1):
        super().__init__()
        self.down_layers = torch.nn.ModuleList(
            [
                torch.nn.Conv2d(in_channels, 32, kernel_size=5, padding=2),
                torch.nn.Conv2d(32, 64, kernel_size=5, padding=2),
                torch.nn.Conv2d(64, 64, kernel_size=5, padding=2),
            ]
        )
        self.up_layers = torch.nn.ModuleList(
            [
                torch.nn.Conv2d(64, 64, kernel_size=5, padding=2),
                torch.nn.Conv2d(64, 32, kernel_size=5, padding=2),
                torch.nn.Conv2d(32, out_channels, kernel_size=5, padding=2),
            ]
        )
        self.act = torch.nn.SiLU()  # The activation function
        self.downscale = torch.nn.MaxPool2d(2)
```

```
        self.upscale = torch.nn.Upsample(scale_factor=2)

    def forward(self, x):
        h = []
        for i, l in enumerate(self.down_layers):
            x = self.act(l(x))  # Through the layer and the activation function
            if i < 2:  # For all but the third (final) down layer:
                h.append(x)  # Storing output for skip connection
                x = self.downscale(x)  # Downscale ready for the next layer

        for i, l in enumerate(self.up_layers):
            if i > 0:  # For all except the first up layer
                x = self.upscale(x)  # Upscale
                x += h.pop()  # Fetching stored output (skip connection)
            x = self.act(l(x))  # Through the layer and the activation function

        return x
```

The diffusion model is then trained using a simple mean squared error loss `torch.nn.MSELoss()` between the network's prediction on an input corrupted with random noise `noisy_x` and the input itself.

```
def train(unet, data, epochs=20):
    opt = torch.optim.Adam(unet.parameters())
    loss_fn = torch.nn.MSELoss()
    for epoch in range(epochs):
        for x, y in data:
            x = x.to(device) # GPU

            noise_amount = torch.rand(x.shape[0]).to(device) # Pick random noise amounts
            noisy_x = corrupt(x, noise_amount)

            pred = unet(noisy_x)
            loss = loss_fn(pred, x)

            opt.zero_grad()
            loss.backward()
            opt.step()
    return autoencoder
```

To sample using the trained model, we start with random noise. With each sampling step, we add the model's prediction and feed it back into the model. A simple sampling procedure is shown below.

```
def sample(unet, n_steps)
    x = torch.rand(8, 1, 64, 64).to(device)  # Start from random
    for i in range(n_steps):
        with torch.no_grad():
            pred = unet(x)  # Predict the denoised x0
        mix_factor = 1 / (n_steps - i)
        x = x * (1 - mix_factor) + pred * mix_factor
    return x
```

It is worth noting that this toy example introduces a lot of simplification compared to the theoretical formulation above. In general, more advanced U-net architectures are employed, and the diffusion process is usually not linear but governed by a more elaborate noise scheduler dependent on the time step t. Additionally, the time step t is provided during the forward pass to condition the model on the amount of noise added. As described above, changing the training objective to predict the noise instead of the mean (or image) is beneficial. Lastly, there is a wide choice of sampling strategies that can improve sampling time and output quality [47, 50–52]. Considerable improvements were achieved when running the diffusion process in the latent space [17]. These so-called latent diffusion models, like Stable Diffusion, employ pretrained autoencoders to convert between the pixel and latent space. Applying the diffusion model only to low-dimensional representations allows the generation of high-quality samples while using only a fraction of resources for training and inference. Although the U-net architecture is the current go-to model for the reverse diffusion process, other architectures can also be employed. For example, transformer-based models show promising results and might improve scalability in the future [53]. In conclusion, diffusion models are both tractable and flexible, demonstrating state-of-the-art performance on a plethora of tasks and data types. However, the iterative sampling process (or inference) is considerably slower than that of other generative models, such as generative adversarial networks.

The mathematical descriptions have been adapted from [4, 54] which are highly recommended sources for further reading. For an extensive tutorial on diffusion models for imaging and vision, see [55] and consider [56] on how to tune them.

8.6 Transformers

Another prominent network architecture propelled to fame through tools like BERT [57] or ChatGPT [58] are transformers [59]. The transformer is a network architecture designed to process sequential data (common in natural language processing) in parallel, exploiting self-attention mechanisms and thus competing with recurrent neural networks and their variations. The two key components of a transformer are

- the positional encoding,
- and the self-attention mechanism.

The positional encoding allows the sequential input $\boldsymbol{X} = (\boldsymbol{x}_0, \boldsymbol{x}_1, \ldots, \boldsymbol{x}_w)$ to be fed into the network not sequentially, as in recurrent neural networks, but in parallel, by simultaneously providing a positional encoding to the network. In the simplest sense, $\boldsymbol{x}_0$ would receive the integer encoding 0, while $\boldsymbol{x}_1$ receives the integer 1, i.e., their indices. In practice, the encoding has the same dimension N^e as each sequential entry $\boldsymbol{x}_i \in \mathbb{R}^{N^e}$ and is thus encoded using more complex functions, which allow for unique encodings that capture relative distances between positions, such as trigonometric functions, as presented in [59]. The input can thus be fed to the network independently of the sequential order while retaining the positional information.

The second key ingredient, the self-attention mechanism, enables the network to selectively focus on different parts of the input sequence to establish context. Consider the example of language processing, where an input is a sentence $\boldsymbol{X} \in \mathbb{R}^{N^w \times N^e}$ consisting of words $(\boldsymbol{x}_0, \boldsymbol{x}_1, \ldots, \boldsymbol{x}_w)$. For the sentence: "*I eat a lot of pasta.*", every word has an attention with respect to the other words. As eating is closely related to pasta, we would expect greater attention of the word "*eat*" toward "*pasta*" than to the other words.

On a technical level, the self-attention is encapsulated within a self-attention layer, taking an embedded input $\boldsymbol{X}$ and providing an output z for the next layer. First, the similarity of the inputs $\boldsymbol{x}_0, \boldsymbol{x}_1, \ldots, \boldsymbol{x}_n$ is determined by querying each input $\boldsymbol{x}_i$ and comparing it to a key, i.e., another input $\boldsymbol{x}_j$. This is achieved by identifying all queries $\boldsymbol{Q}$ and all keys $\boldsymbol{K}$ via linear transformations with learnable weights $\boldsymbol{W}^q$, $\boldsymbol{W}^k$ of size $N^e \times N^s$.

$$\boldsymbol{Q} = \boldsymbol{X}\boldsymbol{W}^q \in \mathbb{R}^{N^w \times N^s} \tag{8.17}$$

$$\boldsymbol{K} = \boldsymbol{X}\boldsymbol{W}^k \in \mathbb{R}^{N^w \times N^s} \tag{8.18}$$

Next the similarity $\boldsymbol{S}$ is obtained by the dot-product

$$\boldsymbol{S} = \boldsymbol{Q}\boldsymbol{K}^\intercal \in \mathbb{R}^{N^w \times N^w}, \tag{8.19}$$

which is a matrix of size $N^w \times N^w$ representing the attention of the input with respect to each other, i.e., the self-attention. An example for the sentence "*I eat a lot of pasta.*" is provided in Fig. 8.9. The values are only exemplary but highlight the attention between the words "*eat*" and "*pasta*", or between "*I*" and "*eat*".

The similarity is then applied to the values $\boldsymbol{V}$ obtained through a linear transformation of the inputs $\boldsymbol{X}$ with the learnable weights $\boldsymbol{W}^v \in \mathbb{R}^{N^e \times N^s}$[3]

$$\boldsymbol{V} = \boldsymbol{X}\boldsymbol{W}^v \in \mathbb{R}^{N^w \times N^s}. \tag{8.20}$$

[3] The second dimension of W^v may also be chosen differently than N^s.

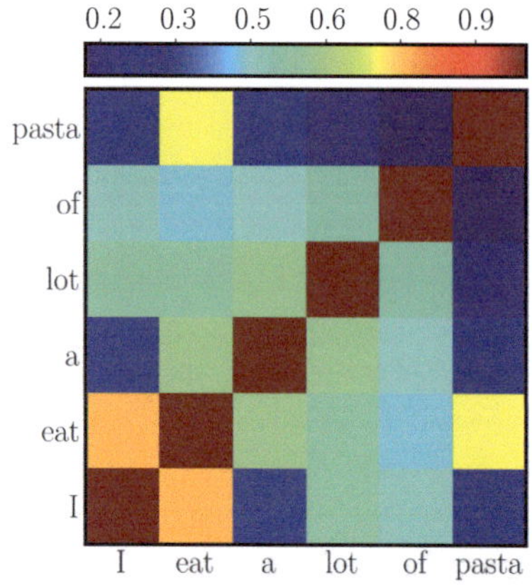

Fig. 8.9 Similarity $\boldsymbol{S}$ plotted in terms of attention on the example of "*I eat a lot of pasta*". Note that the values shown are only representative in order to illustrate the concept and thus relatively arbitrary

However, first, the similarities are normalized by the square root of the key and query dimension N^s and passed through the softmax activation, such that the sum of each queried similarity, i.e., each row, yields one. This results in the following expression for the self-attention layer.

$$\tilde{z} = \text{softmax}\left(\frac{\boldsymbol{Q}\boldsymbol{K}^\top}{\sqrt{N^s}}\right)\boldsymbol{V} \in \mathbb{R}^{N^w \times N^s}. \tag{8.21}$$

The problem with this representation is that the model is limited in capturing multiple patterns in the input sequence simultaneously. However, using multiple self-attention layers simultaneously, referred to as multi-headed attention, allows the model to jointly capture the attention from different representation subspaces. Thus the attention equation, Eq. (8.21) is applied multiple times with different weights, yielding multiple outputs $\tilde{\boldsymbol{Z}} = (\tilde{z}_0, \tilde{z}_1, \ldots, \tilde{z}_h)$. These are then combined by performing another linear transformation with $\boldsymbol{W}^z$, resulting in the final output of the self-attention layer[4]

$$z = \tilde{\boldsymbol{Z}}\boldsymbol{W}^z \in \mathbb{R}^{N^w \times N^e}. \tag{8.22}$$

The multi-headed attention is then in combination with feed-forward layers applied within an encoder and decoder, where the encoder acts solely on the input data $\boldsymbol{X}$, and the decoder acts on both the encoded data and the outputs $\hat{\boldsymbol{Y}}$ produced up to that point. These are then concatenated and fed through additional multi-headed attention layers. The first transformer architecture proposed by [59] was designed for language translation and is illustrated in Fig. 8.10. Here additional details, such as the layer normalization (layer-norm) [60], leading to better-conditioned gradients helping the optimization, and residual connections [61] (indicated in gray), improving the information flow and the model's ability to learn more complex relationships, are provided. Take note that the output from the last layer-norm, denoted as $\boldsymbol{H}^{(L)}$,

[4] Again the second dimension of z may be chosen differently than the embedding N^e.

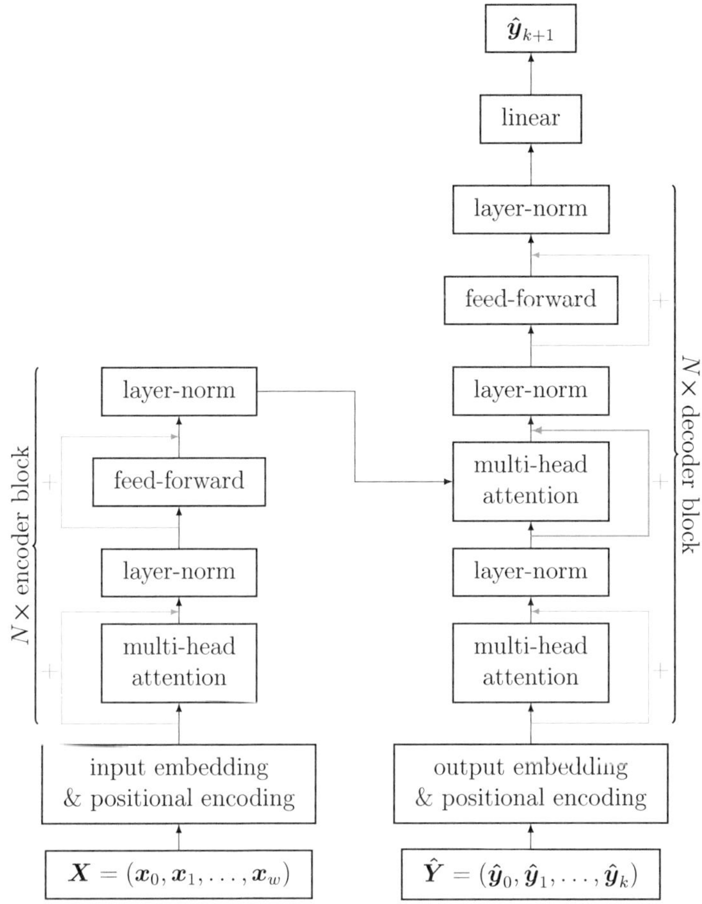

Fig. 8.10 A conceptual transformer architecture for language translation [59] predicting the next output $\hat{\boldsymbol{y}}_{k+1}$. Both the input $\boldsymbol{X}$ and outputs obtained up to now $\hat{\boldsymbol{Y}}$ are fed through multi-headed self-attention layers, layer normalization, and feed-forward layers. Each encoder block comprises two sub-layers. The decoder blocks have an additional third sub-layer performing multi-headed attention over the output of the last encoder block. Additionally, residual connections indicated in gray skip over the attention and feed-forward layers. Adapted from [59]

has dimensions $N^{w+k} \times N^e$. This differs from the final output $\boldsymbol{y}_{k+1}$, which has a dimension of N^e. In the case of sequence-to-sequence prediction, i.e., mapping $\boldsymbol{X}$ to $\hat{\boldsymbol{Y}}$, only the kth column vector of $\boldsymbol{H}^{(L)}$, which has a dimension of N^e, is forwarded to the final linear layer. An alternative approach involves an aggregation operation, such as pooling across the dimension N^{w+k}, to handle the dimensionality discrepancy.

Transformers have successfully been applied in various domains, including natural language processing [57, 58, 62], computer vision [63], and more, showcasing their versatility and effectiveness in handling different types of data.[5]

Different tasks require alternations of the encoder-decoder architecture depicted in Fig. 8.10. Text classification, extractive question answering (from the information provided in the input), or named entity recognition (to extract structured information from unstructured text) are best handled by only the encoder, e.g., achieved by BERT [57]. Text generation, autoregressive question answering (without required context in input), unstructured summarization, or dialogue systems (i.e., chatbots) are tackled by decoder-only systems, such as GPT [58, 70]. Encoder-decoder architectures, such as BART [71], T5 [72], and BigBird [73], are suitable for machine translation, structured summarization, data to text, or speech processing. However, all of the models are based on the basic building blocks of the transformer architecture [59], relying on the self-attention mechanism and positional encoding.

For a more extensive illustration of the transformer architecture, we refer the reader to [74], while a detailed mathematical description can be found in [75]. A practical hands-on guide is [76], allowing you to solve your own problems with transformers (with a focus on natural language processing).

8.7 Applications in Computational Mechanics

8.7.1 Data Generation

The most straightforward application of generative models in computational mechanics is the generation of new data based on existing examples. This has been demonstrated in [35, 77–81] for microstructures in [82] for velocity models used in full waveform inversion, and in [83] for optimized structures using generative adversarial networks. Variational autoencoders have also been used to model the crossover operation in evolutionary algorithms to create new designs from parent designs [84]. Applications of diffusion models for microstructure generation can be found in [85–87].

[5] The resultant models, such as Bert [57] and ChatGPT [58], provide tools for code generation, comprehension of complex topics, and text generation, as, e.g., used during the creation of this book. All of these may accelerate the pace of progress within science, including computational mechanics. On a more speculative level, the transformer architecture might, when integrated with other artificial intelligence methods, lead to artificial general intelligence (AGI), as speculated in [64]. Artificial general intelligence is a model that has surpassed human abilities and could solve numerous problems within science and computational mechanics. Through recursive self-improvement cycles [65], the network would be able to improve itself, possibly outperforming humans. The technical requirements of a model to self-improve are access to its own architecture and the ability to improve on itself, essentially a form of code generation. These aspects are discussed in books such as [66, 67] and are one of the key concerns of the field of artificial intelligence safety [66–69], where the alignment of the cost function of the artificial general intelligence with humanity is of interest (alignment problem).

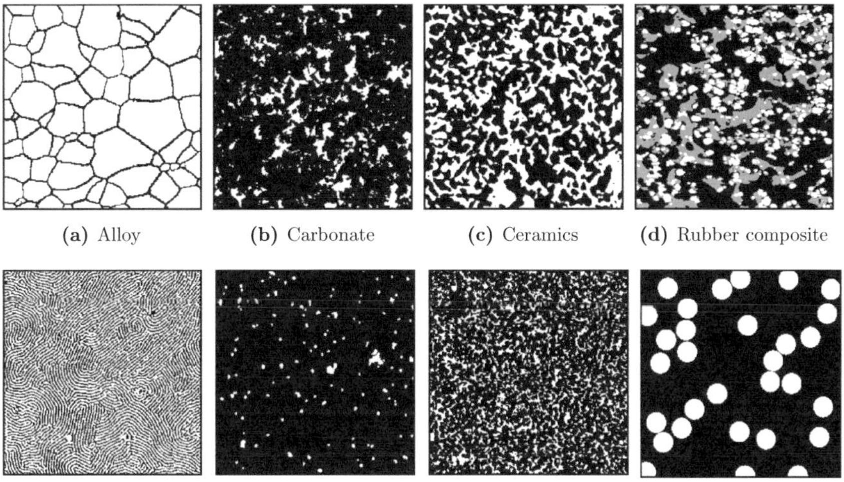

(a) Alloy (b) Carbonate (c) Ceramics (d) Rubber composite

(e) Copolymer (f) Polymer composite (g) Sandstone (h) Fiber composite

Fig. 8.11 Exemplary microstructures that serve as motivating examples of what can be created with generative modeling. The structures are taken from [88] and are derived from real microstructures [89], except from **(h)**, which was generated with the accompanying code

Microstructures pose a unique challenge due to their inherent three-dimensional nature, while often, only two-dimensional reference images are available. This has led to the development of specialized architectures that are capable of creating three-dimensional structures from representative two-dimensional slices [90–92]. The approach typically involves treating three-dimensional voxel data as a sequence of two-dimensional slices of pixels. Sequences of images are predicted from individual slices, ultimately forming a three-dimensional microstructure. In [93], a recurrent neural network is applied to a two-dimensional reference image, yielding an additional dimension and consequently creating a three-dimensional structure. The recurrent neural network is applied to the latent vector inside an encoder-decoder architecture, such that the recurrent neural network's inputs and outputs are relatively small. Similarly, [94, 95] apply a transformer [59] to the latent vector. An alternative formulation using variational autoencoder generative adversarial networks is presented in [96] to reconstruct three-dimensional voxel models of porous media from two-dimensional images.

The generated datasets can subsequently be leveraged to train surrogate models, as demonstrated in [83, 97–99] where convolutional neural networks were used to verify the physical properties of designs, and in the study by [100] on the homogenization of microstructures with convolutional neural networks. Similarly, [82, 101] generate realistic material distributions, such as velocity distributions, to train an inverse operator for full waveform inversion.

In the following examples, we demonstrate the generation of microstructures using a variational autoencoder and a diffusion model. We stick with generating two-dimensional microstructures for simplicity and faster training, but the same principles can be applied to three- or higher-dimensional data.

8.7.1.1 Data Generation with Variational Autoencoders

To demonstrate the generative power and latent space structuring of the variational autoencoder, we apply it to a simple yet illustrative example: generating basic geometric shapes—circles, triangles, squares, and plus signs. Despite being a toy example, it is well suited to showcasing how a variational autoencoder can learn to represent and interpolate between distinct structural features in a low-dimensional latent space. We created a synthetic dataset comprising binary images of the four shapes with a resolution of 128×128 pixels. Each shape is rendered at a random position to introduce variability, allowing the variational autoencoder to learn a more generalized representation of each shape category. An example of the generated shapes is shown in Fig. 8.12.

The variational autoencoder architecture for this example follows the implementation outlined in Sect. 8.2. Instead of fully connected layers, the encoder employs convolutional layers followed by fully connected layers that output the mean $\boldsymbol{\mu}(\boldsymbol{x})$ and standard deviation $\boldsymbol{\sigma}(\boldsymbol{x})$ vectors of the latent distribution. To enable the visualization of the latent space, the encoder network compresses the input images into two latent dimensions. The decoder mirrors this architecture, using transposed convolutional layers to reconstruct the images from the latent space. The variational autoencoder is trained for 100 epochs using the combined reconstruction and Kullback-Leibler divergence loss.

After training, we can visualize the latent space by encoding the test set images and plotting the resulting latent vectors. As shown in Fig. 8.13a, the variational autoencoder organizes the shapes into distinct clusters corresponding to each shape category. This clustering demonstrates that the variational autoencoder has learned to represent the essential features of each shape in the latent space.

One of the strengths of the variational autoencoder is its ability to generate new data by sampling from the smooth latent space. In our example, we can generate

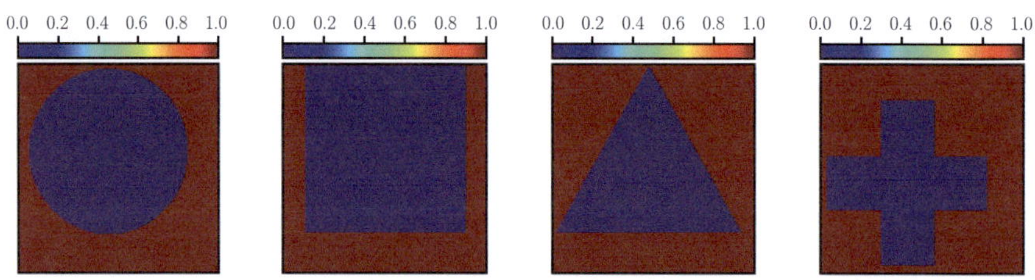

(a) circle (b) square (c) triangle (d) cross

Fig. 8.12 Sample images from the synthetic shapes dataset used for training the variational autoencoder. The dataset includes circles, squares, triangles, and plus signs with varying positions

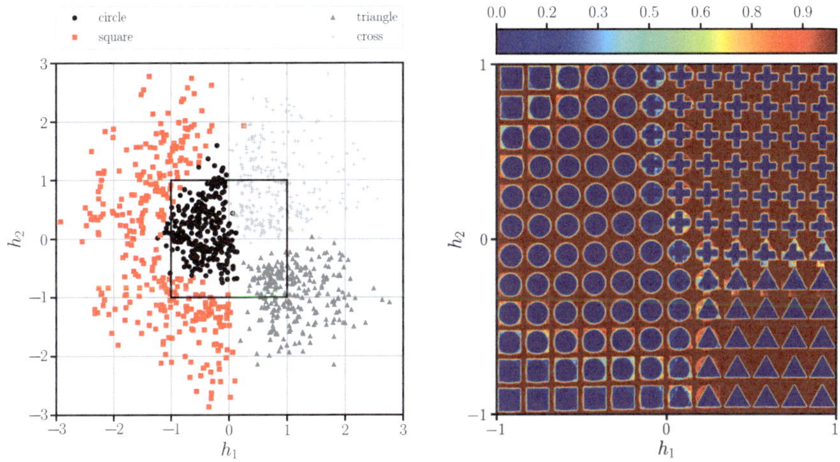

(a) Visualization of the two-dimensional latent space after training the variational autoencoder on the shapes dataset illustrated in Figure 8.12. Each point represents an encoded image, colored according to its shape category. The variational autoencoder clusters similar shapes together, reflecting the learned structural features

(b) Interpolation in the latent space between different shapes. The images are generated by decoding points along a line between two latent vectors corresponding to different shapes. The smooth transition indicates that the variational autoencoder has learned a continuous latent representation of the shape structures

Fig. 8.13 Variational autoencoder for a basic shape generation task

transitions between shapes by interpolating between points in the latent space as illustrated in Fig. 8.13b. This interpolation showcases the smoothness of the learned latent manifold and the variational autoencoder's capacity to capture continuous variations in shape structures. In practice, the reconstruction loss and the Kullback-Leibler loss are weighted with a hyperparameter or even a trainable parameter like in the β-variational autoencoder (β-VAE) [102]. Balancing the loss terms and, thus, finding the right trade-off between a structured latent space and the desired output, is key. However, as a consequence of adding the Kullback-Leibler loss term, samples generated by variational autoencoders are often perceived as noisy or blurry. This led to the development of perceptual losses with the aim of increasing the fidelity of the outputs [22].

This example with simple shapes illustrates the fundamental capabilities of variational autoencoders, laying the groundwork for more advanced applications in microstructure generation and materials design. The expressive capacity of our exemplary variational autoencoder is somehow limited due to the extensive compression from 128×128 pixels into only two latent dimensions.[6] It is common to choose a latent vector of higher dimensionality. To deepen the understanding of generative variational autoencoders, Exercise E.28 explores the example code of the variational autoencoder that was used to produce the figures in this section.

[6] A larger latent space would have prevented the visualization in Fig. 8.13a.

8.7.1.2 Data Generation with Diffusion Model

To showcase the generative capabilities of diffusion models in microstructure modeling, we consider a simplified representation of a fiber composite material, as illustrated in Fig. 8.11h. This microstructure consists of a 64×64 pixel image containing five fibers randomly distributed within a homogeneous material. The objective is to train a diffusion model that can generate new, realistic instances of such fiber composites. For this example, we adopt the simplified U-net architecture, the corruption process, and the training loop introduced in Sect. 8.5. This simplified U-net takes as input a noisy version of the microstructure image $\boldsymbol{x}^t$ at a given time step t and predicts the denoised input $\boldsymbol{x}^0$. The forward diffusion process incrementally adds noise to the original image $\boldsymbol{x}^0$, resulting in a sequence of increasingly noisy images $\boldsymbol{x}^1, \boldsymbol{x}^2, \ldots, \boldsymbol{x}^T$. Figure 8.14a illustrates this noising process for our fiber composite example. During training we randomly sample varying amounts of noise for each batch.

In practice, the choice of the variance schedule β^t significantly affects the performance of the diffusion model. While a simple linear schedule may suffice for basic examples, more complex schedules—such as cosine or quadratic schedules—have been shown to improve model convergence and sample quality [47]. These schedules control how the noise level increases over time, allowing for a better balance between preserving image structure in early time steps and achieving a nearly isotropic Gaussian distribution at the final time step T.

Figure 8.14b visualizes the networks's prediction on the corrupted training data after training the diffusion model. The predictions show that the network learned to reconstruct the desired fiber composite microstructure, even for very noisy samples. However, predicting the desired output from complete random noise is very challenging. Usually, new outptus are generated by starting from random Gaussian noise

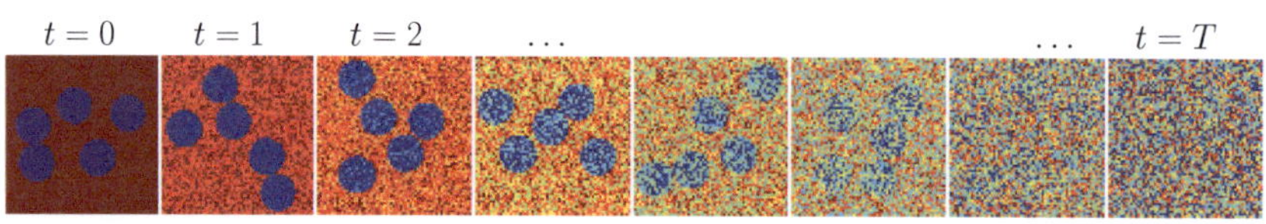

(a) Illustration of the forward diffusion process for multiple training samples with varying amount of noise added

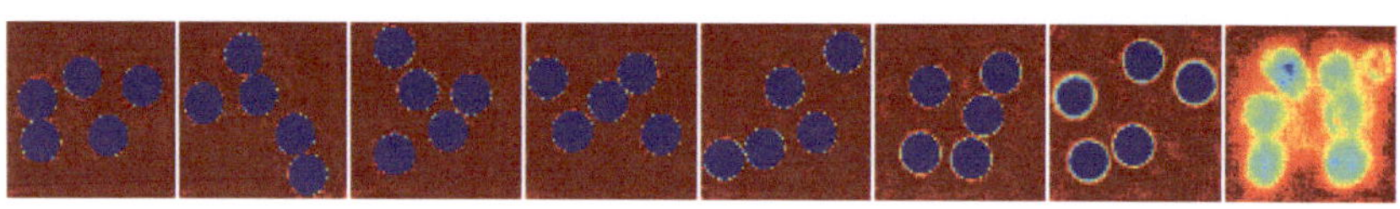

(b) Visualization of the learned inversed diffusion process for the corresponding training samples

Fig. 8.14 Forward diffusion process and the network's reconstruction for multiple training samples with varying amount of noise added. To showcase the training sample space, different samples are illustrated at the different steps t. All training samples, however, undergo the forward diffusion process during training

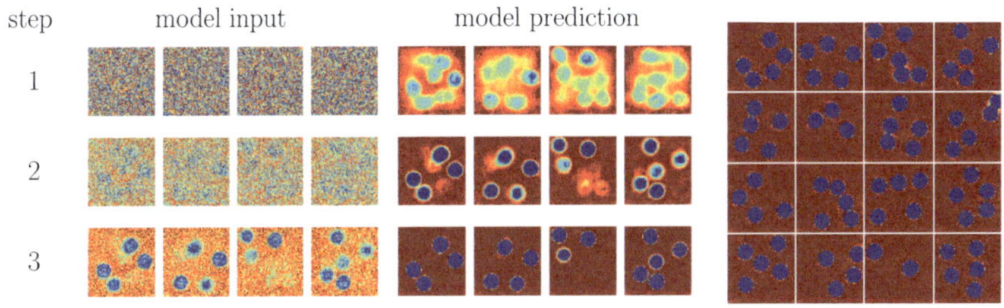

(a) Sampling with three reverse diffusion steps

(b) Sampling results after 40 reverse diffusion steps

Fig. 8.15 Comparison of generated microstructures using different numbers of reverse diffusion steps. Increasing the number of steps during sampling improves the quality and detail of the generated images but also increases computational cost

$\boldsymbol{x}^T$ and iteratively applying the reverse diffusion process. The number of reverse diffusion steps used during sampling influences both the quality of the generated images and the inference time. Figure 8.15 compares the reconstruction results when using three steps versus 40 steps in the reverse diffusion process. With only three steps, the model produces mediocre results. Many fibers remain blurry, and the sharp edges are distorted. As the number of steps increases to 40, the generated images become more accurate and closely resemble the training data, better capturing the round fiber shapes and their spatial distribution. However, we can see multiple generated samples that exhibit more or less than five enclosed fibers. These hallucinations are a common observation in diffusion models and are the subject of ongoing research [103].

This trade-off between reconstruction quality and inference speed is a crucial consideration when deploying diffusion models in practice. While more steps generally lead to higher-quality samples, they also increase computational cost and sampling time. Recent research has focused on developing more efficient sampling methods to reduce the number of required steps without compromising sample quality [50–52].

Despite much simplification, this example demonstrated how diffusion models can be applied to generate fiber composite microstructures. In conclusion, diffusion models offer a powerful and flexible framework for generative tasks such as microstructure generation, material design, and beyond. As diffusion models continue to advance, incorporating techniques such as latent space diffusion [17] and alternative architectures like transformers [53], their applicability and efficiency in computational mechanics are expected to improve further.

8.7.2 Generative Design & Design Optimization

Within generative design, the generator can also be considered as a reparametrization of the design space that reduces the number of design variables. With autoencoders, the latent vector serves as the design parameter [104, 105], which is then optimized.[7] Similarly, [107] find that point cloud autoencoders [108–110] are advantageous as geometric dimensionality reduction tools (potentially combined with performance features) for efficiently exploring the design space. In the context of generative adversarial networks, the optimization task is aimed at the random input $\boldsymbol{h}$ provided to the generator. This approach is demonstrated in various studies, such as ship hull design parameterized by NURBS surfaces [111], airfoil shapes expressed with Bézier curves [112, 113], structural optimization [114], and full waveform inversion [115]. For optimization, variational autoencoder generative adversarial networks are particularly important, as the generative adversarial network ensures high-quality designs, while the autoencoder ensures well-behaving gradients. This was shown for microstructure optimization in [116].

An important requirement for generative design is diversity. Achieving design diversity involves ensuring that the generated data spans the entire design space. For this, the cost function can be extended, as presented in [117], using determinantal point processes [118] or in [111] with a space-filling term [119].

Other strategies focus explicitly on promoting design diversity. This involves identifying novel designs via a novelty score [120]. The novelty within these designs is segmented and used to modify the generative adversarial network using methods outlined in [121]. An alternative approach proposed by [122] quantifies creativity and maximizes it. This is achieved by performing a classification in pre-determined categories by the discriminator. If the classification is unsuccessful, the design must lie outside the categories and is therefore deemed creative. Thus, the generator then seeks to minimize the classification accuracy.

However, some applications necessitate a resemblance to prior designs due to factors such as aesthetics [123] or manufacturability [124]. In [123], a pixel-wise L^1-distance to previous designs is included in the loss.[8] A complete workflow with generative design enforcing the resemblance of previous designs and surrogate model training for the quantification of mechanical properties is described in [125]. Another option is the use of style transfer techniques [106], which in [126] is incorporated into a conventional topology optimization scheme [128] as a constraint in the loss. These are tools with the purpose of incorporating vague constraints based on previous designs for topology optimization.

Generative adversarial networks can also be applied to inverse problems, as presented in [129] for full waveform inversion. The generator predicts the material distribution, which is used in a differentiable simulation, providing the forward solution

[7] It is worth noting that to ensure physically meaningful designs, a style transfer technique can be implemented [106]. Here, the training data is perceived as a style, and the Gram matrices' difference, characterizing the distribution of visual patterns or textures in the generated designs, is minimized.

[8] Similarly, this loss can be used to filter out designs that are too similar.

in the form of a seismogram. The discriminator attempts to distinguish between the seismogram indirectly coming from the generator and the measured seismograms. The underlying material distribution is determined through gradient descent.

Lastly, diffusion models have been applied to inverse design [130], where the diffusion model learns the relationship between load-displacement curves and corresponding structures (including their deformation).

8.7.3 Anomaly Detection

Deep learning-based anomaly detection shows the potential to outperform classical outlier detection methods in computational mechanics. As anomaly detection is a typical pattern recognition task, it has been extensively explored with machine learning [131], thereby forming the current state-of-the-art in anomaly detection.

One of the most successful research directions for this task is the application of generative approaches [2]. The general idea is that if a generative approach—trained on normal data, i.e., data without anomalies—is unable to reproduce a sample, then the sample must be an anomaly. A highlight of these generative approaches is that only normal training data is required. Thus, the potentially rare anomaly can be unknown prior to application, e.g., in non-destructive testing.

At first, the generative model is trained on structures without flaws. During evaluation, the structures to be tested are then fed through the neural network. In the case of an autoencoder, as in [132], it is fed through the encoder and decoder. For a generative adversarial network, as discussed, e.g., in [133–135], the input of the generator is optimized to fit the output as well as possible. The mismatch in reconstruction then provides a spatially dependent measure of where an anomaly, i.e., the defect, is located.

Another approach is to use the discriminator directly, as presented in [36]. If a flawed specimen is given to the discriminator, it will be categorized as fake, as it was not part of the undamaged structures during training. The discriminator can also be used to check if the domain of application of a surrogate model is valid. Trained on the same training data as the surrogate model, the discriminator estimates the dissimilarity between the data to be tested and the training data. For large discrepancies, the discriminator detects that the surrogate model becomes invalid.[9]

An important use case of anomaly detection in engineering is online monitoring of structures during manufacturing or deployment by analyzing time series signals from measurement devices [136]. Alternatively, image data from techniques such as computed tomography can be considered. This is especially valuable in the context of additive manufacturing, where manufacturing processes are less consistent—leading to large discrepancies between as-designed and as-built structures [32–34]. To speed

[9] Note, however, that the discriminator does not guarantee an accurate assessment of the validity of the surrogate model.

up the certification process of additively manufactured structures, anomaly detection is considered for image data obtained, e.g., by computed tomography.

8.7.3.1 Autoencoders for Anomaly Detection

Of the three generative approaches, applying autoencoders to anomaly detection is the most intuitive. The idea of an autoencoder [11, 13–15] is to encode an input into a lower dimensional latent vector and subsequently decode this latent vector as an output resembling the input. In other words, the aim is to learn a compression or dimensionality reduction. A smaller latent vector results in greater compression. As seen in Sect. 8.2, variational autoencoders [19, 20] extend this idea beyond a dimensionality reduction technique to a generative method. By randomly sampling and decoding latent vectors—from a distribution matching the latent vectors of the training dataset—novel yet similar samples are obtained. Repurposing this idea to identify samples that cannot be created by the autoencoder enables anomaly detection in two ways:

- Given a sample as input to the autoencoder, an outlier can be identified in the latent space through a comparison with the training data's distribution in the latent space [137, 138].
- If the autoencoder is unable to reconstruct the input sample, the sample must be an outlier [132, 137, 139–141].

The latter approach is particularly advantageous for image data since the mismatch between the input and output provides a pixel-wise error, i.e., an image and an explanation of where the anomaly originates.

To illustrate this, consider a simple microstructure consisting of five randomly placed, non-overlapping circles (possibly representing a fiber composite, as seen in Fig. 8.11h), as depicted in Fig. 8.16a. Using an autoencoder trained with 180 samples, the test sample from Fig. 8.16a can be reconstructed; see Fig. 8.16b. The pixel-wise squared difference, i.e., the reconstruction error, is shown in Fig. 8.16c. However, given an anomaly, e.g., a microstructure with only four circles and one square; see Fig. 8.16d, the reconstruction in Fig. 8.16e results in a microstructure with five circles. In the pixel-wise squared reconstruction error provided in Fig. 8.16f, the mismatch in the square's reconstruction reveals the outlier.

Statistical evaluation is possible by looking at additional test samples. To this end, normal and anomalous data with an increasing number of squares instead of circles are considered. Each dataset consists of 128 samples. Figure 8.17 illustrates a histogram of the mean squared errors between autoencoder input and output. The red bars indicate the normal data, i.e., microstructures with only circles, which are clearly separated from the anomalous data points.

This toy example illustrates how autoencoders can be employed to detect anomalies in an interpretable manner, which can similarly be achieved with generative adversarial networks and diffusion models. Although these methods result in

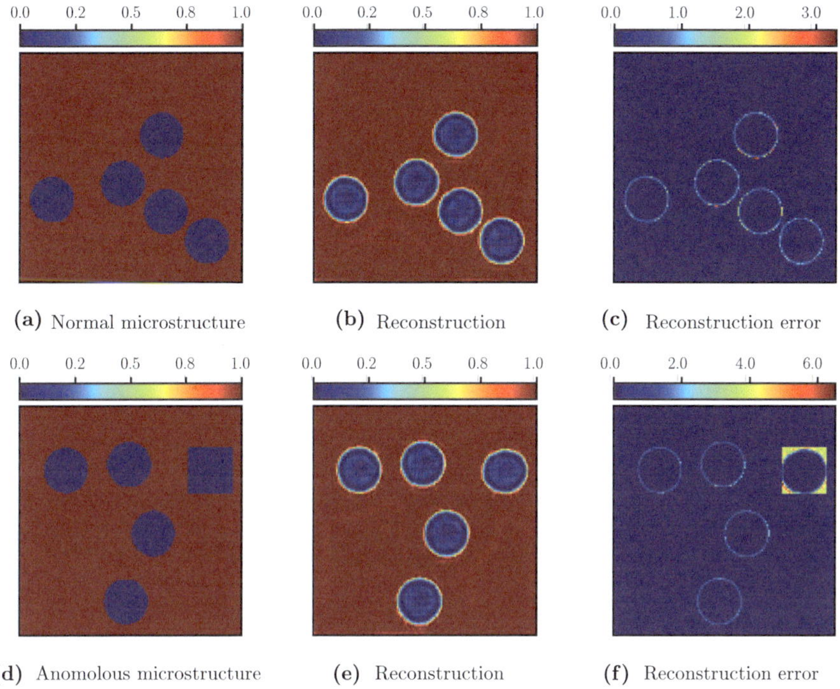

Fig. 8.16 Autoencoder reconstruction with normal data (top) and anomalous data (bottom). Given a microstructure as input, the autoencoder attempts a reconstruction, whose quality is assessed by the pixel-wise squared difference, i.e., the reconstruction error

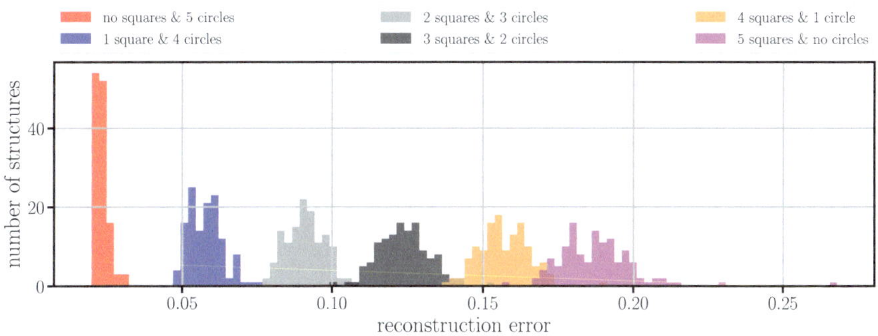

Fig. 8.17 Comparison of the distributions of the normal and anomalous data in terms of the mean squared error between the autoencoder input and output

more complicated anomaly detection approaches, generative adversarial networks and diffusion models are advantageous due to their superior capabilities as data generators—and therein the potential for better anomaly detectors.

8.7.3.2 Generative Adversarial Networks for Anomaly Detection

Both the generator and discriminator can be used for anomaly detection [133, 135, 142, 143]. Instead of discarding the discriminator after training, it can be used directly as an anomaly detector. The discriminator cannot distinguish between generated and real data due to their similarity [127, 144, 145]. It can, however, still identify data samples dissimilar to the real data (including the generated data).

To demonstrate the discriminative power, we introduce another toy example. We trained the DC-GAN architecture introduced in Sect. 8.3 to generate circles in a domain of 64×64 pixels. Training samples and fake samples generated by the generator are presented in Fig. 8.18.

Figure 8.19 shows the generator and discriminator loss during training. Even this very trivial example gives an idea of how difficult training a generative adversarial network can be. In this example, the discriminator became too good too fast, preventing the generator from learning to generate reasonable samples. Heuristics have shown that reducing the number of parameters in the discriminator can help improve training stability[10] [146]. In contrast to the symmetric architecture proposed by [28], the feature maps' size in the discriminator network's convolutional layers was halved.

Now, let us consider a faulty sample in the form of a square shape that we can use to test the discriminative power of the discriminator. Figure 8.18c shows a very high discrimination score (≈ 1) for a circle and a very low score (≈ 0) for the other shapes, which the network has never seen before during training. Thus, the faulty samples, i.e., the square, are clearly identified by the discriminator. This same approach of using the discriminator as an anomaly detector has been applied to more

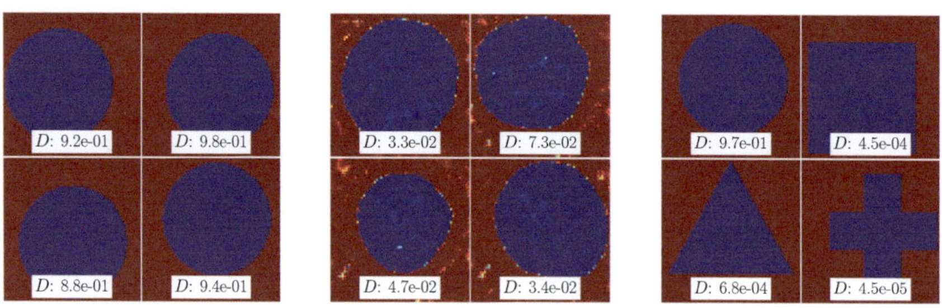

(a) Real samples used from the training set

(b) Fake samples generated by the generator

(c) Shapes unknown to the discriminator

Fig. 8.18 Real samples, fake samples generated by the generator, and shape types not known to the discriminator together with the outptut of the descriminator D. The descriminator recognizes the circle shapes used during training indicated by discrimination scores close to 1. The fake samples by the generator are detected, but with less confidence than the shapes that are not from the training data distribution

[10] A practice that seems deprecated after the introduction of Wasserstein generative adversarial networks [31].

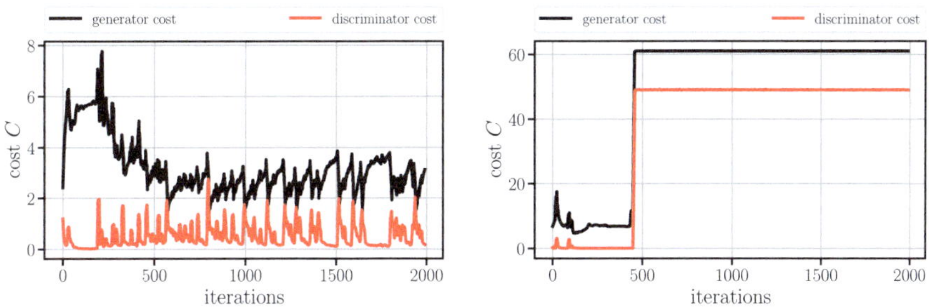

(a) Convergence of cost functions (b) Exploding gradients during GAN training

Fig. 8.19 Achieving convergence of the generator and discriminator cost is key to succesful GAN training. However, if the discriminator becomes too strong too quickly the gradients of the generator may explode. A high learning rate or poor hyperparameter tuning can be other reasons for divergence of the discriminator and generator cost

complex lattice structures seen in Fig. 8.6 [36], where subpar lattices (in terms of elastic stiffness) were identified.

Alternatively, the generator can be utilized similarly to how the autoencoder was used. Instead of subtracting input from output, the difference between the output and the sample of interest is minimized using the generative adversarial input, i.e., the random noise as optimization variables.[11] These are constrained to still represent random noise after the optimization. The pixel-wise squared difference between the optimized output and the sample provides an interpretable reconstruction error, as presented in Fig. 8.16 for the autoencoder.

8.7.3.3 Diffusion Models for Anomaly Detection

Similar to generative adversarial networks, a sample can be subtracted from a diffusion network's output, and the input, i.e., the random noise, is optimized under constraints to reduce the mismatch. Again, the reconstruction error can be used as seen in Fig. 8.16 for the autoencoder. In practice, diffusion models are frequently used to identify anomalies in image data, e.g., for detecting lesions and tumors in the brain [148, 149] or faulty parts in industry [150]. However, it was shown that diffusion models can be adapted to handle a variety of data types [151–153]. For example, Mikuni and Nachman introduced a permutation-invariant density estimator based on diffusion models for particle physics data, which effectively identifies anomalies like jets with low likelihood under the background-only hypothesis [154]. Diffusion provides a promising alternative to traditional methods and deep learning techniques in the context of unsupervised and semi-supervised anomaly detection, as well as in engineering applications.

[11] Optimization during evaluation can be avoided by utilizing a variational autoencoder generative adversarial network [147], which relies on an autoencoder architecture for the generator.

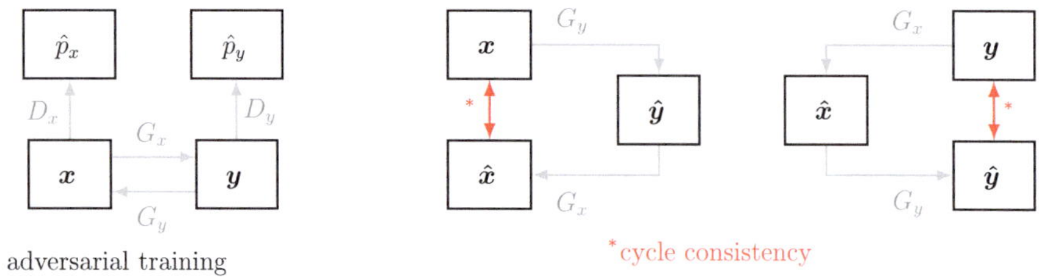

Fig. 8.20 CycleGAN. Two generators, G_x and G_y, are trained to correspondingly generate $\boldsymbol{x}$ from $\boldsymbol{y}$ and $\boldsymbol{y}$ from $\boldsymbol{x}$, while two discriminators, D_x and D_y, judge the outputs. The cycle consistency ensures that applying the two distinct generators sequentially yields the original data, thereby bypassing the need for paired data

8.7.4 Conditional Generation

As stated earlier, generative adversarial networks can take specific inputs to dictate the nature of the output. The key difference to data-driven surrogate models from Sect. 10.1.1 is that generative adversarial networks provide a tool to generate multiple outputs given the same conditional input. They are thus applicable to problems with multiple solutions, such as design optimization or data generation.

Examples of conditional generation are rendered cars from car sketches [155], hierarchical shape generation [156], where the child shape considers its parent shape and topology optimization with predictions of optimal structures from initial fields, e.g., strain energy, of the unoptimized structure [157, 158]. Physical properties can also be used as input. After generation, the properties are computed by a differentiable solver and incorporated in the loss. This was, e.g., presented in [159] for airplane shapes and in [160] for inverse homogenization. For full waveform inversion, [161] trained a conditional generative adversarial network with seismograms as input to predict the corresponding velocity distributions. A similar effort was made by [162] with CycleGANs [163] to circumvent the need for paired data. Here, one generator generates a seismogram $\hat{\boldsymbol{y}} = G_y(\boldsymbol{x})$ and another a corresponding velocity distribution $\hat{\boldsymbol{x}} = G_x(\boldsymbol{y})$. The predictions are judged by two separate discriminators. Additionally, a cycle-consistency loss ensures that a prediction from a prediction, i.e., $G_y(\hat{\boldsymbol{x}})$ or $G_x(\hat{\boldsymbol{y}})$, matches the initial input $\boldsymbol{x}$ or $\boldsymbol{y}$. This cycle-consistency loss ensures that the learned transformations preserve the essential features and structures of the original seismograms or velocity distributions when they are transformed from seismogram to velocity distribution and back again. The conceptual methodology of CycleGANs is visualized in Fig. 8.20.

Coarse-to-fine mappings, as previously discussed in Sect. 10.2.4, can also be learned by generative adversarial networks. This was, for example, demonstrated in topology optimization, where a conditional generative adversarial network refines coarse designs obtained from classical optimizations [157, 164] or convolutional neural network predictions [166]. For temporal problems, such as fluid flows, the temporal coherence between time steps poses an additional challenge. Temporal coherence can be ensured by a second discriminator, which receives three consecutive frames of either the generator or the real data and decides if they are real or generated. The method is referred to as tempoGAN [167].

Lastly, the generative process of the diffusion model can be conditioned. For example, conditional score-based diffusion models for Bayesian inference were employed in mechanics inverse problems, specifically inferring spatially varying material properties from noisy mechanical measurements [168]. Another study presented a Bayesian generative framework utilizing probabilistic diffusion models to efficiently generate spatiotemporal turbulence, overcoming limitations of traditional simulations and deterministic deep learning models by unifying unconditional and conditional sampling strategies and accommodating diverse conditioning scenarios [169].

8.7.5 Surrogate Modeling

Surrogate modeling plays a pivotal role in computational mechanics by providing efficient approximations of complex simulations, thereby accelerating the analysis of highly non-linear systems. Recently, transformer models have shown to potentially outperform existing methods in this area (Sect. 10.1.1). Geneva and Zabaras pioneered the application of self-attention transformer models for modeling physical dynamics, utilizing Koopman dynamics for physics-inspired embeddings of high-dimensional systems and discussing the relation between self-attention and numerical time-integration [170]. Their work demonstrated the capability of transformers to handle high-dimensional partial differential equation problems, marking a significant step in surrogate modeling of physical systems.

Building upon this foundation, researchers have explored the use of transformers in fluid dynamics and turbulence modeling [171]. For instance, some studies have combined variational autoencoders and transformers to develop compact and near-orthogonal reduced-order models for chaotic fluid flows [165], resulting in interpretable models that capture underlying flow dynamics and outperform traditional prediction models. Others have applied transformer architectures to predict future velocity fields in turbulent flows [172], aiming to reduce the computational expense of time-marching in computational fluid dynamics (CFD) simulations. These models achieve high predictive accuracy while offering significant computational savings, demonstrating the potential of transformers to accelerate complex simulations when coupled with traditional solvers. Additionally, transformer-based models have been employed to embed partial differential equations directly into the learning architecture. An example is the physics-informed token transformer (PITT), which tokenizes partial differential equations and incorporates partial derivatives to learn an analytically driven numerical update operator, making the model aware of underlying physical processes. These advancements showcase the versatility of transformers in surrogate modeling, enabling efficient and accurate solutions to high-dimensional and complex systems [173]. In the realm of biochemistry, transformer models have also been applied to molecular dynamics simulations, further highlighting their broad applicability in modeling complex systems [174].

In summary, transformer models are an exciting addition to surrogate modeling within computational mechanics, providing efficient and (potentially) accurate tools for simulating highly non-linear systems and possibly transforming the way complex simulations are performed.

Exercises

E.27 Autoencoder (C)

1. Generate another anomalous shape and verify that the anomaly detection works. A notebook for data generation and anomaly detection is provided in `exercise 27 .ipynb`.
2. Modify a standard autoencoder to create a Denoising Autoencoder (DAE) that learns to recover the original input from a corrupted version, enhancing robustness and preventing overfitting. Create a corruption function $\mathcal{M}_\mathcal{D}$ that adds noise to the input by either setting random values to zero (masking), adding Gaussian noise, or applying salt-and-pepper noise. Corrupt each input sample $\boldsymbol{x}^{(i)}$ to produce $\tilde{\boldsymbol{x}}^{(i)}$. Define the DAE Loss Function as:

$$\mathcal{L}_{\text{DAE}}(\boldsymbol{\Theta}, \boldsymbol{\phi}) = \frac{1}{m} \sum_{i=1}^{m} (\boldsymbol{x}^{(i)} - f_{\boldsymbol{\Theta}}(g_{\boldsymbol{\phi}}(\tilde{\boldsymbol{x}}^{(i)})))^2$$

   where $f_{\boldsymbol{\Theta}}(g_{\boldsymbol{\phi}}(\tilde{\boldsymbol{x}}^{(i)}))$ is the reconstructed output from the corrupted input $\tilde{\boldsymbol{x}}^{(i)}$, and $\boldsymbol{x}^{(i)}$ is the original input. Implement the Denoising Autoencoder and train the model to minimize $\mathcal{L}_{\text{DAE}}$, focusing on reconstructing the clean input from the noisy input. Experiment with various types and levels of noise (e.g., Gaussian, masking) and observe the effects on model robustness and denoising capability.
3. Which technique is similar to masking the inputs and is now widely used in the training of deep neural networks?

E.28 Variational Autoencoder (C)

1. Replace the linear layers in the variational autoencoder with convolutional layers. Use the variational autoencoder framework provided in `exercise 28 .ipynb` as a starting point. Plot the latent space and the generated images.
2. How does the latent space look like when `latent_dim = 3`?. Use the variational autoencoder framework provided in `exercise 28 .ipynb` as a starting point. Plot the latent space and the generated images.

3. What happens, when you train on the fiber composite dataset `normalData128.pt`? Is `latent_dim = 2` still sufficient?
4. Modify the variational autoencoder by introducing a disentanglement constraint using a hyperparameter β [102]. This constraint promotes a latent space where each variable independently captures a unique generative factor. Extend Sect. 8.2 to include the parameter β as follows

$$C = \|\boldsymbol{x} - \hat{\boldsymbol{x}}\|^2 + \beta \cdot \mathrm{KL}\left(\mathcal{N}\left(\boldsymbol{\mu}(\boldsymbol{x}), \boldsymbol{\sigma}\right), \mathcal{N}(\boldsymbol{0}, \boldsymbol{I})\right)$$

Note, that $\beta > 1$ encourages disentanglement, and $\beta = 1$ reduces to the standard variational autoencoder. Experiment with varying β to observe effects on disentanglement and reconstruction quality. If you want to improve the VAE further you can make β a trainable parameter (see [175]).

E.29 Generative Adversarial Network (C)

1. The main issue with GANs is the training stability, especially with deep networks. Heuristics have shown that adding noise to the discriminator can incnrease the robustness of the training process [176]. Add continuous noises onto the inputs of the discriminator and see if this can improve the training stability. Use the GAN framework provided in `exercise 29 .ipynb` as a starting point.
2. The so called Wasserstein GAN [31] was introduced to tackle the common problems with training instabilities. After reading the original paper try to extend the exisitng framework provided in `exercise 29 .ipynb` using the Wasserstein loss.
3. Try to futher improve the training stability by implemmenting the Wasserstein GAN with gradient penalty (WGAN-GP) [177] or experiment with combining autoencoder and GAN as proposed by [40].

E.30 Diffusion Models (C)

1. Familiarize yourself with the diffusion model code to generate fiber microstructures. Apply the simple diffusion model to other datasets composed of microstructures as seen in Fig. 8.11. See `exercise 30 .ipynb` for more details.
2. What happens if you combine the datasets of two or more different microstructures?

References

1. Lyle, Regenwetter, Amin Heyrani Nobari, and Faez Ahmed. 2022. Deep generative models in engineering design: A review. en., Feb 2022. arXiv:2110.10863 [cs, stat]. arXiv: 2110.10863.
2. David, Foster. 2023. *Generative deep learning: Teaching machines to paint, write, compose, and play*. eng. 2nd ed. OCLC: 1378390519. Sebastopol, CA: O'Reilly Media, Incorporated. ISBN: 978-1-09-813414-3.
3. Ivan, Kobyzev, Simon J. D. Prince, and Marcus A. Brubaker. 2021. Normalizing flows: An introduction and review of current methods. In: *IEEE Transactions on Pattern Analysis and Machine Intelligence 43.11*. arXiv:1908.09257 [cs, stat], pp. 3964–3979. ISSN: 0162-8828, 2160-9292, 1939-3539. https://doi.org/10.1109/TPAMI.2020.2992934.
4. Lilian, Weng. *What Are Diffusion Models?*, July 2021.
5. Alexey, Dosovitskiy et al. 2021. *An image is worth 16x16 words: Transformers for image recognition at scale*. arXiv:2010.11929 [cs]. June 2021. https://doi.org/10.48550/arXiv.2010.11929.
6. Gokul, Yenduri et al. 2023. *Generative pre-trained transformer: A comprehensive review on enabling technologies, potential applications, emerging challenges, and future directions*, May 2023. https://doi.org/10.48550/arXiv.2305.10435. arXiv: 2305.10435.
7. Staphord, Bengesi et al. 2024. Advancements in generative AI: A comprehensive review of GANs, GPT, autoencoders, diffusion model, and transformers. *IEEE Access 12*: 69812–69837. ISSN: 2169-3536. https://doi.org/10.1109/ACCESS.2024.3397775.
8. Emma, Yann Zhang et al. 2023. From turing to transformers: A comprehensive review and tutorial on the evolution and applications of generative transformer models. *Science 5.4*: 46. ISSN: 2413-4155. https://doi.org/10.3390/sci5040046.
9. Qian, Chen et al. 2023. Using the transformer model for physical simulation: An application on Transient Thermal Analysis for 3D Printing Process Simulation, Oct 2023. en.
10. Anni, Li et al. 2023. *Induced Generative Adversarial Particle Transformers*, Dec 2023. https://doi.org/10.48550/arXiv.2312.04757 `10.48550/arXiv.2312.04757`. arXiv: 2312.04757.
11. Hinton, G. E., and R. R. Salakhutdinov. Reducing the dimensionality of data with neural networks. en. *Science, 313.5786*: 504–507. ISSN: 0036-8075, 1095-9203. https://doi.org/10.1126/science.1127647.
12. Le Cun, Y., and Françoise Fogelman-Soulié. 1987. Modèles connexionnistes de l'apprentissage. fre. *Intellectica 2.1*: 114–143. https://doi.org/10.3406/intel.1987.1804.
13. Bourlard, H., and Y. Kamp. 1988. Auto-association by multilayer perceptrons and singular value decomposition. en. *Biological Cybernetics 59.4*: 291–294. ISSN: 1432-0770. https://doi.org/10.1007/BF00332918.
14. Geoffrey, E. Hinton, and Richard Zemel. 1993. Autoencoders, minimum description length and helmholtz free energy. In: *Advances in Neural Information Processing Systems*, vol. 6. Morgan-Kaufmann.
15. Shuangshuang, Chen and Wei Guo. (2023). Auto-Encoders in deep learning—A review with new perspectives. en. *Mathematics 11.8*: 1777. ISSN: 2227-7390. https://doi.org/10.3390/math11081777.
16. Olaf, Ronneberger, Philipp Fischer, and Thomas Brox. 2015. U-Net: Convolutional networks for biomedical image segmentation. en. In: *Medical Image Computing and Computer-Assisted Intervention—MICCAI 2015*. Ed. by Nassir, Navab et al. Lecture Notes in Computer Science, 234–241. Cham: Springer International Publishing. ISBN: 978-3-319-24574-4. https://doi.org/10.1007/978-3-319-24574-4_28.
17. Robin, Rombach et al. 2022. High-Resolution image synthesis with latent diffusion models. In: *2022 IEEE/CVF conference on computer vision and pattern recognition (CVPR)*, 10674–10685, June 2022. New Orleans, LA, USA: IEEE. ISBN: 978-1-66546-946-3. DOIurlhttps://doi.org/10.1109/CVPR52688.2022.01042.
18. Aditya, Ramesh et al. 2021. *Zero-Shot Text-to-Image Generation*, Feb 2021. https://doi.org/10.48550/arXiv.2102.12092.

19. Danilo, Jimenez Rezende, Shakir Mohamed, and Daan Wierstra. *Stochastic Backpropagation and Approximate Inference in Deep Generative Models*, May 2014. arXiv:1401.4082 [cs, stat]. https://doi.org/10.48550/arXiv.1401.4082.
20. Diederik, P. Kingma, and Max Welling. 2022. *Auto-Encoding Variational Bayes*, Dec 2022. arXiv:1312.6114 [cs, stat]. https://doi.org/10.48550/arXiv.1312.6114.
21. Kullback, S., and R. A. Leibler. 1951. On information and sufficiency. en.*The Annals of Mathematical Statistics 22.1*: 79–86. ISSN: 0003-4851. https://doi.org/10.1214/aoms/1177729694.
22. Xianxu, Hou et al. 2016. *Deep Feature Consistent Variational Autoencoder*, Oct 2016. https://doi.org/10.48550/arXiv.1610.00291. arXiv: 1610.00291.
23. Adam, Roberts et al. 2018. *A Hierarchical Latent Vector Model for Learning Long-Term Structure in Music*, Mar 2018. https://doi.org/10.48550/arXiv.1803.05428. arXiv: 1803.05428.
24. Petar, Veličković. 2018. *PetarV-/TikZ*, Dec 2018.
25. John, F. Nash, 1950. Equilibrium points in n-person games. en. In: *Proceedings of the National Academy of Sciences*, vol. 36.1, 48–49, Jan 1950. ISSN: 0027-8424, 1091-6490. https://doi.org/10.1073/pnas.36.1.48.
26. Ian, Goodfellow, Yoshua Bengio, and Aaron Courville. 2016. *Deep Learning*. MIT Press. ISBN: 0-262-03561-8.
27. Ian, Goodfellow et al. 2014. Generative adversarial nets. In: *Advances in Neural Information Processing Systems*, vol. 27. Curran Associates, Inc.
28. Alec, Radford, Luke Metz, and Soumith Chintala. *Unsupervised Representation Learning with Deep Convolutional Generative Adversarial Networks*, Jan 2016. arXiv:1511.06434 [cs]. https://doi.org/10.48550/arXiv.1511.06434.
29. Tim, Salimans et al. 2016. Improved techniques for training GANs. In: *Advances in Neural Information Processing Systems*, vol. 29. Curran Associates, Inc.
30. Akash, Srivastava et al. 2017. VEEGAN: Reducing mode collapse in GANs using implicit variational learning. In: *Proceedings of the 31st international conference on neural information processing systems*. NIPS'17. Red Hook, NY, USA: Curran Associates Inc., Dec 2017, pp 3310–3320. ISBN: 978-1-5108-6096-4.
31. Martin, Arjovsky, Soumith Chintala, and Léon Bottou. *Wasserstein GAN*, Dec 2017. arXiv:1701.07875 [cs, stat]. https://doi.org/10.48550/arXiv.1701.07875.
32. Korshunova, N. et al. 2020. Image-based material characterization of complex microarchitectured additively manufactured structures. In: *Computers & Mathematics with Applications*. High-Order Finite Element and Isogeometric Methods 2019 80.11, 2462–2480, Dec 2020. ISSN: 0898-1221. https://doi.org/10.1016/j.camwa.2020.07.018.
33. Korshunova, N. et al. 2021. Image-based numerical characterization and experimental validation of tensile behavior of octet-truss lattice structures. In: *Additive Manufacturing 41*: 101949. ISSN: 2214-8604. https://doi.org/10.1016/j.addma.2021.101949.
34. Korshunova, N. et al. 2021. Bending behavior of octet-truss lattice structures: Modelling options, numerical characterization and experimental validation. en. *Materials & Design 205*: 109693. ISSN: 02641275. https://doi.org/10.1016/j.matdes.2021.109693.
35. Alexander, Henkes, and Henning Wessels. 2022. Three-dimensional microstructure generation using generative adversarial neural networks in the context of continuum micromechanics. en. *Computer Methods in Applied Mechanics and Engineering 400*: 115497. ISSN: 0045-7825. https://doi.org/10.1016/j.cma.2022.115497.
36. Alexander, Henkes et al. 2024. Generative adversarial networks enable outlier detection and property monitoring for additive manufacturing of complex structures [dataset], 2, July 2024. en. https://doi.org/10.17632/zbbrbzrdtd.2.
37. Mehdi, Mirza, and Simon Osindero. 2014. *Conditional Generative Adversarial Nets*. arXiv:1411.1784 [cs, stat]. https://doi.org/10.48550/arXiv.1411.1784.
38. Xi, Chen et al. 2016. InfoGAN: Interpretable representation learning by information maximizing generative adversarial nets. In: *Proceedings of the 30th international conference on neural information processing systems*, 2180–2188, Dec 2016. NIPS'16. Red Hook, NY, USA: Curran Associates Inc. ISBN: 978-1-5108-3881-9.

39. John, S. Bridle, Anthony, J .R. Heading, and David, J. C. MacKay. 1991. Unsupervised classifiers, mutual information and 'phantom targets'. In: *Proceedings of the 4th international conference on neural information processing systems*, 1096–1101, Dec 1991. NIPS'91. San Francisco, CA, USA: Morgan Kaufmann Publishers Inc. ISBN: 978-1-55860-222-9.
40. Anders, Boesen Lindbo Larsen et al. 2016. Autoencoding beyond pixels using a learned similarity metric. In: *Proceedings of the 33rd international conference on international conference on machine learning*, vol. 48, 1558–1566. ICML'16. New York, NY, USA: JMLR.org, June 2016.
41. Zongwei, Zhou et al. 2018. *UNet++: A nested U-Net architecture for medical image segmentation*, July 2018. arXiv:1807.10165 [cs, eess, stat]. https://doi.org/10.48550/arXiv.1807.10165.
42. Nahian, Siddique et al. 2021. U-Net and its variants for medical image segmentation: A review of theory and applications. *IEEE Access 9*: 82031–82057. ISSN: 2169-3536. https://doi.org/10.1109/ACCESS.2021.3086020.
43. Jascha, Sohl-Dickstein et al. 2015. Deep unsupervised learning using nonequilibrium thermodynamics. en. In: *Proceedings of the 32nd international conference on machine learning*, 2256–2265, June 2015. PMLR.
44. Yang, Song, and Stefano Ermon. 2020. *Generative Modeling by Estimating Gradients of the Data Distribution*, Oct 2020. https://doi.org/10.48550/arXiv.1907.05600. arXiv: 1907.05600.
45. Deepanway Ghosal et al. *Text-to-Audio Generation Using Instruction-Tuned LLM and Latent Diffusion Model*. May 2023. DOI: https://doi.org/10.48550/arXiv.2304.13731. arXiv: 2304.13731.
46. Jonathan, Ho, A. Jain, and P. Abbeel. 2020. Denoising diffusion probabilistic models. In: *Advances in neural information processing systems*, vol. 33, 6840–6851. Curran Associates, Inc.
47. Alex, Nichol, and Prafulla Dhariwal. 2021. *Improved Denoising Diffusion Probabilistic Models*, Feb 2021. arXiv:2102.09672 [cs, stat]. https://doi.org/10.48550/arXiv.2102.09672.
48. Haohe, Liu et al. 2023. *AudioLDM: Text-to-Audio generation with latent diffusion models*, Sept 2023. https://doi.org/10.48550/arXiv.2301.12503. arXiv: 2301.12503.
49. Zach, Evans et al. 2024. *Fast Timing-Conditioned Latent Audio Diffusion*, May 2024. arXiv: 2402.04825.
50. Jiaming, Song, Chenlin Meng, and Stefano Ermon. 2022. *Denoising Diffusion Implicit Models*, Oct 2022. https://doi.org/10.48550/arXiv.2010.02502. arXiv: 2010.02502.
51. Tim, Salimans, and Jonathan Ho. 2022. *Progressive distillation for fast sampling of diffusion models*, June 2022. https://doi.org/10.48550/arXiv.2202.00512. arXiv: 2202.00512.
52. Yang, Song et al. 2023. *Consistency Models*, May 2023. https://doi.org/10.48550/arXiv.2303.01469. arXiv: 2303.01469.
53. William, Peebles, and Saining Xie. 2023. *Scalable Diffusion Models with Transformers*, Mar 2023. https://doi.org/10.48550/arXiv.2212.09748. arXiv: 2212.09748.
54. Inga, Strümke, and Helge Langseth. *Lecture Notes in Probabilistic Diffusion Models*, Dec 2023. https://doi.org/10.48550/arXiv.2312.10393. arXiv: 2312.10393 [cs].
55. Stanley, H. Chan. *Tutorial on diffusion models for imaging and vision*, Sept 2024. arXiv: 2403.18103 2403.18103.
56. Tero, Karras et al. 2022. *Elucidating the design space of diffusion-based generative models*, Oct 2022. https://doi.org/10.48550/arXiv.2206.00364. arXiv: 2206.00364.
57. Jacob, Devlin et al. 2019. *BERT: Pre-training of deep bidirectional transformers for language understanding*, May 2019. arXiv:1810.04805 [cs]. https://doi.org/10.48550/arXiv.1810.04805.
58. OpenAI. 2023. *GPT-4 Technical Report*, Mar 2023. arXiv:2303.08774 [cs]. https://doi.org/10.48550/arXiv.2303.08774.
59. Ashish, Vaswani et al. 2017. Attention is all you need. In: *Advances in neural information processing systems*, vol. 30. Curran Associates, Inc.
60. Jimmy, Lei Ba, Jamie Ryan Kiros, and Geoffrey E. Hinton. 2016. *Layer Normalization*, July 2016. arXiv:1607.06450 [cs, stat]. https://doi.org/10.48550/arXiv.1607.06450.

61. Alex, Graves. 2014. *Generating sequences with recurrent neural networks*, June 2014. arXiv:1308.0850 [cs]. https://doi.org/10.48550/arXiv.1308.0850.
62. Yonghui, Wu et al. 2016. Google's neural machine translation system: Bridging the gap between Human and Machine Translation, Sept 2016.
63. Salman, Khan et al. 2022. Transformers in vision: A survey. *ACM Computing Surveys 54.10*s: 200:1–200:41. ISSN: 0360-0300. https://doi.org/10.1145/3505244.
64. Sébastien, Bubeck et al. 2023. *Sparks of artificial general intelligence: Early experiments with GPT-4*, Apr 2023. arXiv:2303.12712 [cs]. https://doi.org/10.48550/arXiv.2303.12712.
65. Irving, John Good. 1966. Speculations Concerning the First Ultraintelligent Machine**Based on talks given in a Conference on the Conceptual Aspects of Biocommunications, Neuropsychiatric Institute, University of California, Los Angeles, Oct 1962; and in the Artificial Intelligence Sessions of the Winter General Meetings of the IEEE, Jan 1963 [1, 46]. The first draft of this monograph was completed in Apr 1963, and the present slightly amended version in May 1964. I am much indebted to Mrs. Euthie Anthony of IDA for the arduous task of typing. en. In: *Advances in Computers*. Ed. by Franz, L. Alt and Morris Rubinoff, vol. 6, 31–88 Jan 1966. Elsevier. https://doi.org/10.1016/S0065-2458(08)60418-0.
66. Nick, Bostrom. 2014. *Superintelligence: Paths, dangers, strategies*, 1st ed. OCLC: ocn881706835. Oxford: Oxford University Press. ISBN: 978-0-19-967811-2.
67. Max, Tegmark. 2017. *Life 3.0: Being human in the age of artificial intelligence*. 1st ed. New York: Alfred A. Knopf. ISBN: 978-1-101-94659-6 978-1-101-97031-7.
68. Dario, Amodei et al. *Concrete problems in AI safety*, July 2016. arXiv:1606.06565 [cs]. https://doi.org/10.48550/arXiv.1606.06565.
69. Jessica, Taylor et al. 2020. Alignment for advanced machine learning systems. en. In: *Ethics of Artificial Intelligence*, 342–382, Sept 2020. Oxford University Press. ISBN: 978-0-19-090503-3 978-0-19-090507-1. https://doi.org/10.1093/oso/9780190905033.003.001310.1093/oso/9780190905033.003.0013.
70. Tom, B. Brown et al. 2020. Language models are few-shot learners. In: *Proceedings of the 34th international conference on neural information processing systems*, pp 1877–1901, Dec 2020. NIPS '20. Red Hook, NY, USA: Curran Associates Inc. ISBN: 978-1-71382-954-6.
71. Mike, Lewis et al. 2019. *BART: Denoising sequence-to-sequence pre-training for natural language generation, translation, and comprehension*, Oct 2019. arXiv:1910.13461. https://doi.org/10.48550/arXiv.1910.13461.
72. Colin, Raffel et al. 2020. Exploring the limits of transfer learning with a unified text-to-text transformer. In: *Journal of Machine Learning Research 21.1*: 140:5485–140:5551. ISSN: 1532-4435.
73. Manzil, Zaheer et al. 2020. Big bird: transformers for longer sequences. In: *Proceedings of the 34th international conference on neural information processing systems*, 17283–17297, Dec 2020. NIPS '20. Red Hook, NY, USA: Curran Associates Inc. ISBN: 978-1-71382-954-6.
74. Jay, Alammar. 2018. *The illustrated transformer*, July 2018.
75. Richard, E. Turner. 2024. *An introduction to transformers*, Feb 2024. arXiv: 2304.10557 2304.10557.
76. Lewis, Tunstall, Leandro von Werra, and Thomas Wolf, 2022. *Natural language processing with transformers: Building language applications with Hugging Face*, 1st ed. OCLC: on1266359932. Sebastopol, CA: O'Reilly Media. ISBN: 978-1-09-810324-8 978-1-09-813679-6.
77. Lukas, Mosser, Olivier Dubrule, and Martin J. Blunt. 2017. Reconstruction of three-dimensional porous media using generative adversarial neural networks. en. *Physical Review E 96.4*: 043309. ISSN: 2470-0045, 2470-0053. https://doi.org/10.1103/PhysRevE.96.043309.
78. Junxi, Feng et al. 2019. Reconstruction of porous media from extremely limited information using conditional generative adversarial networks. *Physical Review E 100.3*: 033308. https://doi.org/10.1103/PhysRevE.100.033308.
79. Reza, Shams et al. 2020. Coupled generative adversarial and auto-encoder neural networks to reconstruct three-dimensional multi-scale porous media. en. *Journal of Petroleum Science and Engineering 186*: 106794. ISSN: 09204105. https://doi.org/10.1016/j.petrol.2019.106794.

80. Pengfei, Xia, Hualin Bai, and Ting Zhang. 2022. Multi-scale reconstruction of porous media based on progressively growing generative adversarial networks. en. *Stochastic Environmental Research and Risk Assessment 36.11*: 3685–3705. ISSN: 1436-3240, 1436-3259. https://doi.org/10.1007/s00477-022-02216-z.
81. Felix, Pütz et al. 2020. Generating input data for microstructure modelling: A deep learning approach using generative adversarial networks. In: *Materials 13.19*: 4236. ISSN: 1996-1944. https://doi.org/10.3390/ma13194236.
82. Mauricio, Araya-Polo, Stuart Farris, and Manuel Florez. 2019. Deep learning-driven velocity model building workflow. en. *The Leading Edge 38.11*: 872a1–872a9. ISSN: 1070-485X, 1938-3789. https://doi.org/10.1190/tle38110872a1.1.
83. Sharad, Rawat, and M. H. Herman Shen. 2019. *A novel topology design approach using an integrated deep learning network architecture*, Jan 2019. https://doi.org/10.48550/arXiv.1808.02334.
84. Kentaro, Yaji, Shintaro Yamasaki, and Kikuo Fujita. 2022. Data-driven multifidelity topology design using a deep generative model: Application to forced convection heat transfer problems. en. *Computer Methods in Applied Mechanics and Engineering 388*: 114284. ISSN: 0045-7825. https://doi.org/10.1016/j.cma.2021.114284.
85. Kang-Hyun, Lee, and Gun Jin Yun. *Microstructure reconstruction using diffusion-based generative models*, Jan 2023. arXiv:2211.10949 [cond-mat, physics:physics]. https://doi.org/10.48550/arXiv.2211.10949.
86. Christian, Düreth et al. 2023. Conditional diffusion-based microstructure reconstruction. en. In: *Materials Today Communications 35*: 105608. ISSN: 2352-4928. https://doi.org/10.1016/j.mtcomm.2023.105608.
87. Nikolaos, N. Vlassis, and WaiChing Sun. 2023. Denoising diffusion algorithm for inverse design of microstructures with fine-tuned nonlinear material properties. en. *Computer Methods in Applied Mechanics and Engineering 413*: 116126. ISSN: 0045-7825. https://doi.org/10.1016/j.cma.2023.116126.
88. Paul, Seibert et al. 2022. Microstructure characterization and reconstruction in python: MCRpy. en. *Integrating Materials and Manufacturing Innovation 11.3*: 450–466. ISSN: 2193-9772. https://doi.org/10.1007/s40192-022-00273-4.
89. Xiaolin, Li et al. 2018. A transfer learning approach for microstructure reconstruction and structure-property predictions. en. *Scientific Reports 8.1*: 13461. ISSN: 2045-2322. https://doi.org/10.1038/s41598-018-31571-710.1038/s41598-018-31571-7.
90. Junxi, Feng et al. 2020. An end-to-end three-dimensional reconstruction framework of porous media from a single two-dimensional image based on deep learning. en. *Computer Methods in Applied Mechanics and Engineering 368*: 113043. ISSN: 0045-7825. https://doi.org/10.1016/j.cma.2020.113043.
91. Steve, Kench, and Samuel J. Cooper. 2021. Generating three-dimensional structures from a two-dimensional slice with generative adversarial network-based dimensionality expansion. en. *Nature Machine Intelligence 3.4*:299–305. ISSN: 2522-5839. https://doi.org/10.1038/s42256-021-00322-1.
92. Yang, Li, Pengpeng Jian, and Guanghui Han. 2022. Cascaded progressive generative adversarial networks for reconstructing three-dimensional grayscale core images from a single two-dimensional image. *Frontiers in Physics 10*. ISSN: 2296-424X.
93. Fan, Zhang et al. 2022. 3D-PMRNN: Reconstructing three-dimensional porous media from the two-dimensional image with recurrent neural network. en. *Journal of Petroleum Science and Engineering 208*: 109652. ISSN: 09204105. https://doi.org/10.1016/j.petrol.2021.109652.
94. Qiang, Zheng, and Dongxiao Zhang. 2022. RockGPT: Reconstructing three-dimensional digital rocks from single two-dimensional slice with deep learning. en. *Computational Geosciences 26.3*: 677–696. ISSN: 1573-1499. https://doi.org/10.1007/s10596-022-10144-8.
95. Johan, Phan et al. 2022. Size-invariant 3D generation from a single 2D rock image. en. *Journal of Petroleum Science and Engineering 215*: 110648. ISSN: 0920-4105. https://doi.org/10.1016/j.petrol.2022.110648.

96. Fan, Zhang et al. 2021. Slice-to-voxel stochastic reconstructions on porous media with hybrid deep generative model. en. *Computational Materials Science 186*: 110018. ISSN: 0927-0256. https://doi.org/10.1016/j.commatsci.2020.110018.
97. Sharad, Rawat and MH Herman Shen. 2019. Application of adversarial networks for 3D structural topology optimization, 2019–01–0829, Apr 2019. https://doi.org/10.4271/2019-01-0829.
98. Sharad, Rawat, and M.-H. Herman Shen. 2019. *A novel topology optimization approach using conditional deep learning*, Jan 2019. https://doi.org/10.48550/arXiv.1901.04859.
99. Herman Shen, M.-H., and Liang Chen. 2019. *A New CGAN technique for constrained topology design optimization*, Apr 2019. https://doi.org/10.48550/arXiv.1901.07675.
100. Henning, Wessels et al. 2022. Computational homogenization using convolutional neural networks. en. In: *Current trends and open problems in computational mechanics*. Ed. by Fadi, Aldakheel et al., 569–579. Cham: Springer International Publishing. ISBN: 978-3-030-87312-7. https://doi.org/10.1007/978-3-030-87312-7sps55.
101. Lukas, Mosser, Olivier Dubrule, and Martin J. Blunt. 2020. Stochastic seismic waveform inversion using generative adversarial networks as a geological prior. en. *Mathematical Geosciences 52.1*: 53–79. ISSN: 1874-8953. https://doi.org/10.1007/s11004-019-09832-6.
102. Higgins, I. et al. 2016. Beta-VAE: Learning basic visual concepts with a constrained variational framework. In: *International conference on learning representations*, Nov 2016.
103. Sumukh, K. Aithal et al. 2024. *Understanding hallucinations in diffusion models through mode interpolation*, Aug 2024. https://doi.org/10.48550/arXiv.2406.09358. arXiv: 2406.09358.
104. Tinghao, Guo et al. 2018. An indirect design representation for topology optimization using variational autoencoder and style transfer. In: *2018 AIAA/ASCE/AHS/ASC structures, structural dynamics, and materials conference*. AIAA SciTech Forum. American Institute of Aeronautics and Astronautics, Jan 2018. https://doi.org/10.2514/6.2018-0804.
105. Praveen, S. Vulimiri et al. 2021. Integrating geometric data into topology optimization via neural style transfer. en. *Materials 14.16*: 4551. ISSN: 1996-1944. https://doi.org/10.3390/ma14164551.
106. Leon, Gatys, Alexander Ecker, and Matthias Bethge. 2016. A neural algorithm of artistic style. en. *Journal of Vision 16.12*: 326. ISSN: 1534-7362. https://doi.org/10.1167/16.12.326.
107. Nivesh, Dommaraju et al. 2023. Evaluation of geometric similarity metrics for structural clusters generated using topology optimization. en. *Applied Intelligence 53.1*: 904–929. ISSN: 1573-7497. https://doi.org/10.1007/s10489-022-03301-0.
108. Charles, R. Qi et al. 2017. *PointNet: Deep learning on point sets for 3D classification and segmentation*, Apr 2017. https://doi.org/10.48550/arXiv.1612.00593.
109. Panos, Achlioptas et al. 2018. Learning representations and generative models for 3D point clouds. en. In: *Proceedings of the 35th International Conference on Machine Learning*. PMLR, July 2018, pp. 40–49.
110. Yaoqing, Yang et al. 2018. FoldingNet: Point Cloud Auto-Encoder via Deep Grid Deformation, 206–215.
111. Shahroz, Khan et al. 2023. ShipHullGAN: A generic parametric modeller for ship hull design using deep convolutional generative model. en. *Computer Methods in Applied Mechanics and Engineering 411*: 116051. ISSN: 0045-7825. https://doi.org/10.1016/j.cma.2023.116051.
112. Qiuyi, Chen et al. 2022. Inverse design of two-dimensional airfoils using conditional generative models and surrogate log-likelihoods. en. In: *Journal of Mechanical Design 144.2*: 021712. ISSN: 1050-0472, 1528-9001. https://doi.org/10.1115/1.4052846.
113. Wei, Chen, and Mark Fuge. 2021. *B'ezierGAN: Automatic generation of smooth curves from interpretable low-dimensional parameters*, Jan 2021. arXiv:1808.08871 [cs, stat]. https://doi.org/10.48550/arXiv.1808.08871.
114. Wei, Chen, and Faez Ahmed. 2021. MO-PaDGAN: Reparameterizing engineering designs for augmented multi-objective optimization. en. *Applied Soft Computing 113*: 107909. ISSN: 1568-4946. https://doi.org/10.1016/j.asoc.2021.107909.
115. Alan, Richardson. 2018. *Generative adversarial networks for model order reduction in seismic full-waveform inversion*, June 2018. arXiv:1806.00828 [physics]. https://doi.org/10.48550/arXiv.1806.00828.

116. Yichi, Zhang et al. 2023. *DA-VEGAN: Differentiably augmenting VAE-GAN for microstructure reconstruction from extremely small data sets*, Feb 2023. arXiv:2303.03403 [cs]. https://doi.org/10.48550/arXiv.2303.03403.
117. Wei, Chen, and Faez Ahmed. 2021. PaDGAN: Learning to generate high-quality novel designs. en. *Journal of Mechanical Design 143.3*: 031703. ISSN: 1050-0472, 1528-9001. https://doi.org/10.1115/1.4048626.
118. Alex, Kulesza, and Ben Taskar. 2012. Determinantal point processes for machine learning. *Foundations and Trends & reg; in Machine Learning 5.2-3*: 123–286. arXiv:1207.6083 [cs, stat]. ISSN: 1935-8237, 1935-8245. https://doi.org/10.1561/2200000044.
119. Bates, S. J., J. Sienz, and D. S. Langley. 2003. Formulation of the Audze–Eglais Uniform Latin Hypercube design of experiments. en. In: *Advances in Engineering Software, 34.8*: 493–506. ISSN: 0965-9978. https://doi.org/10.1016/S0965-9978(03)00042-5.
120. Amin, Heyrani Nobari, Muhammad Fathy Rashad, and Faez Ahmed. 2021. CreativeGAN: Editing generative adversarial networks for creative design synthesis. In: *Volume 3A: 47th design automation conference (DAC)*. Virtual, Online: American Society of Mechanical Engineers, Aug 2021, V03AT03A002. ISBN: 978-0-7918-8538-3. https://doi.org/10.1115/DETC2021-68103.
121. David, Bau et al. 2020. *Rewriting a Deep Generative Model*. arXiv:2007.15646 [cs]. https://doi.org/10.48550/arXiv.2007.15646.
122. Ahmed, Elgammal et al. 2017. *CAN: Creative adversarial networks, generating "Art" by learning about styles and deviating from style norms*, June 2017. arXiv:1706.07068 [cs]. https://doi.org/10.48550/arXiv.1706.07068.
123. Sangeun, Oh et al. 2019. Deep generative design: Integration of topology optimization and generative models. *Journal of Mechanical Design 141.11*, 111405. ISSN: 1050-0472, 1528-9001. https://doi.org/10.1115/1.4044229.
124. Michael, Greminger. 2020. Generative adversarial networks with synthetic training data for enforcing manufacturing constraints on topology optimization. In: *Volume 11A: 46th design automation conference (DAC)*. Virtual, Online: American Society of Mechanical Engineers, Aug 2020, V11AT11A005. ISBN: 978-0-7918-8400-3. https://doi.org/10.1115/DETC2020-22399.
125. Soyoung, Yoo et al. 2021. Integrating deep learning into CAD/CAE system: Generative design and evaluation of 3D conceptual wheel. en. In: *Structural and Multidisciplinary Optimization 64.4*: 2725–2747. ISSN: 1615-1488. https://doi.org/10.1007/s00158-021-02953-9.
126. Weisheng, Zhang et al. 2023. Machine-learning assisted topology optimization for architectural design with artistic flavor. en. In: *Computer Methods in Applied Mechanics and Engineering 413*: 116041. ISSN: 0045-7825. https://doi.org/10.1016/j.cma.2023.116041.
127. Panpan, Zheng et al. 2019. One-Class adversarial nets for fraud detection. In: *Proceedings of the AAAI Conference on Artificial Intelligence*, vol. 33.01, 1286–1293, July 2019. ISSN: 2374-3468, 2159-5399. https://doi.org/10.1609/aaai.v33i01.33011286.
128. Martin, P. Bendsøe, and O. Sigmund. 2023. *Topology optimization: Theory, methods, and applications*. Berlin, New York: Springer. ISBN: 978-3-540-42992-0.
129. Fangshu, Yang, and Jianwei Ma. 2023. FWIGAN: Full-Waveform inversion via a physics-informed generative adversarial network. en. *Journal of Geophysical Research: Solid Earth 128.4*: e2022JB025493. ISSN: 2169-9313, 2169-9356. https://doi.org/10.1029/2022JB025493.
130. Jan-Hendrik, Bastek, and Dennis M. Kochmann. 2023. Inverse design of nonlinear mechanical metamaterials via video denoising diffusion models. en. *Nature Machine Intelligence 5.12*: 1466–1475. ISSN: 2522-5839. https://doi.org/10.1038/s42256-023-00762-x.
131. Guansong, Pang et al. 2022. Deep learning for anomaly detection: A review. en. *ACM Computing Surveys 54.2*: 1–38. ISSN: 0360-0300, 1557-7341. https://doi.org/10.1145/3439950.
132. Simon, Hawkins et al. 2002. Outlier detection using replicator neural networks. en. In: *Data Warehousing and Knowledge Discovery*, Ed. by Yahiko, Kambayashi, Werner Winiwarter, and Masatoshi Arikawa. Lecture Notes in Computer Science, pp 170–180. Berlin, Heidelberg: Springer. ISBN: 978-3-540-46145-6. https://doi.org/10.1007/3-540-46145-0_17.

133. Thomas, Schlegl et al. 2017. Unsupervised anomaly detection with generative adversarial networks to guide marker discovery. en. In: *Information Processing in Medical Imaging*, Ed. by Marc, Niethammer et al., Lecture Notes in Computer Science. Cham: Springer International Publishing, 2017, pp. 146–157. ISBN: 978-3-319-59050-9. https://doi.org/10.1007/978-3-319-59050-9_12.
134. Houssam, Zenati et al. 2019. *Efficient GAN-Based anomaly detection*, May 2019. arXiv:1802.06222 [cs, stat]. https://doi.org/10.48550/arXiv.1802.06222.
135. Thomas, Schlegl et al. 2019. f-AnoGAN: Fast unsupervised anomaly detection with generative adversarial networks. en. *Medical Image Analysis 54*: 30–44. ISSN: 1361-8415. https://doi.org/10.1016/j.media.2019.01.010.
136. Sebastian, Schmidl, Phillip Wenig, and Thorsten Papenbrock. 2022. Anomaly detection in time series: A comprehensive evaluation. en. In: *Proceedings of the VLDB endowment*, vol. 15.9, pp. 1779–1797, May 2022. ISSN: 2150-8097. https://doi.org/10.14778/3538598.3538602.
137. Dan, Xu et al. 2017. Detecting anomalous events in videos by learning deep representations of appearance and motion. In: *Computer vision and image understanding*. Image and video understanding in big data, vol. 156, 117–127, Mar 2017. ISSN: 1077-3142. https://doi.org/10.1016/j.cviu.2016.10.010.
138. Radu, Tudor Ionescu et al. 2019. Object-Centric auto-encoders and dummy anomalies for abnormal event detection in video. In: *2019 IEEE/CVF conference on computer vision and pattern recognition (CVPR)*, 7834–7843, June 2019. ISSN: 2575-7075. https://doi.org/10.1109/CVPR.2019.00803.
139. Chong, Zhou. and Randy C. Paffenroth. 2017. Anomaly detection with robust deep autoencoders. en. In: *Proceedings of the 23rd ACM SIGKDD international conference on knowledge discovery and data mining*, 665–674, Aug 2017. Halifax NS Canada: ACM. ISBN: 978-1-4503-4887-4. https://doi.org/10.1145/3097983.3098052.
140. Weining, Lu et al. 2017. Unsupervised sequential outlier detection with deep architectures. *IEEE Transactions on Image Processing*, vol. 26.9, 4321–4330, Sept 2017. ISSN: 1057-7149, 1941-0042. https://doi.org/10.1109/TIP.2017.2713048.
141. Tung, Kieu et al. 2019. Outlier detection for time series with recurrent autoencoder ensembles. In: *Proceedings of the 28th international joint conference on artificial intelligence*, 2725–2732, Aug 2019. IJCAI'19. Macao, China: AAAI Press. ISBN: 978-0-9992411-4-1.
142. Houssam, Zenati et al. 2018. Efficient GAN-Based anomaly detection. https://doi.org/10.48550/ARXIV.1802.06222.
143. Xuan, Xia et al. 2022. GAN-based anomaly detection: A review. en. *Neurocomputing 493*: 497–535. ISSN: 09252312. https://doi.org/10.1016/j.neucom.2021.12.093.
144. Mohammad, Sabokrou et al. 2018. Adversarially learned one-class classifier for novelty detection. In: *2018 IEEE/CVF conference on computer vision and pattern recognition*, pp 3379–3388, June 2018. Salt Lake City, UT: IEEE. ISBN: 978-1-5386-6420-9. https://doi.org/10.1109/CVPR.2018.00356.
145. Phuc, Cuong Ngo et al. 2019. Fence GAN: Towards better anomaly detection. In: *2019 IEEE 31st international conference on tools with artificial intelligence (ICTAI)*, pp 141–148, Nov 2019. ISSN: 2375-0197. https://doi.org/10.1109/ICTAI.2019.00028.
146. Christopher, Beckham. 2021. *Training GANs the Right Way*, June 2021.
147. Samet, Akcay, Amir Atapour-Abarghouei, and Toby P. Breckon. 2019. GANomaly: Semi-supervised anomaly detection via adversarial training. en. In: *Computer Vision–ACCV 2018*, Ed. by Jawahar, C. V. et al., vol. 11363, 622–637. Cham: Springer International Publishing. ISBN: 978-3-030-20892-9 978-3-030-20893-6. https://doi.org/10.1007/978-3-030-20893-6_39.
148. Julia, Wolleb et al. 2022. *Diffusion Models for Medical Anomaly Detection*, Oct 2022. https://doi.org/10.48550/arXiv.2203.04306. arXiv: 2203.04306.
149. Walter, H. L. Pinaya et al. 2022. *Fast unsupervised brain anomaly detection and segmentation with diffusion models*, June 2022. https://doi.org/10.48550/arXiv.2206.03461. arXiv: 2206.03461.

150. Arian, Mousakhan, Thomas Brox, and Jawad Tayyub. 2023. *Anomaly detection with conditioned denoising diffusion models*, Dec. 2023. https://doi.org/10.48550/arXiv.2305.15956. arXiv: 2305.15956.
151. Victor, Livernoche et al. 2023. *On diffusion modeling for anomaly detection*, Oct 2023. https://doi.org/10.48550/arXiv.2305.18593. arXiv: 2305.18593.
152. He, Wang et al. 2023. Odd: One-Class anomaly detection via the diffusion model. In: *2023 IEEE international conference on image processing (ICIP)*, pp 3000–3004, Oct 2023. https://doi.org/10.1109/ICIP49359.2023.10222162.
153. Yuhang, Chen et al. 2023. *ImDiffusion: Imputed diffusion models for multivariate time series anomaly detection*, Nov 2023. https://doi.org/10.48550/arXiv.2307.00754. arXiv: 2307.00754.
154. Vinicius, Mikuni, and Benjamin Nachman. 2024. High-Dimensional and permutation invariant anomaly detection. *SciPost Physics 16.3*: 062. ISSN: 2542-4653. https://doi.org/10.21468/SciPostPhys.16.3.062.
155. Sreedhar, Radhakrishnan et al. 2018. Creative intelligence – automating car design studio with generative adversarial networks (GAN). en. In: *Machine learning and knowledge extraction*, Ed. by Andreas, Holzinger et al. Lecture Notes in Computer Science, pp 160–175. Cham: Springer International Publishing. ISBN: 978-3-319-99740-7. https://doi.org/10.1007/978-3-319-99740-7_11.
156. Wei, Chen. and Mark Fuge. 2019. Synthesizing designs with interpart dependencies using hierarchical generative adversarial networks. en. *Journal of Mechanical Design 141.11*: 111403. ISSN: 1050-0472, 1528-9001. https://doi.org/10.1115/1.4044076.
157. Zhenguo, Nie et al. 2020. TopologyGAN: Topology optimization using generative adversarial networks based on physical fields over the initial domain, Mar. 2020. en. https://doi.org/10.48550/arXiv.2003.04685.
158. Nathan, Hertlein et al. 2021. Generative adversarial network for early-stage design flexibility in topology optimization for additive manufacturing. en. *Journal of Manufacturing Systems 59*: 675–685. ISSN: 0278-6125. https://doi.org/10.1016/j.jmsy.2021.04.007.
159. Amin, Heyrani Nobari, Wei (Wayne) Chen, and Faez Ahmed. 2021. RANGE-GAN: Design synthesis under constraints using conditional generative adversarial networks. en. *Journal of Mechanical Design*, 1–16. ISSN: 1050-0472, 1528-9001. https://doi.org/10.1115/1.4052442.
160. Jun, Wang et al. 2022. IH-GAN: A conditional generative model for implicit surface-based inverse design of cellular structures. en. In: *Computer methods in applied mechanics and engineering*, vol. 396, 115060, June 2022. ISSN: 0045-7825. https://doi.org/10.1016/j.cma.2022.115060.
161. Duque, L. et al. 2019. Automated velocity estimation by deep learning based seismic-to-velocity mapping. en. *European association of geoscientists and engineers*, vol. 2019, 1–5, June 2019. https://doi.org/10.3997/2214-4609.201901523.
162. Yu-Qing, Wang et al. 2022. Seismic impedance inversion based on cycle-consistent generative adversarial network. en. *Petroleum Science 19.1*: 147–161. ISSN: 1995-8226. https://doi.org/10.1016/j.petsci.2021.09.038.
163. Jun-Yan, Zhu et al. 2020. *Unpaired image-to-image translation using cycle-consistent adversarial networks*, Aug 2020. arXiv:1703.10593 [cs]. https://doi.org/10.48550/arXiv.1703.10593.
164. Baotong, Li et al. 2019. Non-iterative structural topology optimization using deep learning. en. *Computer-Aided Design 115*: 172–180, Oct 2019. ISSN: 0010-4485. https://doi.org/10.1016/j.cad.2019.05.038.
165. Alberto, Solera-Rico et al. 2024. β-Variational autoencoders and transformers for reduced-order modelling of fluid flows. *Nature Communications 15.1*: 1361. ISSN: 2041-1723. https://doi.org/10.1038/s41467-024-45578-4.
166. Yonggyun, Yu et al. 2019. Deep learning for determining a near-optimal topological design without any iteration. en. In: *Structural and Multidisciplinary Optimization 59.3*: 787–799. ISSN: 1615-1488. https://doi.org/10.1007/s00158-018-2101-5.

167. You, Xie et al. 2018. tempoGAN: A temporally coherent, volumetric GAN for super-resolution fluid flow. *ACM Transactions on Graphics 37.4*: 95:1–95:15. ISSN: 0730-0301. https://doi.org/10.1145/3197517.3201304.
168. Agnimitra, Dasgupta et al. 2025. Conditional score-based diffusion models for solving inverse elasticity problems. *Computer Methods in Applied Mechanics and Engineering 433*: 117425. ISSN: 0045-7825. https://doi.org/10.1016/j.cma.2024.117425.
169. Han, Gao et al. 2024. Bayesian conditional diffusion models for versatile spatiotemporal turbulence generation. *Computer Methods in Applied Mechanics and Engineering 427*: 117023. ISSN: 0045-7825. https://doi.org/10.1016/j.cma.2024.117023.
170. Nicholas, Geneva. and Nicholas Zabaras. 2022. Transformers for modeling physical systems. *Neural Networks 146*: 272–289. ISSN: 0893-6080. https://doi.org/10.1016/j.neunet.2021.11.022.
171. Han, Gao, Sebastian Kaltenbach, and Petros Koumoutsakos. 2024. Generative learning for forecasting the dynamics of high-dimensional complex systems. *Nature Communications 15.1*: 8904. ISSN: 2041-1723. https://doi.org/10.1038/s41467-024-53165-w.
172. Rakesh, Sarma et al. 2024. Prediction of turbulent boundary layer flow dynamics with transformers. *Mathematics 12.19*: 2998. ISSN: 2227-7390. https://doi.org/10.3390/math12192998.
173. Cooper, Lorsung, Zijie Li, and Amir Barati Farimani. 2024. Physics informed token transformer for solving partial differential equations. *Machine Learning: Science and Technology 5.1*: 015032. ISSN: 2632-2153. https://doi.org/10.1088/2632-2153/ad27e3.
174. Thorben Frank, J. et al. 2024. A euclidean transformer for fast and stable machine learned force fields. *Nature Communications 15.1*: 6539. ISSN: 2041-1723. https://doi.org/10.1038/s41467-024-50620-6.
175. Christopher, P. Burgess et al. 2018. *Understanding Disentangling inβ-VAE*, Apr 2018. https://doi.org/10.48550/arXiv.1804.03599. arXiv: 1804.03599 [stat].
176. Lilian, Weng. *From GAN to WGAN*, Apr 2019. https://doi.org/10.48550/arXiv.1904.08994. arXiv: 1904.08994.
177. Ishaan, Gulrajani et al. 2017. *Improved training of wasserstein GANs*, Dec 2017. https://doi.org/10.48550/arXiv.1704.00028 arXiv: 1704.00028

Chapter 9
Inverse Problems and Deep Learning

Abstract Inverse problems address the problem of identifying models from data. The models to be identified may be differential equations, and the data may be solutions of the differential equations, which are obtained through observations. Inverse problems often require innovative approaches for accurate solutions. This chapter introduces deep learning-based methods for inverse problems, including physics-informed neural networks, iterative forward solvers (adapted from adjoint optimization), and data-driven approaches. Applications in nondestructive testing and topology optimization highlight the practical utility of these techniques. The discussions are aimed at equipping researchers with tools to address complex, data-constrained engineering challenges in model discovery.

Inverse problems are a class of problems often encountered in physics and engineering applications. As already introduced in Sect. 4.3, instead of solving a differential equation, the goal is to estimate the underlying differential equation from an observed solution $\tilde{u}(x, t)$, which is considered the ground truth. Typically, this is achieved by making a prediction $\hat{u}(x, t)$ based on a differential equation and then adjusting the differential equation and the prediction $\hat{u}(x, t)$ in an effort to better fit the ground truth. The problem can be simplified if the form of the differential equation is already known, and thereby, only the parameters λ of the differential equation have to be estimated. These problems are often ill-posed, meaning the solution is not unique. Furthermore, there is a lack of general and efficient solution strategies, which is why inverse problems are an active field of research within which deep learning methods provide a rich toolbox.

9.1 Basic Methodology

Despite their young age, there is already a wide variety of approaches to address inverse problems with deep learning. However, they are often related and typically fall into one of the following three categories.

L. Herrmann et al., *Deep Learning in Computational Mechanics*,
https://doi.org/10.1007/978-3-031-89529-6_9

- Physics-Informed Neural Networks
- Iterative Forward Solvers
- Data-Driven Solvers

The degree of enforcement of the underlying physics decreases from top to bottom for these three approaches. The physics-informed neural networks rely fully on the underlying partial differential equation in the sense that neural networks are used to discretize the partial differential equation. The solution is then sought by simultaneously minimizing the residual of the partial differential equation and the residual of the measurement data. Iterative forward solvers solve the forward problem and minimize the residual of the measurement data in an alternating fashion. Classical laws of physics are not used directly in the iterative forward solver strategy but are employed to solve the forward problem. Traditionally, iterative forward solvers have been employed without neural networks but have recently been enhanced by deep learning, as seen in tasks like topology optimization. Finally, data-driven solvers rely entirely on a labeled dataset to learn the inverse quantities from a set of measurements. Thus, the corresponding laws of physics are fully removed from this approach.

9.1.1 Physics-Informed Neural Networks

Physics-informed neural networks in the form that [82] conceived them have already been applied to simple inverse problems in Sect. 4.3. In this case, the underlying differential equation is known, and the goal is to estimate the corresponding parameters λ that describe the observed data best. This is achieved by predicting the parameters λ and the solution $u(x, t)$ with a neural network. The residual of the differential equation and the measurement loss (between prediction $\hat{u}$ and measurement of the ground truth $\tilde{u}$) are minimized during the training process.

The problem can, however, also be expressed in a more general sense: find the differential equation that best describes the observed data. An approach that addresses this more general problem is the PDE-net, by [83, 84]. In the PDE-net, the same parametrized non-linear differential equation, Eq. (4.1) from Chap. 4 is considered.

$$\frac{\partial u}{\partial t} = \mathcal{N}[u; \lambda], \ x \in \Omega, \ t \in \mathcal{T}. \tag{9.1}$$

The objective is to identify the non-linear operator $\mathcal{N}[u; \lambda] = f(x, u, \lambda, \nabla u, \nabla^2 u, ...)$, which is a function of various differential operators with respect to x applied to $u(x, t)$ and λ. A neural network is trained to predict the non-linear function f and the differential operators. The differential operators are expressed as discrete approximations using convolutional filters. This requires a convolutional neural network architecture as the one described in Sect. 3.9.1. Once correctly trained, these filters are equivalent to a finite difference approximation. These filters are often encountered in image processing, where the image gradients are, e.g., computed

Table 9.1 Convolutional filters as differential operator approximators compared to their corresponding finite difference approximation

Finite difference approximation	Convolutional filter
$f'(x) = \frac{f(x+\Delta x)-f(x)}{\Delta x}$	$\frac{1}{\Delta x}[-1, 1]$
$f''(x) = \frac{f(x+\Delta x)-2f(x)+f(x-\Delta x)}{\Delta x^2}$	$\frac{1}{\Delta x^2}[1, -2, 1]$

with Sobel filters. More information can be found in textbooks like [1]. To illustrate that a finite difference stencil can be expressed with filters, two example filters with their corresponding finite difference approximation are presented in Table 9.1. The shown filters must be applied to a discretized domain in one dimension with a stride $s = 1$.

This technique enables the prediction of $\mathcal{N}[u; \lambda]$. With $\mathcal{N}[u; \lambda]$ and the previous time step $u(x, t_{n+1})$, the next time step $u(x, t_{n+1})$ is obtained by integrating Eq. (9.1) with respect to time. In [83], this is achieved with a forward Euler scheme and results in a prediction of the discretized solution $\hat{u}(x, t)$. The loss is defined by the mean squared error between the prediction $\hat{u}(x, t)$ and the observation of the ground truth $\tilde{u}(x, t)$. An approximation of the non-linear differential operator from Eq. (9.1) is thus found by minimizing this loss.

The fundamental concepts of this approach have also been used for forward problems. The adaptation of the numerical differentiation using convolutional filters instead of automatic differentiation was, e.g., applied by [85]. Additionally, the usage of the time-stepping scheme to enforce the underlying physics was applied by [2] to solve differential equations.

These approaches are already very close to the iterative forward solvers, which will be discussed in Sect. 9.1.2. However, they still closely combine the neural networks with the differential equation with a dedicated classical discretization.

An alternative approach to solving Eq. (9.1) presented by [3] is to use two neural networks. One network predicts the solution $\hat{u}(x, t)$, while the other predicts the non-linear differential operator $\hat{\mathcal{N}}[u; \lambda]$. The input to the network for the prediction of the solution $\hat{u}(x, t)$ are the coordinates x and the time t. The second network for the differential operator can require coordinates x, time t, the previous prediction $\hat{u}(x, t)$, and all expected derivatives of the prediction $\nabla\hat{u}, \nabla^2\hat{u}, \ldots$. The loss is defined by the residual of the predicted differential equation and the mean squared error between the predicted solution $\hat{u}(x, t)$ and the observations $\tilde{u}(x, t)$. The training of the two networks may occur simultaneously or sequentially, depending on the problem. One downside of this approach is that the differential equation is expressed through the neural network. Thus, extracting an analytical expression is not possible as the network acts as a black box.

It depends strongly on the problem, which of these approaches is viable. However, the three key ingredients to all of the different variants of physics-informed neural networks can be summarized by the following three points:

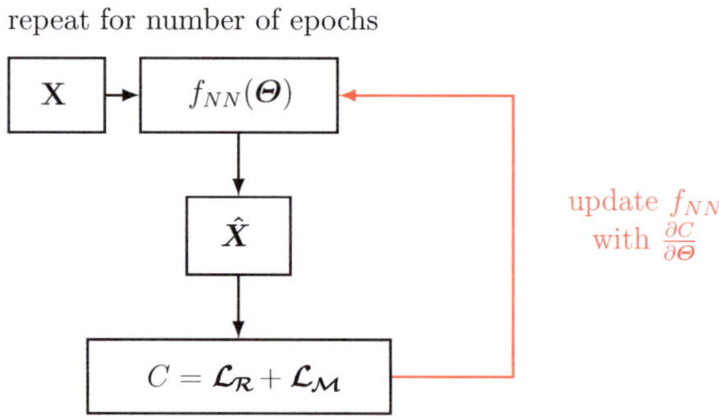

Fig. 9.1 Flowchart of physics-informed neural networks for inverse problems

- prediction of the differential equation and solution,
- enforcement of the physics through minimization of a residual of the differential equation or a time-stepping scheme,
- minimization of the measurement loss.

This is also outlined in the flowchart in Fig. 9.1, where the training loop is illustrated. Generally, a single neural network or multiple neural networks will predict $\hat{\boldsymbol{X}}$ from an input $\boldsymbol{X}$. The contents of the input vector vary depending on the specific approach. It might consist of the coordinates x and time t, solutions at previous time steps, or even a combination of solutions u and their gradients ∇u. The output $\hat{\boldsymbol{X}}$ consists of the predicted solution $\hat{u}$ and the prediction of the differential equation, which can be in the form of inverse parameters $\hat{\lambda}$ or the evaluation of the parametrized differential equation defined in terms of the non-linear operator $\hat{\mathcal{N}}$. The cost function C comprises two terms. The first term is the physics-informed loss $\mathcal{L}_{\mathcal{R}}$, i.e., the residual of the differential equation, which has been discussed thoroughly in Chap. 4. The second term is the measurement error $\mathcal{L}_{\mathcal{M}}$, defined as the mean squared error between the measurement $\tilde{u}$ and the forward solution $\hat{u}$ defined as:

$$\mathcal{L}_{\mathcal{M}} = \frac{1}{m_{\mathcal{M}}} \sum_{i}^{m_{\mathcal{M}}} (\tilde{u}_i - \hat{u}_i)^2. \tag{9.2}$$

The gradient of the cost function with respect to the network parameters $\boldsymbol{\Theta}$ is subsequently used to train the network.

9.1.2 *Iterative Forward Solvers*

An alternative is the iterative forward solver. Iterative forward solvers are based on conventional approaches that do not make use of deep learning directly but rely on similar concepts such as gradient-based optimization employed in, e.g., full waveform inversion [4] or topology optimization [5]. Here, the forward problem is solved (with forward solver S) iteratively for each gradient descent step. However, also gradient-free approaches [93], such as genetic algorithms/evolution strategies/differential evolution, particle swarm optimization/ant colony optimization, simulated annealing, simplex methods, or even Bayesian optimization, iteratively solve (using S) the forward problem to arrive at the set of inverse parameters or optimized designs. As gradient-based approaches are connected more directly to deep learning and are generally more promising [6], this chapter focuses on gradient-based optimization. As illustrated in Fig. 9.2, a gradient-based optimization loop can be enhanced in multiple ways. The main modifications concern

- the forward solver,
- the gradient computation,
- the ansatz of the inverse quantity.

The forward solver can, for example, be replaced by a differentiable neural network surrogate—reducing computation times during forward solves and easing the gradient computation through the differentiability of the neural network. A surrogate can similarly be employed during the gradient computation to provide either the gradient estimate or an enhanced update, emulating multiple gradient descent steps. Both of these approaches are essentially data-driven solvers (see Sect. 9.1.3), applied within a larger context, i.e., the optimization loop. Lastly, neural networks can be used as ansatz for the inverse quantity λ, which—as significantly different from data-driven solvers—will be discussed in detail in the following.

9.1.2.1 Neural Network Ansatz in Iterative Solvers

Typically, in classical solvers, the parameters λ are discretized. This can, e.g., be achieved with a traditional ansatz, such as known from the finite element method, where a combination of shape functions $N_i(x)$ and coefficients λ_i are used to express a field quantity, in this case, $\hat{\lambda}(x)$.

$$\hat{\lambda}(x) \approx \sum_i \lambda_i N_i(x) \tag{9.3}$$

The shape functions are known, while the coefficients λ_i have to be estimated. The preliminary approximation of the coefficients $\hat{\lambda}$ can subsequently be utilized to solve

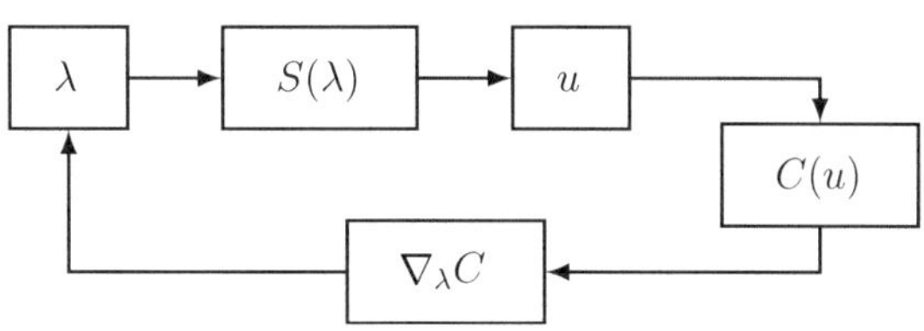

Fig. 9.2 Gradient-based optimization. Adapted from [7]

the forward problem. The forward solution $\hat{u}$ together with the observed data $\tilde{u}$ is used to compute a misfit functional $\mathcal{F}(\tilde{u}, \hat{u})$. A basic example of a misfit functional is the mean squared error as previously shown in Eq. (9.2).

To determine a good estimate of the inverse solution, the misfit functional has to be minimized. Thus, the inverse problem is posed as an optimization problem. For this task, gradient-based optimizers (as those discussed in Sect. 2.8) are commonly employed. Here the parameters λ and the forward solution $\hat{u}$ are estimated to compute the misfit functional. Subsequently, the gradient $\nabla\mathcal{F}$ of the misfit functional with respect to the coefficients λ_i must be computed. The negative gradient $-\nabla\mathcal{F}$ is then used to update the coefficients λ_i with a step size α to realize a minimization.

$$\lambda_i^{n+1} = \lambda_i^n - \alpha\nabla\mathcal{F} \tag{9.4}$$

These steps are repeated until convergence is reached. More sophisticated and efficient optimization algorithms exist, such as the conjugate gradient method or Newton methods; see, e.g., [8]. However, the concept of updating the coefficients through a computation of the gradients remains the same.

This approach is only feasible if the computation of the gradient of the misfit functional is possible. One way of estimating the gradient is through adjoint sensitivity methods [9]; see, e.g., [10] for an introduction to formulating the adjoint problem for inverse problems.

One notices that this optimization approach is similar to how neural networks are employed, where the parameters of the neural network $\boldsymbol{\Theta}$ are modified to minimize a cost function C. Thus, the neural network parameters $\boldsymbol{\Theta}$ act analogously to the coefficients λ of the finite element ansatz in Eq. (9.3). From this, the idea arises to replace the estimation of the parameters $\hat{\lambda}$ with a neural network. With this neural network ansatz (as an alternative to a traditional ansatz, like the finite element ansatz from Eq. (9.3)) for the inverse quantities, the universal function approximation property [11] can be exploited.

$$\hat{\lambda}(x) \approx \lambda_{NN}(x; \boldsymbol{\Theta}) \tag{9.5}$$

The forward problem is solved using classical approaches, such as the finite element or finite difference method to predict the solution $\hat{u}$, with which the misfit functional, i.e., the cost function, is computed. The neural network is then trained by minimizing the cost function with the gradient descent methods discussed in Chap. 3. Note that the approach is very similar to conventional approaches. The differences lie only in the ansatz for the inverse quantity and the computation of the gradient of the misfit functional, commonly achieved through automatic differentiation in deep learning.

A general outline of the approach is given in Fig. 9.3. Similar to the physics-informed neural networks, a neural network $f_{NN}(\boldsymbol{\Theta})$ predicts an output $\hat{\boldsymbol{X}} \setminus \hat{u}$ from an input $\boldsymbol{X}$. Again, the contents of the input and output may vary slightly, but it is important to note that the output does not include the forward solution $\hat{u}$, as it is found in the next step with a forward solver. Finally, the cost function is evaluated, consisting

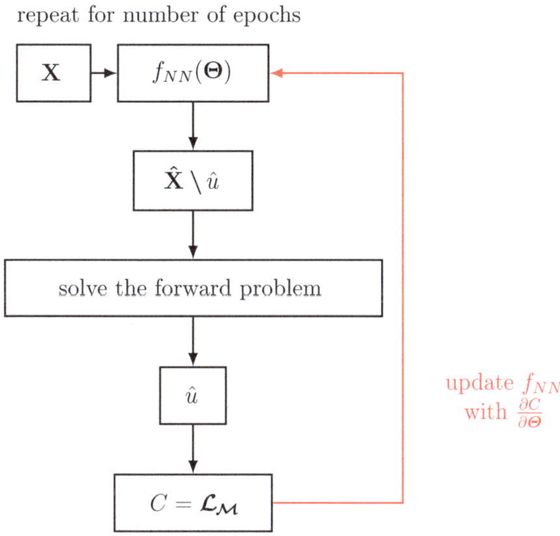

Fig. 9.3 Flowchart of the iterative forward solver for inverse problems using a neural network ansatz for the inverse quantity $\hat{X} \setminus \hat{u}$, from which the forward solution is computed $\hat{u}$. Note that neural networks can be incorporated in different ways in the iterative forward solver. Moreover, a linear ansatz, such as standard shape functions or a voxel representation, can replace the neural network, which recovers a standard gradient-based optimization

only of the measurement loss $\mathcal{L}_{\mathcal{M}}$ from Eq. (9.2). The network is then optimized with the gradient of the cost function with respect to the network parameters $\boldsymbol{\Theta}$.

9.1.2.2 Motivation for a Neural Network Ansatz

To illustrate the benefit of a neural network ansatz within an optimization framework, a toy example is considered [12]. The Rosenbrock function

$$g(y_1, y_2) = (1 - y_1)^2 + 100(y_2 - y_1^2)^2 \tag{9.6}$$

with y_1 and y_2 as design variables, is minimized directly (referred to as a linear ansatz, as the map between design variables and optimization function is linear) and an indirect neural network ansatz. Specifically a fully connected neural network f_{NN}, parametrized by $\boldsymbol{\Theta}$, predicts $\hat{y}_1$ and $\hat{y}_2$ from a random input vector $\boldsymbol{\xi}$.

$$\begin{pmatrix} \hat{y}_1 \\ \hat{y}_2 \end{pmatrix} = f_{NN}(\boldsymbol{\xi}; \boldsymbol{\Theta}) \tag{9.7}$$

Note that the aim of the neural network is simply a reparametrization of the design space (y_1, y_2) to the (higher-dimensional) neural network parameters $\boldsymbol{\Theta}$. The random input vector $\boldsymbol{\xi}$ remains constant during optimization and prediction. The input is, in principle, not needed and only an artifact of the neural network architecture, which requires an input. Thus, no real mapping is learned, but only an optimum for a single problem is identified. As the neural network is used as ansatz for only this one

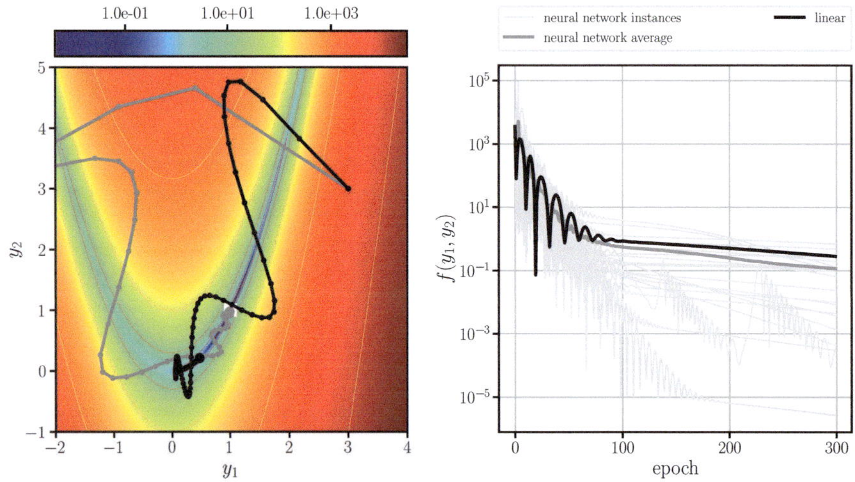

(a) Optimization path after 300 epochs (b) Cost histories

Fig. 9.4 Rosenbrock function optimized with neural network ansatz and directly on the coordinates (linear ansatz) using Adam. Note that both parametrizations will converge to the optimum given a sufficient number of epochs. Reprinted from [12]

problem, the approach is not machine learning, as this would require a subsequent application to yet unseen problems.

The employed neural network architecture consists of 625 trainable parameters, i.e., the optimized problem is overparametrized. Starting from an initial guess $y_1^{(0)} = 3$, $y_2^{(0)} = 3$ for both parametrizations, an optimization is performed with Adam [13][1] for 300 epochs. Figure 9.4a showcases a representative optimization path where the neural network ansatz (gray) clearly outperforms the linear ansatz (black). Since different neural network initializations can be obtained with the same initial guess $y_1^{(0)} = 3$, $y_2^{(0)} = 3$, multiple neural network instantiations are compared to the linear ansatz in Fig. 9.4b. From these cost histories, it becomes clear that a neural network parametrization can outperform the linear ansatz. This is, however, not always guaranteed, as seen by some neural network instances with worse local optima after 300 epochs. This behavior is even worse for other optimization benchmarks, such as Rastrigrin, Ackley, or Lévi [12]. Nevertheless, this toy example illustrates the potential of an overparametrization achieved through a neural network ansatz.

[1] Currently, the advantage of a neural network ansatz has only been observed in combination with Adam [12].

9.1.3 Data-Driven Solvers

Data-driven solvers do not take the underlying physics into account. Instead, they rely solely on labeled datasets during training. That is, in the dataset, every example consists of a matrix $\boldsymbol{X}$ and a corresponding label or target vector $\tilde{\boldsymbol{y}}$, as has been described in Chap. 3. The loss is defined as the misfit between the prediction vector obtained with the neural network $\hat{\boldsymbol{y}}$ and the target vector $\tilde{\boldsymbol{y}}$. After training, the neural network can be applied to unseen data. In the context of applications for physics, the residual of the differential equation can still be used to quantify the quality of a prediction after training.

Data-driven solvers, as the one seen in the learning strain distributions example in Sect. 3.9.1.2, are the most common approach in computer science. However, they are not always viable in the field of physics and engineering, as they require large datasets. However, if a large dataset exists or it is possible to generate one cheaply, they are well suited for inverse problems. The idea is that the network learns to predict the parameters $\hat{\lambda}$ of a differential equation from the ground truth solution u which can, for example, be measurements $\tilde{u}$.

The flowchart in Fig. 9.5 provides an overview of the approach. The neural network predicts $\hat{\boldsymbol{y}}$ from the input $\boldsymbol{X}$, which is subsequently used to compute the loss $\mathcal{L}_{\mathcal{D}}$ expressed in terms of an L^2 norm where only one sample is used at a time.

$$\mathcal{L}_{\mathcal{D}} = ||\hat{\boldsymbol{y}} - \tilde{\boldsymbol{y}}||_2^2 \tag{9.8}$$

This process is repeated for each sample i, and the loss term of each sample is added to the cost function C. Finally, the cost function is normalized by the number of samples $m_{\mathcal{D}}$, and its gradient $\partial C/\partial \boldsymbol{\Theta}$ with respect to the network weights $\boldsymbol{\Theta}$ is used to update the network.

Variations of these approaches have been demonstrated for the wave equation [14–23, 86, 94] and topology optimization [24–33]. As this approach is highly flexible and relatively simple it has been deployed widely. Note that this approach is not limited to only inverse problems but can also be employed for forward problems with minor modifications, as previously seen in Sect. 3.9.1.2. A variety of examples in computational mechanics is provided by [34]. The advantage of this approach is that the task of training neural networks with labeled data is very common and relatively well understood, which shifts the task of considering the underlying physics and numerical algorithms to tuning neural networks. A great resource for this task are [35, 36].

9.2 Ultrasonic Nondestructive Testing

Ultrasonic nondestructive testing is used to identify flaws in components without incurring further damage. This is achieved with ultrasonic pulses and sensors that then receive the corresponding signal. If a flaw perturbs the signal, the sensor will

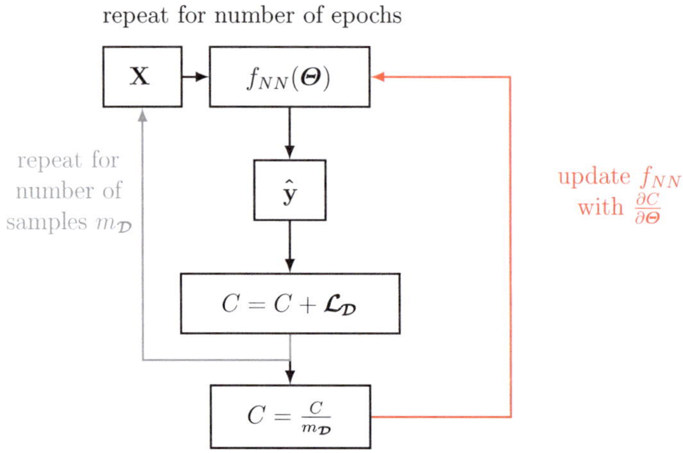

Fig. 9.5 Flowchart of the data-driven solver

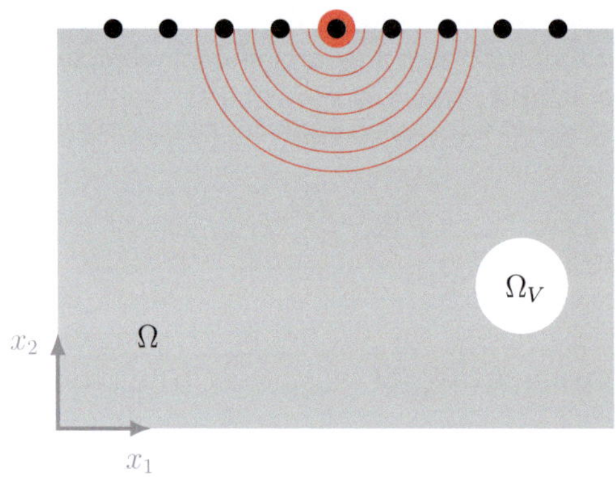

Fig. 9.6 Ultrasonic nondestructive testing on a two-dimensional rectangular domain Ω with a void Ω_V. The signal emitted by the source (red circle) is recorded at the sensors (black circles) over a time period. Adapted from [37]

observe this. The general methodology is visualized for a two-dimensional specimen in Fig. 9.6, where a signal is emitted by a source located at the red circle. The sensors are depicted as black circles.

A typical pulse used as source is a sine burst illustrated in Fig. 9.7. A typical spatio-temporal response, referred to as wave field, in a damaged two-dimensional specimen is illustrated in Fig. 9.8. The pulse is applied at the upper edge and travels in time towards the lower edge. The wave field is visibly perturbed when it reaches the hole at the center. Measurements of the wave field in terms of signals can typically be obtained at the boundaries in order to determine the damage inside a specimen by solving the inverse problem.

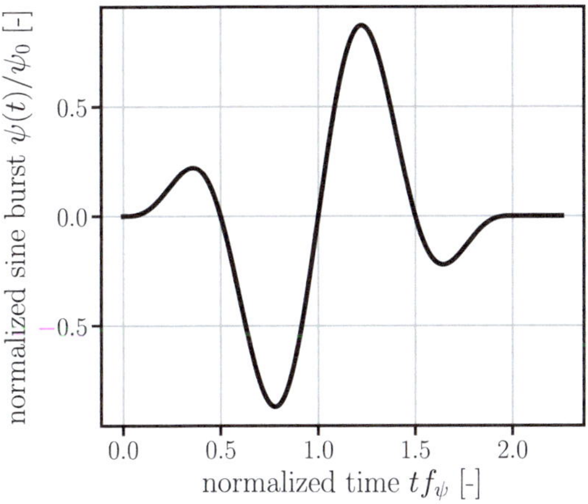

Fig. 9.7 Sine burst from Eq. (9.12) with $n_c = 2$ cycles. Adapted from [37]

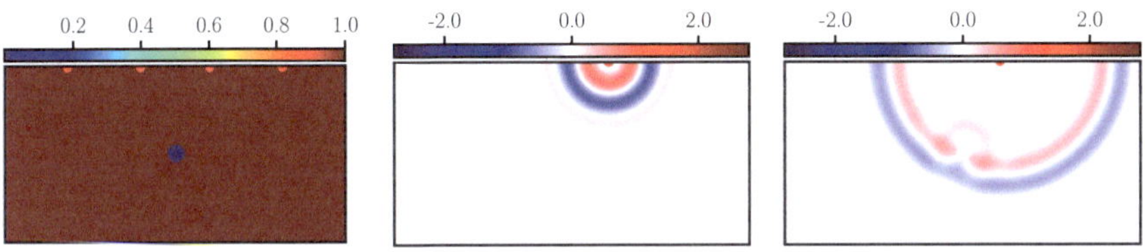

(a) Material distribution, where zero signifies a hole

(b) wave field $u(x_1, x_2, t)$ at $t =$ 0.00375µs

(c) wave field $u(x_1, x_2, t)$ at $t =$ 0.0075µs

Fig. 9.8 The wave field initiated by the sine burst and perturbed by the centered circular hole. Adapted from [37]

This allows for the identification of a flaw. The shape and position of the flaw can be computed from the sensor data with procedures such as the full waveform inversion [4, 38], which considers the problem as a non-linear optimization problem, where a misfit functional between prediction and measurement is minimized. The gradient required for the optimization is usually found via an adjoint sensitivity approach.

Alternatively, deep learning approaches can be applied. Physics-informed neural networks for nondestructive testing are discussed in Sect. 9.2.3. An iterative approach relying on a neural network ansatz is considered in Sect. 9.2.4, and a data-driven approach adapted from [39, 94] is shown in Sect. 9.2.5. However, before proceeding to deep learning, the fundamentals are covered in terms of the governing differential equation for full waveform inversion, which is the acoustic (or scalar) wave equation[2]

[2] Although the scalar wave equation is an approximation of the underlying physics in solids, it is popular for method development in full waveform inversion due to the reduced computational complexity; see, e.g. [37].

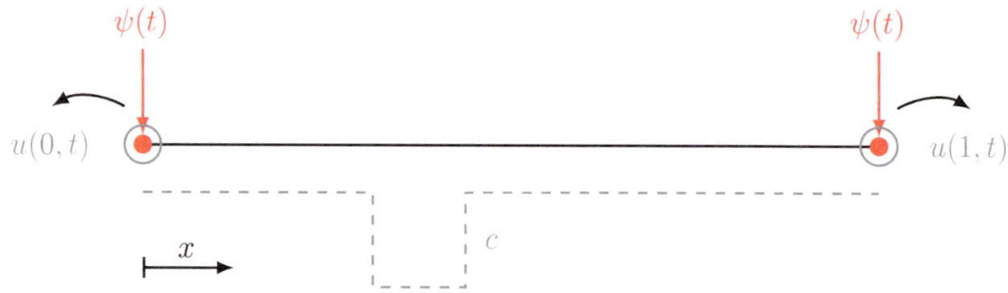

Fig. 9.9 Setup for the ultrasonic nondestructive test. The goal is to infer the wave speed c from the measurements $u(0, t)$ and $u(L, t)$ induced by the two sine bursts applied separately at $x = 0$ and $x = L$

(Sect. 9.2.1) and an efficient solution technique, where we use the finite difference method (Sect. 9.2.2).

9.2.1 Acoustic Wave Equation

The linear response[3] of a specimen subjected to a pulse such as a sine burst can be described by the wave equation presented in one dimension in Eq. (9.9) acting on the spatio-temporal domain $\Omega \times \mathcal{T}$. In addition, boundary conditions (Eq. (9.10)) on Γ_N as well as initial conditions (Eq. (9.11)) are required. The derivation can be found in textbooks such as [40]. The two-dimensional wave equation was used to compute the response presented in Fig. 9.8.

$$\frac{\partial^2 u(x,t)}{\partial t^2} - \frac{\partial}{\partial x}\left(c^2(x)\frac{\partial u(x,t)}{\partial x}\right) = p(x,t) \qquad \text{on } \Omega \times \mathcal{T} \tag{9.9}$$

$$\boldsymbol{n} \cdot \frac{\partial u}{\partial x} = 0 \qquad \text{on } \Gamma_N \times \mathcal{T} \tag{9.10}$$

$$u(x,0) = u_t(x,0) = 0 \qquad \text{on } \Omega \tag{9.11}$$

Here u is the wave field (related to wave pressures) that sensors can detect, and p is the source term, which typically consists of a local pulse such as the sine burst $\psi(x, t)$ given by Eq. (9.12). The amplitude ψ_0, the frequency ω, and the number of sinusoidal bursts n_c are the parameters controlling the pulse. Using a Dirac delta $\delta(x)$, the location x_s of the pulse ψ is specified within the source term $p(x, t)$.

$$\psi(t) = \begin{cases} \psi_0 \sin(\omega t)\sin^2(\frac{\omega t}{2n_c}), & \text{for } 0 \leq t \leq \frac{2\pi n_c}{\omega} \\ 0, & \text{for } \frac{2\pi n_c}{\omega} < t. \end{cases} \tag{9.12}$$

$$p(x,t) = \psi(t)\delta(x - x_s) \tag{9.13}$$

[3] Neglecting shear waves.

Table 9.2 Problem configuration of the one-dimensional wave equation example illustrated in Fig. 9.9

Physical parameters	L	c_0	ω	ψ_0	n_c
	0.02 mm	6000 m/s	$500 \cdot 2\pi$ kHz	10^{14} Pa/s^2	2
Numerical parameters	N_1	N_t	Δt		
	117	700	$2 \cdot 10^{-8}$ s		

The most important quantity for nondestructive testing is the spatially varying wave speed $c(x)$. The wave speed c is a material property related to the Young's modulus E and the mass density ρ. It can thereby be used to identify damaged material, cracks, or other flaws. For more sophisticated formulations of the wave equation for nondestructive testing; see [41].

We will focus on the one-dimensional wave equation in the following. Two-dimensional examples are provided in later sections. The boundary conditions will always be homogeneous Neumann boundary conditions as denoted in Eq. (9.10). Both initial conditions, the initial wave field, and its temporal derivative will also be homogeneous as presented in Eq. (9.11). This also corresponds to a realistic testing scenario, where the specimen will be free at the boundaries and undisturbed before the source is applied. The hypothetical test setup used for the following sections is illustrated in Fig. 9.9. A spatial domain of $\Omega = [0, L]$ and a temporal domain of $\mathcal{T} = [0, N_t \cdot \Delta t]$ will be considered, where N_t is the number of time steps and Δt is the time step size. Two pulses at $x_s = 0$ and $x_s = L$ are applied separately, resulting in two sets of measurements $\tilde{u}$. Sensors are applied at both boundaries $x = 0$ and $x = L$ to capture the full temporal response. It will be assumed that the unflawed specimen has a homogeneous wave speed of $c(x) = c_0$. A flaw is assumed to only reduce the wave speed to an interval between $c_0/60$ and c_0.[4] This limits the search space and thereby helps during optimization. To avoid scaling issues, the data is normalized in the range between -1 and 1, as neural networks can handle the interval $[-1, 1]$ well. The specific physical and numerical parameters used throughout the examples in the following sections are summarized by Table 9.2 and lead to a representative spatio-temporal wave field in Fig. 9.10.

9.2.2 *Finite Difference Method*

The wave equation can be solved numerically with the finite difference method, which is going to be used for the iterative forward solver in Sect. 9.2.4 and for the data generation in Sect. 9.2.5. The method is, therefore, briefly explained. To this end,

[4] This choice is made to guarantee that the wave can propagate within a reasonable amount of time for low wave speeds. The upper bound for the wave speed is employed to ensure numerical stability. The choice is arbitrary and can be adapted to the specific problem at hand.

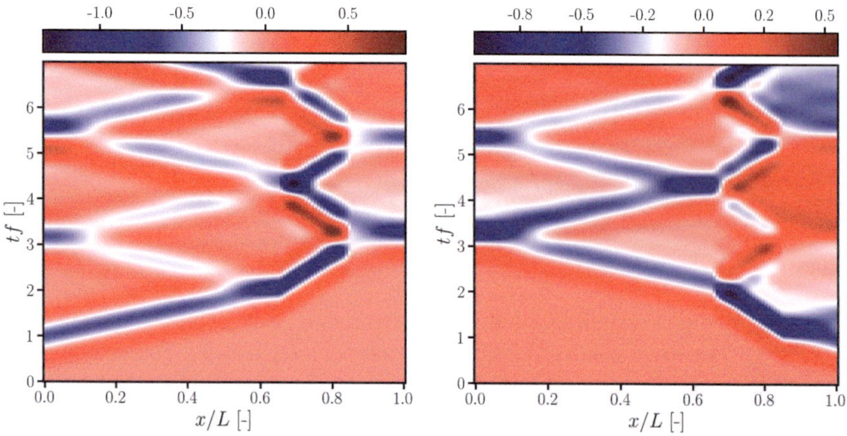

(a) Sine burst applied on left boundary **(b)** Sine burst applied on right boundary

Fig. 9.10 Representative one-dimensional wave field $u(x, t)$ evolving over time and perturbed by a defect

the one-dimensional wave equation (Eq. (9.9)) together with homogeneous Neumann boundary conditions (Eq. (9.9)) and homogeneous initial conditions (Eq. (9.11)) will be considered. The following derivation is adapted from [42], where more in-depth material on the finite difference method can be found.

The first step is discretization, where the domain is described by a two-dimensional grid consisting of $N_1 + 1$ points in the spatial dimension and $N_t + 1$ points in the temporal domain. Thereby the total number of grid points is $(N_1 + 1) \times (N_t + 1)$.

$$x_i = i\Delta x, i = 0, 1, ..., N_1 \tag{9.14}$$

$$t_n = n\Delta t, n = 0, 1, ...N_t \tag{9.15}$$

The discretized wave field will, from now on, be referred to as u_i^n, where i, n refer to the grid point (x_i, t_n). The same applies for the source term p_i^n and the wave speed c_i, which, in contrast, is time-invariant.

The essential ingredient in the finite difference method is the finite difference approximation of the gradients. Here, second-order central difference approximations will be used for the temporal second-order derivatives.

$$\frac{\partial^2 u_i^n}{\partial t^2} \approx \frac{u_i^{n+1} - 2u_i^n + u_i^{n-1}}{\Delta t^2} \tag{9.16}$$

Additionally, first-order central difference approximations will be employed for the initial conditions, boundary conditions, and spatial gradient term of the wave equation.

$$\frac{\partial u_i^n}{\partial t} \approx \frac{u_i^{n+1} - u_i^{n-1}}{2\Delta t} \tag{9.17}$$

$$\frac{\partial u_i^n}{\partial x} \approx \frac{u_{i+1}^n - u_{i-1}^n}{2\Delta x} \tag{9.18}$$

The next step towards a finite difference scheme is replacing the derivatives in the differential equation (Eq. (9.9)) with the numerical approximations. The temporal second-order derivative $\partial^2 u(x,t)/\partial t^2$ can directly be replaced with Eq. (9.16). The second term of Eq. (9.9) requires an additional treatment, where $c^2(x)\partial u(x,t)\partial x$ is temporarily substituted by $v(x,t)$:

$$\frac{\partial}{\partial x}\left(c^2(x)\frac{\partial u(x,t)}{\partial x}\right) = \frac{\partial v(x,t)}{\partial x}. \tag{9.19}$$

The spatial derivative can be approximated with the first-order central difference approximation from Eq. (9.18) using half grid spacings, i.e., $i - 1/2, i + 1/2$:

$$\frac{\partial v_i^n}{\partial x} \approx \frac{v_{i+\frac{1}{2}}^n - v_{i-\frac{1}{2}}^n}{\Delta x}. \tag{9.20}$$

Relying on the substitution in Eq. (9.19), the first-order derivatives within $v_{i-\frac{1}{2}}^n$ and $v_{i+\frac{1}{2}}^n$ can again be approximated by first-order central difference approximations with half grid spacings:

$$v_{i-\frac{1}{2}}^n = c_{i-\frac{1}{2}}^2 \frac{\partial u_{i-\frac{1}{2}}^n}{\partial x} \approx c_{i-\frac{1}{2}}^2 \frac{u_i^n - u_{i-1}^n}{\Delta x}, \tag{9.21}$$

$$v_{i+\frac{1}{2}}^n = c_{i+\frac{1}{2}}^2 \frac{\partial u_{i+\frac{1}{2}}^n}{\partial x} \approx c_{i+\frac{1}{2}}^2 \frac{u_{i+1}^n - u_i^n}{\Delta x}. \tag{9.22}$$

The wave speed between grid points $c_{i-\frac{1}{2}}^2, c_{i+\frac{1}{2}}^2$ can be obtained by interpolation, specifically by computing the mean. A possible choice is the arithmetic mean $c_{i+\frac{1}{2}}^2 \approx \frac{1}{2}(c_i^2 + c_{i+1}^2)$, but for functions exhibiting jumps,[5] the harmonic mean is preferred [42].

$$c_{i-\frac{1}{2}}^2 \approx 2\left(\frac{1}{c_{i-1}^2} + \frac{1}{c_i^2}\right)^{-1} \tag{9.23}$$

$$c_{i+\frac{1}{2}}^2 \approx 2\left(\frac{1}{c_i^2} + \frac{1}{c_{i+1}^2}\right)^{-1} \tag{9.24}$$

[5] As we are dealing with defects in materials, jumps in the material distribution and thereby the wave speed are to be expected.

Using the harmonic mean from Eqs. (9.23) and (9.24) together with the finite difference approximations of $v_{i-\frac{1}{2}}, v_{i+\frac{1}{2}}$ from Eqs. (9.21) and (9.22) in Eq. (9.20), yields

$$\frac{\partial}{\partial x}\left(c_i^2\frac{\partial u_i^n}{\partial x}\right) = \frac{\partial v_i^n}{\partial x} \approx \frac{c_{i+\frac{1}{2}}^2(u_{i+1}^n - u_i^n) - c_{i-\frac{1}{2}}^2(u_i^n - u_{i-1}^n)}{\Delta x^2}$$
$$\approx 2\frac{\left(\frac{1}{c_i^2}+\frac{1}{c_{i+1}^2}\right)^{-1}(u_{i+1}^n - u_i^n) - \left(\frac{1}{c_{i-1}^2}+\frac{1}{c_i^2}\right)^{-1}(u_i^n - u_{i-1}^n)}{\Delta x^2}. \tag{9.25}$$

Substituting the finite difference approximations of the temporal second-order derivative (Eq. (9.16)) and the spatial derivative term (Eq. (9.25)) into Eq. (9.9), results in the following discretized equation:

$$\frac{u_i^{n+1}-2u_i^n+u_i^{n-1}}{\Delta t^2} - 2\frac{\left(\frac{1}{c_i^2}+\frac{1}{c_{i+1}^2}\right)^{-1}(u_{i+1}^n-u_i^n)-\left(\frac{1}{c_{i-1}^2}+\frac{1}{c_i^2}\right)^{-1}(u_i^n-u_{i-1}^n)}{\Delta x^2} = p_i^n, \tag{9.26}$$

which can be rewritten as an explicit formula depending only on previous time steps, i.e., $n-1$ and n.

$$u_i^{n+1} = -u_i^{n-1} + 2u_i^n + \Delta t^2 p_i^n$$
$$+2\left(\frac{\Delta t}{\Delta x}\right)^2\left[\left(\frac{1}{c_i^2}+\frac{1}{c_{i+1}^2}\right)^{-1}(u_{i+1}^n - u_i^n) - \left(\frac{1}{c_{i-1}^2}+\frac{1}{c_i^2}\right)^{-1}(u_i^n - u_{i-1}^n)\right] \tag{9.27}$$

To declutter the notation, $q_i^n = \partial v_i^n/\partial x$ is introduced, yielding the following shorter expression for the finite difference scheme.

$$u_i^{n+1} = -u_i^{n-1} + 2u_i^n + \Delta t^2 p_i^n + \Delta t^2 q_i^n \tag{9.28}$$

The derivation is analogous in higher dimensions with additional terms of the form q_i^n but defined with respect to the additional spatial dimensions. To describe the numerical stability of the scheme, the Courant number $C_i = c_i\frac{\Delta t}{\Delta x}$ can be considered. For explicit schemes, its square C_i^2 should usually be lower than 1 to ensure numerical stability. In the case of heterogeneous materials, the maximum value of C_i^2 is typically considered.

The derived scheme (Eq. (9.28)) is a multi-step time-stepping scheme, i.e., the two previous time steps are required to compute the next time step. Therefore, special care has to be taken at the first time step. The first time step can be estimated as,

$$u_i^1 = -u_i^{-1} + 2u_i^0 + \Delta t^2 p_i^0 + \Delta t^2 q_i^0 \tag{9.29}$$

where u_i^{-1} is unknown. However, the homogeneous initial conditions (Eq. (9.11)) can be used together with the approximation of the first order temporal derivative from Eq. (9.17) to derive an expression for u_i^{-1}:

$$0 = \frac{\partial u_i^0}{x} = \frac{u_i^1 - u_i^{-1}}{2\Delta t}, \tag{9.30}$$

$$u_i^{-1} = u_i^1, \tag{9.31}$$

which can be inserted in Eq. (9.29) and results in the special scheme for the first time step.

$$2u_i^1 = 2u_i^0 + \Delta t^2 p_i^0 + \Delta t^2 q_i^0 \tag{9.32}$$

Using the homogeneous initial conditions from Eq. (9.11) and assuming $p_i^0 = 0$ at the first time step, the second time step must also be $u_i^1 = 0$ at all grid points.

A similar problem arises for the boundary grid points, as u_i^{n+1} depends on u_{i-1}^n and u_{i+1}^n, as seen in Eq. (9.27). Consider the left boundary with $i = 0$, which leads to a dependency on u_{-1}^n for the computation of the next time step u_0^{n+1}. To resolve the problem, the boundary conditions (Eq. (9.10)) have to be considered. Specifically, the homogeneous Neumann boundary conditions are considered, which through the first-order finite difference approximation (Eq. (9.18)) can be rewritten as

$$0 = \frac{\partial u_i^n}{\partial x} = \frac{u_{i+1}^n - u_{i-1}^n}{2\Delta x} \quad \text{on } \Gamma_N, \tag{9.33}$$

$$u_{i-1}^n = u_{i+1}^n. \tag{9.34}$$

Thus in the previous example, the value of u_{-1}^n can be estimated with u_1^n. Extending the grid $x_i, i = 0, 1, \ldots, N_1$ with one grid point at each boundary yields the grid x_i with $i = -1, 0, \ldots, N_1, N_1 + 1$ points. Thereby, the boundary conditions can be incorporated efficiently. The additional grid points are referred to as ghost cells, as they are non-physical and only employed for implementational reasons. The finite difference scheme (Eq. (9.28)) is then applied on the extended grid with $N_1 + 3$ gridpoints, where after each time step, the ghost cells are updated with

$$u_{-1}^n = u_1^n, \tag{9.35}$$

$$u_{N_1+1}^n = u_{N_1}^n, \tag{9.36}$$

leading to a fulfillment of the homogeneous Neumann boundary conditions from Eq. (9.10) at $i = 1$ and $i = N_1$.

This concludes the derivation of the finite difference scheme, which is to be used in Sects. 9.2.4 and 9.2.5. The scheme is summarized by Algorithm 14.

Algorithm 14 Finite difference scheme for the wave equation Equation (9.9) with homogeneous initial conditions Equation (9.11) and homogeneous Neumann boundary conditions Equation (9.10)

Require: discretized grid Δx, Δt, N_1, N_t, source term p_i^n, wave speed c_i^2
  First two time steps from homogeneous initial conditions
  $u_i^0 = 0, u_i^1 = 0$
  Initialization of first two time steps
  for all time steps **do**
    Update all internal grid points for time step $n+1$
    $$u_i^{n+1} = -u_i^{n-1} + 2u_i^n + \Delta t^2 p_i^n$$
    $$+2\left(\frac{\Delta t}{\Delta x}\right)^2 \left[\left(\frac{1}{c_i^2} + \frac{1}{c_{i+1}^2}\right)^{-1} (u_{i+1}^n - u_i^n) - \left(\frac{1}{c_{i-1}^2} + \frac{1}{c_i^2}\right)^{-1} (u_i^n - u_{i-1}^n)\right]$$
    Update ghost cells for time step $n+1$
    $u_{-1}^{n+1} = u_1^{n+1}$
    $u_{N_1+1}^{n+1} = u_{N_1}^{n+1}$
  end for

9.2.3 Physics-informed Neural Networks

As previously discussed in Sect. 4.3, physics-informed neural networks can be applied to inverse problems. It is, however, the weakest of the three methodologies discussed in this chapter because of its prohibitively challenging training. For general inverse problems, the physics-informed concept is only viable in conjunction with a data-driven solver, where it can be employed as a regularization term to punish unphysical solutions. Inverse problems with full domain knowledge of the forward solution exhibit an exception [7, 37] and is possibly a reasonable choice for non-linear problems. The problem with partial domain knowledge is that the physics-informed neural network must learn both the forward solution and inverse quantities, which is a highly complex task leading to excessive computation times, which in the two-dimensional case of full waveform inversion are at least ~2000 times larger than the iterative solver, as shown in [37] and discussed in more detail in Sect. 9.3.1 in the context of topology optimization.

Therefore, full domain knowledge, i.e., sensors recording the entire domain $\Omega \times \mathcal{T}$, is considered in this section for physics-informed neural networks. This has the advantage that u does not have to be learned, but the recorded signals $\tilde{u}$ can directly be used in the residual computation. As the measurements are only available at discrete points, the residual computation is conducted with numerical differentiation. Specifically, finite differences are used. In fact, the update rule from Eq. (9.27) can directly be rewritten as the partial differential equation residual:

$$C(\tilde{\boldsymbol{u}}; \boldsymbol{p}) = \frac{1}{2(N_t+1)(N_1+1)} \sum_{n=0}^{N_t} \sum_{i=0}^{N_1} \left(\tilde{u}_i^{n+1} + \tilde{u}_i^{n-1} - 2\tilde{u}_i^n - \Delta t^2 p_i^n + \Delta t^2 \hat{q}_i^n\right)^2. \quad (9.37)$$

Note that the cost $C(\tilde{\boldsymbol{u}}_j)$ in Eq. (9.37) is defined with respect to a single observation $\tilde{\boldsymbol{u}}_j$ corresponding to a specific source $\boldsymbol{p}_j$. A cumulative cost function is employed in the case of multiple measurements with different sources, as in the setup in Fig. 9.9 with $M = 2$ measurements.

$$C = \frac{1}{M} \sum_{j=1}^{M} C(\tilde{\boldsymbol{u}}_j; \boldsymbol{p}_j) \tag{9.38}$$

Boundary conditions, initial conditions, and the measurement mismatch do not have to be enforced, as these are already fulfilled by the measured signal $\tilde{u}$. The prediction of the spatial derivative term $\hat{q}_i^n$ is based on measurements $\tilde{u}$ and the wave speed $\hat{c}(\boldsymbol{\Theta})$ predicted by a neural network:

$$\hat{q}_i^n \left(\tilde{u}_{i-1}^n, \tilde{u}_i^n, \tilde{u}_{i+1}^n; \hat{c}_{i-1}(\boldsymbol{\Theta}), \hat{c}_i(\boldsymbol{\Theta}), \hat{c}_{i+1}(\boldsymbol{\Theta}) \right). \tag{9.39}$$

The neural network parameters $\boldsymbol{\Theta}$ are updated by differentiating the cost function C with respect to the parameters $\boldsymbol{\Theta}$.

Here, three different network architectures are considered:

- linear ansatz,
- fully connected neural network,
- convolutional neural network.

The linear ansatz is technically not a neural network but consists of piece-wise constant patches[6] and is deployed as a reference. The fully connected neural network simply predicts the wave speed distribution $c(x)$ from coordinates x. In contrast, the convolutional neural network has a generator structure, which upscales and manipulates random noise sampled from a Gaussian distribution. For more in-depth information on the network architectures, the reader is referred to Exercise E.32 for which the corresponding code is made available.

The resulting inversions are presented in Fig. 9.11. The linear ansatz leads to spurious oscillations resulting from the residual's finite difference approximation. The numerically evaluated spatial gradients reduce to zero over the oscillations. However, if averaged the identified wave speed distribution is correct. The neural networks smoothen the solutions and, therefore, lack these artifacts. Notably, the convolutional neural network converges much faster, whereas the fully connected neural network takes much longer to converge to a reasonable result. It takes about 10 000 epochs instead of 1 000 to converge to a similar result. Caution is advised in conclusions on network architectures, as performance strongly depends on the hyperparameters. A convolutional neural network is, however, expected to perform best for the task at hand, as the one-dimensional wave speed is a discrete grid, i.e., a one-dimensional image.

[6] Note, that this results in a linear problem (due to the linear partial differential equation and linear ansatz) and could also be solved directly with a linear regression without gradient-based iterations.

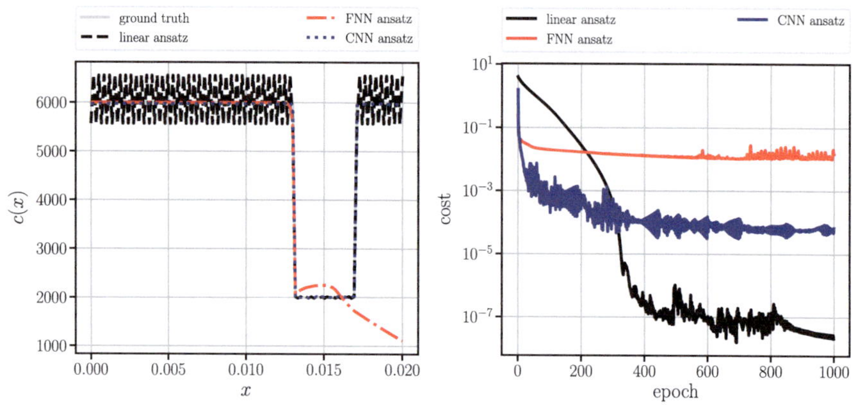

(a) Wave speed prediction (b) Training history

Fig. 9.11 One-dimensional full waveform inversion example using three different ansatz functions with a physics-informed neural network and full-domain knowledge

9.2.4 *Iterative Forward Solver with Neural Network Ansatz*

Unlike the physics-informed neural network, the iterative forward solver is also viable for inverse problems with partial domain knowledge. To illustrate this, only measurements at the boundaries $x = 0$ and $x = L$ are taken, i.e., $\tilde{u}(0, t)$, $\tilde{u}(L, t)$. The main concept is demonstrated on the one-dimensional case from Fig. 9.9. However, as full waveform inversion works best in two and three dimensions, a two-dimensional case is also illustrated, better underlining the power of a neural network ansatz.

9.2.4.1 One-Dimensional Case

The aim of the iterative solver is to identify $c(x)$ given a set of boundary measurements $\tilde{u}(0, t)$, $\tilde{u}(L, t)$. The iterative solver utilizes a neural network to predict the wave speed $\hat{c}(x; \boldsymbol{\Theta})$. To improve on this (initially random) prediction, training has to occur. Here the prediction $\hat{c}(x; \boldsymbol{\Theta})$ is fed into the finite difference scheme from Algorithm 14 to estimate the discretized wave field $\hat{u}(x_i, t_n; \hat{c})$ on the entire spatio-temporal domain $\Omega \times \mathcal{T}$ up to time step N_t. The predicted wave field is then compared to the boundary measurements $\tilde{u}(0, t)$, $\tilde{u}(L, t)$ via a misfit functional, which is used as the cost function:

$$C(\tilde{\boldsymbol{u}}; \boldsymbol{p}) = \int_{\mathcal{T}} \Big(\big(\hat{u}(0, t; \hat{c}, \boldsymbol{p}) - \tilde{u}(0, t)\big)^2 + \big(\hat{u}(L, t; \hat{c}, \boldsymbol{p}) - \tilde{u}(L, t)\big)^2 \Big) d\mathcal{T}. \tag{9.40}$$

As elaborated in Sect. 9.2.3, multiple measurements $\tilde{\boldsymbol{u}}_j$ obtained with M different sources $\boldsymbol{p}_j$ can be considered. This leads to the following cumulative cost function C:

$$C = \frac{1}{M}\sum_{j=1}^{M} C(\tilde{\boldsymbol{u}}_j; \boldsymbol{p}_j) \tag{9.41}$$

To aid the learning procedure, the cost function can be extended by prior information on the wave speed, such as, e.g., the wave speed values at the boundaries, which typically are undamaged and thereby known.

Subsequently, the gradient of the cost function is computed to minimize the cost function C. The gradient computation can be conducted via automatic differentiation through the forward solver, which requires a differentiable solver and can be costly. Alternatively, the adjoint sensitivity method can be employed to compute the forward solver's sensitivity $\partial C/\partial c$ with respect to the wave speed. Via the chain rule, the sensitivities can be propagated to the neural network parameters $\boldsymbol{\Theta}$ to compute the gradient $\partial C/\partial \boldsymbol{\Theta}$.

$$\frac{\partial C}{\partial \boldsymbol{\Theta}} = \frac{\partial C}{\partial \hat{c}}\frac{\partial \hat{c}}{\partial \boldsymbol{\Theta}} \tag{9.42}$$

This is the most efficient method and makes it possible to treat the solver as a black box; see [37] for details and Exercise E.33 for an implementation.

The gradient is subsequently used to minimize the cost function C with Adam. The described steps are repeated until the prediction $\hat{c}$ is sufficiently accurate. A summary of the steps is provided in Algorithm 15.

Algorithm 15 Iterative forward solver: Training an iterative solver to find the wave-speed c given a measurement at the boundary $u(0,t)$, $u(1,t)$

Require: boundary measurements $\tilde{u}(0,t)$ and $\tilde{u}(1,t)$, source term $p(x,t)$, network parameters
  define network architecture
  initialize network parameters $\boldsymbol{\Theta}$: weights $\{\boldsymbol{W}^l\}_{l=1}^L$ and biases $\{\boldsymbol{b}^l\}_{l=1}^L$ for all layers L
  set hyperparameters for the Adam optimizer (epochs, learning rate α, ...)

for all epochs **do**
  Predict $\hat{c}$ with the neural network
  Compute $\hat{u}(x_i, t_n)$ from $\hat{c}$ with the finite difference method ▷ cf. Algorithm 14
  Compute cost function C ▷ cf. Equation (9.40)
  Compute gradient $\frac{\partial C}{\partial \boldsymbol{\Theta}}$ ▷ Automatic differentiation/adjoint method cf. Equation (9.42)
  update parameters: $\boldsymbol{\Theta} \leftarrow \boldsymbol{\Theta} - \alpha\frac{\partial C}{\partial \boldsymbol{\Theta}}$ ▷ Adam
end for

As in Sect. 9.2.3, a linear ansatz, a fully connected neural network, and a convolutional neural network are employed for the prediction of the wave speed $\hat{c}$. Again, details are provided in Exercise E.33. The recovered wave speeds and the training history are presented in Fig. 9.12. The neural network discretizations converge much

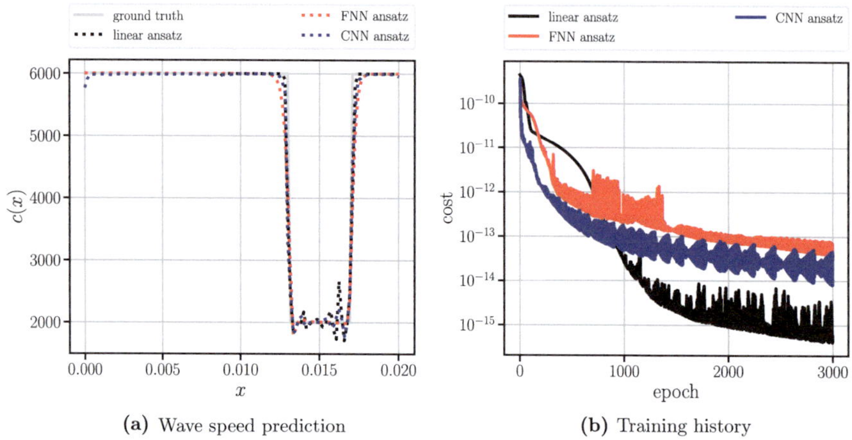

Fig. 9.12 One-dimensional full waveform inversion example using three different ansatz functions with an iterative forward solver and only partial domain knowledge

faster. However, the reached minima are worse than with the linear ansatz. Nevertheless, when considering the recovered wave speed, the neural networks predict smoother solutions with a small loss in sharpness visible at the corner of the jump. The difference in quality between the two neural networks is minor.

9.2.4.2 Two-Dimensional Case

Full waveform inversion is a better-posed problem in higher dimensions, as the multi-dimensional waves are able to transmit more information on the defect. Therefore, the full waveform inversion results from [37] are presented in Fig. 9.13 after 50 epochs,[7] where a circular hole is to be recovered. To detect voids, a specialized formulation of the scalar wave equation from [41] is employed:

$$\gamma(x)\rho_0\frac{\partial^2 u(x,t)}{\partial t^2} - \nabla\cdot\Big(\gamma(x)\rho_0 c_0^2 \nabla u(x,t)\Big) = p(x,t) \tag{9.43}$$

Instead of scaling the wave speed c, as in Eq. (9.9), the density ρ is scaled using an indicator function $\gamma_0 \leq \gamma(x) \leq 1$, where $0 < \gamma_0 \ll 1$ is used as lower bound to ensure numerical stability. The formulation is more suitable for the detection of voids [41, 43] and referred to as ρ-scaling.

The inversion with a linear ansatz is compared with one using a neural network. The linear ansatz results in oscillatory artifacts, slightly masking the defect, while the neural network leads to a smooth recovery without loss of sharpness. These

[7] This again demonstrates how much easier it is to perform a full waveform inversion in higher dimensions. The one-dimensional case took about 3 000 epochs.

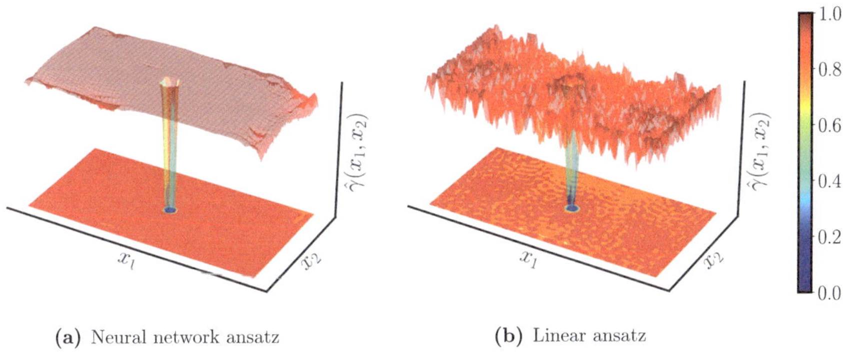

Fig. 9.13 Surface plot of the predicted material field after 50 epochs. The training was performed with a neural network and linear ansatz using automatic differentiation. Figure reprinted from [37]

Fig. 9.14 Comparison of the predicted material field after 50 epochs with noisy measurements. The level of noise is quantified by β. Figure reprinted from [37]

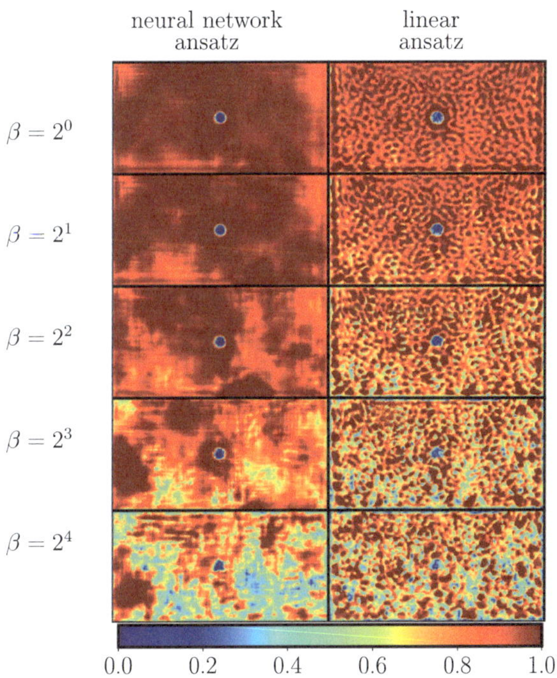

results have independently been reported for other inverse problems [44–52, 89]. The neural network acts as a regularizer. The reason herefore is connected to the overparametrization enabled by the neural network [12] and makes it possible to identify better local minima.

The effect becomes even more prominent in the presence of noise, as illustrated in Fig. 9.14. Here β is the level of noise added to the measurements $\tilde{u}$.

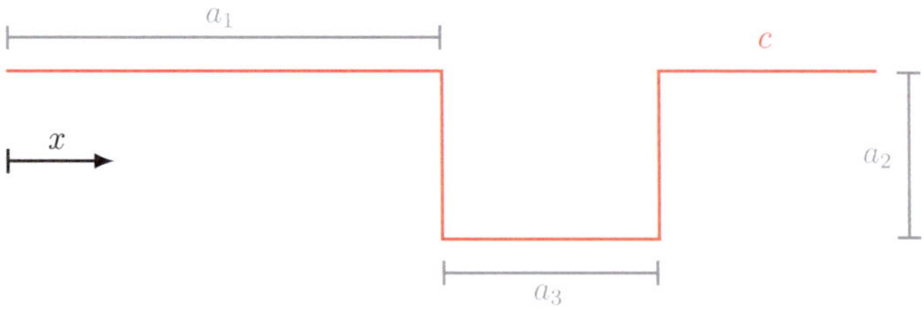

Fig. 9.15 The parametrization of the wave speed c with the parameters a_1, a_2 and a_3. In the undamaged domain, i.e., left to and right to the flaw, a wave speed of $c = c_0$ is assumed

9.2.5 Data-Driven Solver

In contrast to the previous methods, the data-driven solver relies on a dataset to learn the inversion. To reduce the space of possible solutions and thereby the necessary size of the dataset, the problem is simplified with a parametrization of the wave speed distribution $c(x)$. Here a single wave speed distribution $c(x)$ is described by three parameters a_1, a_2 and a_3, as depicted in Fig. 9.15. The location of the defect is controlled by a_1 and a_3, while its amplitude is defined by a_2.

$$c(x) = \begin{cases} c_0 \text{ if } x < a_1 \text{ or } x > a_3 \\ c_0 - a_2 \text{ else} \end{cases} \tag{9.44}$$

The first step is the creation of a dataset. To this end, $m_{\mathcal{D}} = 10\,000$ different wave speed distributions $\tilde{\boldsymbol{c}} = \{\tilde{c}_i(x)\}_{i=1}^{m_{\mathcal{D}}}$ are generated with the three parameters. These are fed through the finite difference scheme from Algorithm 14 to estimate the corresponding wave fields $\tilde{\boldsymbol{u}} = \{\tilde{u}_i(0,t), \tilde{u}_i(L,t)\}_{i=1}^{m_{\mathcal{D}}}$. The final dataset $\tilde{\boldsymbol{c}}, \tilde{\boldsymbol{u}}$ is thus composed of $m_{\mathcal{D}}$ samples with either a wave speed sampled at $N_1 = 117$ grid points or the parameters a_1, a_2, a_3 defining $c(x)$ indirectly and four measurements (two boundaries for two separate source signals) with $N_t = 700$ time steps each. To ease the learning process, $c(x)$ is learned indirectly, i.e., the parametrization a_1, a_2, a_3 is learned.

However, predicting the three parameters from the measurements $\tilde{u}$ with 4×700 values (two sensors $\times$ two sources $\times$ 700 time steps) is challenging. It can be achieved by using a convolutional neural network, first reducing the data and simultaneously extracting the most essential information. However, a simpler approach is to reduce the dimensionality of the data through a linear dimensionality reduction technique, such as a singular value decomposition, as discussed in Sect. 6.1.1. The singular value decomposition yields an orthogonal transformation matrix $\boldsymbol{V}$, which can be truncated $\bar{\boldsymbol{V}}$, yielding a low-dimensional representation $\bar{\tilde{u}}$:

$$\bar{\tilde{u}} = \tilde{u} \cdot \bar{\boldsymbol{V}}^{\mathsf{T}}, \tag{9.45}$$

$$\tilde{u} \approx \bar{\tilde{u}} \cdot \bar{\boldsymbol{V}}. \tag{9.46}$$

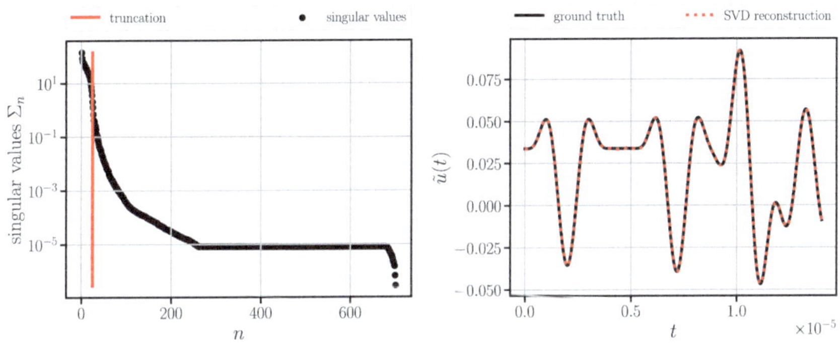

(a) Singular values (b) Reconstruction at a truncation level of 25

Fig. 9.16 Singular value decomposition of the signal $\tilde{u}$

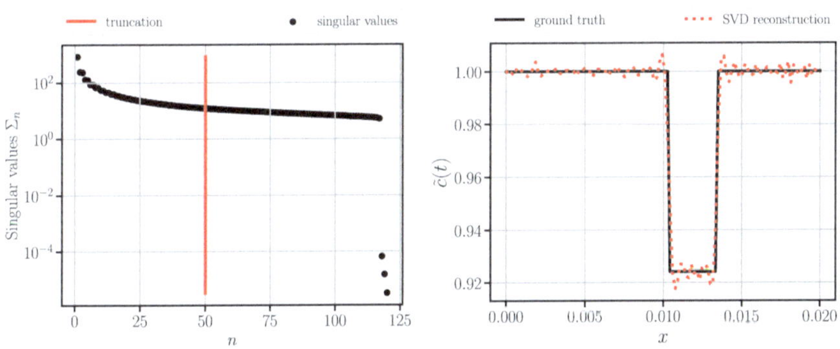

(a) Singular values (b) Reconstruction at a truncation level of 50

Fig. 9.17 Singular value decomposition of the wave speed c

By performing a singular value decomposition and considering the singular values Σ_n, a reasonable degree of truncation can be determined; see Fig. 9.16a. At a truncation level of 25—thus reducing the dimensionality from 700 to 25 with Eq. (9.45)—the reconstruction (Eq. (9.46)) captures the main features of the measurement, as seen in Fig. 9.16. Thereby the measured signals $\tilde{u}$ can be reduced from 4×700 to 4×25, i.e., effectively 100 values

If the low-dimensional space of the wave speed distribution, i.e., the parametrization a_1, a_2, a_3, were unknown, the singular value decomposition could similarly be applied. However, due to the non-linear connection between the parameters and the wave speed, the singular value decomposition cannot determine a suitable low-dimensional space, as demonstrated in Fig. 9.17 with a truncation level of 50. Instead, a non-linear dimensionality reduction technique, such as an autoencoder, could be employed.

To highlight the benefit of the transformation to a low-dimensional space, a convolutional neural network acting on the full measurements $\tilde{u}$ is compared to a fully connected neural network acting on the reduced measurements $\bar{\tilde{u}}$. The fully connected neural network consists of five hidden layers with 100 neurons each, yielding 40 703 parameters, while the convolutional neural network has 27 905.

Not all 10 000 cases are used for the training. Instead, the data is split into a training and validation set, as explained in Sect. 2.2. Here, 90% of the data is used to train the neural network. Both networks predict the parameters $\hat{a}_1, \hat{a}_2, \hat{a}_3$. The cost function C can then be computed using the dataset of size $m_{\mathcal{D}}$.

$$C = \frac{1}{m_{\mathcal{D}}} \sum_{i=1}^{m_{\mathcal{D}}} \Big((\tilde{a}_{1_i} - \hat{a}_{1_i})^2 + (\tilde{a}_{2_i} - \hat{a}_{2_i})^2 + (\tilde{a}_{3_i} - \hat{a}_{3_i})^2 \Big) \tag{9.47}$$

The cost function is subsequently minimized with an Adam optimizer, whereby the prediction is improved. After the training, the neural network can be applied to unseen data to predict how an unknown specimen is flawed. To assess the quality of this generalization ability, the validation set is used. The validation set is fed through the neural network, and the cost function is computed. This can already be done during training to spot overfitting early and thus stop the training process, as in early stopping regularization. Examples of the predictions in the validation set are shown in Fig. 9.18a.

Additionally, the training history with the training and validation costs are provided in Fig. 9.18b. The training was performed for 400 epochs. If the training continues, overfitting would occur without regularization, where the training cost continues decreasing, and the validation cost starts to increase, as previously discussed with Fig. 2.6. Furthermore, it is seen that the fully connected neural network benefits from the prior transformation to a reduced state.

The examples from the validation set show that the approach seems less accurate than the iterative approach described in Sect. 9.2.4. However, the network still captures the essential features of the flaw, which might be sufficient in practical circumstances. Furthermore, in the absence of a ground truth, the quality of each prediction can still be assessed using the residual of the differential equation. Importantly, the data-driven approach is significantly faster than the iterative approach if a large amount of predictions have to be made. The main cost comes from the one-time offline training, which is comparable to solving a single case with the iterative solver (neglecting the data acquisition effort). After this training, predictions can be made extremely fast, i.e., as online predictions. Thus, if quantity is important and accuracy (and reliability) is not important, the data-driven solver is favorable.

An alternative to training a neural network is to use the training data as a lookup table and pick the closest case to the sample under investigation. This is illustrated in Fig. 9.18a, where the material distribution with the smallest mean squared error between the considered sample and all of the training data is selected. This yields similar results but is more expensive in terms of storage. The neural network requires only 40 701 parameters, while the lookup table is based on the entire training data

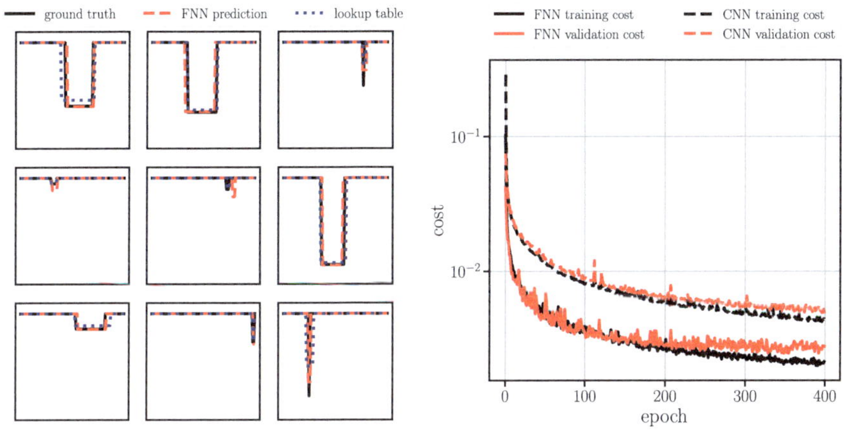

(a) Predictions of the validation set (b) Training history

Fig. 9.18 One-dimensional full waveform inversion example using two different neural network architectures in a data-driven solver. The convolutional neural network predicts the parametrization of the wave speed directly from the measurements, while the fully connected neural network relies on a prior transformation to a reduced space

consisting of $9000 \cdot (2 \cdot 2 \cdot 700 + 3) = 25\,227\,000$ floating point values. For larger problems, the lookup table becomes unviable, while a neural network evaluation remains efficient.

9.2.5.1 Transfer Learning

To increase the reliability of the data-driven approach, it can be combined with the iterative solver from Sect. 9.2.4, as proposed in [53–55]. First, the network can be trained as a data-driven solver. Subsequently, it can be employed as a pretrained ansatz in the iterative forward solver, thus reducing the number of epochs to find an inversion and guarantee a correct inversion. This establishes a middle-ground between data-driven approaches, which are fast but unreliable, and the reliable iterative forward solver with a higher training cost. The procedure relies on the concept of transfer learning [56–58] (transferring knowledge from a prior training to the current problem at hand) and is summarized in Fig. 9.19, where a data-driven solver yields pretrained weights $\boldsymbol{\Theta}^p$ to be used in the iterative forward solver.

To demonstrate the concept, a two-dimensional case is considered.[8] As in the previous two-dimensional case in Sect. 9.2.4.2, the ρ-scaling formulation from Eq. (9.43) is employed. Thus, the goal is the identification of the indicator function $\gamma(x)$ over the two-dimensional spatial domain Ω. The defects are assumed to be ellipsoidal and parametrized according to Fig. 9.20, yielding five parameters per sample: the

[8] The results can be reproduced by the code from [96].

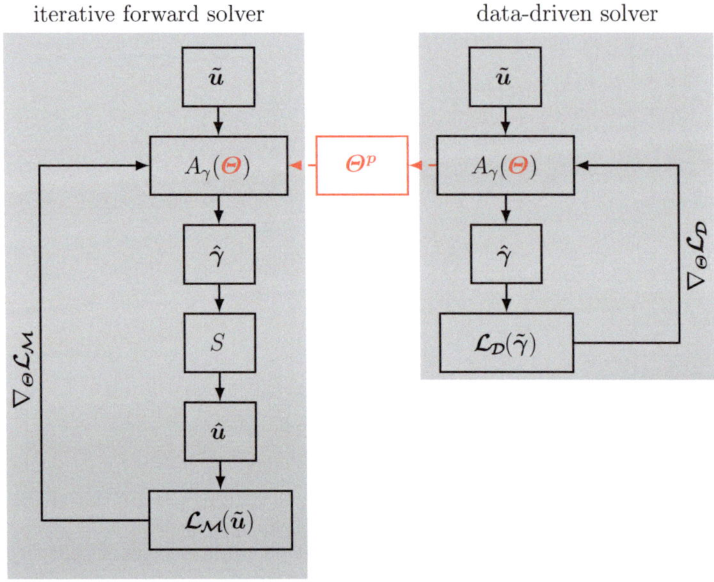

Fig. 9.19 Schematic of the transfer learning procedure applied to the iterative forward solver. First, the neural network is pretrained with a data-driven solver, resulting in a pretrained set of weights $\boldsymbol{\Theta}^p$. Subsequently, the pretrained weights are employed within the iterative forward solver, reducing the number of iterations. Figure adapted from [54]

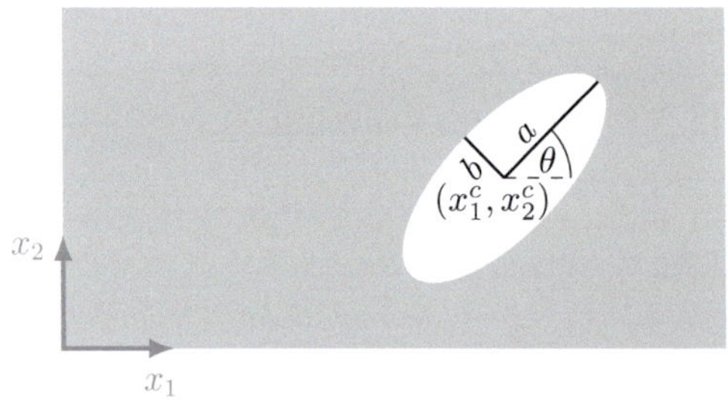

Fig. 9.20 Ellipsoidal parametrization for the dataset used by the data-driven solver

origin (x_1^c, x_2^c), the axes a, b, and the rotation θ. For each sample, the material field $\tilde{\gamma} = \gamma(\boldsymbol{x})$ is generated and used to compute the sensor response $\tilde{\boldsymbol{u}}$, yielding the labeled dataset $\{\tilde{\gamma}_j, \tilde{\boldsymbol{u}}_j\}_{j=1}^m$ with m samples. In total, 2 000 samples are generated, where a 90%/10% training validation split is employed.

The quality of the prediction $\hat{\gamma}$ for the data-driven solver is assessed with a mean squared error integrated over the domain Ω (adapted from Eq. (9.47), where the parametrization coefficients were used instead of the entire material field):

$$\mathcal{L}_\mathcal{D} = \frac{1}{2m_\mathcal{D}} \sum_{i=1}^{m_\mathcal{D}} \int_\Omega \Big(\hat{\gamma}_i(\boldsymbol{\Theta}; \boldsymbol{x}) - \tilde{\gamma}_i(\boldsymbol{x})\Big)^2 d\Omega. \tag{9.48}$$

In contrast, the iterative forward solver relies on a variation of Eq. (9.40), a spatio-temporal integral over the mismatch between measured and simulated wave fields:

$$\mathcal{L}_{\mathcal{M}} = \frac{1}{2}\int_T \int_\Omega \sum_{i=1}^{m_{\mathcal{M}}} \Big(\hat{u}(\hat{\gamma}(\boldsymbol{\Theta}; \boldsymbol{x}), t) - \tilde{u}(\boldsymbol{x}_i, t)\Big)^2 \delta(\boldsymbol{x} - \boldsymbol{x}_i) d\Omega dT. \tag{9.49}$$

The data is transformed before applying a neural network to the measurement data $\tilde{u}$. The difficulty lies in transforming the measurement data $\tilde{u}$ of size number of sensors × number of sources × number of time steps N_t to an output $\hat{\gamma}$ with the grid dimensions $N_1 \times N_2$. One elegant encoding is to use the measurement cost gradient $\nabla_{\gamma_0}\mathcal{L}_{\mathcal{M}}$ with respect to an intact field $\gamma_0(x) = 1$. This contains the most important information of the measurements $\tilde{u}$ and simultaneously results in a dimensionality corresponding to the grid dimensions $N_1 \times N_2$. The added benefit is that the flaws can already be partially identified in the gradient $\nabla_{\gamma_0}\mathcal{L}_{\mathcal{M}}$, as shown for an example in Fig. 9.21. Thus, the neural network only has to amplify the effect and remove the noisy artifacts. As a neural network, a U-net [59] with 784 265 trainable parameters is chosen, which was invented for segmentation tasks, where identically sized images are mapped to each other.

The data-driven solver is trained for 50 epochs using Adam and mini-batch gradient descent with a batch size of 32. The corresponding learning curve is depicted in Fig. 9.22a. From the non-increasing validation cost, it can be seen that overfitting is avoided. Predictions of the validation set are given in Fig. 9.22b, where decent predictions can be seen. The essence of the flaws, i.e., size and location, is captured, while the exact shape is not.

To capture a more accurate shape of the flaws, the network can be retrained with the iterative solver for each case. The pretrained weights $\boldsymbol{\Theta}$ are therefore transferred to the neural network in the iterative forward solver. Averaged training histories are provided in Fig. 9.23. Figure 9.23 shows that the cost for the pretrained neural network starts much lower and is only overtaken at around 30 epochs. The cost

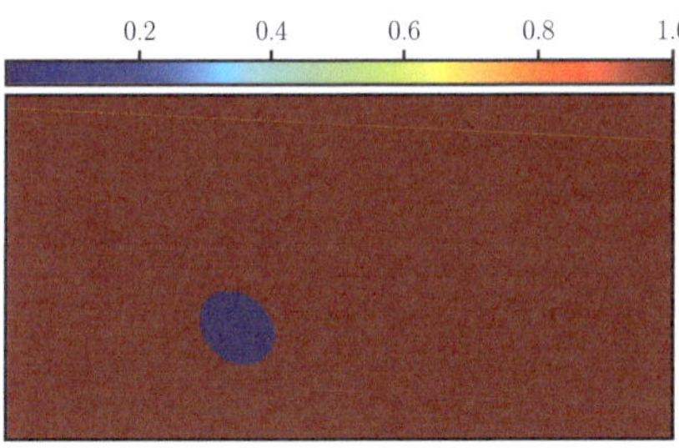

(a) Material distribution γ

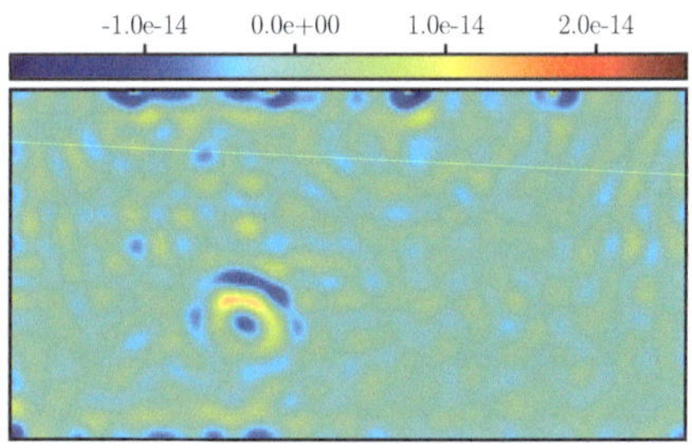

(b) Initial gradient $\nabla_\gamma C$

Fig. 9.21 The initial gradient $\nabla_\gamma C$ provides an indication of the defect. Thereby it enables an encoding of the wave signal $\tilde{u}$ with the same shape of the dimensionality of the material distribution γ

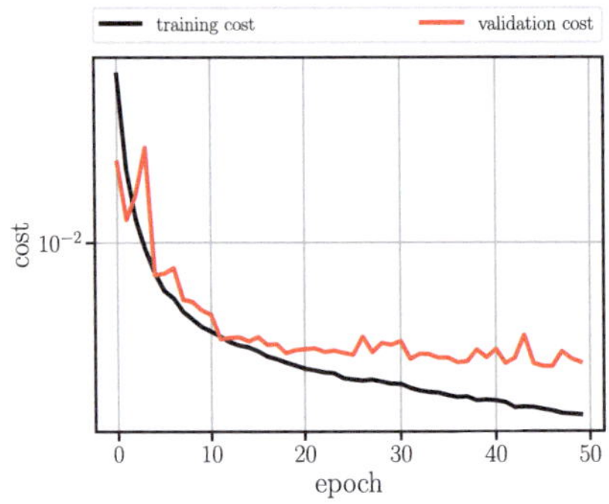

(a) Data-driven training history

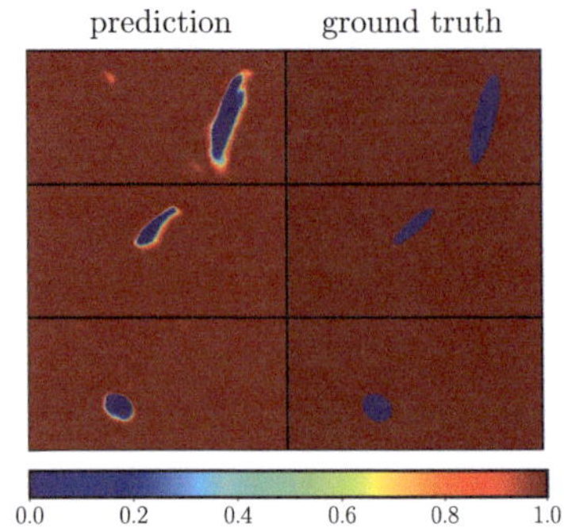

(b) Validation predictions with trained network

Fig. 9.22 Data-driven solver as pretrained neural network for an iterative solver

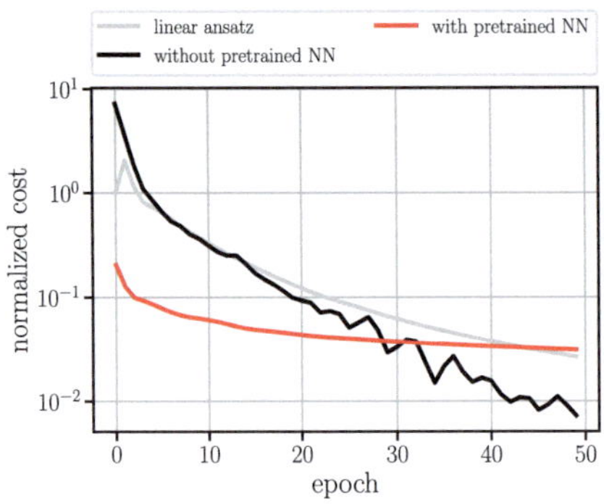

(a) Cost history

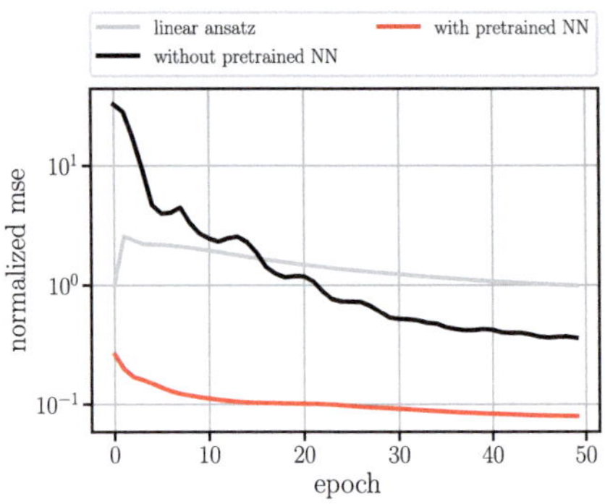

(b) MSE history

Fig. 9.23 Training histories averaged over ten samples

does not decrease further due to the less favorable statistical distribution of the neural network weights caused by the pretraining. In contrast, a neural network without pretraining and proper weight initialization is able to reduce the cost much more efficiently. This difference can be reduced by increasing the L^2 regularization while increasing the network size to retain the same capacity. It is, however, also interesting to study the history of the mean squared error of the material field γ in Fig. 9.23b. Again, the pretrained neural network starts at a much lower value and, for this measure, also stays lower for the entire 50 epochs. This can be attributed to the pretraining being conducted using the mean squared error, such that minor artifacts, such as background noise as still seen in Fig. 9.13, are fully removed. This becomes especially evident when comparing the evolution of the material distribution over the first 40 epochs during training, presented in Fig. 9.24. Here, the pretrained neural network shows no background noise whatsoever, whereas even the pretrained version has minor artifacts at the boundaries. Furthermore, it is seen how the rough shape of

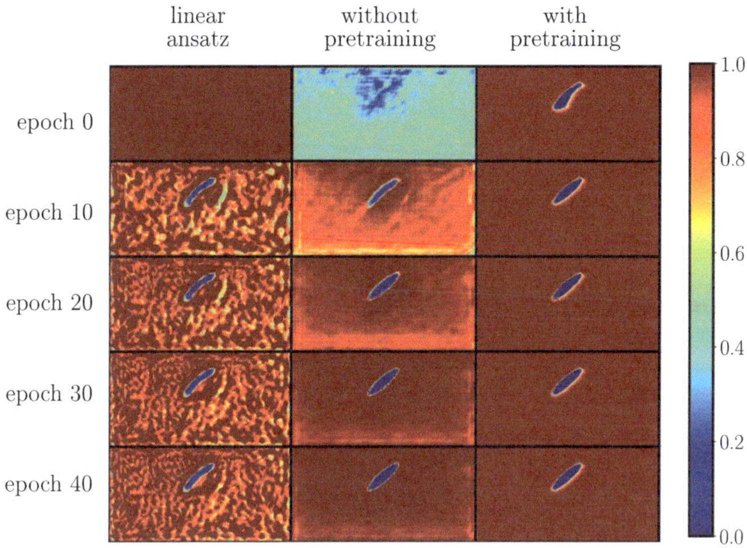

Fig. 9.24 Prediction histories starting at epoch 0 until epoch 40. From left to right, the piece-wise linear ansatz, the neural network without pretraining, and the neural network with pretraining

the pretrained neural network is adjusted within the first ten epochs, whereupon the network accurately captures the true flaw.

Lastly, we test the iterative forward solver on test cases that do not follow the same ellipsoidal parametrization defined in Fig. 9.20. Specifically, a rectangular defect (Fig. 9.25a), multiple circular voids (Fig. 9.25b), and a void derived from a CT-scan from a rock core sample [87] (Fig. 9.25c) are considered. Figure 9.26 shows the corresponding evolution of the material field prediction during the first 40 epochs. The defect from test case 2 from Fig. 9.25b is almost immediately recovered with only minor adjustments in the later epochs. For the other two test cases, the defect is roughly identified and subsequently adjusted to better represent the ground truth. For test case 3 in Fig. 9.25c, the defect is identified better with the pretraining, while in test case 1 from Fig. 9.25a, the neural network without pretraining overtakes the pretrained neural network at around 40 epochs. For reference, the closest material distributions of the training dataset are shown in Fig. 9.25d, f for the three test cases. To this end, the mean squared error of the measurement signals $\tilde{u}$ was employed, yielding a similar lookup table as previously used in Sect. 9.2.5. As can be seen in Fig. 9.25d, f, the location of the flaws is captured by the data, while the exact outline is not. Furthermore, the lookup table is more expensive than the neural network evaluation and has a larger memory footprint.

While the hyperparameters varied for the neural network with and without pretraining, they were kept consistent throughout all validation and test samples presented in Figs. 9.24 and 9.26. This is also one reason for the poor performance of the

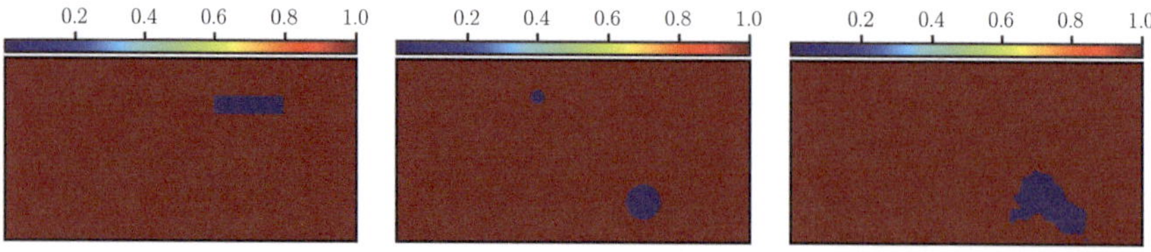

(a) Test case 1 (b) Test case 2 (c) Test case 3

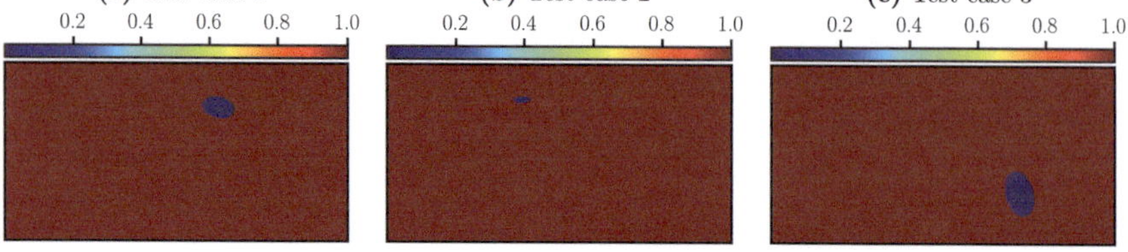

(d) Closest material distribution to test case 1 (e) Closest material distribution to test case 2 (f) Closest material distribution to test case 3

Fig. 9.25 Test cases, that do not come from the ellipsoidal parametrization. Below the closest material distributions of the training dataset are provided. The closeness was determined by the mean squared error between the measurement signals $\tilde{u}$

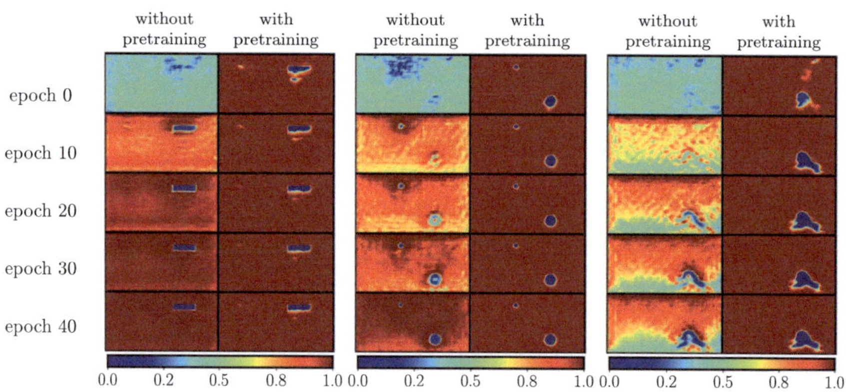

(a) Test case 1 (b) Test case 2 (c) Test case 3

Fig. 9.26 Prediction histories for the three test cases starting at epoch 0 until epoch 40. Left and right are correspondingly the neural network ansatz without and with pretraining

network without pretraining in test case 3 (Fig. 9.26c). A different set of hyperparameters would lead to a better convergence for the specific test case. To exploit this, a per-sample hyperparameter tuning would be required, which thereby increases the cost of training. Instead, transfer learning also reduces the sensitivity with respect to the choice of hyperparameters, i.e., it increases the method's robustness.

Finally, another remark on fairness has to be made. The U-net was selected as neural network architecture, as it is one of the state-of-the-art architectures for data-driven learning for image-to-image mappings, and yields impressive results, as seen in Fig. 9.22. However, for the iterative forward solver, only a single image has to be generated. Therefore, [37] employed a generator architecture inspired from GANs[9] (as also used in Sect. 9.2.4.2), which leads to a more favorable convergence than when using a U-net. Thus, the network without pretraining can also be improved slightly by employing a different network architecture. A better and more thorough comparison is conducted in [55]. Nevertheless, the advantage of a pretrained neural network as ansatz in the iterative solver remains clear.

9.3 Topology Optimization

Topology optimization [60, 61] is a mathematical method used to optimize the material layout within a given design space for a structure or component. By iteratively redistributing material, this technique aims to improve performance based on predefined objectives $\mathcal{L}_{\mathcal{O}}$, such as minimizing weight or maximizing stiffness. Constraints like material properties, loading conditions, and design boundaries guide the process. The result is an optimized structure that meets performance criteria while using material efficiently, such as those illustrated in Figs. 9.32 or 9.35b.

Unlike inverse problems and the example of nondestructive testing, topology optimization does not infer the differential equation parameters λ from observations $\tilde{u}(x, t)$, but comes up with new material distributions, i.e., spatially distributed parameters λ using an objective $\mathcal{L}_{\mathcal{O}}$. Importantly, given a specific problem, a ground truth of optimal parameters λ is unknown prior to the iterative optimization and not necessarily unique, i.e., ill-posed.

However, topology optimization represents a major field in computational mechanics and relies on similar techniques as inverse problems. Therefore, the three enhancement methodologies from Sect. 9.1 are briefly covered in the context of topology optimization.

9.3.1 *Physics-Informed Neural Networks*

As mentioned in Sect. 9.1.1, physics-informed neural networks are incredibly challenging for inverse problems without full-field data, which is impossible in the context of topology optimization, as both the optimal parameters λ and corresponding solution u are unknown. Empirically, the problem was investigated with full waveform inversion with physics-informed neural networks being at least $\sim$2000 slower than iterative solvers for simple toy problems. However, the challenges of physics-informed neural networks persist for other inverse problems and topology optimization.

[9] The implementation with the generator structure can be found in [91].

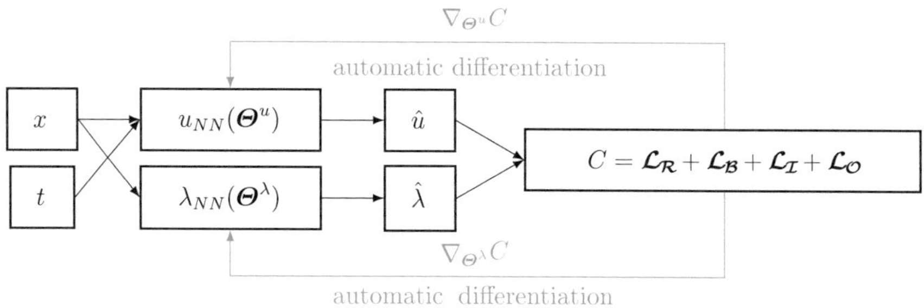

Fig. 9.27 Prediction and training of physics-informed neural networks for topology optimization (and inverse problems). Predictions of the forward solution $\hat{u}$ and the differential equation coefficients $\hat{\lambda}$ are obtained by the neural networks $u_{NN}(\boldsymbol{\Theta}^u)$ and $\lambda_{NN}(\boldsymbol{\Theta}^\lambda)$. During training, the weights $\boldsymbol{\Theta}^u$, $\boldsymbol{\Theta}^\lambda$ are updated with the gradients of the cost function C obtained by automatic differentiation. Modified from [37]

Consider the training scheme in Fig. 9.27, where two neural networks u_{NN} and λ_{NN}, correspondingly predict the solution $\hat{u}$ and the material parameters $\hat{\lambda}$, whose quality is assessed by the cost function C comprised of the residual of the differential equation $\mathcal{L}_\mathcal{R}$, boundary $\mathcal{L}_\mathcal{B}$, and initial conditions $\mathcal{L}_\mathcal{I}$; see Chap. 4, and the topology optimization's objective function $\mathcal{L}_\mathcal{O}$ (which in the case of an inverse problem is the measurement mismatch, i.e., $\mathcal{L}_\mathcal{O} = \mathcal{L}_\mathcal{M}$; see Eq. (9.2)). As discussed in Chap. 4, the cost function C is differentiated with respect to the neural network parameters $\boldsymbol{\Theta}^u$, $\boldsymbol{\Theta}^\lambda$ and optimized with gradient-based optimization.

This suggested nested optimization problem is extremely challenging. Different idealized scenarios preventing convergence can be thought of. One scenario is that given a prediction of the parameters $\hat{\lambda}$, the optimizer focuses on the forward solution $\hat{u}$, which it successfully improves by minimizing $\mathcal{L}_\mathcal{R} + \mathcal{L}_\mathcal{B} + \mathcal{L}_\mathcal{I}$. Now, modifying $\hat{\lambda}$—to reduce $\mathcal{L}_\mathcal{O}$—leads to an increase of $\mathcal{L}_\mathcal{R} + \mathcal{L}_\mathcal{B} + \mathcal{L}_\mathcal{I}$, thereby preventing further overall improvement. These effects can be slightly reduced through learnable weightings of the loss terms [62, 92], i.e., attention, but remain a prohibitive issue for solving topology optimization or inverse problems with physics-informed neural networks.

As proposed in [37], by exchanging the neural network $u_{NN}(\boldsymbol{\Theta}^u)$ with a forward solver S, as illustrated in Fig. 9.28, one arrives at adjoint optimization with a neural network ansatz $\lambda_{NN}(\boldsymbol{\Theta})$ (or alternatively a classical (linear) ansatz $\lambda_N(\boldsymbol{\Theta})$ with shape functions $N(x)$ using Eq. (9.3)) for the material parameters λ and automatic differentiation [63] for the gradient computation. In the representative example of full waveform inversion, this simple change reduces the computational effort from $22.2 - 33.3$ h to 125 s [37]. Further replacing the gradient computation with the adjoint sensitivity method [9, 10] instead of automatic differentiation further reduces the computation time. The ansatz of the material λ does not affect the computation time, but the solution quality, as seen in Sects. 9.2.4 and 9.3.3.

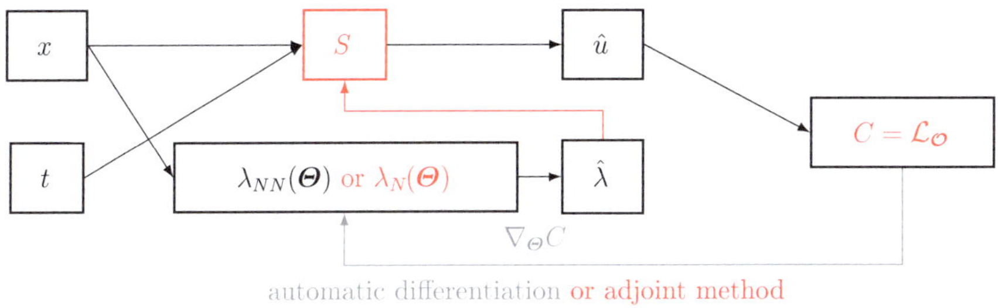

Fig. 9.28 Physics-informed neural network with non-trainable forward operator. The modification of the original physics-informed neural network from Fig. 9.27 is marked in red. Instead of learning the forward problem, a non-trainable forward operator S with $\hat{\lambda}$ as additional input is used. Simultaneously, this leads to a simplification of the cost functional C, as $\mathcal{L}_{\mathcal{R}}$, $\mathcal{L}_{\mathcal{B}}$, $\mathcal{L}_{\mathcal{I}}$ are approximately satisfied due to S. Modified from [37]

Thus, physics-informed neural networks in their original form [3] are not viable for topology optimization and inverse problems with partial domain knowledge. However, a workaround is to optimize the two neural networks separately, i.e., to use the forward neural network $u_{NN}(\boldsymbol{\Theta}^u)$ as forward solver, as, e.g., seen in [64, 88]. This only makes sense if the physics-informed neural network is, on average, faster than established forward solvers for the problem at hand, which, as seen in Chaps. 4 and 5, is also unlikely. However, within an optimization, the physics-informed neural network acting as a forward solver does not have to relearn the slightly different solution from scratch. Instead, it can rely on the learned solution from the previous optimization iteration, i.e., by using transfer learning [56–58], as seen in [65, 66]. Nevertheless, the solution quality remains subpar compared to classical solvers; see Chaps. 4 and 5, making physics-informed neural networks unsuitable for topology optimization.

9.3.2 Data-Driven Solvers

Also data-driven solvers [24–33] are questionable in the context of topology optimization, as, e.g., heavily criticized in [67]. One reason is a more expensive data acquisition, as each optimal design, i.e., label, requires a full optimization (consisting of many forward solves), in contrast to a single forward simulation in the nondestructive testing example from Sect. 9.2.5. Thus, the breakeven threshold is increased drastically. Furthermore, the patterns encountered in topology optimization are more challenging than in other machine learning tasks. Woldseth et al. [67] illustrate this by changing the orientation of one boundary condition by 0.01°, significantly altering the optimal design. This stands in contrast not only with image classification of cats and dogs but also with the full waveform inversion example from Sect. 9.2.5. Lastly, the optimized measures $\mathcal{L}_{\mathcal{D}}$ for training data-driven solvers (cf.

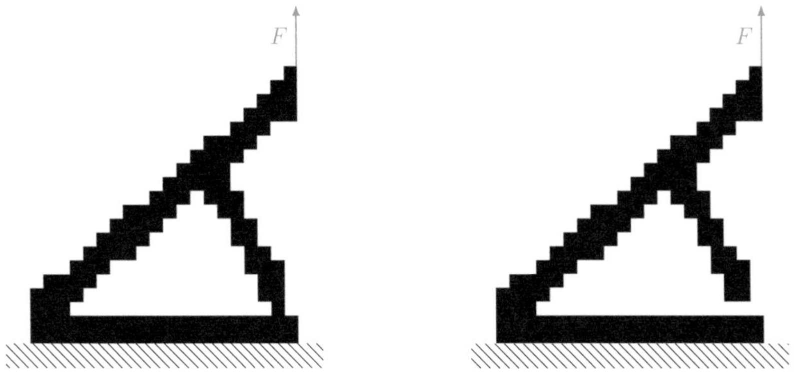

(a) Without disconnected pixel, $c = \sim 351$ (b) With disconnected pixel, $c = \sim 4\,729$

Fig. 9.29 Exemplary structure illustrating the effect of a single pixel that greatly affects the compliance (~ 351 versus $\sim 4\,729$), but not the mean squared between the two images ($2.5 \cdot 10^{-3}$). Adapted from [67]

Eq. (9.8)), like the mean squared error do not align with the topology optimization objectives $\mathcal{L}_{\mathcal{O}}$. Consider Fig. 9.29, where a single disconnected pixel leads to only a minor change in the mean squared error while majorly influencing the structure's integrity—quantifiable through its compliance c,[10] which is ~ 13.5 times greater due to the disconnection. These are the main hurdles preventing data-driven solvers from proper topology optimization.

9.3.3 *Iterative Forward Solver with Neural Network Ansatz*

Similar to Sect. 9.2.4, where neural networks were solely utilized for the parametrization of the inverse parameters λ in full waveform inversion (or other inverse problems [45, 47, 68–70]), neural network parametrizations can be employed for topology optimization [12, 44, 48, 50, 51, 71–73, 89, 90]. Due to the interesting optimization toy example from Sect. 9.1.2.2, this is explored for two topology optimization problems with increasing complexity, namely, the well-known compliance problem (Sect. 9.3.3.1), and acoustic topology optimization (Sect. 9.3.3.2).

9.3.3.1 Compliance Optimization

Compliance optimization will be studied on the well-known Messerschmitt-Bolkow-Blohm (MBB) beam, where Fig. 9.30 depicts half of the symmetric beam. The symmetry axis is located along the x_2-axis. The aim is to distribute (a limited amount of)

[10] Assuming a 1×1 domain with 20×20 elements, $E = 10\,000$, $\nu = 0.3$ and $F = 100$.

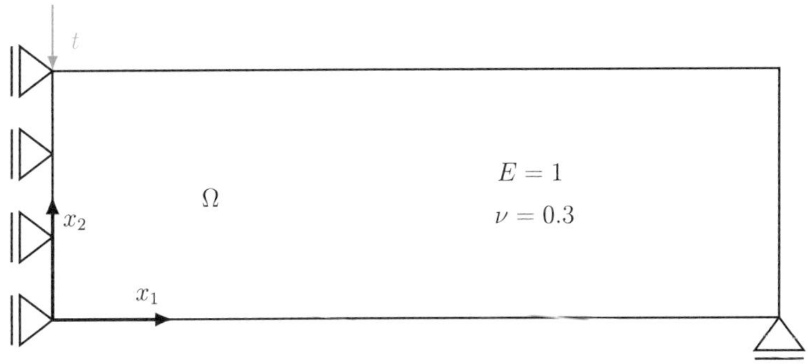

Fig. 9.30 Half of the symmetric Messerschmitt-Bolkow-Blohm beam with symmetry boundary conditions

solid material within the domain Ω, such that the compliance

$$c = \boldsymbol{F}^\mathsf{T}\boldsymbol{u} \tag{9.50}$$

is minimal (equivalent to a maximal stiffness). Linear elasticity is considered with the finite element method, where $\boldsymbol{F}$ are the discretized applied forces and $\boldsymbol{u}$ are the corresponding displacement degrees of freedom obtained by solving

$$\boldsymbol{K}\boldsymbol{u} = \boldsymbol{F}. \tag{9.51}$$

The global stiffness matrix $\boldsymbol{K}$ depends on the constitutive matrix $\boldsymbol{C}_e(x_1, x_2)$ of all N elements through the local stiffness matrices $\boldsymbol{K}_e$:

$$\boldsymbol{K} = \sum_{e=1}^{N} \boldsymbol{K}_e(\boldsymbol{C}_e). \tag{9.52}$$

The constitutive matrix can be parameterized in terms of the constitutive matrix of the solid material $\boldsymbol{C}_e^0$ through the spatially varying density, i.e., the indicator $\gamma(x_1, x_2)$

$$\boldsymbol{C}_e = \gamma(x_1, x_2)^p \boldsymbol{C}_e^0, \qquad p > 1. \tag{9.53}$$

The density $\gamma(x_1, x_2)$ describes, where material is placed ($\gamma = 1$) and where not ($\gamma = 0$).[11] Gray values between zero and one are unphysical and must be avoided. To this end, the exponent p from Eq. 9.53 penalizes material that is neither zero nor one. It is a hyperparameter, typically chosen as $p \geq 3$.

Assuming a constant material distribution through a finite element domain Ω_e, i.e., $\gamma(x_1, x_2) = \gamma_e$ for $x_1, x_2 \in \Omega_e$, the stiffness matrix from Eq. (9.52) can be

[11] In practice, $\gamma = 0$ is approximated by a small non-zero value $0 < \epsilon \ll 1$ to avoid numerical problems.

reformulated as

$$\boldsymbol{K} = \sum_{e=1}^{N} \gamma_e^p \boldsymbol{K}_e. \tag{9.54}$$

With the established stiffness matrix parametrization (Eq. (9.54)) and the state equation (Eq. (9.51)), the optimization problem can be formulated. The cost function C is the compliance

$$C = c = \boldsymbol{F}^\intercal \boldsymbol{u}. \tag{9.55}$$

The cost function C is to be minimized with respect to the densities γ_e under several constraints, namely the volume constraint with volume fraction $\tilde{V}$

$$V = \sum_{e=1}^{N} v_e \gamma_e \leq \tilde{V} \tag{9.56}$$

and the acceptable range of the element material density

$$\epsilon \leq \gamma_e \leq 1, e = 1, \dots, N. \tag{9.57}$$

These constraints are best taken into account through constrained optimization algorithms, such as optimality criterion methods (see, e.g., [61]), the method of moving asymptotes [74], or the globally convergent method of moving asymptotes [75]. Alternatively—in order to allow a neural network ansatz, the problem can be transformed to an unconstrained optimization problem via penalization of the constraints, as seen in [51, 71, 72, 89]

$$\tilde{C} = c + \kappa \left(\sum_{e=1}^{N} v_e \gamma_e - \tilde{V} \right)^2. \tag{9.58}$$

The constraint's importance is handled through the parameter κ, which is typically increased during the optimization to ensure fulfillment of the constraints at the end while avoiding initial overconstraining of the problem. Equation (9.57) is ensured by clamping the values γ_e in the range $[\epsilon, 1]$ after each gradient descent update.

With the augmented cost function (Eq. (9.58)), i.e., the unconstrained optimization problem, the gradient descent algorithms from Sect. 2.8.1 can readily be applied. Only the sensitivity with respect to the densities γ_e

$$\frac{\partial \tilde{C}}{\partial \gamma_e} = -p\gamma_e^{p-1} \boldsymbol{u}^\intercal \boldsymbol{K}_e \boldsymbol{u} + \left(\sum_{e=1}^{N} v_e \gamma_e - \tilde{V} \right) v_e \tag{9.59}$$

is needed.

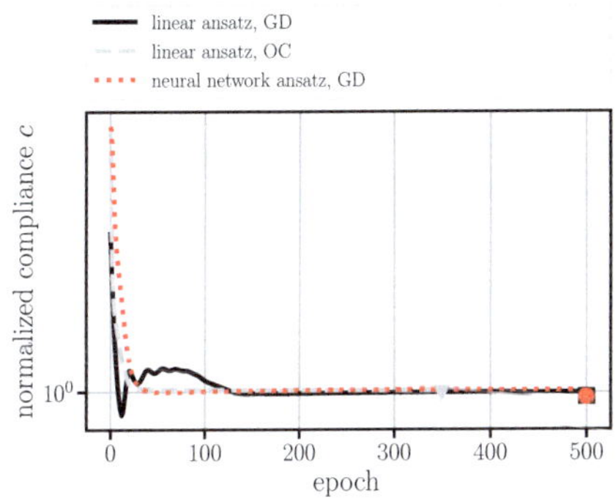

(a) Compliance normalized by the best achieved compliance ~187.7 (before thresholding)

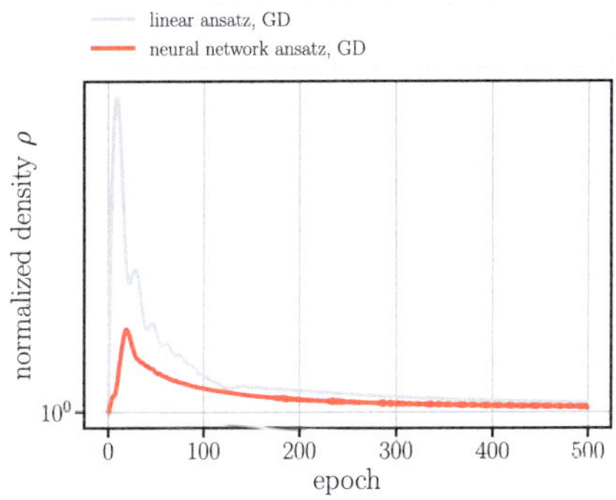

(b) Density normalized by $V = 0.5$

Fig. 9.31 Learning history of the three topology optimization approaches (gradient descent (GD) with linear and neural network ansatz and the optimality criterion (OC)) for the compliance problem from Fig. 9.30

Table 9.3 Attained compliance and density values

	Linear ansatz, GD	Neural network ansatz, GD	Linear ansatz, OC
c before thresholding	~187.7	~190.3	~189.3
c after thresholding	~182.1	~181.5	~185.2
V after thresholding	0.502	0.500	0.500

Conventionally, the densities γ_e are optimized directly[12] (linear ansatz). However, as announced, a neural network may be employed to predict $\hat{\gamma}_e$ (neural network ansatz). The impact of this reparametrization is illustrated by Fig. 9.31 showcasing the learning histories for a volume fraction of $\tilde{V} = 0.5$. Initially, the neural network improves faster than the linear ansatz with gradient descent, which can be seen for both compliance and density evolution. As the epochs progress, the difference in performance decreases and the attained designs (see Fig. 9.31 for the thresholded designs[13]) have similar performances, summarized in Table 9.3. Furthermore, the advantage during the early stages of optimization is diminished by considering a state-of-the-art optimizer for the problem at hand, namely the optimality criterion method (Fig. 9.32).

Nevertheless, Table 9.3 indicates that the neural network achieves a slightly better design than the linear ansatz. However, this result depends on the neural network initialization and should therefore be assessed statistically. By performing multiple

---

[12] After a filtering [76] to avoid mesh-dependent solutions with too small and, thereby, unmanufacturable solutions. Here, a filter radius twice as large as a single finite element is employed for all investigations.

[13] After the optimization, gray values persist and are removed by thresholding to a binary design. The markers indicate the compliances after thresholding in Fig. 9.31a.

(a) With linear ansatz, $c = {\sim}182.1$ after thresholding (b) With neural network ansatz, $c = {\sim}181.5$ after thresholding

Fig. 9.32 Optimized structures after 500 epochs of gradient descent

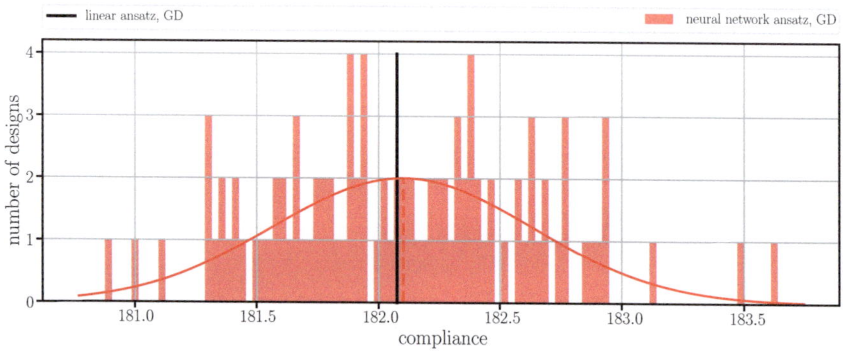

Fig. 9.33 Statistical evaluation of the compliance values using 100 samples for the neural network ansatz, relying on different initializations

optimizations with different neural network initializations (by varying the random seed), Fig. 9.33 is obtained, namely the distribution of the achieved compliances in the form of a histogram. Although the neural network outperforms the linear ansatz in some cases, the performance is, on average, similar. Furthermore, the variation in design quality ($< 1.5\%$) is limited.

Thus, a neural network ansatz is not beneficial (nor detrimental) in the compliance problem, which is in line with the observations from [48, 50, 51]. However, as seen in Sect. 9.2.4, for more complex optimization problems, the neural network ansatz is more promising [12, 73]. Therefore, we move on to the more challenging problem of acoustic topology optimization [77, 78], which is more prone to end in premature local optima.

9.3.3.2 Acoustic Topology Optimization

Figure 9.34 provides a possible application of acoustic topology optimization (or acoustic design), in which a ceiling in Ω_c is to be optimized to reduce the wave pressure u in a rectangular domain Ω_s. The wave pressures in the entire domain Ω

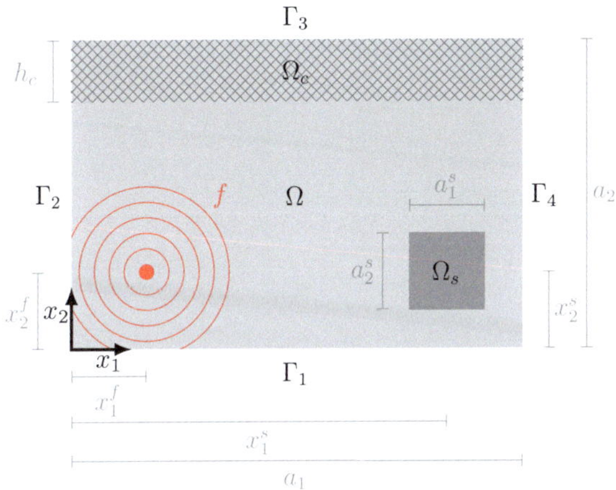

Fig. 9.34 Topology optimization of a ceiling Ω_c of height h_c in order to suppress the sound pressure in the domain Ω_s from a harmonic source p at x_1^f, x_2^f with frequency f. The suppressed domain is rectangular with dimensions a_s, b_s and centered at x_1^s, x_2^s. The domain Ω with dimensions $a \times b$, including Ω_c and Ω_s, is governed by the acoustic wave equation. The Neumann boundary conditions consist of $\Gamma_{Nx_1} = \Gamma_1 \cup \Gamma_3$ and $\Gamma_{Nx_2} = \Gamma_2 \cup \Gamma_4$. Reproduced from [12]

are induced by a harmonic source p with angular frequency $2\pi f = \omega$ [77]. Due to the periodic excitation, the problem is governed by the Helmholtz equation, given in a normalized form

$$\nabla \cdot (\tilde{\rho}^{-1} \nabla u(x)) + i\tilde{\omega}\eta_d \tilde{\kappa}^{-1} u(x) + \tilde{\omega}^2 \tilde{\kappa}^{-1} u(x) = p(x). \tag{9.60}$$

The angular excitation frequency $\tilde{\omega} = \omega / \sqrt{\kappa_1/\rho_1}$ is normalized by the mass density ρ_1, and bulk modulus κ_1 of air. The normalized inverse mass density $\tilde{\rho}^{-1}(\gamma(x))$, and bulk modulus $\tilde{\kappa}^{-1}(\gamma(x))$ are parametrized by an indicator $0 \leq \gamma \leq 1$, as

$$\tilde{\rho}^{-1}(\gamma) = 1 + \gamma \left(\frac{\rho_1}{\rho_2} - 1 \right), \tag{9.61}$$

$$\tilde{\kappa}^{-1}(\gamma) = 1 + \gamma \left(\frac{\kappa_1}{\kappa_2} - 1 \right), \tag{9.62}$$

where ρ_2 is the mass density, and κ_2 the bulk modulus of the solid material—which in this study is aluminum; see [12] for more details on the simulation setup including the material parameters. The distribution of the material is defined by the spatially distributed indicator $\gamma(x)$, which represents air when $\gamma(x) = 0$ and solid material when $\gamma(x) = 1$. In addition to Eq. (9.60), reflecting boundary conditions, i.e., homogeneous Neumann boundary conditions are assumed

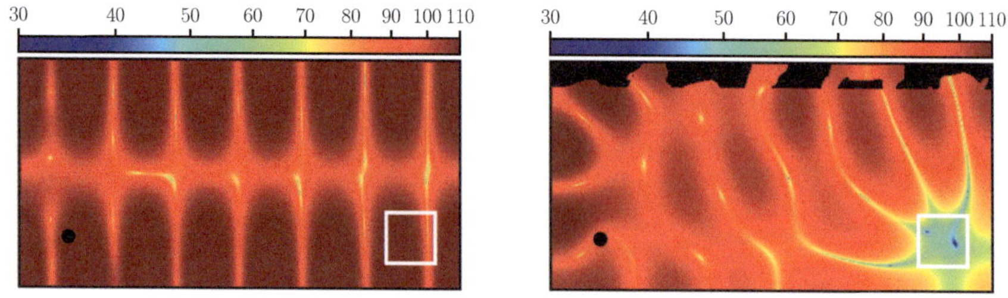

(a) Without optimized ceiling, $L_p = 110.0$ dB (b) With optimized ceiling, $L_p = 61.3$ dB

Fig. 9.35 Sound pressure level distribution in dB at $f = 69.43$ Hz without and with an optimized ceiling. Reproduced from [12]

$$\frac{\partial u(x)}{\partial x_1} = 0 \quad \text{on } \Gamma_{N x_2} \times \mathcal{T}, \tag{9.63}$$

$$\frac{\partial u(x)}{\partial x_2} = 0 \quad \text{on } \Gamma_{N x_1} \times \mathcal{T}. \tag{9.64}$$

The quality of a design given by $\gamma(x)$ is determined by the following cost function

$$C = \frac{1}{\int_{\Omega_s} d\Omega_s} \int_{\Omega_s} |u|^2 d\Omega_s, \tag{9.65}$$

which, as an unconstrained optimization problem, can be minimized via a gradient descent approach. Here, Adam [13] is utilized. An exemplary optimization is illustrated in Fig. 9.35, where the sound pressure level in the area Ω_s is reduced from $L_p = 110.0$ dB to $L_p = 61.3$ dB by an optimized ceiling.

To employ the neural network ansatz, minor pretraining is necessary to ensure that filtering [76] and projection [79–81] work adequately. The pretraining (based on only four samples) ensures that the initial guesses are already close to a binary 0/1 design; see [12] for more details.

Relying on this pretrained neural network ansatz, a comparison with the linear ansatz is conducted for the frequencies $f = 21.33$ Hz, $f = 34.39$ Hz, $f = 57.22$ Hz, $f = 69.43$ Hz, $f = 76.30$ Hz, $f = 95.37$ Hz, $f = 141.78$ Hz, and $f = 166.56$ Hz.[14] In a generalization case—relying on a different problem setup for the pretraining—improvements of correspondingly 13.6 dB, 24 dB, 10.1 dB, 2.1 dB, 29.6 dB, 22.8 dB, 1.4 dB, and 1.9 dB are obtained. As different designs can be obtained through different initializations of the neural network and the linear ansatz, a statistical assessment is provided to ensure that the neural network is truly consistently beneficial. By performing 30 different pretrainings with the same four training samples and using 30 different homogeneous initializations in the range [0, 1] for the linear ansatz, the histograms from Fig. 9.36 are obtained. The sound pressure levels achieved with the neural network ansatz are indicated in blue, while red stands for the linear ansatz. For

[14] The results can be reproduced with the code from [95].

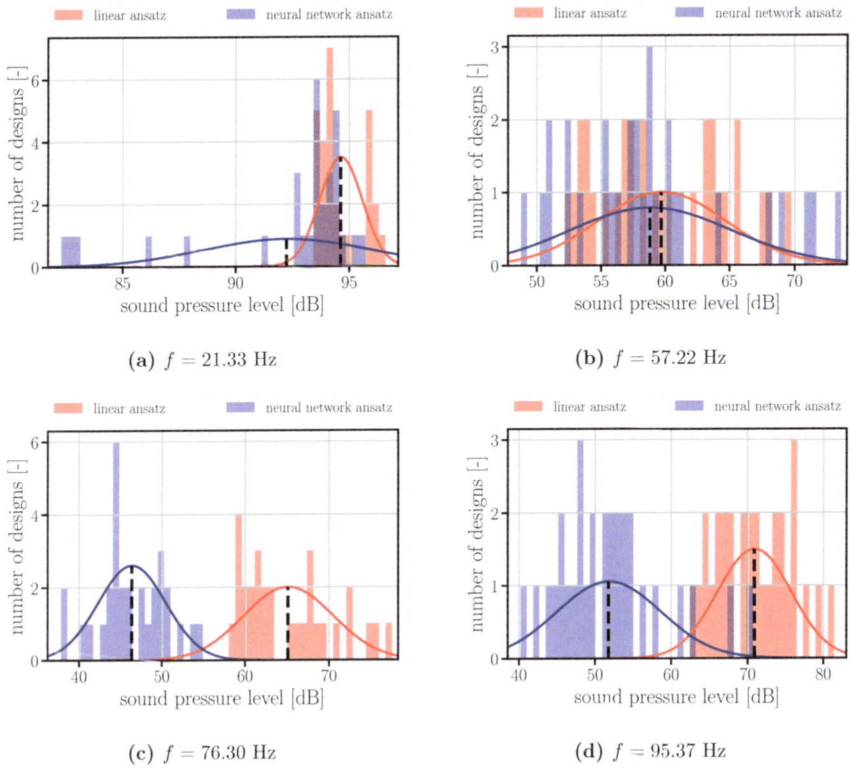

Fig. 9.36 Statistical evaluation of the local optima quality with 30 samples for the linear and neural network ansatz—relying on different initializations. In the linear case, we used 30 homogeneous initial guesses equally spaced between 0 and 1, while the neural network relies on 30 different pretrainings. Reproduced from [12]

$f = 21.33$ Hz and $f = 57.22$ Hz, given in Figs. 9.36a and 9.36b, the distributions mostly overlap. However, in both cases, the best designs are obtained with the neural network—and in Fig. 9.36a, outliers in the neural network distribution represent an improvement of $\sim$10 dB over the linear ansatz. Figure 9.36c, d with $f = 76.30$ Hz, and $f = 95.37$ Hz demonstrate a clear separation in design quality distribution. This also corresponds to the cases with the greatest improvement through a neural network ansatz. The overlaps of the distributions are minimal, and in these scenarios, the neural network ansatz consistently outperforms the linear ansatz [12]. This showcases the potential of a neural network ansatz within the unconstrained optimization problem of acoustic topology optimization.

Exercises

E.31 Neural Network Ansatz on Optimization Benchmarks (C)

This exercise revisits the example from Sect. 9.1.2.2. Four basic optimization benchmarks, the Rosenbrock, Rastrigrin, Ackley, and Levy functions, are to be optimized with a linear ansatz and a neural network ansatz. To this end, an implementation is provided in `exercise 31 .ipynb` with which the functions can be minimized. Study and understand the implementation. Six optimizers are provided, namely gradient descent, gradient descent with momentum, AdaGrad, RMSprop, Adam, and L-BFGS (cf. Sect. 2.8), which can be selected via `selectOptimizer`. Investigate the performance of the linear and neural network ansatz for the different optimization benchmarks (interchangeable through `benchmark`) using the six optimizers. Perform statistical analyses to determine when the neural network ansatz is beneficial and when it is not.

E.32 Physics-Informed Neural Networks for Inverse Problems (C)

A physics-informed neural network for one-dimensional full waveform inversion with full domain knowledge is provided in `exercise 32 .ipynb`. The exact problem setup is defined in `measurement1DFWI/settings.csv`. Your goal is to study the implementation and compare the three ansatz spaces: a piece-wise constant ansatz (linear), a fully connected neural network, and a convolutional neural network. These can be specified with `selectModel`, whereupon tuned (ansatz-specific) hyperparameters are selected. Which ansatz converges the fastest, and which approach is the most robust? If an approach is unsuccessful, try increasing the number of epochs.

E.33 Iterative Forward Solver (C)

An iterative forward solver for one-dimensional full waveform inversion is presented in `exercise 33 .ipynb`. The problem setup is again defined in `measurement1DFWI/settings.csv`. Two measurements are performed. A source is emitted at each of the two boundaries. The sensor locations are specified with `sensorPositions = (0, −1)`. In this case, the first and last indices, i.e., the boundaries, are selected. Again, the ansatz space can be changed. Compare the effect of the three ansatz spaces. Try changing the hyperparameters and observe if the inversion process improves or not. Lastly, study the training loop. What are the main differences from the previously encountered training loops and why is it more expensive?

E.34 Data-Driven Solver (C)

Lastly, a data-driven solver implementation is studied. Again, the problem specifications are provided in `dataset1DFWI/settings.csv`.

1. To this end, a dataset has to be generated first. Run the `exercise 34 _1.ipynb` file, which produces random material distributions and solves the wave equation with the finite difference implementation.
2. Before training the data-driven solver, we want to reduce the dimensionality of the measurements using singular value decomposition. The measurements currently have the shape: number of measurements × number of sensors × number of sources × number of recorded time steps. The model order reduction is performed with `exercise 34 _2.ipynb`. Is it more reasonable to use the reduced basis of the wave velocity distribution $c(x)$ or of the wave field measurements $u(x, t)$?
3. Finally, the data-driven solver is to be trained with `exercise 34 _3.ipynb`. Two options are provided. The first is a fully connected neural network, which relies on the reduced basis for the wave field measurements and the parametrization of the wave velocity from Fig. 9.15. The second option is a convolutional neural network. You can look up the exact neural network architectures in the implemented classes `FNN` and `CNN`. Compare the two approaches. Why is it easier to train the fully connected neural network despite all the advantages of a convolutional neural network? Perform a hyperparameter optimization in order to improve the predictions. The results can further be improved by increasing the number of samples in the `dataset1DFWI/settings.csv`. Remember to rerun the data generation and model order reduction.

E.35 Finite Difference Method with ρ-Scaling (P)

Consider the modified formulation of the scalar wave equation (Eqs. (9.9)) in (9.43). Modify the finite difference scheme presented in Algorithm 14, such that ρ-scaling is taken into account. Provide an explicit expression in the style of Eq. (9.27) with one spatial dimension.

References

1. Rafael, C. Gonzalez, and Richard E. Woods. 2018. *Digital image processing*. eng. 4th edition, global edition. New York, NY: Pearson. ISBN: 978-1-292-22304-9.
2. Nicholas, Geneva, and Nicholas Zabaras. 2020. Modeling the dynamics of PDE systems with physics-constrained deep auto-regressive networks. *Journal of Computational Physics 403*: 109056. tex.ids= Geneva, Geneva2020a arXiv: 1906.05747. ISSN: 00219991. https://doi.org/10.1016/j.jcp.2019.109056.

3. Maziar, Raissi. 2018. *Deep hidden physics models: Deep learning of nonlinear partial differential equations.*
4. Andreas, Fichtner, and Florian Bleibinhaus. 2011. Full seismic waveform modelling and inversion. eng. In: *Advances in geophysical and environmental mechanics and mathematics*. Berlin, Heidelberg: Springer. ISBN: 978-3-642-15807-0 978-3-642-15806-3.
5. Martin, P. Bendsøe, and O. Sigmund. 2003. *Topology optimization: Theory, methods, and applications*. Berlin, New York: Springer. ISBN: 978-3-540-42992-0.
6. Ole, Sigmund. 2011. On the usefulness of non-gradient approaches in topology optimization. en. In: *Structural and Multidisciplinary Optimization 43.5*: 589–596. ISSN: 1615-1488. https://doi.org/10.1007/s00158-011-0638-7.
7. Leon, Herrmann, and Stefan Kollmannsberger. 2024. Deep learning in computational mechanics: a review. en. In: *Computational Mechanics*, Jan 2024. ISSN: 1432-0924. https://doi.org/10.1007/s00466-023-02434-4.
8. Jorge, Nocedal, and Stephen J. Wright. 2006. *Numerical optimization*, 2nd ed. Springer series in operations research. OCLC: ocm68629100. New York: Springer. ISBN: 978-0-387-30303-1.
9. Plessix, R.-E. 2006. A review of the adjoint-state method for computing the gradient of a functional with geophysical applications. en. *Geophysical Journal International 167.2*: 495–503. ISSN: 0956540X, 1365246X. https://doi.org/10.1111/j.1365-246X.2006.02978.x.
10. Dan, Givoli. 2021. A tutorial on the adjoint method for inverse problems. en. *Computer Methods in Applied Mechanics and Engineering 380*: 113810. ISSN: 00457825. https://doi.org/10.1016/j.cma.2021.113810.
11. Kurt, Hornik, Maxwell Stinchcombe, and Halbert White. 1989. Multilayer feedforward networks are universal approximators. en. In: *Neural Networks 2.5*: 359–366. tex.ids= Hornik1989a. ISSN: 0893-6080. https://doi.org/10.1016/0893-6080(89)90020-8.
12. Leon, Herrmann et al. 2024. *Neural networks for generating better local optima in topology optimization [Software]*, July 2024. https://doi.org/10.5281/zenodo.12806412.
13. Diederik, P. Kingma, and Jimmy Ba. 2017. Adam: A method for stochastic optimization, Jan 2017. arXiv:1412.6980 [cs].
14. Yue, Wu, Youzuo Lin, and Zheng Zhou. 2018. Inversionet: Accurate and efficient seismic-waveform inversion with convolutional neural networks. en. In: *SEG technical program expanded abstracts 2018*. Anaheim, California: Society of Exploration Geophysicists, 2096–2100, Aug 2018. https://doi.org/10.1190/segam2018-2998603.1.
15. Wenlong, Wang, Fangshu Yang, and Jianwei Ma. 2018. Velocity model building with a modified fully convolutional network. en. In: *SEG technical program expanded abstracts 2018*. Anaheim, CA: Society of Exploration Geophysicists, pp. 2086–2090, Aug 2018. https://doi.org/10.1190/segam2018-2997566.1.
16. York, Zheng et al. 2019. Applications of supervised deep learning for seismic interpretation and inversion. en. *The Leading Edge 38.7*: 526–533. ISSN: 1070-485X, 1938-3789. https://doi.org/10.1190/tle38070526.1.
17. Mauricio, Araya-Polo, Stuart Farris, and Manuel Florez. 2019. Deep learning-driven velocity model building workflow. en. *The Leading Edge 38.11*: 872a1–872a9. ISSN: 1070-485X, 1938-3789. https://doi.org/10.1190/tle38110872a1.1.
18. Vishal Das et al. 2019. Convolutional neural network for seismic impedance inversion. en. *GEOPHYSICS 84.6*: R869–R880. ISSN: 0016-8033, 1942-2156. https://doi.org/10.1190/geo2018-0838.1.
19. Wenlong, Wang, and Jianwei Ma. 2020. Velocity model building in a crosswell acquisition geometry with image-trained artificial neural networks. en. *GEOPHYSICS 85.2*: U31–U46. ISSN: 0016-8033, 1942-2156. https://doi.org/10.1190/geo2018-0591.1.
20. Bangyu, Wu et al. 2020. Seismic impedance inversion using fully convolutional residual network and transfer learning. *IEEE Geoscience and Remote Sensing Letters 17.12*: 2140–2144. ISSN: 1558-0571. https://doi.org/10.1109/LGRS.2019.2963106.

21. Min, Jun Park, and Mauricio D. Sacchi. 2020. Automatic velocity analysis using convolutional neural network and transfer learning. en. *GEOPHYSICS 85.1*: V33–V43. ISSN: 0016-8033, 1942-2156. https://doi.org/10.1190/geo2018-0870.1.
22. Jiaxing, Ye, and Nobuyuki Toyama. 2022. Automatic defect detection for ultrasonic wave propagation imaging method using spatio-temporal convolution neural networks. en. *Structural Health Monitoring 21.6*: 2750–2767. ISSN: 1475-9217, 1741-3168. https://doi.org/10.1177/14759217211073503.
23. Jing, Rao et al. 2023. Quantitative reconstruction of defects in multi-layered bonded composites using fully convolutional network-based ultrasonic inversion. en. *Journal of Sound and Vibration 542*: 117418. ISSN: 0022-460X. https://doi.org/10.1016/j.jsv.2022.117418.
24. Qiyin, Lin et al. 2018. Investigation into the topology optimization for conductive heat transfer based on deep learning approach. en. *International Communications in Heat and Mass Transfer 97*: 103–109. ISSN: 0735-1933. https://doi.org/10.1016/j.icheatmasstransfer.2018.07.001.
25. Yonggyun, Yu et al. 2019. Deep learning for determining a near-optimal topological design without any iteration. en. *Structural and Multidisciplinary Optimization 59.3*: 787–799. ISSN: 1615-1488. https://doi.org/10.1007/s00158-018-2101-5.
26. Diab, W. Abueidda, Seid Koric, and Nahil A. Sobh. 2020. Topology optimization of 2D structures with nonlinearities using deep learning. *Computers & Structures 237*: 106283. ISSN: 00457949. https://doi.org/10.1016/j.compstruc.2020.106283.
27. Keigo, Nakamura, and Yoshiro Suzuki. 2020. *Deep learning-based topological optimization for representing a user-specified design area*, Apr 2020. https://doi.org/10.48550/arXiv.2004.05461.
28. Yiquan, Zhang et al. 2020. *A deep Convolutional Neural Network for topology optimization with strong generalization ability*, Mar 2020. arXiv:1901.07761 [cs, stat]. https://doi.org/10.48550/arXiv.1901.07761.
29. Shuai, Zheng, Zhenzhen He, and Honglei Liu. 2021. Generating three-dimensional structural topologies via a U-Net convolutional neural network. en. *Thin-Walled Structures 159*: 107263. ISSN: 0263-8231. https://doi.org/10.1016/j.tws.2020.107263.
30. Shuai, Zheng et al. 2021. Accurate and real-time structural topology prediction driven by deep learning under moving morphable component-based framework. en. *Applied Mathematical Modelling 97*: 522–535. ISSN: 0307-904X. https://doi.org/10.1016/j.apm.2021.04.009.
31. Dalei, Wang et al. 2022. A deep convolutional neural network for topology optimization with perceptible generalization ability. *Engineering Optimization 54.6*: 973–988. ISSN: 0305-215X. https://doi.org/10.1080/0305215X.2021.1902998.
32. Jun, Yan et al. 2022. Deep learning driven real time topology optimisation based on initial stress learning. en. *Advanced Engineering Informatics 51*: 101472. ISSN: 1474-0346. https://doi.org/10.1016/j.aei.2021.101472.
33. Junhyeon, Seo, and Rakesh K. Kapania. 2023. Topology optimization with advanced CNN using mapped physics-based data. en. *Structural and Multidisciplinary Optimization 66.1*: 21. tex.ids= Seo2023. ISSN: 1615-1488. https://doi.org/10.1007/s00158-022-03461-0.
34. Genki, Yagawa, and Atsuya Oishi. 2021. *Computational mechanics with neural networks*. en. Lecture Notes on Numerical Methods in Engineering and Sciences. Cham: Springer International Publishing. ISBN: 978-3-030-66110-6 978-3-030-66111-3. https://doi.org/10.1007/978-3-030-66111-3.
35. Ian, Goodfellow, Yoshua Bengio, and Aaron Courville. 2016. *Deep learning*. MIT Press. ISBN: 0-262-03561-8.
36. Aurélien, Géron. 2023. *Hands-on machine learning with Scikit-Learn, Keras and TensorFlow: Concepts, tools, and techniques to build intelligent systems*. 3rd ed. OCLC: on1346503549. Beijing: O'Reilly Media, Inc. ISBN: 978-1-09-812597-4.
37. Leon, Herrmann et al. 2023. On the use of neural networks for full waveform inversion. *Computer Methods in Applied Mechanics and Engineering 415*, 116278. ISSN: 0045-7825. https://doi.org/10.1016/j.cma.2023.116278.
38. Robert, Seidl. 2018. Full waveform inversion for nondestructive ultrasonic testing. Doctoral dissertation. Technische Universität München.

39. Oishi, A. et al. 2001. Neural network-based inverse analysis for defect identification with Laser Ultrasonics. *Research in Nondestructive Evaluation 13.2*: 79–96. tex.ids= Oishi2001. ISSN: 1432-2110. https://doi.org/10.1007/s00164-001-0010-1.
40. Salsa, S. 2015. *Partial differential equations in action: From modelling to theory*, 2nd ed. Unitext volume 86. OCLC: ocn910078747. Cham, New York: Springer. ISBN: 978-3-319-15092-5.
41. Tim, Bürchner et al. 2023. Immersed boundary parametrizations for full waveform inversion. en. *Computer Methods in Applied Mechanics and Engineering 406*: 115893. ISSN: 00457825. https://doi.org/10.1016/j.cma.2023.115893.
42. Hans, Petter Langtangen, Svein Linge, and Springer International Publishing AG. 2018. *Finite difference computing with PDEs a modern software approach*. English. OCLC: 1049590683. ISBN: 978-3-319-85666-7.
43. Daniel, Rabinovich, Dan Givoli, and Eli Turkel. 2023. Single-field identification of inclusions and cavities in an elastic medium. en. *International Journal for Numerical Methods in Engineering*, nme.7364. ISSN: 0029-5981, 1097-0207. https://doi.org/10.1002/nme.7364.
44. Stephan, Hoyer, Jascha Sohl-Dickstein, and Sam Greydanus. 2019. *Neural reparameterization improves structural optimization*. https://doi.org/10.48550/arXiv.1909.04240.
45. Kailai, Xu, and Eric Darve. 2019. *The neural network approach to inverse problems in differential equations*, Jan 2019. https://doi.org/10.48550/arXiv.1901.07758.
46. Philipp, Holl, Vladlen Koltun, and Nils Thuerey. 2020. *Learning to control PDEs with differentiable physics*, Jan 2020. arXiv:2001.07457 [physics, stat]. https://doi.org/10.48550/arXiv.2001.07457.
47. Jens, Berg and Kaj Nyström. 2021. Neural networks as smooth priors for inverse problems for PDEs. en. *Journal of Computational Mathematics and Data Science 1*: 100008. tex.ids= Berg2021a. ISSN: 27724158. https://doi.org/10.1016/j.jcmds.2021.100008.
48. Liang, Chen, and Mo-How Herman Shen. 2021. A new topology optimization approach by physics-informed deep learning process. *Advances in Science, Technology and Engineering Systems Journal 6.4*: 233–240. ISSN: 24156698, 24156698. https://doi.org/10.25046/aj060427.
49. Alex, Halle, Lucio Flavio Campanile, and Alexander Hasse. 2021. An artificial intelligence–assisted design method for topology optimization without pre-optimized training data. en. *Applied Sciences 11.19*: 9041. ISSN: 2076-3417. https://doi.org/10.3390/app11199041.
50. Hao, Deng, and Albert C. To. 2020. Topology optimization based on deep representation learning (DRL) for compliance and stress-constrained design. en. *Computational Mechanics 66.2*: 449–469. ISSN: 1432-0924. https://doi.org/10.1007/s00466-020-01859-5.
51. Aaditya, Chandrasekhar, and Krishnan Suresh. TOuNN: Topology optimization using neural networks. en. *Structural and Multidisciplinary Optimization 63.3*: 1135–1149. ISSN: 1615-1488. https://doi.org/10.1007/s00158-020-02748-4.
52. Aaditya, Chandrasekhar, and Krishnan Suresh. *Length scale control in topology optimization using fourier enhanced neural networks*, Sept 2021. https://doi.org/10.48550/arXiv.2109.01861.
53. Ana, P. O. Muller et al. 2023. Deep pre-trained FWI: Where supervised learning meets the physics-informed neural networks. en. *Geophysical Journal International 235.1*: 119–134. ISSN: 0956-540X, 1365-246X. https://doi.org/10.1093/gji/ggad215.
54. Stefan, Kollmannsberger, Divya Singh, and Leon Herrmann. 2023. Transfer learning enhanced full waveform inversion. In: *2023 IEEE/ASME International Conference on Advanced Intelligent Mechatronics (AIM)*, 866–871, June 2023. Seattle, WA, USA: IEEE. ISBN: 978-1-66547-633-1. https://doi.org/10.1109/AIM46323.2023.10196158.
55. Divya, Shyam Singh et al. 2024. *Accelerating Full Waveform Inversion By Transfer Learning*, Aug 2024. arXiv:2408.00695 [cs]. https://doi.org/10.48550/arXiv.2408.00695.
56. Sinno, Jialin Pan and Qiang Yang. 2010. A survey on transfer learning. *IEEE Transactions on Knowledge and Data Engineering 22.10*: 1345–1359. ISSN: 1558-2191. https://doi.org/10.1109/TKDE.2009.191.
57. Jason, Yosinski et al. 2014. How transferable are features in deep neural networks? In: *Proceedings of the 27th international conference on neural information processing systems* 2, 3320–3328, Dec 2014. NIPS'14. Cambridge, MA, USA: MIT Press.

58. Peng, Yan et al. 2024. A comprehensive survey of deep transfer learning for anomaly detection in industrial time series: Methods, applications, and directions. *IEEE Access 12*: 3768–3789. ISSN: 2169-3536. https://doi.org/10.1109/ACCESS.2023.3349132.
59. Olaf, Ronneberger, Philipp Fischer, and Thomas Brox. 2015. U-Net: Convolutional networks for biomedical image segmentation. en. In: *Medical Image Computing and Computer-Assisted Intervention—MICCAI 2015*. Ed. by Nassir, Navab et al. Lecture Notes in Computer Science. Cham: Springer International Publishing, 234–241. ISBN: 978-3-319-24574-4. https://doi.org/10.1007/978-3-319-24574-4_28.
60. O. Sigmund. 2001. A 99 line topology optimization code written in Matlab. en. *Structural and Multidisciplinary Optimization 21.2*: 120–127. ISSN: 1615-1488. https://doi.org/10.1007/s001580050176.
61. Martin, P. Bendsøe and Ole Sigmund. 2004. *Topology Optimization*. en. Berlin, Heidelberg: Springer. ISBN: 978-3-642-07698-5 978-3-662-05086-6. https://doi.org/10.1007/978-3-662-05086-6.
62. Yatin, Nandwani et al. 2019. A primal dual formulation for deep learning with constraints. In: *Advances in Neural Information Processing Systems*, vol 32. Curran Associates, Inc.
63. Atılım, Gunes Baydin et al. 2018. Automatic differentiation in machine learning: A survey. en. *Journal of Machine Learning Research*, 43.
64. Hyogu, Jeong et al. 2023. A complete Physics-Informed Neural Network-based framework for structural topology optimization. en. *Computer Methods in Applied Mechanics and Engineering 417*: 116401. ISSN: 00457825. https://doi.org/10.1016/j.cma.2023.116401.
65. Somdatta, Goswami et al. 2019. Transfer learning enhanced physics informed neural network for phase-field modeling of fracture. en, July 2019. arXiv:1907.02531 [cs, stat]. tex.ids= Goswami2019a.
66. Junyan, He et al. 2023. Deep energy method in topology optimization applications. en. *Acta Mechanica 234.4*: 365–1379. ISSN: 1619-6937. https://doi.org/10.1007/s00707-022-03449-3.
67. Rebekka, V. Woldseth et al. 2022. On the use of artificial neural networks in topology optimisation. en. *Structural and Multidisciplinary Optimization 65.10*: 294. ISSN: 1615-1488. https://doi.org/10.1007/s00158-022-03347-1.
68. Qinglong, He, and Yanfei Wang. 2021. Reparameterized full-waveform inversion using deep neural networks. en. *GEOPHYSICS 86.1*: V1–V13. ISSN: 0016-8033, 1942-2156. https://doi.org/10.1190/geo2019-0382.1.
69. Weiqiang, Zhu et al. 2022. Integrating deep neural networks with full-waveform inversion: Reparameterization, regularization, and uncertainty quantification. en. *GEOPHYSICS 87.1*: R93–R109. ISSN: 0016-8033, 1942-2156. https://doi.org/10.1190/geo2020-0933.1.
70. Peng, Jiang et al. 2024. Full waveform inversion based on inversion network reparameterized velocity. en. *Geophysical Prospecting 72.1*: 52–67. ISSN: 0016-8025, 1365-2478. https://doi.org/10.1111/1365-2478.13292.
71. Aaditya, Chandrasekhar, and Krishnan Suresh. 2022. Approximate length scale filter in topology optimization using fourier enhanced neural networks. *Computer-Aided Design 150*: 103277. ISSN: 0010-4485. https://doi.org/10.1016/j.cad.2022.103277.
72. Connor, N. Mallon et al. 2024. *Neural level set topology optimization using unfitted finite elements*, Feb 2024. arXiv:2303.13672 [cs].
73. Suryanarayanan, Manoj Sanu, Alejandro M. Aragon, and Miguel A. Bessa. 2024. *Neural topology optimization: The good, the bad, and the ugly*, July 2024. arXiv:2407.13954 [cs, math]. https://doi.org/10.48550/arXiv.2407.13954.
74. Krister, Svanberg. 1987. The method of moving asymptotes—a new method for structural optimization. en. *International Journal for Numerical Methods in Engineering 24.2*: 359–373. ISSN: 0029-5981, 1097-0207. https://doi.org/10.1002/nme.1620240207.
75. Krister, Svanberg. 2002. A class of globally convergent optimization methods based on conservative convex separable approximations. en. *SIAM Journal on Optimization 12.2*: 555–573. ISSN: 1052-6234, 1095-7189. https://doi.org/10.1137/S1052623499362822.
76. Sigmund, O., and J. Petersson. 1998. Numerical instabilities in topology optimization: A survey on procedures dealing with checkerboards, mesh-dependencies and local minima". en. *Structural optimization 16.1*: 68–75. ISSN: 1615-1488. https://doi.org/10.1007/BF01214002.

77. Maria, B. Dühring, Jakob S. Jensen, and Ole Sigmund. 2008. Acoustic design by topology optimization. en. *Journal of Sound and Vibration 317.3-5*: 557–575. ISSN: 0022460X. https://doi.org/10.1016/j.jsv.2008.03.042.
78. Rasmus, E. Christiansen et al. 2015. Creating geometrically robust designs for highly sensitive problems using topology optimization: Acoustic cavity design. en. *Structural and Multidisciplinary Optimization 52.4*: 737–754. ISSN: 1615-147X, 1615-1488. https://doi.org/10.1007/s00158-015-1265-5.
79. Guest, J. K., J. H. Prévost, and T. Belytschko. 2004. Achieving minimum length scale in topology optimization using nodal design variables and projection functions. en. *International Journal for Numerical Methods in Engineering 61.2*: 238–254. ISSN: 0029-5981, 1097-0207. https://doi.org/10.1002/nme.1064.
80. Fengwen, Wang, Boyan Stefanov Lazarov, and Ole Sigmund. 2011. On projection methods, convergence and robust formulations in topology optimization. en. *Structural and Multidisciplinary Optimization 43.6*: 767–784. ISSN: 1615-1488. https://doi.org/10.1007/s00158-010-0602-y.
81. Lei, Li, and Kapil Khandelwal. 2015. Volume preserving projection filters and continuation methods in topology optimization. *Engineering Structures 85*: 144–161. ISSN: 0141-0296. https://doi.org/10.1016/j.engstruct.2014.10.052.
82. M. Raissi, P. Perdikaris, and G.E. Karniadakis. 2019. Physics-informed neural networks: A deep learning framework for solving forward and inverse problems involving nonlinear partial differential equations. en. *Journal of Computational Physics 378*: 686–707. ISSN: 00219991. https://doi.org/10.1016/j.jcp.2018.10.045.
83. Zichao, Long, et al. 2018. PDE-Net: Learning PDEs from data. en. In: *Proceedings of the 35th international conference on machine learning*, 3208–3216, July 2018. PMLR.
84. Zichao, Long, Yiping Lu, and Bin Dong. 2019. PDE-Net 2.0: Learning PDEs from data with a numeric-symbolic hybrid deep network. en. *Journal of Computational Physics 399*: 108925. ISSN: 0021-9991. https://doi.org/10.1016/j.jcp.2019.108925.
85. Yinhao, Zhu et al. 2019. Physics-Constrained deep learning for high-dimensional surrogate modeling and uncertainty quantification without labeled data. en. *Journal of Computational Physics 394*. arXiv: 1901.06314, pp. 56–81. ISSN: 00219991. https://doi.org/10.1016/j.jcp.2019.05.024.
86. Shucai, Li et al. 2020. Deep-Learning inversion of seismic data. In: *IEEE transactions on geoscience and remote sensing*, vol 58.3, Mar 2020, 2135–2149. ISSN: 0196-2892, 1558-0644. https://doi.org/10.1109/TGRS.2019.2953473.
87. Hug, L. et al. 2022. A three-field phase-field model for mixed-mode fracture in rock based on experimental determination of the mode II fracture toughness. en. *Engineering with Computers 38.6*: 5563–5581. ISSN: 1435-5663. https://doi.org/10.1007/s00366-022-01684-9.
88. Hyogu, Jeong et al. 2023. A physics-informed neural network-based topology optimization (PINNTO) framework for structural optimization. en. *Engineering Structures 278*: 115484. ISSN: 01410296. https://doi.org/10.1016/j.engstruct.2022.115484.
89. Aaditya, Chandrasekhar, and Krishnan Suresh. 2021. Multi-Material topology optimization using neural networks. en. *Computer-Aided Design 136*:103017. ISSN: 0010-4485. https://doi.org/10.1016/j.cad.2021.103017.
90. Gawel, Kus, and Miguel A. Bessa. 2024. Gradient-free neural topology optimization. https://doi.org/10.48550/ARXIV.2403.04937.
91. Leon, Herrmann et al. 2023. On the use of neural networks for full waveform inversion [Software]. en., 1, Jan 2023. https://doi.org/10.17632/7kps2hnj6g.1.
92. Levi, McClenny, and Ulisses Braga-Neto. 2022. *Self-Adaptive Physics-Informed Neural Networks using a Soft Attention Mechanism*, Apr 2022. arXiv:2009.04544 [cs, stat].
93. Charles, Audet, and Warren Hare. 2017. *Derivative-Free and Blackbox Optimization*. 1st ed. 2017. Springer Series in Operations Research and Financial Engineering. Cham: Springer International Publishing: Imprint: Springer. ISBN: 978-3-319-68913-5.

94. Fangshu, Yang and Jianwei Ma. 2019. Deep-learning inversion: A next-generation seismic velocity model building method. *Geophysics 84.4*.
95. Leon, Herrmann et al. 2024. On neural networks for generating better local optima in topology optimization. en. *Structural and Multidisciplinary Optimization 67.11*: 192. ISSN: 1615-1488. https://doi.org/10.1007/s00158-024-03908-6.
96. Divya, Shyam Singh et al. 2024. *Accelerating full waveform inversion by transfer learning [Software]*, Aug 2024. https://doi.org/10.5281/zenodo.13150916.

Chapter 10
Methodological Overview of Deep Learning in Computational Mechanics

Abstract As conclusion of this book, it is time to broaden the view. The rapid growth of deep learning research, including within the field of computational mechanics, has resulted in an extensive and diverse body of literature. To help researchers identify key concepts and promising methodologies within this field, this chapter provides an overview of deep learning in computational mechanics. Five main categories are identified and explored: simulation substitution, simulation enhancement, discretizations as neural networks, generative approaches, and deep reinforcement learning. This chapter focuses on presenting the methods used in deep learning in computational mechanics rather than their specific applications in computational mechanics, thereby enabling researchers to explore this field more effectively. The discussed concepts are explained as simple as possible. This renders the content accessible to both beginners who worked through the previous chapters, as well as experts. This makes it possible to gain an overview of the current state-of-the-art and thereby obtain ideas for future research directions.

This overview chapter provides a summary of the contents provided in the previous chapters and places them into the greater context of other deep learning methodologies used in computational mechanics that were not covered in detail in this book. This should aid the reader in navigating the scientific literature on the topic. To this end, we introduce the following taxonomy:

- **simulation substitution** (Sect. 10.1)
  - **data-driven modeling** (Sect. 10.1.1)
  - **physics-informed learning** (Sect. 10.1.2)
- **simulation enhancement** (Sect. 10.2)
- **discretizations as neural networks** (Sect. 10.3)
- **generative approaches** (Sect. 10.4)
- **deep reinforcement learning** (Sect. 10.5).

This chapter is a modified version of the publication *Deep Learning in Computational Mechanics: a review* [1].

L. Herrmann et al., *Deep Learning in Computational Mechanics*,
https://doi.org/10.1007/978-3-031-89529-6_10

Simulation substitution replaces the entire simulation with a surrogate model, which are deep neural networks in the context of deep learning. The model can be trained with supervised learning, which purely relies on labeled data and therefore is referred to as **data-driven modeling**. The generalization errors of these models can be reduced by **physics-informed learning**. Here, physics constraints are imposed on the learnable space such that only physically admissible solutions are learned.

Simulation enhancement instead only replaces components of the simulation chain, while the remaining parts are still handled by classical methods. Approaches within this category are strongly linked to their respective applications and will, therefore, be presented in the context of their specific use cases. Both data-driven and physics-informed approaches will be discussed.

Treating **discretizations as neural networks** is achieved by constructing a discretization from the basic building blocks of neural networks, i.e., linear transformations and non-linear activation functions. Thereby, techniques within deep learning frameworks—such as automatic differentiation, gradient-based optimization, and efficient GPU-based parallelization—can be leveraged to improve classical simulation techniques.

Generative approaches create new content based on a dataset. The goal is not, however, to recreate the data, but to generate statistically similar data. This is useful in exploring the design space or enhancing a dataset to train surrogate models.

Finally, in **deep reinforcement learning**, an agent learns how to interact with an environment in order to maximize rewards provided by the environment. In the case of deep reinforcement learning, the agent is modeled with neural networks. In the context of computational mechanics, the environment is modeled by the governing physical equations. Reinforcement learning provides an alternative to gradient-based optimization, which is useful when gradient information is not available.

The unique proposed taxonomy arises from a methodological viewpoint, instead of an application [2–15], or problem [16] oriented perspective. However, parallels can be drawn to the challenges identified in [16] and proposed areas of investigation in machine learning. Similarly, the distinction between machine learning enhanced[1] and substitution by machine learning models is made. Additionally, challenges such as robustness, explainability, and handling of complex and high-dimensional data are highlighted. Also, the separation between physics-informed learning and data-driven modeling is made by [16], as well as by [17]. Interestingly, older reviews [18, 19] arrived at similar categories, additionally including neural networks as means of more efficient implementations, i.e., discretizations as neural networks. Only the last two proposed categories, generative approaches, and deep reinforcement learning, have not been spotlighted as methodologies within reviews of computational mechanics. But these are by now well-established within the machine learning community [20–23] and sufficiently distinct to be treated separately.

---

[1] A further interesting distinction is made between inner (within a forward simulation) and outer loop enhancements (using multiple forward simulations, e.g., within an optimization).

10.1 Simulation Substitution

In the field of computational mechanics, numerical procedures are developed to solve or find partial differential equations. A generic partial differential equation can be written as

$$\mathcal{N}[u; \lambda] = 0, \qquad \text{on } \Omega \times \mathcal{T}, \tag{10.1}$$

where a non-linear operator $\mathcal{N}$ acts on a solution $u(x, t)$ of a partial differential equation as well as its coefficients $\lambda(x, t)$[2] in the spatio-temporal domain $\Omega \times \mathcal{T}$. In the forward problem, the solution $u(x, t)$ is to be computed, while the inverse problem considers either the non-linear operator $\mathcal{N}$ or its coefficients $\lambda(x, t)$ as unknowns.

A further distinction is made between methods treating the temporal dimension t as a continuum, as in space-time approaches [24] (Sects. 10.1.1.1 and 10.1.2.1),[3] or in discrete sequential time steps, as in time-stepping procedures (Sects. 10.1.1.2 and 10.1.2.2). For simplicity, but without loss of generality, time-stepping procedures will be presented on partial differential equations with a first order derivative with respect to time:

$$\frac{\partial u}{\partial t} = \mathcal{N}^{\mathcal{T}}[u; \lambda], \qquad \text{on } \Omega \times \mathcal{T} \tag{10.2}$$

with the non-linear operator $\mathcal{N}^{\mathcal{T}}$. Another task in computational mechanics is the forward modeling and identification of systems of ordinary differential equations. For this, we will consider systems of the following form:

$$\frac{d\boldsymbol{x}(t)}{dt} = \boldsymbol{f}(\boldsymbol{x}(t)). \tag{10.3}$$

Here, $\boldsymbol{x}(t)$ are the time-dependent degrees of freedom and $\boldsymbol{f}$ is the right-hand side defining the system of equations.[4] Both the forward problem of computing $\boldsymbol{x}(t)$ and the inverse problem of identifying $\boldsymbol{f}$ will be discussed in the following.

10.1.1 Data-Driven Modeling

Data-driven modeling relies entirely on labeled data $\tilde{x}$, $\tilde{y}$. The neural network learns the mapping between $\tilde{x}$ and $\tilde{y}$ with $\hat{y}_i = f_{NN}(x_i; \boldsymbol{\Theta})$. Thereby, an interpolation to yet unseen data points is established. A data-driven loss $\mathcal{L}_{\mathcal{D}}$, such as the mean squared error, for example, can be used as cost function C, as seen in Chap. 3.

---

[2] In case of the bar equation, the partial differential equation coefficients could be the cross-sectional stiffness $EA(x)$ or/and the distributed load $p(x)$.

[3] Static problems without time-dependence can only be treated by the space-time approaches.

[4] Note that a spatial discretization of the partial differential equation Eq. 10.2 can also be written as a system of ordinary differential equations.

$$C = \mathcal{L}_{\mathcal{D}} = \frac{1}{m_D} \sum_{i=1}^{m_{\mathcal{D}}} ||\hat{y}(\tilde{x}_i) - \tilde{y}_i||_2^2 \tag{10.4}$$

10.1.1.1 Space-Time Approaches

To declutter the notation, but without loss of generality, the temporal dimension t is dropped in this section, as it is possible to treat it like any other spatial dimension x in the scope of these methods. The goal of the upcoming methods is to either learn a forward operator $\hat{u} = S[\lambda; x]$, an inverse operator for the coefficients $\hat{\lambda} = I[u; x]$, or an inverse operator for the non-linear operator $\hat{\mathcal{N}} = O[u; \lambda; x]$.[5] The methods will be explained using the forward operator, but they apply analogously to the inverse operators. Only the inputs and outputs differ.

The solution prediction $\hat{u}_i$ at coordinate x_i or $\hat{\boldsymbol{u}}_i$ on the entire domain Ω is made based on a set of unknown coefficients $\boldsymbol{\lambda}_i$. The cost function C is formulated analogously to Eq. 10.4:

$$C = \mathcal{L}_{\mathcal{D}} = \frac{1}{m_{\mathcal{D}}} \sum_{i=1}^{m_{\mathcal{D}}} ||\hat{u}_i - \tilde{u}_i||_2^2. \tag{10.5}$$

Fully Connected Neural Networks

The simplest procedure is to approximate the operator S with a fully connected neural network S_{FNN} (cf. Sect. 3.1).

$$\hat{u}(x) = S_{FNN}(\boldsymbol{\lambda}; x; \boldsymbol{\Theta}) \tag{10.6}$$

Example applications are flow classification [25, 26], fluid flow in turbomachinery [27], dynamic beam displacements from previous measurements [28], wall velocity predictions in turbulence [29], heat transfer [30], prediction of source terms in turbulence models [31], full waveform inversion [32–34], and topology optimization based on moving morphable bars [35]. The approach is however limited to simple problems, as an abundance of data is required. To this end, several improvements have been proposed and presented in the following subsections.

Image-To-Image Mapping

One downside of the application of fully connected neural networks to problems in computational mechanics is that they often need to learn spatial relationships with respect to x from scratch. Convolutional neural networks (cf. Sect. 3.9.1) inherently account for these spatial relationships due to their kernel-based structure, incorporating the corresponding inductive bias. Therefore, image-to-image mappings using convolutional neural networks have been proposed, where an image, i.e., a uniform grid (see Fig. 10.1) of the coefficients $\boldsymbol{\lambda}$, is used as input.

[5] Note that u might only be partially known on the domain Ω for inverse problems.

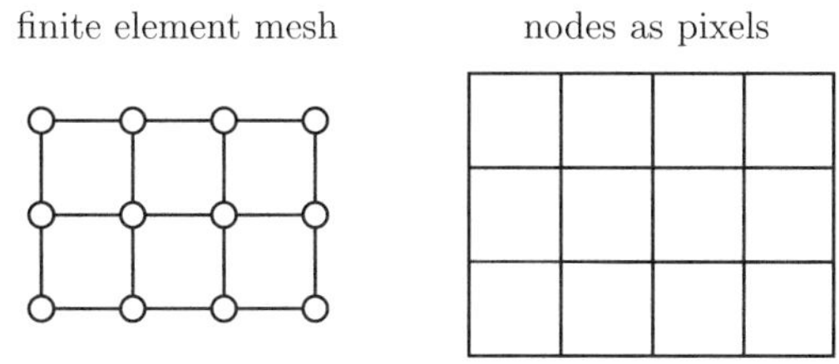

Fig. 10.1 Representation of nodes of a Cartesian grid as pixels in an image. Adapted from [36]

$$\hat{\boldsymbol{u}} = S_{CNN}(\boldsymbol{\lambda}; \boldsymbol{\Theta}) \tag{10.7}$$

This results in a prediction of the solution $\hat{\boldsymbol{u}}$ throughout the entire image, i.e., the domain, as explored in Sect. 3.9.1.2 by learning pixelized strain distributions.

Applications include pressure and velocity predictions around airfoils [37–40], stress predictions from geometries and boundary conditions [41, 42], steady flow predictions [43], detection of manufacturing features [44, 45], full waveform inversion [46–57], and topology optimization [58–67]. An important choice in the design of the learning algorithm is the encoding of the input data. In the case of geometries and boundary conditions, binary representations are the most straightforward approach. These are, however, challenging for convolutional neural networks, as discussed in [43]. Signed distance functions [43] or simulations on coarse grids provide superior alternatives. For inverse problems, an initial forward simulation of an initial guess of the inverse field can be used to encode the desired boundary conditions [62, 65–67]. Another possibility for convolutional neural networks is a decomposition of the domain. The mapping can be performed on the full domain [68], smaller subdomains [69], or even individual pixels [70]. In the latter two cases, interfaces require special treatment.

The disadvantage of convolutional neural network mappings is being constrained to uniform grids on rectangular domains. This can be circumvented by using graph neural networks (cf. Sect. 3.9.2) acting on graph data, e.g., meshes, such as in [71–73], or point cloud-based neural networks [74, 75] acting on point cloud data, such as in [76]. Just as convolutional neural networks, graph neural networks operate on the invariant structural elements of the data, which for graph neural networks are edges connecting vertices instead of pixels aligned on a structured grid for convolutional neural networks. In fact, graph neural networks can be regarded as a generalization of convolutional neural networks since they can handle a broader class of data structures, i.e., graphs (including images).[6] This comes at the cost of less efficient implementations when compared to pure convolutional neural networks.

[6] For an in-depth treatment of the inner workings of graph neural networks; see [77].

Model Order Reduction Encoding

Independent of the neural network architecture, learning can be aided by applying the neural network to a lower-dimensional space that is able to capture the data. For complex problems, mappings e to low-dimensional spaces (also referred to as latent space or latent vector) $\boldsymbol{h}$ can be identified with model order reduction techniques; see Sect. 6.1. Thus, in the case of simulation substitution, a low-dimensional encoding $\boldsymbol{h}^\lambda = e(\boldsymbol{\lambda})$ of $\boldsymbol{\lambda}$ (sampled on all sample points x) is identified. This is provided as input to a neural network to predict the solution field $\boldsymbol{h}^u$ in a reduced latent space. The full solution field $\boldsymbol{u}$ (on all sample points $\boldsymbol{x}$) is obtained in a decoding $d = e^{-1}$ step. The prediction is given as

$$\hat{\boldsymbol{u}} = d(\hat{\boldsymbol{h}}^u) = d\big(S_{NN}(\boldsymbol{h}^\lambda; \boldsymbol{\Theta})\big) = d\Big(S_{NN}\big(e(\boldsymbol{\lambda}); \boldsymbol{\Theta}\big)\Big). \tag{10.8}$$

The dimensional reduction can, e.g., be performed with singular value decomposition, principal components analysis [78, 79] (cf. Sect. 6.1), as proposed in [80], proper orthogonal decomposition [81], or reduced manifold learning [82]. These techniques have been applied to learning aortic wall stresses [83], arterial wall stresses [84], flow velocities in viscoplastic flow [85], and the inverse problem of identifying unpressurized geometries from pressurized geometries [86]. Currently, the most impressive results in data-driven surrogate modeling are achieved with model order reduction encodings combined with neural networks [87, 88], which can be combined with most other methodologies presented in this chapter.

Another dimensionality reduction technique are autoencoders [89], where e and d are modeled by neural networks.[7] These are treated in detail in Chap. 8 and enable non-linear encodings. An early investigation is presented in [90], where proper orthogonal decomposition is related to neural networks. Application areas are the prediction of designs of acoustic scatterers from the reduced latent space [91], or mappings from dynamic responses of bridges to damage [92]. Furthermore, it has to be stated that many of the image-to-image mapping techniques rely on neural network architectures inspired by autoencoders, such as U-nets [93, 94].

Neural Operators

The most recent trend in surrogate modeling with neural networks are neural operators [95], which map between function spaces instead of selective functions of those spaces.[8] Neural operators rely on the extension of the universal approximation theorem [96] to non-linear operators [97]. The two most prominent neural operators are DeepONets[9] [98] and Fourier neural operators [99].

---

[7] Note that the autoencoder is modified, as it does not perform an identity mapping. Nonetheless, the idea of mapping to a reduced latent state is exploited.

[8] In practice, this involves representing function spaces by evaluating them on a set of functions, analogous to evaluating functions at different sampling points to obtain data vectors. In addition, this set of functions is evaluated at sampling points, yielding data vectors.

[9] Originally proposed in [97] with shallow neural networks.

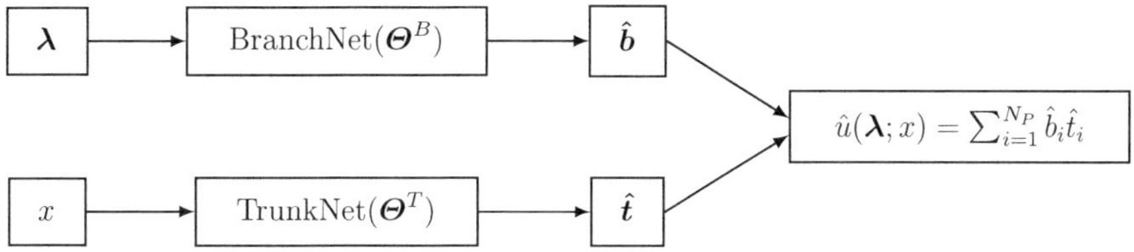

Fig. 10.2 DeepONet, operator learning via prediction of the basis functions $\hat{\boldsymbol{t}}$ and the corresponding coefficients $\hat{\boldsymbol{b}}$ [98]

DeepONet

In DeepONets [98], illustrated in Fig. 10.2, the task of predicting the operator $\hat{u}(\boldsymbol{\lambda}; x)$ is split up into two sub-tasks:

- the prediction of N_P basis functions $\hat{\boldsymbol{t}}(x)$ (TrunkNet),
- the prediction of the corresponding N_P problem-specific coefficients $\hat{\boldsymbol{b}}(\boldsymbol{\lambda})$ (BranchNet).

The basis is predicted by the TrunkNet with parameters $\boldsymbol{\Theta}^T$ via an evaluation at coordinates x. The coefficients are estimated from the coefficients $\boldsymbol{\lambda}$ using the BranchNet parametrized by $\boldsymbol{\Theta}^B$ and, thus, specific to the problem being solved. Taking the scalar product of the evaluated basis and the coefficients yields the solution prediction $\hat{u}(\boldsymbol{\lambda}; x)$.[10]

$$\hat{\boldsymbol{t}}(x) = F_{FNN}^T(x; \boldsymbol{\Theta}^T) \tag{10.9}$$

$$\hat{\boldsymbol{b}}(\boldsymbol{\lambda}) = F_{FNN}^B(\boldsymbol{\lambda}; \boldsymbol{\Theta}^B) \tag{10.10}$$

$$\hat{u}(x) = \hat{\boldsymbol{b}}(\boldsymbol{\lambda}) \cdot \hat{\boldsymbol{t}}(x) \tag{10.11}$$

Applications can be found in [100, 102–113]. DeepONets have also been extended with physics-informed loss functions [114–116].

Fourier Neural Operators

Fourier neural operators [99] predict the solution $\hat{\boldsymbol{u}}$ on a uniform grid $\boldsymbol{x}$ from the spatially varying coefficients $\boldsymbol{\lambda} = \lambda(\boldsymbol{x})$. As the aim is to learn a mapping between different functions (from a function space), sampled on the entire domain, non-local mappings can be performed at each layer [117]. For example, mappings such as integral kernels [118, 119], Laplace transformations [120], and Fourier transforms [99] can be employed. These transformations enhance the non-local expressivity of the neural network [117], where Fourier transforms are particularly favorable due to the computational efficiency achievable through fast Fourier transforms.

The Fourier neural operator, as illustrated in Fig. 10.3, consists of Fourier layers, where linear transformations $\boldsymbol{K}$ are performed after Fourier transforms $\mathcal{F}$ along the

[10] For a scalar solution. An extension to vector-valued solution quantities is straightforward.

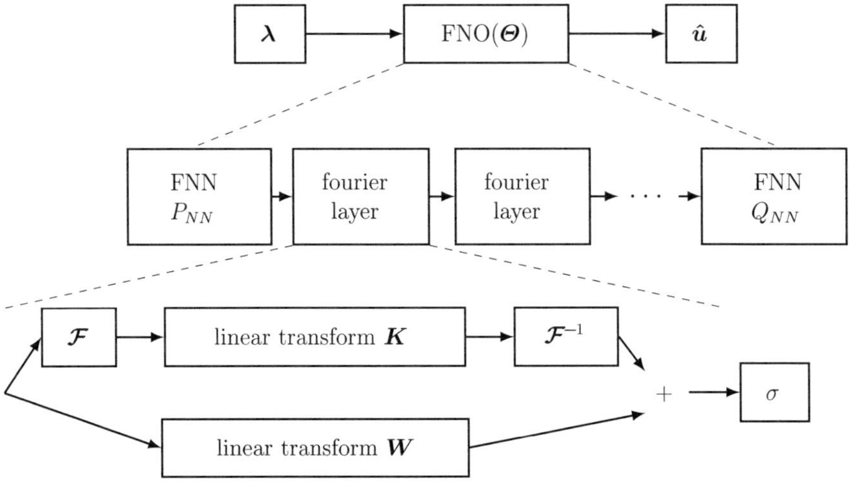

Fig. 10.3 Fourier neural operator, operator learning in the Fourier space [99]

spatial dimensions x. Subsequently, an inverse Fourier transform $\mathcal{F}^{-1}$ is applied, which is added to the output of a linear transformation $\boldsymbol{W}$ performed outside the Fourier space. Thus, the Fourier transform can be skipped by the neural network. The final step is an activation function σ. The manipulations within a Fourier layer to predict the next activation on the uniform grid $\boldsymbol{a}^{(j+1)}(\boldsymbol{x})$ can be written as

$$\boldsymbol{a}^{(j+1)}(\boldsymbol{x}) = \sigma\left(\boldsymbol{W}\boldsymbol{a}^{(j)}(\boldsymbol{x}) + \boldsymbol{b} + (\mathcal{F}^{-1}\Big[\boldsymbol{K}\mathcal{F}\big[\boldsymbol{a}^{(j)}(\boldsymbol{x})\big]\Big]\right), \tag{10.12}$$

where $\boldsymbol{b}$ is the bias. Both the linear transformations $\boldsymbol{K}$, $\boldsymbol{W}$ and the bias $\boldsymbol{b}$ are learnable and thereby part of the parameters $\boldsymbol{\Theta}$. Multiple Fourier layers can be employed, typically used in combination with an encoding network P_{NN} and a decoding network Q_{NN}.

Applications can be found in [121–131]. An extension relying on the attention mechanisms of transformers [132] is presented in [133]. Analogously to DeepONets, Fourier neural operators have been combined with physics-informed loss functions [134].

Neural Network Approximation Power

Despite the advancements in neural network architectures,[11] neural network surrogates struggle to learn solutions of general partial differential equations. Typically, successes have only been achieved for parametrized partial differential equations with relatively small parameter spaces—or in cases where accuracy, reliability, or

[11] Including architectures specifically designed to solve partial differential equations.

generalization were disregarded. It has, however, been shown—both for simple architectures such as fully connected neural networks [135, 136] as well as for advanced architectures such as DeepONets [137]—that neural networks possess an excellent theoretical approximation power which can capture solutions of various partial differential equations. Currently, there are two obstacles that impede the identification of sufficiently good optima with these desirable neural network parameter spaces [135]:

- training data: generalization error,
- training algorithm: optimization error.

A lack of sufficient training data leads to poor generalization. This might be alleviated through faster data generation using, e.g., faster and specialized classical methods [138], or improved sampling strategies, i.e., finding the minimum number of required data points distributed in a specific manner to train the surrogate. Additionally, current training algorithms only converge to local optima. Research into improved optimization algorithms, such as current trends in computing better initial weights [139] and thereby better local optima, attempts to reduce the optimization error. At the same time, training times are reduced drastically increasing the competitiveness. Another problem is that most known optimization algorithms have been designed to search in linear spaces. While that is a challenging endeavor, improving the optimization algorithms to search in non-linear spaces would be needed.

10.1.1.2 Time-Stepping Procedures

For the time-stepping procedures, we will consider Eqs. (10.2) and (10.3) in the following.

Recurrent Neural Networks

The simplest approach to modeling time series data is by using fully connected neural networks to predict the next time step t_{i+1} from the current time step t_i:

$$\hat{u}(x, t_{i+1}) = S_{FNN}\big(x, t_i; u(x, t_i); \boldsymbol{\Theta}\big). \tag{10.13}$$

However, this approach lacks the ability to capture the temporal dependencies between different time steps, as each input is treated independently and without considering more than just the previous time step. Incorporating the sequential nature of the data can be achieved directly with recurrent neural networks; see Sect. 3.9.6. Recurrent neural networks maintain a hidden state which captures information from the previous time steps, to be used for the next time step prediction. By unrolling the recurrent neural network, the entire time-history can be predicted.

$$\{\hat{u}(x, t_2), \hat{u}(x, t_3), \dots, \hat{u}(x, t_N)\} = S_{RNN}(x; u(x, t_1); \boldsymbol{\Theta}) \tag{10.14}$$

As previously seen in Sect. 3.9.6, shortcomings of recurrent neural networks, such as their tendency to struggle with learning long-term dependencies due to the problem of vanishing or exploding gradients, have been addressed by more sophisticated architectures such as long short-term memory networks (LSTM) [140], gated recurrent unit networks (GRU) [141] (see Sect. 3.9.6), and transformers [132] introduced in Sect. 8.6 (see [142] for a recent contribution on transformers for thermal analysis in additive manufacturing). The concept of recurrent units has also been combined with other architectures, as demonstrated for convolutional neural networks [143] and graph neural networks [71, 72, 144–148].

Further applications of recurrent neural networks are full waveform inversion [149–151], high-dimensional chaotic systems [152], fluid flow [153, 154], fracture propagation [73], sensor signals in non-linear dynamic systems [155, 156], and settlement field predictions induced by tunneling [157], which was extended to damage prediction in affected structures [158, 159]. Recurrent neural networks are often combined with reduced order model encodings [160], where the dynamics are predicted on the reduced latent space, as demonstrated in [161–167]. Further variations employ classical time-stepping schemes on the reduced latent space obtained by autoencoders [168, 169].

Dynamic Mode Decomposition

Another approach that was formulated for system dynamics, i.e., Eq. 10.3 is dynamic mode decomposition (DMD) [170, 171]. The aim of DMD is to identify a linear operator $\boldsymbol{A}$ that relates two successive snapshot matrices with n time steps $\boldsymbol{X} = [\boldsymbol{x}(t_1), \boldsymbol{x}(t_2), \ldots, \boldsymbol{x}(t_n)]^\intercal$, $\boldsymbol{X}' = [\boldsymbol{x}(t_2), \boldsymbol{x}(t_3), \ldots, \boldsymbol{x}(t_{n+1})]^\intercal$:

$$\boldsymbol{X}' \approx \boldsymbol{A}\boldsymbol{X}. \tag{10.15}$$

To solve this, the problem is reframed as a regression task. The operator $\boldsymbol{A}$ is approximated by minimizing the Frobenius norm of the difference between $\boldsymbol{X}'$ and $\boldsymbol{A}\boldsymbol{X}$. This minimization can be performed using the Moore-Penrose pseudoinverse $\boldsymbol{X}^\dagger$ (see, e.g., [172]):

$$\boldsymbol{A} = \arg\min_{\boldsymbol{A}} ||\boldsymbol{X}' - \boldsymbol{A}\boldsymbol{X}||_F = \boldsymbol{X}'\boldsymbol{X}^\dagger. \tag{10.16}$$

Once the operator is identified, it can be used to propagate the dynamics forward in time, approximating the next state $\boldsymbol{x}(t_{i+1})$ using the current state $\boldsymbol{x}(t_i)$:

$$\boldsymbol{x}(t_{i+1}) \approx \boldsymbol{A}\boldsymbol{x}(t_i). \tag{10.17}$$

This framework, is however, only valid for linear dynamics. DMD can be extended to handle non-linear systems through the application of Koopman operator theory [173]. According to Koopman operator theory, it is possible to represent a non-linear system as a linear one by using an infinite-dimensional Koopman operator $\mathcal{K}$ that acts on a transformed state $e(\boldsymbol{x}(t_i))$:

$$g(\boldsymbol{x}(t_{i+1})) = \mathcal{K}[e(\boldsymbol{x}(t_i))]. \tag{10.18}$$

In theory, the Koopman operator $\mathcal{K}$ is an infinite-dimensional linear transformation. In practice, however, finite-dimensional approximations are employed. This approach is, for example utilized in the extended DMD [174], where the regression from Eq. 10.16 is performed on a higher-dimensional state $\boldsymbol{h}(t_i) = e(\boldsymbol{x}(t_i))$ relying on a dictionary of orthonormal basis functions $\boldsymbol{h}(t_i) = \boldsymbol{\psi}(\boldsymbol{x}(t_i))$. Alternatively, the dictionary can be learned using neural networks, i.e., $\hat{\boldsymbol{\psi}}(\boldsymbol{x}) = \psi_{NN}(\boldsymbol{x}; \boldsymbol{\Theta})$, as demonstrated in [175, 176]. The neural network is trained by minimizing the mismatch between predicted state $\boldsymbol{\psi}(\hat{\boldsymbol{x}}(t_{i+1})) = \boldsymbol{A}\hat{\boldsymbol{\psi}}(\boldsymbol{x}(t_i))$ (Eq. 10.17) and the true state in the dictionary space. Orthogonality is not required and therefore not enforced.

$$C = \frac{1}{2n}\sum_{i=1}^{n} ||\hat{\boldsymbol{\psi}}(\boldsymbol{x}(t_{i+1})) - \boldsymbol{A}\hat{\boldsymbol{\psi}}(\boldsymbol{x}(t_i))||_2^2 \tag{10.19}$$

When the dictionary is learned, the state predictions can be reconstructed using the Koopman mode decomposition, as explained in detail in [175].

Alternatively, the mapping to the augmented state can be performed with autoencoders, which at the same time allows for a direct map back to the original space [177–180]. Thus, an encoder learns a reduced latent space $\hat{\boldsymbol{h}}(\boldsymbol{x}) = e_{NN}(\boldsymbol{x}; \boldsymbol{\Theta}^e)$ and a decoder learns the inverse mapping $\hat{\boldsymbol{x}}(\boldsymbol{h}) = d_{NN}(\boldsymbol{h}; \boldsymbol{\Theta}^d)$. The networks are trained using three losses: the autoencoder reconstruction loss $\mathcal{L}_{\mathcal{A}}$, the linear dynamics loss $\mathcal{L}_{\mathcal{R}}$, and the future state prediction loss $\mathcal{L}_{\mathcal{F}}$.

$$\mathcal{L}_{\mathcal{A}} = \frac{1}{n+1}\sum_{i=1}^{n+1} ||\boldsymbol{x}(t_i) - d_{NN}(e_{NN}(\boldsymbol{x}(t_i); \boldsymbol{\Theta}^e); \boldsymbol{\Theta}^d)||_2^2 \tag{10.20}$$

$$\mathcal{L}_{\mathcal{R}} = \frac{1}{n}\sum_{i=1}^{n} ||e_{NN}(\boldsymbol{x}(t_{i+1}); \boldsymbol{\Theta}^e) - \boldsymbol{A}e_{NN}(\boldsymbol{x}(t_i); \boldsymbol{\Theta}^e)||_2^2 \tag{10.21}$$

$$\mathcal{L}_{\mathcal{F}} = \frac{1}{n}\sum_{i=1}^{n} ||\boldsymbol{x}(t_{i+1}) - d_{NN}(\boldsymbol{A}e_{NN}(\boldsymbol{x}(t_i); \boldsymbol{\Theta}^e); \boldsymbol{\Theta}^d)||_2^2 \tag{10.22}$$

$$C = \kappa_{\mathcal{A}}\mathcal{L}_{\mathcal{A}} + \kappa_{\mathcal{R}}\mathcal{L}_{\mathcal{R}} + \kappa_{\mathcal{F}}\mathcal{L}_{\mathcal{F}} \tag{10.23}$$

The cost function C is composed of a weighted sum of the loss terms $\mathcal{L}_{\mathcal{A}}, \mathcal{L}_{\mathcal{R}}, \mathcal{L}_{\mathcal{F}}$ and weighting terms $\kappa_{\mathcal{A}}, \kappa_{\mathcal{R}}, \kappa_{\mathcal{F}}$. Furthermore, [179] allows $\boldsymbol{A}$ to vary depending on the state. This is achieved by predicting the eigenvalues of $\boldsymbol{A}$ with an auxiliary network and constructing the matrix from these.

10.1.1.3 Active Learning & Transfer Learning

Finally, an important machine learning technique independent of the neural network architecture and applicable to both space-time and time-stepping approaches is active learning [181]. Instead of precomputing a labeled dataset, data is only provided when the prediction quality of the neural network is insufficient. Furthermore, the data is not chosen arbitrarily, but only in the vicinity of the failed prediction. In computational mechanics, the prediction of the neural network can be assessed with an error indicator. For an insufficient result, the results of a classical simulation are used to retrain the neural network. Over time, the neural network estimates improve in the respective domain of application. Due to the error indicator and the classical simulations, the predictions are reliable. Examples for active learning in computational mechanics can be found in [182–184].

Another technique, transfer learning [185, 186] (seen in the context of full waveform inversion in Sect. 9.2.5.1), aims at accelerating the neural network training. Here, the neural network is first trained on a similar task. Subsequently, it is applied to the task of interest—where it converges faster than an untrained neural network. Applications in computational mechanics can be found in [55, 187, 188].

10.1.2 Physics-Informed Learning

In supervised learning, as discussed in Sect. 10.1.1, the quality of prediction strongly depends on the amount of training data. Acquiring data in computational mechanics may be expensive. To reduce the amount of required data, constraints enforcing the physics have been proposed. Two main approaches exist [17, 189]. Physical laws such as conservation or compatibility, can be enforced by modifying the cost function through a penalty term punishing unphysical predictions, thus acting as a regularizer, as seen in Chaps. 4 and 5. The prominent modifications are recapitulated in the upcoming section. Alternatively, physical laws can be enforced by construction, i.e., by reducing the learnable space to a physically meaningful space and is often referred to as physics-aware, physics-augmented, physics-inspired or physics-enhanced learning, as seen in the context of material modeling in Chap. 7 with physics-aware neural networks (PANNs). This approach is highly specific to its application, as, e.g., seen in Chap. 7 for material modeling. It will be further explored in Sect. 10.2, due to its direct connection to specific applications. A brief coverage of physics-informed learning via penalization is provided in Sect. 10.1.2.3.

Both concepts can be found in overview publications, where [17] defines four overarching methodologies: (i) augmentation of training data using prior knowledge, (ii) modification of the model, i.e., enforcement by construction, (iii) enhancement of the learning algorithm with regularization terms, i.e., enforcing constraints through the cost function, and (iv) checking the final estimate and thereby discarding physical violations (using, e.g., error indicators). The two most prominent methodologies, i.e., modifying the cost function and enforcement by construction are similarly mentioned in [189], which correspondingly refers to them as physics-informed and physics-augmented. Further variations in terminology can be found in [143, 190], who refer to physics-informed neural networks for multiple solutions as physics-constrained deep learning, or [191] using the term physics-enhanced neural networks for neural networks enforcing the physics by construction. Due to the many names within the relatively new and interconnected field, in this chapter we cover the variations under the overarching term of physics-informed learning.

10.1.2.1 Space-Time Approaches

Once again and without loss of generality, the temporal dimension t is dropped to declutter the notation. However, in contrast to Sect. 10.1.1.1, the following methods are not equally applicable to forward and inverse problems. Thus, the prediction of the solution $\hat{u}$, the partial differential equation coefficients $\hat{\lambda}$, and the non-linear operator $\mathcal{N}$ are treated separately.

Differential Equation Solving with Neural Networks

The concept of solving partial differential equations[12] with neural networks was first proposed in the 1990s [192–194], but was recently popularized by the so-called physics-informed neural networks (PINNs) [195], introduced in Chaps. 4 and 5 (see [196–198] for recent review articles and SciANN [199], SimNet [200], DeepXDE [201] for libraries).

The common idea is the parametrization of the differential equation's solution or coefficients by a neural network, which is subsequently trained by minimizing the differential equation's residual. This can be using the strong form of the differential equation as for physics-informed neural networks (Chap. 4), or the weak form as for variational physics-informed neural networks (Sect. 5.1.4), weak adversarial networks (Sect. 5.1.5), and the deep energy method/deep Ritz method (Sect. 5.1.1). The difference between the latter three approaches acting on the weak form essentially boils down to the choice of test functions, as elaborated in Chap. 5. Noteworthy variations of physics-informed neural networks operating on the strong form are the deep least-squares method [202] and the deep Galerkin method [203]. Instead of focusing on the residuals at individual collocation points as in physics-informed neural networks, these methods consider the L^2-norm of the residuals integrated over the domain. However, as seen in Chap. 5, operating on the weak form is typically preferred due to lower computation times per epoch and potentially faster convergence. The difficulty herein lies in the accurate numerical integration of the weak form, which is currently evaluated with Gaussian quadrature, Monte Carlo integration methods [204, 205], sparse grid quadratures [206], or specialized schemes [207]. A multitude of extensions to the physics-informed neural network methodology exist,

[12] Typically, a single solution to a partial differential equation is obtained. If the partial differential equation is parametrized, multiple solutions can be obtained.

of which the most prominent were outlined in Sect. 5.2. A broader overview is provided below, but can be skipped if there is no further interest in physics-informed neural networks. For in-depth reviews; see [196–198].

Learning Multiple Solutions

Currently, physics-informed neural networks are mainly employed to learn a single solution. As the training effort exceeds the solving effort of classical solvers, the viability of physics-informed neural networks is questionable [208]. However, physics-informed neural networks can also be employed to learn multiple solutions. This is achieved by providing the parametrization of the partial differential equation, i.e., λ as an additional input to the network, as discussed in Sect. 10.1. This enables a cheap prediction stage without retraining for new solutions.[13] One possible example for this is [209], where different geometries are captured in terms of point clouds and processed with point cloud-based neural networks [74].

Boundary Conditions

The enforcement of the boundary conditions through a penalty term $\mathcal{L}_{\mathcal{B}}$ in physics-informed neural networks leads to an unbalanced optimization, due to the competing loss terms $\mathcal{L}_{\mathcal{R}}, \mathcal{L}_{\mathcal{B}}, \mathcal{L}_{\mathcal{D}}$; see Chap. 4.[14] One remedy is to modify the neural network output f_{FNN} by multiplication of a function, such that the Dirichlet boundary conditions are satisfied a priori, i.e., $\mathcal{L}_{\mathcal{B}} = 0$, as demonstrated in [210, 211].

$$\hat{u}(x) = G(x) + D(x) f_{FNN}(x; \boldsymbol{\Theta}^u) \tag{10.24}$$

Here, $G(x)$ is a smooth interpolation of the boundary conditions, and $D(x)$ is a signed distance function that is zero at the boundary. For Neumann boundary conditions, [212] propose to predict u and its derivatives $\partial u/\partial x$ with separate networks, such that the Neumann boundary conditions can be enforced strongly by modifying the derivative network. This requires an additional constraint, ensuring that the derivative predictions match the derivative of u. For complex domains, $G(x)$ and $D(x)$ cannot be found analytically. Therefore, [210] use neural networks to learn $G(x)$ and $D(x)$ in a supervised manner by prescribing either the boundary values or zero at the boundary and restricting the values within the domain to be non-zero. Similarly [213] proposed using radial basis function networks for $G(x)$, where $D(x) = 1$ is assumed. The radial basis function networks are determined by solving a linear system of equations

[13] Importantly, the training would be without training data and would only require a definition of the parametrized partial differential equation. Currently, this is only possible for simple partial differential equations with small parameter spaces.

[14] Consider, for instance, a training procedure in which the partial differential equation loss $\mathcal{L}_{\mathcal{R}}$ is first minimal, such that the partial differential equation is fulfilled. Without fulfilment of the boundary conditions, the solution is not unique. However, the neural network struggles to modify the current boundary values without violating the partial differential equation loss and thereby increasing the total cost function C. The neural network is thus stuck in a bad local minimum. Similar scenarios can be formulated for a too rapid minimization of the other loss terms.

constructed with the boundary conditions. On uniform grids, strong enforcement can be achieved through specialized convolutional neural network kernels [166] with constant padding terms for Dirichlet boundary conditions and ghost cells for Neumann boundary conditions. Constrained backward propagation [214] has also been proposed to guarantee the enforcement of boundary conditions [215, 216].

Another possibility is to introduce weighting terms $\kappa_{\mathcal{R}}, \kappa_{\mathcal{B}}, \kappa_{\mathcal{D}}$ for each loss term. These are either hyperparameters, or they are learned during the optimization with attention mechanisms [217–219]. This is achieved by performing a minimax optimization with respect to all weighting terms $\boldsymbol{\kappa} = \{\kappa_{\mathcal{R}}, \kappa_{\mathcal{B}}, \kappa_{\mathcal{D}}\}$

$$\min_{\boldsymbol{\Theta}} \max_{\boldsymbol{\kappa}} C. \tag{10.25}$$

Expanding on this idea, each collocation point used for the loss terms can be considered an individual equality constraint [220, 221]. Therefore, a weighting term $\kappa_{\mathcal{R}_i}$ is allocated for each collocation point x^i, as illustrated for the partial differential equation loss $\mathcal{L}_{\mathcal{R}}$ from Eq. 4.13 in Chap. 4

$$\mathcal{L}_{\mathcal{R}} = \frac{1}{m_{\mathcal{R}}} \sum_{i=1}^{m_{\mathcal{R}}} \kappa_{\mathcal{R}_i} ||\hat{r}[u(x^i); \lambda(x^i)]||_2^2. \tag{10.26}$$

This has the added advantage that greater emphasis is assigned on more important collocation points, i.e., points which lead to larger residuals. This approach is strongly related to the approaches relying on the augmented Lagrangian method [222] and competitive physics-informed neural networks [223], where an additional neural network models the penalty weights $\kappa(x) = K_{FNN}(x; \boldsymbol{\Theta}^{\kappa})$. This is similar to weak adversarial networks, but instead formulated using the strong form.

Ansatz

Another prominent topic is the question of which ansatz to choose. The type of ansatz is, for example, determined by different neural network architectures (see [224] for a comparison) or combinations with classical ansatz formulations, as seen in Sect. 5.2.5. Instead of using fully connected neural networks, some authors [143, 190] employ convolutional neural networks to exploit the spatial structure of the data. Irregular geometries can be handled by embedding the structure in a rectangular domain using binary encodings [225] or signed distance functions [43, 226]. Another option are coordinate transformations into rectangular grids [227]. The convolutional neural network requires a full-grid discretization, meaning that the coordinates x are analytically independent of the prediction $\hat{u} = S_{CNN}$. Thus, the gradients of u are not obtained with automatic differentiation, but with numerical differentiation, i.e., finite differences. Alternatively, the output of the convolutional neural network can represent coefficients of an interpolation, as proposed under the name spline-physics-informed neural networks [228] using Hermite splines. This again allows for an automatic differentiation. This is similarly applied for irregular geometries in [229], where graph neural networks are used in combination with a piecewise polynomial basis. Using a classical basis has the added advantage that Dirichlet boundary conditions can be satisfied exactly. A further variation is the approximation of the coefficients of classical bases with fully connected neural networks. This is shown with B-splines in [230] in the sense of isogeometric analysis [231]. This was similarly done for piecewise polynomials in [232]. However, instead of simply minimizing the partial differential equation residual directly (as in physics-informed neural networks), the finite element discretization [233, 234] is exploited. The loss $\mathcal{L}_{\mathcal{F}}$ can thus be formulated in terms of the non-linear stiffness matrix $\boldsymbol{K}$, the force vector $\boldsymbol{F}$, and the degrees of freedom $\boldsymbol{u}$.

$$\mathcal{L}_{\mathcal{F}} = ||\boldsymbol{K}(\boldsymbol{u})\boldsymbol{u} - \boldsymbol{F}||_2^2 \tag{10.27}$$

In the forward problem, $\boldsymbol{u}$ is approximated by a fully connected neural network, whereas for the inverse problem a fully connected neural network predicts $\boldsymbol{K}$. Similarly, [235, 236] map a neural network onto a finite element space by using the neural network evaluations at nodal coordinates as the corresponding basis function coefficients. This also allows a straightforward strong enforcement of Dirichlet boundary conditions, as demonstrated in [36] with convolutional neural networks. The nodes are represented as pixels (see Fig. 10.1).

Prior information on the solution can be incorporated through a feature layer [237]. If, for example, it is known that the solution is composed of trigonometric functions, a feature layer with trigonometric functions can be applied after the input layer. Thus, known features are given to the neural network directly to aid the learning. Without known features, the task can also be modified to improve learning. Inspired by adaptivity from finite elements, refinements are progressively learned by additional layers of the neural network [238]; see Fig. 10.4. Thus, a coarse solution $\boldsymbol{u}_1$ is learned to begin with, then refined to $\boldsymbol{u}_2$ by an additional layer, which again is refined to $\boldsymbol{u}_3$ until the deepest refinement level is reached.

Domain Decomposition

To improve the scalability of physics-informed neural networks to more complex problems, several domain decomposition methods have been proposed. One approach are *hp*-variational physics-informed neural networks [239], where the domain is decomposed into patches. Piecewise polynomial test functions are defined on each patch separately, while the solution is approximated by a globally acting neural network. This enables a separate numerical integration of each patch, improving its accuracy.

In an alternative formulation, one neural network can be used per subdomain. This was proposed as conservative physics-informed neural networks [240], where conservation laws are enforced at the interface to ensure continuity. Here, the discrepancies between both solution and flux were penalized at the interface in a least squares manner. The advantages of this approach are twofold: Firstly, parallelization is possible [241] and, secondly, adaptivitiy can be introduced. Shallower networks can be employed for smooth solutions and deeper networks for more complex solutions. The approach was generalized for any partial differential equation in the context

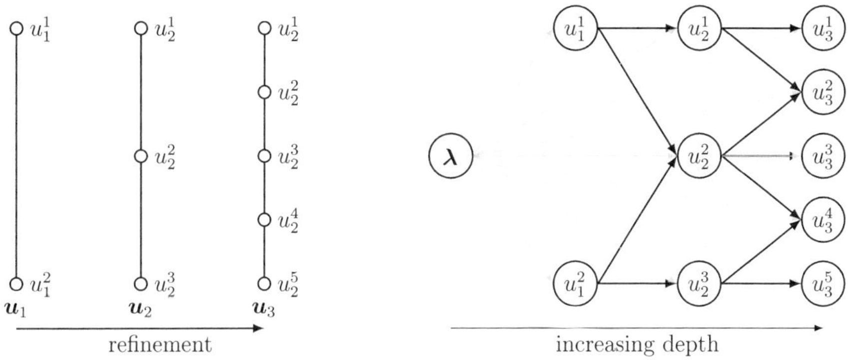

Fig. 10.4 Refinement expressed with neural networks in terms of neural network depth. Thick black lines indicate non-learnable connections and gray lines indicate learnable connections. Each added layer is composed of a projection from the coarser level and a correction obtained through the learnable connection

of extended physics-informed neural networks [242]. Here, the interface condition is formulated in terms of the difference in both the residual and the solution.

Acceleration Methods

Analogously to supervised learning, as discussed in Sect. 10.1.1, transfer learning can be applied to physics-informed neural networks [243] as, e.g., demonstrated in phase-field fracture [244] or topology optimization [245, 246]. These are very suitable problems since crack and displacement fields evolve with mostly local changes in phase-field fracture. For topology optimization, only minor updates are expected between each optimization iteration [245, 246], where [246] achieved computational times comparable to conventional SIMP[15] approaches [247, 248].

The poor performance of physics-informed neural networks in their original form can also be improved with better sampling strategies. In importance sampling [249, 250], the collocation point density is proportional to the value of the cost function. Alternatively, residual-based adaptive refinement [201] adds collocation points in the vicinity of areas with a higher cost function.

Another essential ingredient for neural networks is to normalize the inputs, outputs, and loss terms [251, 252]. For time-dependent problems, it is possible to use time-dependent normalization [253] to ensure that the solution is always in the same range regardless of the time step.

Furthermore, the cost function can be enhanced by including the derivative of the residual [254] as well. The derivative should also be minimized, as both the residual and its derivative should be zero at the correct solution. However, a general problem in the cost function formulation persists. The cost function should correspond to the norm of the error, which is not necessarily the case. This means that a reduction in the cost does not necessarily yield an improvement in the accuracy of solution. The

[15] Solid isotropic material with penalization.

error norm can be expressed in terms of the H^{-1}-norm, which, according to [255], can efficiently be computed on rectangular domains with Fourier transforms. Thus, the H^{-1}-norm can directly be used as a cost function and minimized.

Another aspect is to take care of an efficient numerical differentiation, which is advantageous for the residual of the partial differential equation [256], as automatic differentiation may be erroneous due to spurious oscillations between collocation points. Thus, numerical differentiation enforces regularity, which was exploited in [256] by coupling automatic differentiation and numerical differentiation to retain the advantages of automatic differentiation.

Further specialized modifications to neural network architectures have been proposed. Adaptive activation functions [257] have shown acceleration in convergence. Extreme learning machines [258, 259] remove the need for iterations altogether. All layers are randomly initialized in extreme learning machines, and only the last layer is learnable. Without a non-linear activation function, the parameters are found with a least-squares regression. This was demonstrated for physics-informed neural networks in [260]. Instead of only learning the last layer, the problem can be split into a non-linear and a linear regression problem, which are solved separately [261], such that the full expressivity of neural networks is retained.

Applications To Forward Problems

Physics-informed neural networks have been applied to various partial differential equations (see [196–198] for an overview). In computational mechanics, forward problems can, for example, be found in solid mechanics [251, 262, 263], fluid mechanics [264–271], and thermomechanics [272, 273]. Currently, physics-informed neural networks do not outperform classical solvers such as the finite element method [208, 274] in terms of speed for a given accuracy of engineering relevance. In the author's experience and judgement, this is especially the case for forward problems even if the extensions mentioned above are employed. Often, the mentioned gains compared to classical forward solvers disregard the training effort and only report evaluation times [275].

Incorporating large parts of the solution in the form of measurements with the data-driven loss $\mathcal{L}_{\mathcal{D}}$ improves the performance of physics-informed neural networks, which thereby can become a viable method in some cases. Yet, [276] states that data-driven methods outperform physics-informed neural networks. Thus physics-informed neural networks should not be regarded as a replacement for data-driven methods, but rather as a potential regularization technique for data-driven methods to reduce the generalization error.

Applications To Inverse Problems

However, physics-informed neural networks are in particular useful for (non-linear) inverse problems with full domain knowledge, i.e., the solution is available throughout the entire domain. This has, for example, been shown for the identification of material properties [252, 277–280]. By contrast, for inverse problems with only partial knowledge, the applicability of physics-informed neural networks is limited [281], as both forward and inverse solution have to be learned simultaneously, as seen in Chap. 9. Most applications therefore limit themselves to simpler inversions such as size and shape optimization. Examples are published, e.g., in [262, 282–287]. Exceptions that deal with the identification of entire fields can be found in full waveform inversion [288], topology optimization [289], elasticity, and the heat equation [290].

Inverse Problems

Physics-informed neural networks are capable of discovering governing equations by either learning the operator $\mathcal{N}$ or the coefficients λ. The resulting operator is, however, not always interpretable, and in the case of identification of the coefficients, the underlying partial differential equation is assumed. To discover interpretable operators, one can apply sparse regression approaches [291]. Here, potential differential operators are assumed as an input to the non-linear operator

$$\hat{\mathcal{N}}\left[x, u, \frac{\partial u}{\partial x}, \frac{\partial^2 u}{\partial x^2}, \dots\right] = 0. \tag{10.28}$$

Subsequently, a neural network learns the corresponding coefficients using observed solutions inserted into Eq. 10.28. The evaluation of the differential operators is achieved through automatic differentiation by first interpolating the solution with a neural network. Sparsity is ensured with an L^1-regularization.

A more sophisticated and complete framework is AI-Feynman [292]. Sequentially, dimensional analysis, polynomial regression, and brute force search algorithms are applied to identify fundamental laws in the data. If unsuccessful, a neural network interpolates the data, which can thereby be queried for symmetry and separability. The identification of symmetries leads to a reduction in variables, i.e., a reduction of the input space. In the case of separability, the problem is decomposed into two subproblems. The reduced problems or subproblems are iteratively fed through the framework until an equation is identified. AI-Feynman has been successfully applied to 100 equations from the Feynman lectures [293].

10.1.2.2 Time-Stepping Procedures

Again Eqs. 10.2 and 10.3 will be considered for the time-stepping procedures.

Physics-Informed Neural Networks

In the spirit of domain decomposition, parareal physics-informed neural networks [294] split up the temporal domain in subdomains $[t_i < t_{i+1}]$. A rough estimate of the solution u is provided by a conjugate gradient solver on a simplified form of the partial differential equation starting from t_0. Physics-informed neural networks are then independently applied in each subdomain to correct the estimate. Subsequently, the conjugate gradient solver is applied again, starting from t_1. This process is repeated until all time steps have been traversed. A closely related approach can be found in [295], where a physics-informed neural network is retrained on successive time segments. It is however ensured that previous time steps are kept fulfilled through a data-driven loss term for time segments that were already learned.

Another approach are the discrete-time physics-informed neural networks [195], which consider the temporal dimension in a discrete manner. The differential equation from Eq. 10.2 is discretized with the Runge-Kutta method with q stages [296]:

$$u^{n+c_i} = u^n + \Delta t \sum_{j=1}^{q} a_{ij} \mathcal{N}^{\mathcal{T}}[u^{n+c_j}], \qquad i = 1, \dots, q, \tag{10.29}$$

$$u^{n+1} = u^n + \Delta t \sum_{j=1}^{q} b_j \mathcal{N}^{\mathcal{T}}[u^{n+c_j}], \tag{10.30}$$

where

$$u^{n+c_j}(x) = u(t^n + c_j \Delta t, x), \qquad j = 1, \dots, q. \tag{10.31}$$

A neural network S_{NN} predicts all stages $i = 1, \dots, q$ from an input x:

$$\hat{\boldsymbol{u}} = [\hat{u}^{n+c_1}(x), \dots, \hat{u}^{n+c_q}(x), \hat{u}^{n+1}(x)] = S_{NN}(x; \boldsymbol{\Theta}). \tag{10.32}$$

The cost is then constructed by rearranging Eqs. (10.29) and (10.30).

$$\hat{u}^n = \hat{u}_i^n = \hat{u}^{n+c_i} - \Delta t \sum_{j=1}^{q} a_{ij} \mathcal{N}^{\mathcal{T}}[\hat{u}^{n+c_j}], \qquad i = 1, \dots, q, \tag{10.33}$$

$$\hat{u}^n = \hat{u}_{q+1}^n = \hat{u}^{n+1} - \Delta t \sum_{j=1}^{q} b_j \mathcal{N}^{\mathcal{T}}[\hat{u}^{n+c_j}]. \tag{10.34}$$

The $q+1$ predictions $\hat{u}_i^n$, $\hat{u}_{q+1}^n$ of $\hat{u}^n$ have to match the initial conditions $\tilde{u}^n$, where the mean squared error is used as a loss function to learn all stages $\hat{\boldsymbol{u}}$. The approach has been applied to fluid mechanics [297, 298].

Inverse Problems

As for inverse problems in the space-time approaches, the non-linear operator $\mathcal{N}$ can be learned. For temporal problems, this corresponds to the right-hand side of Eq. 10.2 for partial differential equations and to Eq. 10.3 for systems of ordinary differential equations. The predicted right-hand side can then be used to predict a time series using a classical time-stepping scheme, as proposed in [299]. More sophisticated methods leaning on similar principles are presented in the following. Specifically, we will discuss PDE-net for discovering partial differential equations, SINDy for discovering systems of ordinary differential equations in an interpretable sense (cf. Sect. 6.3), and an approach relying on multistep methods for systems of ordinary differential equations. The multistep approach leads to a non-interpretable, but more expressive approximation of the right-hand side.

PDE-Net

PDE-net [300, 301] is designed to learn both the system dynamics $u(x, t)$ and the underlying differential equation it follows. Given a problem of the form of Eq. 10.2, the right-hand side can be approximated as a function of coordinates and gradients of the solution.

$$\hat{\mathcal{N}}^{\mathcal{T}}\left[x, u, \frac{\partial u}{\partial x}, \frac{\partial^2 u}{\partial x^2}, \ldots\right] \tag{10.35}$$

The operator $\hat{\mathcal{N}}^{\mathcal{T}}$ is approximated by neural networks. The first step involves estimating spatial derivatives using learnable convolutional filters. The filters are designed to adjust their order of approximation based on the fit to the underlying measurements $\tilde{u}$, while the type of gradient is predefined.[16] Thus, the neural network learns how to best approximate spatial derivatives specific to the underlying data. Subsequently, the inputs of $\hat{\mathcal{N}}^{\mathcal{T}}$ are combined with point-wise convolutional neural networks [302] in [300] or a symbolic network in [301]. Both yield an interpretable operator from which the analytical expression can be extracted. In order to construct a loss function, Eqs. (10.2) and (10.35) are discretized using the forward Euler method:

$$u(x, t_{n+1}) = u(x, t_n) + \Delta t \hat{\mathcal{N}}^{\mathcal{T}}\left[x, u, \frac{\partial u}{\partial x}, \frac{\partial^2 u}{\partial x^2}, \ldots\right]. \tag{10.36}$$

This temporal discretization is applied iteratively, and the discrepancy between the derived function and the measured data $\tilde{u}(x, t_n)$ serves as the loss function.

Sparse Identification of Non-linear Dynamical Systems—SINDy

As previously introduced in Sect. 6.3, sparse identification of non-linear dynamical systems (SINDy) [303] deals with the discovery of dynamical systems of the form of Eq. 10.3. The task is posed as a sparse regression problem. Snapshot matrices of the state $\boldsymbol{X} = (\boldsymbol{x}(t_1), \boldsymbol{x}(t_2), \ldots, \boldsymbol{x}(t_n))$ and its time derivative $\dot{\boldsymbol{X}} = (\dot{\boldsymbol{x}}(t_1), \dot{\boldsymbol{x}}(t_2), \ldots, \dot{\boldsymbol{x}}(t_n)]$ are related to one another via candidate functions $\boldsymbol{\Theta}(\boldsymbol{X})$ evaluated at $\boldsymbol{X}$ using unknown coefficients $\boldsymbol{\Xi}$:

$$\dot{\boldsymbol{X}} = \boldsymbol{\Theta}(\boldsymbol{X})\boldsymbol{\Xi}. \tag{10.37}$$

The coefficients $\boldsymbol{\Xi}$ are determined through sparse regression, such as sequential thresholded least squares or LASSO regression. By including partial derivatives, SINDy has been extended to the discovery of partial differential equations [304, 305].

[16] This is enforced through constraints using moment matrices of the convolutional filters.

The expressivity of SINDy can further be increased by a coordinate transformation into a basis (and a corresponding coordinate system) allowing for a simpler representation of the system dynamics. This can be achieved with an autoencoder (consisting of an encoder $e_{NN}(x; \boldsymbol{\Theta}^e)$ and a decoder $d_{NN}(h; \boldsymbol{\Theta}^d)$, as proposed in [306], where the dynamics are learned on the reduced latent space h using SINDy. A simultaneous optimization of the neural network parameters $\boldsymbol{\Theta}^e$, $\boldsymbol{\Theta}^d$ and SINDy parameters Ξ is conducted with gradient descent. The cost is defined in terms of the autoencoder reconstruction loss $\mathcal{L}_{\mathcal{A}}$ and the residual of Eq. 10.37 at both the reduced latent space $\mathcal{L}_{\mathcal{R}}$ and the original space $\mathcal{L}_{\mathcal{F}}$.[17] An L^1-regularization for Ξ promotes sparsity.

$$\mathcal{L}_{\mathcal{A}} = \frac{1}{n}\sum_{i=1}^{n} ||\boldsymbol{x}(t_i) - d_{NN}\big(e_{NN}(\boldsymbol{x}(t_i); \boldsymbol{\Theta}^e); \boldsymbol{\Theta}^d\big)||_2^2 \tag{10.38}$$

$$\mathcal{L}_{\mathcal{R}} = \frac{1}{n}\sum_{i=1}^{n} ||\underbrace{\Big(\nabla_x e_{NN}\big(\boldsymbol{x}(t_i); \boldsymbol{\Theta}^e\big)\Big) \cdot \dot{\boldsymbol{x}}(t_i)}_{\dot{h}} - \boldsymbol{\Theta}\Big(e_{NN}\big(\boldsymbol{x}(t_i); \boldsymbol{\Theta}^e\big)\Big)\Xi||_2^2 \tag{10.39}$$

$$\mathcal{L}_{\mathcal{F}} = \frac{1}{n}\sum_{i=1}^{n} ||\dot{\boldsymbol{x}}(t_i) - \nabla_h d_{NN}\big(\underbrace{e_{NN}(\boldsymbol{x}(t_i); \boldsymbol{\Theta}^e)}_{h}; \boldsymbol{\Theta}^d\big) \cdot \underbrace{\boldsymbol{\Theta}\Big(e_{NN}(\boldsymbol{x}(t_i); \boldsymbol{\Theta}^e)\Big)\Xi}_{\dot{h}} ||_2^2 \tag{10.40}$$

$$C = \kappa_{\mathcal{A}}\mathcal{L}_{\mathcal{A}} + \kappa_{\mathcal{R}}\mathcal{L}_{\mathcal{R}} + \kappa_{\mathcal{F}}\mathcal{L}_{\mathcal{F}} \tag{10.41}$$

As in Eq. 10.23, a weighted cost function with weights $\kappa_{\mathcal{A}}, \kappa_{\mathcal{R}}, \kappa_{\mathcal{F}}$ is employed. The reduced latent space can be exploited for forward simulations of the identified system. By solving the system with classical time-stepping schemes in the reduced latent space, the solution is obtained in the full space through the decoder, as outlined in [307]. Thus, a reduced order model of a previously unknown system is identified. The downside is, that the model no longer is interpretable in the full space.

Multistep Methods

Another approach [308] to learning the system dynamics from Eq. 10.3 is to approximate the right-hand side directly with a neural network $\hat{\boldsymbol{f}}(\boldsymbol{x}_i) = O_{NN}(\boldsymbol{x}_i; \boldsymbol{\Theta}), \boldsymbol{x}_i = \boldsymbol{x}(t_i)$. A residual can be formulated by considering linear multistep methods [296]. In general, these methods take the form:

$$\sum_{m=0}^{M}[\alpha_m \boldsymbol{x}_{n-m} + \Delta t \beta_m \boldsymbol{f}(\boldsymbol{x}_{n-m})] = 0, \tag{10.42}$$

where $M, \alpha_0, \alpha_1, \beta_0, \beta_1$ are parameters specific to a multistep scheme. The scheme can be reformulated with a cost function, given as:

[17] The encoder and decoder are derived with respect to their inputs to estimate the derivatives $\dot{\boldsymbol{x}}, \dot{\boldsymbol{h}}$ using the chain rule.

$$C = \frac{1}{m_{\mathcal{D}} - M + 1} \sum_{n=M}^{m_{\mathcal{D}}} ||\hat{\boldsymbol{y}}_n||_2^2 \tag{10.43}$$

$$\hat{\boldsymbol{y}}_n = \sum_{m=0}^{M} [\alpha_m \boldsymbol{x}_{n-m} + \Delta t \beta_m \hat{\boldsymbol{f}}(\boldsymbol{x}_{n-m})] \tag{10.44}$$

The idea of the method is strongly linked to the discrete-time physics-informed neural network presented in Paragraph "Physics-Informed Neural Networks", where a reformulation of the Runge-Kutta method yields the cost function needed to learn the forward solution.

10.1.2.3 Enforcement of Physics by Construction

Up to this point, this chapter only considered the case where physics are enforced indirectly through penalty terms of the partial differential equation residual. The only exception, and the first example of enforcing physics by construction, was the strong enforcement of boundary conditions [166, 210, 211] by modifying the outputs of the neural network—which led to a fulfillment of the boundary conditions independent of the neural network parameters. For partial differential equations, this can be achieved by manipulating the output, such that the solution automatically obeys fundamental physical laws. Examples for this are, e.g., given in [309], where stream functions are predicted and subsequently differentiated to ensure conservation of mass, the incorporation of symmetries [310], or invariances [311] by using integrity bases [312]. Dynamical systems have been treated by learning the Lagrangian or Hamiltonian with correspondingly Lagrangian neural networks [313–315] and Hamiltonian neural networks [316]; see Sect. 3.9.8. The quantities of interest are obtained through the differentiable neural network and compared to labeled data. Indirectly learning the quantities of interest through the Lagrangian or Hamiltonian improves generalization capabilities by obeying the structure of correspondingly Hamiltonian mechanics or Lagrangians. Similar techniques were observed in Chap. 7 for material modeling. Enforcing the physics by construction is also referred to as physics-constrained learning, as the learnable space is constrained. Note, however, that constraining the learnable space also challenges the learning algorithm, thus potentially rendering convergence more difficult. Therefore, [189] relaxes the requirement of fulfilling the physical laws by introducing a secondary unconstrained network—acting additively on the solution—whose influence is scaled by a hyperparameter. More examples of physics enforcement by construction are provided in the context of simulation enhancement in Sect. 10.2.2.

10.2 Simulation Enhancement

The category of simulation enhancement deals with any deep learning technique that interacts directly with and, thus, improves a component of a classical simulation. This is the most diverse category and will therefore be subdivided into the individual steps of a classical simulation pipeline:

- pre-processing
- physical modeling
- numerical methods
- post-processing.

Both data-driven and physics-informed approaches will be discussed in the following.

10.2.1 Pre-processing

The discussed pre-processing methods are trained in a supervised manner relying on the techniques presented in Sect. 10.1.1 and on labeled data.

10.2.1.1 Data Preparation

Data preparation includes tasks, such as geometry extraction. For instance the detection of cracks from images by means of segmentation [317–319] can subsequently be used in simulations to assess the impact of the identified cracks. Also, convolutional neural networks have been used to prepare voxel data obtained from computed tomography scans; see [320], where scanning artifacts are removed. Similarly neural networks can be employed to enhance measurement data. This was, for example, demonstrated in [321], where the neural network acts as a denoiser for magnetic signals in the scope of nondestructive testing. Similarly, low-frequency extrapolation for full waveform inversion has been performed using neural networks [322–324].

10.2.1.2 Initialization

Instead of preparing the data, the simulation can be accelerated by an initialization. This can, for example, be achieved through initial guesses by neural networks, providing a better starting point for classical iterative solvers [325].[18] A tighter integration is achieved by using a pre-trained [243] neural network ansatz whose parameters are subsequently tweaked by the classical solver, as demonstrated for full waveform inversion in [187].

[18] Here, the initial guess is incorporated through a regularization term.

10.2.1.3 Meshing

Finally, many simulation techniques rely on meshes. This can be achieved indirectly with neural networks, by prediction of mesh density functions [326–330] incorporating either expert knowledge of where small elements are needed, or relying on error estimations. Subsequently, a classical mesh generator is employed. However, neural networks (specifically let-it-grow neural networks [331]) have also been proposed directly as mesh generators [332, 333].

10.2.2 Physical Modeling

Physical models that capture physical phenomena accurately are the core component of mechanics. Deep learning offers three main approaches for physical models. Firstly, a neural network is used as the physical model directly (model substitution). Secondly, an underlying model may be assumed where a neural network determines its coefficients (identification of model parameters). Lastly, the entire model can be identified by a neural network (model identification). In the first approach, the neural network is integrated within the simulation pipeline, while the latter two rely on incorporation of the identified models in a classical sense.

For illustration purposes, the approaches are mostly explained on the example of constitutive models (see [334] for a recent review on constitutive modeling with neural networks). Here, the task is to relate the strain $\boldsymbol{\varepsilon}$ to a stress $\boldsymbol{\sigma}$, i.e., find a function $\boldsymbol{\sigma} = f(\boldsymbol{\varepsilon})$. This can, for example, be used within a finite element framework to determine the element stiffness, as elaborated in [335]. Note that model substitution was discussed in detail in Chap. 7 for material modeling. Here, the perspective is kept broader but shallower.

10.2.2.1 Model Substitution

In model substitution, a neural network f_{NN} replaces the model, yielding the prediction $\hat{\boldsymbol{\sigma}} = f_{NN}(\boldsymbol{\varepsilon}; \boldsymbol{\Theta})$. The quality of the model is assessed with a data-driven cost function (Eq. 10.4) using labeled data $\tilde{\boldsymbol{\sigma}}, \tilde{\boldsymbol{\varepsilon}}$. The approach is applied to a variety of problems, where the key difference lies in the definition of input and output quantities. The same deep learning techniques from data-driven simulation substitution (Sect. 10.1.1) can be employed.

Applications include predictions of stress from strain [335, 336], flow stresses from temperatures, strain rates and strains [337, 338], yield functions [339], crack opening responses from stresses [340], contact stiffness from penetration and contact pressure [341], point of contact from position of neighboring nodes of finite elements [342], or control points of NURBS surfaces [343]. Source terms of simplified equations or coarser discretizations have also been learned for turbulence [31, 344,

345] and the wave equation [346]. Here, the reference—a high-fidelity model—is to be captured in the best possible way by the source term.

Variations also predict the quantity of interest indirectly. For example, strain energy densities W are predicted by neural networks from deformation gradients $\boldsymbol{F}$, and subsequently derived using automatic differentiation to obtain stresses [347, 348]. The approach can also be extended to incorporate uncertainty quantification [349]. By extending the input space with microstructural information, an in-built homogenization is added to the constitutive model [350–352]. Thus, the macroscale simulation considers the microstructure at the integration points in the sense of FE^2 [353, 354], but without an additional finite element computation. Incorporation of microstructures requires a large amount of realistic training data, which can be obtained through generative approaches as discussed in Sect. 10.4. Active learning can reduce the required number of simulations on these geometries [184].

A specialized neural network architecture is employed by [355], where a neural network first estimates invariants I of the deformation gradient $\boldsymbol{F}$ and thereupon predicts the strain energy density, thus mimicking the classical constitutive modeling approach. Another network extension is the use of recurrent neural networks to learn history-dependent models. This was shown in [350, 351, 356, 357] for the prediction of the stress increment from the stress-strain history, the strain energy from the strain energy history [358], and crack patterns based on prior cracks and crystalline orientations [359, 360].

The learned models do not, however, necessarily obey fundamental physical laws. Attempts to incorporate physics as constraints using penalty terms have been made in [361–363]. Still, physical consistency is not guaranteed. Instead, neural network architectures can be chosen such that they satisfy physical requirements by construction. In constitutive modeling, objectivity can be enforced by using only deformation invariants as input [364], and polyconvexity can be enforced through the architecture, such as input-convex neural networks [365–368] or neural ordinary differential equations [364, 369]. It was demonstrated that ensuring fundamental physical aspects such as invariants combined with polyconvexivity delivers a much better behavior for unseen data, especially if the model is used in extrapolation.

Input-convex neural networks [370] enforce the convexity with specialized activation functions such as log-sum-exponential, or softplus functions in combination with constraints on the neural network weights to ensure that they are positive, while neural ordinary differential equations [371] (discussed in Sect. 10.3) approximate the strain energy density derivatives and ensure non-negative values. Alternatively, a mapping from the neural network to a convex function can be defined [372] ensuring a convex function for any neural network output. Related are also thermodynamics-based neural networks [373, 374], e.g., applied to complex microstructures in [375], which by construction obey fundamental thermodynamic laws. Training of these methods can be performed in a supervised manner, relying on stress-strain data, or unsupervised. In the unsupervised setting, the constitutive model is incorporated in a finite element solver, yielding a displacement field for a specific boundary value problem. The computed field, together with measurement data, yields a residual that is referred to as the modified constitutive relation error (mCRE) [376–378], which is minimized to improve the constitutive relation [379, 380]. Instead of formulating the mismatch in terms of displacements, [381, 382] formulate it in terms of boundary forces. For an in-depth overview of constitutive model substitution in deep learning; see [15].

10.2.2.2 Identification of Model Parameters

Identification of model parameters is achieved by assuming an underlying model and training a neural network to predict its parameters for a given input. In the constitutive model example, one might assume a linear elastic model expressed in terms of a constitutive tensor $\boldsymbol{C}$, such that $\boldsymbol{\sigma} = \boldsymbol{C}\boldsymbol{\varepsilon}$. The constitutive tensor can be predicted from the material distribution defined in terms of a heterogeneous elasticity modulus $E(\boldsymbol{x})$ defined throughout the domain

$$\hat{\boldsymbol{C}} = f_{NN}(E(\boldsymbol{x}); \boldsymbol{\Theta}). \tag{10.45}$$

Typical applications are homogenization, where effective properties are predicted from the geometry and material distribution. Examples are convolutional neural network-based homogenizations on computed tomography scans [383, 384], predictions of in-vivo constitutive parameters of aortic walls from its geometry [385], predictions of elastoplastic properties [386] from instrumented indentation results relying on a multi-fidelity approach [387], prediction of stress intensity factors from the geometry in microfabricated microcantilevers [388], estimation of effective bone properties from the boundary conditions and applied stresses within a finite element, and incorporating meso-scale information by training a neural network on representative volume elements [389].

10.2.2.3 Model Identification

Neural network models as a replacement of classical approaches are not interpretable, while only identifying model parameters of known models restricts the models capacity. This gap can be bridged by the identification of models in terms of parsimonious mathematical expressions.

The typical procedure is to pose the problem in terms of candidate functions and to identify the most relevant terms. The methodology was inspired by SINDy [303] and introduced in the framework for efficient unsupervised constitutive law identification and discovery (EUCLID) [390]. The approach is unsupervised, as the stress-strain data is only indirectly available through the displacement field and corresponding reaction forces. The N_I invariants I_i of the deformation gradient $\boldsymbol{F}$ are inserted into a candidate library $Q(\{I_i\}_{i=1}^{N_I})$ containing the candidate functions. Together with the corresponding weights $\boldsymbol{\Theta}$, the strain density W is determined:

$$W(\{I_i\}_{i=1}^{N_I}) = Q^{\mathsf{T}}(\{I_i\}_{i=1}^{N_I})\boldsymbol{\Theta}. \tag{10.46}$$

Through derivation of the strain density W using automatic differentiation, the stresses $\boldsymbol{\sigma}$ are determined. The problem is then cast into the weak form with which the linear momentum balance is enforced. The weak form is then minimized with respect to $\boldsymbol{\Theta}$ using a fixed-point iteration scheme (inspired by [391]), where a L^p-regularization is used to promote sparsity in $\boldsymbol{\Theta}$. Despite its young age, the approach has already been applied to plasticity [392], viscoelasticity [393], combinations [394], and has been extended to incorporate uncertainties through a Bayesian model [395]. Furthermore, the approach has been extended with an ensemble of input-convex neural networks [382], yielding a more accurate, but less interpretable model.

A similar effort was recently carried out by [396, 397], where neural networks are designed to retain interpretability. This is achieved through sparse connections in combination with specialized activation functions representing candidate functions, such that they are able to capture classical forms of constitutive terms. Through the sparse connections in the network and the specialized activation functions, the neural network's weights become physical parameters, yielding an interpretable model. This is best understood by consulting Fig. 10.5, where the strain energy density is expressed as

$$\begin{aligned} \hat{W} =& w_0^1 e^{w_0^0 I_1} + w_1^1 \ln(w_1^0 I_1) + w_2^1 e^{w_2^0 I_1^2} + w_2^1 \ln(w_2^0 I_1^2) + \\ & w_3^1 e^{w_3^0 I_2} + w_4^1 \ln(w_4^0 I_2) + w_5^1 e^{w_5^0 I_2^2} + w_6^1 \ln(w_6^0 I_2^2) \end{aligned} \tag{10.47}$$

with the learnable parameters $\boldsymbol{w}^0$, $\boldsymbol{w}^1$. Differentiating the predicted strain energy density $\hat{W}$ with respect to the invariants I_i yields the constitutive model, relating stress and strain.

10.2.3 Numerical Methods

This subsection describes efforts in which neural networks are used to replace or enhance classical numerical schemes to solve partial differential equations.

10.2.3.1 Algorithm Enhancement

Classical algorithms can be enhanced by neural networks, by learning corrections to commonly arising numerical errors, or by estimating tunable parameters within the algorithm. Corrections have, for example, been used for numerical quadrature [398] in the context of finite elements. Therein, neural networks are used to predict adjustments to quadrature weights and positions from the nodal positions to improve the accuracy for distorted elements. Similarly, neural networks have been applied as correction for strain-displacement matrices for distorted elements [399]. Neural networks have also been employed to provide improved gradient estimates. Specifically,

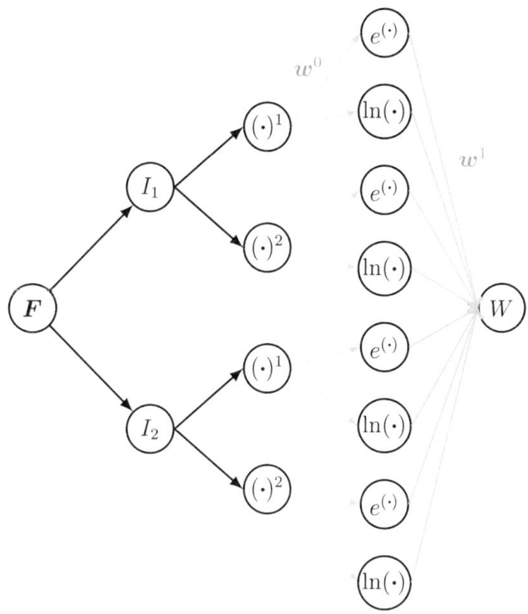

Fig. 10.5 Automated model discovery through a sparsely connected neural network with specialized activation functions acting as candidate functions. The thick black connections are not learnable, while the gray ones represent linearly weighted connections. Figure adapted and simplified from [396]

[400] modify the gradient computation to match a fine scale simulation on a coarse grid:

$$\frac{\partial^n u}{\partial x^n} \approx \sum_i \alpha_i^{(n)} u_i. \tag{10.48}$$

The coefficients α_i are predicted by neural networks from the current coarse solution. Special constraints are imposed on α_i to guarantee accurate derivatives. Another application are specialized strain mappings for damage mechanics embedded within individual finite elements learned by physics-informed neural networks [401]. It has even been suggested to partially replace solvers. For example, [402] replace either the fluid or structural solver by a surrogate model for fluid-structure interaction problems.

Learning tunable parameters was demonstrated for the estimation of the largest possible time step using a recurrent neural network acting at the latent vector of an autoencoder [403]. Also, optimal test functions for finite elements were learned to improve stability [404]. Another approach to learning numerical parameters for simulation is presented in [405], where hyperparameters connected to a similarity-based topology optimization are learned—specifically, an energy scaling factor is predicted from a dissimilarity metric based on a previous topology optimization. These approaches have in common that they spare the user from performing multiple simulations to tune the numerical parameters.

10.2.3.2 Multiscale Methods

Multiscale methods have been proposed to efficiently integrate and resolve systems acting on multiple scales. One approach are the learned constitutive models from Sect. 10.2.2 that incorporate the microstructure. This is essentially achieved through a homogenization at the mesoscale used within a macroscale simulation.

A related approach is element substructuring [406, 407], where superelements mimic the behavior of a conglomerate of classic basic finite elements. In [408], the superelements are enhanced by neural networks, which draw on the boundary displacements to predict the displacements and stresses within the element as well as the reaction forces at the boundary. Through assembly of the reaction forces in the global finite element system, an equilibrium is reached with a Newton-Raphson solver. Similarly, the approach in [409] learns the internal forces from the coarse degrees of freedom of the superelements. These approaches are particularly valuable, as they can seamlessly incorporate history-dependent behavior using recurrent neural networks.

Finally, multiscale analysis can also be performed by first solving a coarse global model with a subsequent local analysis. This is referred to as zooming methods. In [410], a neural network learns the global model and thereby predicts the boundary conditions for the local model. In a similar sense, DeepONets have been applied for the local analysis [411], whereas the global analysis is performed with a finite element solver. Both are conducted in an alternating fashion until convergence is reached.

10.2.3.3 Optimization

Optimization is a fundamental task within computational mechanics and therefore addressed separately. It is not only used to find optimal structures, but also to solve inverse problems, as discussed in Chap. 9. Generally, the task can be formulated as minimizing a cost function C with respect to parameters λ. In computational mechanics, λ is typically fed to a forward simulation $u = S(\lambda)$, yielding a solution u inserted into the cost function C. If the gradients $\nabla_\lambda C$ are available, gradient-based optimization is the state-of-the-art [412], where the gradients are used to update λ. In order to access the gradients, the forward simulation S has to be differentiable. This requirement is, for example, utilized within the branch of deep learning called differentiable physics [413]. Incorporating gradient information from the numerical solver into the neural network improves learning, feedback, and generalization. An overview and introduction to differentiable physics is provided in [413], with applications in [178, 371, 400, 414–416].[19]

The iterative gradient-based optimization procedure was previously illustrated in Fig. 9.2 in Chap. 9. For an in-depth treatment of neural networks in structural optimization; see the recent review [11].

[19] Applications of differentiable physics vary widely and are addressed throughout this work.

Inserting a learned forward operator S, as those discussed in Sect. 10.1.1, into an optimization problem provides two advantages [417–421]. Firstly, a faster forward operator results in faster optimization iterations. Secondly, the gradient computation is simplified, as automatic differentiation through the forward operator S is straightforward in contrast to the adjoint state method [422, 423]. Note however, that for time-stepping procedures, the computational cost might be greater for automatic differentiation, as shown in [281]. Applications include full waveform inversion [281], topology optimization [424–426], and control problems [27, 29, 414].

Similarly, an operator replacing the sensitivity computation can be learned [426–429]. This can be achieved in a supervised manner with precomputed sensitivities to reduce the cost C [426, 428], or by intending to maximize the improvement of the cost function after the gradient update [427, 429]. In [427, 429], an evolutionary algorithm was employed for the general case that the sensitivites are not readily available. Training can adaptively be reintroduced during the optimization phase, if the cost C does not decrease [426], improving the neural network for the specific problem it is handling. Taking this idea to the extreme, the neural network is trained on the initial gradient updates of a specific optimization. Later, solely the neural network delivers the sensitivities [430] with supervised updates every n updates to improve accuracy, where n is a hyperparameter. The ideas of learning a forward operator and a sensitivity operator are combined in [425], where it is pointed out that the sensitivity from automatic differentiation through the learned forward operator can be inaccurate, despite an accurate forward operator.[20] Therefore, an additional loss term is added to the cost function, enforcing the correctness of the sensitivity through labels obtained with the adjoint state method. Alternatively, the sensitivity computation can be enhanced by correcting the sensitivity computation performed on a coarse grid, as proposed in [431] and related to the multiscale techniques discussed in Sect. 10.2.3.2. Here, the adjoint field used for the sensitivity computation is reduced by both a proper orthogonal decomposition, and a coarser discretization. Subsequently, a neural network corrects the coarse estimate through a super-resolution neural network [432]. Similarly, [426, 433] maps the forward solution on a coarse grid to the design variable sensitivity on a fine grid. A similar application is a correction term within a fixed-point iterator, as outlined in [434].

Related to the sensitivity predictions are approaches that directly predict an updated state. The goal is to decrease the total number of iterations. In practice, a combination of predictions and classical gradient-based updates is performed [68–70, 435]. The main variations between the methods in the literature are the inputs and how far the forecasting is performed. In [68], the update is obtained from the current state and gradient, while [70] predicts the final state from the history of initial updates. The history is also considered in [69], but the prediction is performed on subpatches which are then stitched together.

Another option of introducing neural networks to the optimization loop is to use neural networks as an ansatz of λ; see, e.g., [101, 281, 414, 436–443] or Chapter 9.

---

[20] Although automatic differentiation in principle has a high accuracy, oscillations between the sampled points may lead to spurious gradients with regard to the sampled points [207].

In the context of inverse problems [281, 414, 436–440], the neural network acts as a regularizer on a spatially varying inverse quantity $\lambda(x) = I_{NN}(x; \boldsymbol{\Theta})$, providing both smoother and sharper solutions, as seen for full waveform inversion in Sect. 9.2.4. For topology optimization with a neural network parametrization of the density function [101, 441–443], no regularizing effect was observed. It was however possible to obtain a greater design diversity [444] through different initializations of the neural network; see Section 9.3.3.2. Extensions using specialized neural network architectures for implicit representations [445–450] have been presented in the context of topology optimization in [451]. Furthermore, [281, 438, 442] showed how to conduct the gradient computation without automatic differentiation through the solver S. The gradient computation is split up via the chain rule:

$$\nabla_{\boldsymbol{\Theta}} C = \nabla_{\lambda} C \nabla_{\boldsymbol{\Theta}} \lambda. \tag{10.49}$$

The first gradient $\nabla_{\lambda} C$ is computed with the adjoint state method, such that the solver can be treated as a black box. The second gradient $\nabla_{\boldsymbol{\Theta}} \lambda$ is obtained through automatic differentiation. An additional advantage of the neural network ansatz is that, if applied to multiple solutions with a problem specific input, the neural network is trained. Thus, after sufficient inversions, the neural network can be used as predictor, as presented in [452]. The training can also be performed in combination with labeled data, yielding a semi-supervised approach, as demonstrated in [187, 453, 454] and seen by the transfer learning example in Sect. 9.2.5.1.

10.2.4 Post-processing

Post-processing concerns the modification and interpretation of the computed solution. One motivation is to reduce the numerical error of the computed solution. This can for example be achieved with super-resolution techniques relying on specialized convolutional neural network architectures from computer vision [455, 456]. Coarse-to-fine mappings can be obtained in a supervised manner using matching coarse and fine simulations as labeled data, as presented for turbulent flows [432, 457] and topology optimization [458–460]. The mapping is typically performed from coarse-to-fine solution fields, but mappings from a posteriori errors have been proposed as well [461]. Further specialized extensions to the cost function have been suggested in the context of de-homogenization [462].

The methods can analogously be applied to temporal data where the solution is refined at each time step as, e.g., presented with recurrent neural networks as corrector of reduced order models [463]. However, coarse discretizations in dynamical models lead to an error accumulation, that increases with the number of time steps. Thus, a simple coarse-to-fine post-processing at each time step is not sufficient. To this end, [415, 416] apply a correction at each time step before the coarse solver predicts the next time step. As the correction is propagated through the solver, the sensitivities of the solver must be computed to perform the backward propagation. Therefore, a differentiable solver (i.e., differentiable physics) has to be employed. This significantly outperforms the purely supervised approach, where the entire coarse trajectory is applied without corrections in between. The number of steps performed is a hyperparameter, which increases the accuracy but comes with a higher computational effort. This concept is referred to as solver-in-the-loop.

Further variations perform the coarse-to-fine mapping in a patch-based manner, where the interfaces require special treatment [464]. Another approach uses a neural network to map the coarse solution to the closest fine solution stored in a database [465]. The mapping is performed on patches of the domain.

Other post-processing tasks include feature extraction. After a topology optimization, neural networks have been used to extract basic shapes to be used in a subsequent shape optimization [466, 467]. Another aspect that can be ensured through post-processing is manufacturability.

Lastly, adaptive mesh refinement falls under the category of post-processing as well. Closely related to the meshing approaches discussed in Sect. 10.2.1.3, neural networks have been proposed as error indicators [329, 468] that are trained in a supervised manner. The error indicators can subsequently be employed to adapt the mesh based on the error.

10.3 Discretizations as Neural Networks

Neural networks are composed of linear transformations and non-linear functions, which are basic building blocks of most partial differential equation discretizations. Thus, the motivation to construct neural networks utilizing discretizations of partial differential equations are twofold. Firstly, deep learning techniques can hereby be exploited within classical discretization frameworks. Secondly, novel neural network architectures arise, which are more tailored towards many physical problems in computational mechanics but potentially also find their use cases outside of that field.

10.3.1 Finite Element Method

One method are finite element neural networks [469, 470] (see [471–476] for applications), for which we consider the system of equations from a finite element discretization with the stiffness matrix K_{ij}, N degrees of freedom u_j, and the body load p_i:

$$\sum_{j=1}^{N} K_{ij}u_j - p_i = 0, i = 1, 2, \dots, N. \tag{10.50}$$

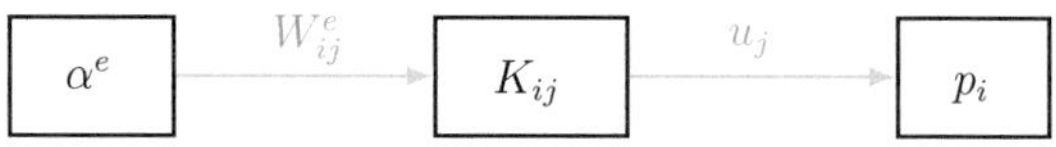

Fig. 10.6 Finite element neural networks, prediction of forces p_i from material coefficients α^e via assembly of global stiffness matrix K_{ij}, and evaluations of equations with the displacements u_j [470]

Assuming constant material properties along an element and uniform elements, a pre-integration of the local stiffness matrix $k^e_{ij} = \alpha^e w^e_{ij}$ can be performed, as, e.g., shown in [477]. The goal is to pull out the material coefficients of the integration, leading to the following assembly of the global stiffness matrix with M elements:

$$K_{ij} = \sum_{e=1}^{M} \alpha^e W^e_{ij} \text{ with } W^e_{ij} = \begin{cases} w^e_{ij} \text{ if } i, j \in e \\ 0 \text{ else} \end{cases} . \tag{10.51}$$

Inserting the assembly into the system of equations from Eq. 10.50 yields

$$\sum_{j=1}^{N} \left(\sum_{e=1}^{M} \alpha^e W^e_{ij} \right) u_j - p_i = 0, i = 1, 2, \ldots, N. \tag{10.52}$$

The nested summation has a similar structure of a fully connected neural network, $a_i^{(l)} = \sigma(z_i^{(l)}) = \sigma(\sum_{j=1}^{N^{(l)}} a_j^{(l-1)} + b_i^{(l)})$, (where $z_i^{(l)} = \sum_{j=1}^{N^{(l)}} a_j^{(l-1)} + b_i^{(l)}$) without activation σ and bias b (see Fig. 10.6):

$$a_i^{(2)} = \sum_{j=1}^{N^{(2)}} W_{ij}^{(1)} a_j^{(1)} = \sum_{j=1}^{N^{(2)}} W_{ij}^{(1)} (\sum_{k=1}^{N^{(1)}} W_{jk}^{(0)} a_k^{(0)}). \tag{10.53}$$

Thus, the stiffness matrix K_{ij} is the hidden layer. In a forward problem, W^e_{ij} are non-learnable weights, while u_j contains a mixture of learnable weights and non-learnable weights coming from the imposed Dirichlet boundary conditions. A loss can be formulated in terms of body load mismatch, as $\frac{1}{2} \sum_{i=1}^{N} (\hat{b}_i - b_i)^2$. In the inverse setting, α^e becomes learnable—instead of u_j, which is then fixed. For partial domain knowledge in the inverse case, u_j becomes partially learnable.

A different approach are the hierarchical deep learning neural networks (HiDeNNs) [478] with extensions in [479–484]. Here, shape functions are treated as neural networks constructed from basic building blocks. Consider, for example, the one-dimensional linear shape functions

$$N_1(x) = \frac{x - x_2^e}{x_1^e - x_2^e} \tag{10.54}$$

$$N_2(x) = \frac{x - x_1^e}{x_2^e - x_1^e}, \tag{10.55}$$

which can be represented as a neural network, as shown in Fig. 10.7, where the weights depend on the nodal positions x_1^e, x_2^e. The interpolated displacement field $u^e(x)$, which is valid in the element domain Ω^e, is obtained by multiplication with the nodal displacements u_1^e, u_2^e, treated as shared neural network weights.

$$u^e(x) = N_1^e(x)u_1^e + N_2^e(x)u_2^e \tag{10.56}$$

They are shared, as the nodal displacements u_1^e, u_2^e are also used for the neighboring elements u^{e-1}, u^{e+1}. Finally the displacement over the entire domain u is obtained by superposition of all elemental displacement fields u^e, which are first multiplied by a step function defined as 1 inside the corresponding element domain Ω^e and 0 outside.

A forward problem is solved with a minimization of the variational loss function, as presented in Sect. 10.2.2 with the nodal values u_i^e as learnable weights. According to [478], this is equivalent to iterative solution procedures employed for large systems of equations in finite elements. The additional advantage is a seamless integration of r-refinement [485–487] (also referred to as adaptive mesh refinement), i.e., the shift of nodal positions to optimal positions by making the nodal positions x_i^e learnable. Special care has to be taken to avoid element inversion, which is handled by an additional loss term. Inverse problems can similarly be solved by using learnable input parameters, as presented for topology optimization [484].

The method has been combined with reduced order modeling techniques [480]. Furthermore, the shape functions have been extended with convolutions [482, 483]. Specifically, a weighting field $W(x)$, i.e., kernel (e.g., radial basis functions) with learnable dilation parameter,[21] is introduced to enhance the finite element space $u^c(x)$ through convolutions, thereby increasing the space's expressivity and continuity:

$$u^c(x) = u^e(x) * W(x). \tag{10.57}$$

This introduces a smoothing effect over the elements and can efficiently be implemented using neural networks and, thereby, obtain a more favorable data-structure to exploit the full parallelization capabilities of GPUs [483]. The enhanced space has been incorporated in the HiDeNN framework. While an independent confirmation is still missing, the authors promise a speedup of several orders of magnitude compared to traditional finite element solvers [484].[22]

[21] The dilation parameter depends on x and thus introduces additional degrees of freedom throughout the domain.

[22] After close examination and exchanges with the authors of [478, 480, 482–484], we have concluded that the current speed-up is mainly attributable to the simultaneously employed reduced order models. Minor improvements in accuracy are possible through the employed r-adaptivity and convolutions, however, accompanied by an increase in computational effort. Furthermore, we have been informed that the HiDeNN methodology is implemented most efficiently in the standard finite element way, i.e., without neural networks, unlike the descriptions in [478, 480, 482–484]. Thus the shape functions $N_i^e(x)$ are implemented in a straightforward manner within automatic differentiation frameworks in order to obtain the derivatives of the solution u with respect to x, and

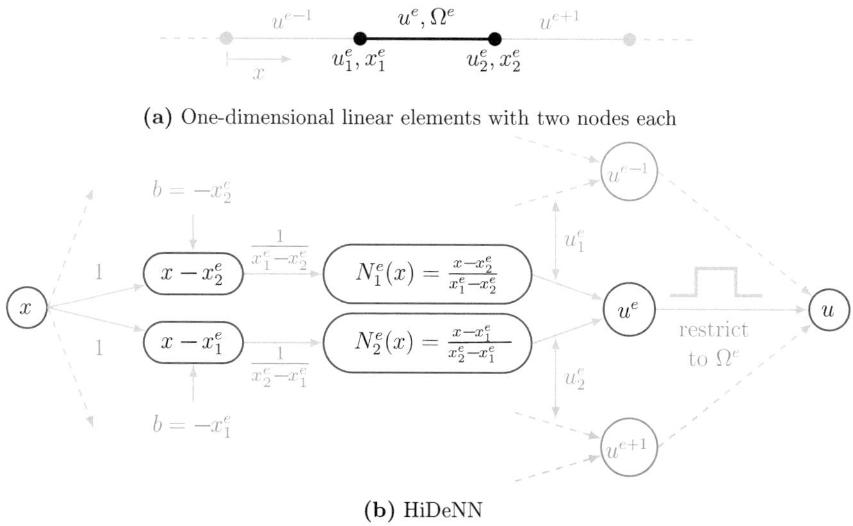

Fig. 10.7 HiDeNN with one-dimensional linear elements [478]

Lastly, another approach related to finite elements was presented as FEA-net [488, 489]. Here, the matrix-vector multiplication of the global stiffness matrix $\boldsymbol{K}$ and solution vector $\boldsymbol{u}$ including the assembly of the global stiffness matrix is replaced by a convolution. In other words, the computation of the force vector $\boldsymbol{f}$ is used to compute the residual $\boldsymbol{r}$.

$$\boldsymbol{r} = \boldsymbol{f} - \boldsymbol{K} \cdot \boldsymbol{u} \tag{10.58}$$

Assuming a uniform mesh with homogeneous material properties, the mesh is defined by the segment illustrated in Fig. 10.8. The degree of freedom u_j only interacts with the stiffness contributions K_i^1, K_i^2, K_{i+1}^1, K_{i+1}^2 of its neighboring elements i and $i+1$. Therefore, the force component f_j acting on node j can be expressed by a convolution:

$$f_j = [K_i^1, K_i^2 + K_{i+1}^1, K_{i+1}^2] * [u_{j-1}, u_j, u_{j+1}] \tag{10.59}$$

This can analogously be applied to all degrees of freedoms, with the same convolution filter $\boldsymbol{W} = [K^1, K^1 + K^2, K^2]$, assuming the same stiffness contributions for each element.

$$\boldsymbol{K} \cdot \boldsymbol{u} = \boldsymbol{W} * \boldsymbol{u} \tag{10.60}$$

---

the loss function with respect to the nodal positions x_i^e and degrees of freedom u^e. If convolutions are not employed, the sensitivities can be precomputed analytically, eliminating the need for automatic differentiation. Hence, the difference to a conventional finite element implementation is that the finite element discretization is solved with gradient descent optimization instead of solving the system of equations directly. This is more expensive but allows for the flexibility of seamlessly (i.e., without further manual derivations) introducing r-adaptivity or convolutions on top of the ansatz space.

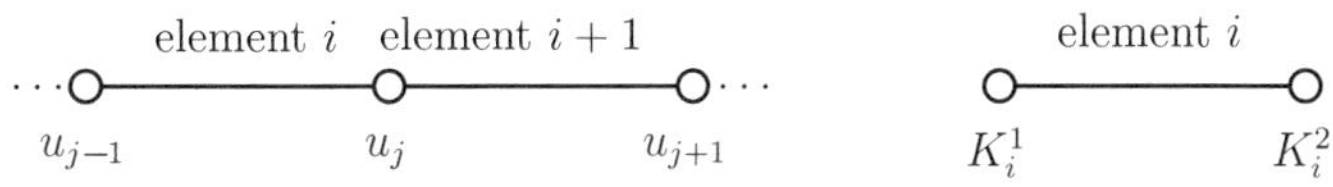

Fig. 10.8 Segment of one-dimensional finite element mesh with degrees of freedom (left). Local element definition with stiffness contributions (right)

The convolution can then be exploited in iterative schemes which minimize the residual $\boldsymbol{r}$ from Eq. 10.58. This saves the effort of constructing and storing the global stiffness matrix. By constructing the filter $\boldsymbol{W}$ as a function of the material properties of the adjacent elements, heterogeneities can be taken into account [489]. If the same iterative solver is employed, FEA-net is able to outperform classical finite elements for non-linear problems on uniform grids.

10.3.2 Finite Difference Method

Similar ideas have been proposed for finite differences [490], as for example employed in [281], where convolutional kernels are used as an implementation of stencils exploiting the efficient neural network libraries with GPU capabilities. Here, the learnable parameters can be the finite difference stencil for inverse problems or the output for forward problems. This has, for example, been presented in the context of full waveform inversion, which is modeled as a recurrent neural network [491, 492]. The stencils are written as convolutional filters and repeatedly applied to the current state and the corresponding inputs. These are the wave field, the material distribution, and the source. The problem can then be regarded as a recurrent neural network. However, it is computationally expensive to perform automatic differentiation throughout the time steps for full waveform inversion, thereby obtaining the sensitivities with respect to γ—both regarding memory and wall clock computational time. A remedy is to combine automatic differentiation with the adjoint state method as in [281, 438, 442] and discussed in Sect. 10.2.3.3.

Taking this idea one step further, the discretized wave equation can be regarded as an analog recurrent neural network [493] where the weights are the material distribution; see Fig. 10.9. Here, a binary material is learned in a trainable region (gray) between source (red) and probing location (black). The input $x(t)$ is encoded as a signal and emitted as source, which is measured at the probing locations $y_i(t)$ as output. By integrating the outputs, a classification of the input can be performed.

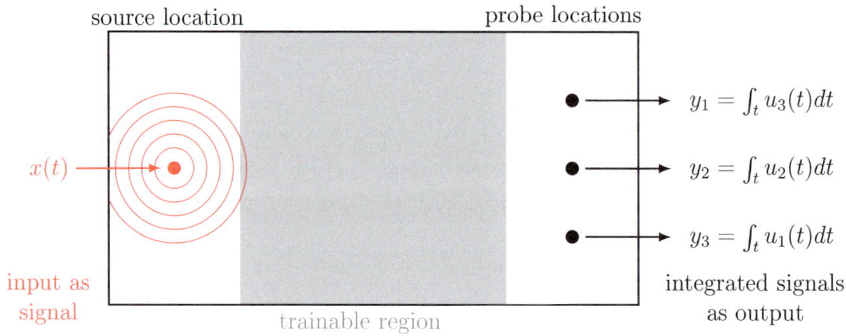

Fig. 10.9 Analog recurrent neural network

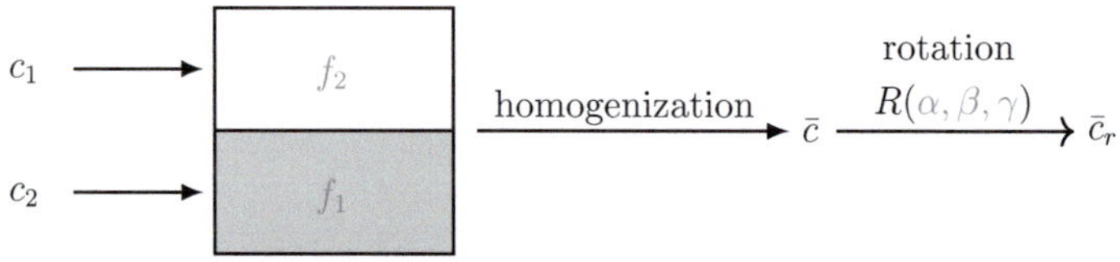

Fig. 10.10 A single building block of the deep material network [494]

10.3.3 Material Discretizations

Deep material networks [494, 495] construct a neural network from a material distribution. An output is constructed from basic building blocks, inspired by analytical homogenization techniques. Given two materials defined in terms of their compliance tensors c_1, c_2, and volume fractions f_1, f_2, an analytical effective compliance tensor $\bar{c}$ is computed. The effective tensor is subsequently rotated with a rotation tensor R, defined in terms of the three rotation angles α, β, γ, yielding a rotated effective tensor $\bar{c}_r$. Thus, the building block takes as input two compliance tensors c_1, c_2 and outputs a rotated effective compliance tensor $\bar{c}_r$, where $f_1, f_2, \alpha, \beta, \gamma$ are the learnable parameters (see Fig. 10.10). By connecting these building blocks, a large network can be created. The network is applied to homogenization tasks of representative volume elements (RVEs) [494, 495], where the material of the phases is varied during evaluation.

Complex microstructures can be built up by combining a multitude of the building blocks presented in Fig. 10.10. In [494], the geometry is hierarchically represented with binary-trees. From stress-strain data from multiple microstructure samples (all with the same base characteristics), material properties of the phases, anisotropic characteristics, or volume fractions can be inferred.

10.3.4 Neural Differential Equations

In a more general setting, neural ordinary differential equations [371] consider the forward Euler discretization of ordinary differential equations. Specifically, recurrent neural networks are viewed as Euler discretizations of continuous transformations [496–498]. Consider the iterative update rule of the hidden states $y_{t+1} = y(t + \Delta t)$ of a recurrent neural network.

$$y_{t+1} = y_t + f(y_t; \boldsymbol{\Theta}) \tag{10.61}$$

Here, f is the evaluation of one recurrent unit in the recurrent neural network. In the limit of the time step size $\lim \Delta t \to 0$, the dynamics of the hidden units y_t can be parametrized by an ordinary differential equation

$$\frac{dy(t)}{dt} = f(y(t), t; \boldsymbol{\Theta}) \tag{10.62}$$

The input to the network is the initial condition $y(0)$, and the output is the solution $y(T)$ at time T. The output of the neural network, $y(T)$, is obtained by solving Eq. 10.62 with a differential equation solver. The sensitivity computation for the weight update is obtained using the adjoint state method [423, 499], as backpropagating through each time step of the solver leads to a high memory cost. This also makes it possible to treat the solver as a black box. Similar extensions to partial differential equations [497] have been proposed by considering recurrent convolutional neural networks with residual connections, where the convolutional neural networks act as spatial gradients.

Similarly, [500] establish a connection between deep residual recurrent neural networks and iterative solvers. Residual connections in neural networks allow information to bypass neural network layers. Consider the estimation of the next state of a partial differential equation with a classical solver $u_{t+1} = u(t + \Delta t) = F[u(t)]$. The residual $r_{t+1} = r(t + \Delta t)$ is determined in terms of the ground truth $\tilde{u}_{t+1}$:

$$r_{t+1} = \tilde{u}_{t+1} - u_{t+1}. \tag{10.63}$$

An iterative correction scheme is formulated with a neural network. The iterations are indicated with the superindex (k).

$$u_{t+1}^{(k+1)} = u_{t+1}^{(k)} + f_{NN}(r_{t+1}^{(k+1)}; \boldsymbol{\Theta}) \tag{10.64}$$

$$r_{t+1}^{(k+1)} = \tilde{u}_{t+1} - u_{t+1}^{(k)} \tag{10.65}$$

Note that the residual connection, i.e., $u_{t+1}^{(k)}$ as directly used in the prediction of $u_{t+1}^{(k+1)}$, allows information to pass past the recurrent unit f_{NN}. A related approach can be found in [501], where an autoencoder iteratively acts on a solution until convergence. In the first iteration, a random initial solution is used as input.

10.4 Generative Approaches

As discussed in depth in Chap. 8, generative approaches aim to model the underlying probability distribution of a dataset to generate new data that resembles the training data. In this category, the most successful approaches are variational autoencoders, generative adversarial networks, diffusion models, flow-based models, and transformers. All of these are covered in detail in Chap. 8 with in-depth descriptions of the most prominent applications in the field of computational mechanics. One area of focus is microstructure generation, which aims to produce a sufficient quantity of realistic training data for surrogate models. The second key application area is generative design, which relies on algorithms to efficiently explore the design space within the constraints established by the designer. Furthermore, anomaly detection is tackled by identifying samples that cannot be reproduced by a generative approach. Recently, also the generative transformer technique has been repurposed for surrogate modeling.

10.5 Deep Reinforcement Learning

In reinforcement learning, an agent interacts with an environment through a sequence of actions a_t, which is illustrated in Fig. 10.11. Upon executing an action a_t, the agent receives an updated state s_{t+1} and reward r_{t+1} from the environment. The agent's objective is to maximize the cumulative reward R_Σ. Generally, the environment can be treated as a black box. This presents an advantage in computational mechanics when differentiable physics are not feasible. However, unlike other gradient-free optimization algorithms, the reinforcement learning algorithm will behave differently, i.e., smarter when encountering a previously seen state again. Thus, the machine learning paradigm is contained in the idea of reinforcement learning, distinguishing it from pure optimization algorithms.

In reinforcement learning, the environment is commonly modeled as a Markov decision process (MDP). This mathematical model is defined by a set of all possible states S, actions A, and associated rewards R. Furthermore, the probability of getting to the next state s_{t+1} from the previous s_t with action a_t is given by $\mathbb{P}(s_{t+1}|s_t, a_t)$. Thus, the environment is not necessarily deterministic. One key aspect of a Markov decision process is the Markov property, stating that future states depend solely on the current state and action, and not the history of states and actions.

The goal of a reinforcement learning algorithm is to determine a policy $\pi(s, a)$ which dictates the next action a_t in order to maximize the cumulative reward R_Σ. The cumulative reward R_Σ is discounted by a discount factor γ^t in order to give more importance to immediate rewards.

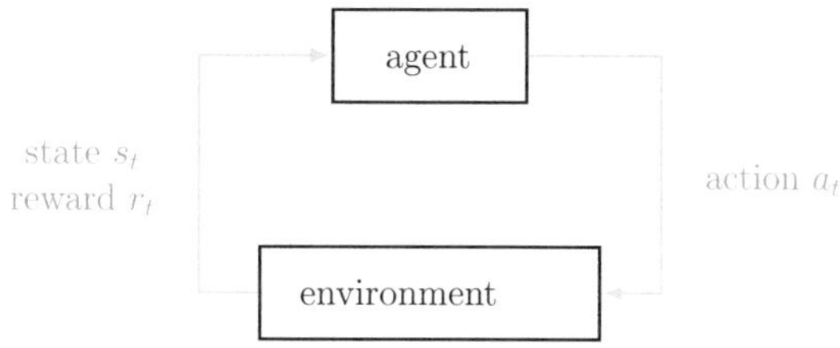

Fig. 10.11 Reinforcement learning in which an agent interacts with an environment with actions a_t, states s_t, and rewards r_t. Figure adapted from [21]

$$R_\Sigma = \sum_{t=0}^{\infty} \gamma^t r_t \tag{10.66}$$

The quality of a policy $\pi(s, a)$ can be assessed by a state-value function $V_\pi(s)$, defined as the expected future reward given the current state s and following the policy π. Similarly, an action-value function $Q_\pi(s)$ determines the expected future reward given the current state s and action a, while subsequently following the policy π. The expected value along a policy π is denoted as $\mathbb{E}_\pi$.

$$V_\pi(s) = \mathbb{E}_\pi\left[R_\Sigma(t)|s\right] \tag{10.67}$$

$$Q_\pi(s, a) = \mathbb{E}_\pi\left[R_\Sigma(t)|s, a\right] \tag{10.68}$$

The optimal value and quality function correspondingly follow the optimal policy:

$$V(s) = \max_\pi V_\pi(s), \tag{10.69}$$

$$Q(s, a) = \max_\pi Q_\pi(s). \tag{10.70}$$

The approaches can be subdivided into model-based and model-free. Model-based methods incorporate a model of the environment. In the most general sense, a probabilistic environment, this entitles the probability distribution of the next state $\mathbb{P}(s_{t+1}|s_t, a_t)$ and of the next reward $\mathbb{R}(r_{t+1}|s_{t+1}, s_t, a_t)$. The model of the environment can be cheaply sampled to improve the policy π with model-free reinforcement learning techniques [502–505] discussed in the sequel (Sects. 10.5.1 and 10.5.2). However, if the model is differentiable, the gradient of the reward can directly be used to update the policy [506–511]. This is identical to the optimization through differentiable physics solvers discussed in Sect. 10.2.3.3. Model-free reinforcement learning techniques can be used to enhance the optimization.

A further distinction is made between policy-based [512–516] and value-based [517–519] approaches. Policy-based methods, such as deep policy networks [172] (Sect. 10.5.1), directly optimize the policy. By contrast, value-based methods, such as deep Q-learning [519] (Sect. 10.5.2) learn the value function from which the optimal policy is selected. Actor-critic methods, such as proximal policy optimization [520] combine the ideas with an actor that performs a policy and a critic that judges its quality. Both can be modeled by neural networks.

10.5.1 Deep Policy Networks

In deep policy networks, the policy, i.e., the mapping of states to actions, is modeled by a neural network $\hat{a} = \pi(s; \boldsymbol{\Theta})$. The quality of the neural network is assessed by the expected cumulative reward R_Σ, formulated in terms of the action-value function $Q(s, a)$.

$$C = R_\Sigma = \mathbb{E}\big[Q(s, a)\big] \tag{10.71}$$

Its gradient (see [172, 513, 515] for a derivation), given as

$$\nabla_{\boldsymbol{\Theta}} R_\Sigma = \mathbb{E}\big[Q(s, a)\nabla_{\boldsymbol{\Theta}} \log\big(\pi(s, a; \boldsymbol{\Theta})\big)\big], \tag{10.72}$$

can be applied within a gradient ascent scheme to learn the optimal policy.

10.5.2 Deep Q-Learning

Deep Q-learning identifies the optimal action-value function $Q(s, a)$ from which the optimal policy is extracted. Q-Learning relies on the Bellman optimality criterion [521, 522]. By separating the reward r_0 at the first step, the recursion formula of the optimal state-value function, i.e., the Bellman optimality criterion, can be established:

$$V(s) = \max_\pi \mathbb{E}_\pi\Big[\sum_{t=0}^{\infty} \gamma^t r_t | s_0 = s\Big] \tag{10.73}$$

$$= \max_\pi \mathbb{E}_\pi\Big[r_0 + \sum_{t=1}^{\infty} \gamma^t r_t | s_1 = s'\Big] \tag{10.74}$$

$$= \max_\pi \mathbb{E}_\pi\big[r_0 + \gamma V(s')\big]. \tag{10.75}$$

Here, s' represents the next state after s. This can be done analogously for the action-value function.

$$Q(s, a) = \max_\pi \mathbb{E}_\pi\big[r_0 + \gamma Q(s', a')\big] \tag{10.76}$$

The recursion enables an update formula, referred to as temporal difference (TD) learning [523, 524]. Specifically, the current estimate $Q^{(m)}$ at state s_t is compared to the more accurate estimate at the next state s_{t+1} using the obtained reward r_t, referred to as the TD target estimate. The difference is the TD error, which in combination with a learning rate α is used to update the function $Q^{(m)}$:

$$Q^{(m+1)}(s_t, a_t) = Q^{(m)}(s_t, a_t) + \alpha \overbrace{\Big(\underbrace{r_t + \gamma \max_a Q(s_{t+1}, a)}_{\text{TD target estimate}} - \underbrace{Q^{(m)}(s_t, a_t)}_{\text{model prediction}}\Big)}^{\text{TD error}}. \tag{10.77}$$

Here, the TD target estimate only looks one step ahead—and is therefore referred to as TD(0). The generalization is called TD(N). In the limit $N \to \infty$, the method is equivalent to Monte Carlo learning, where all steps are performed and a true target is obtained.

Deep Q-learning introduces a neural network for the action-value function $Q(s, a; \boldsymbol{\Theta})$. Its quality is assessed with a loss composed of the mean squared error of the TD error.

$$C = \mathbb{E}\Big[\frac{1}{2}\big(r_t + \gamma \max_a Q(s_{t+1}, a; \boldsymbol{\Theta}) - Q(s_t, a_t; \boldsymbol{\Theta})\big)^2\Big] \tag{10.78}$$

Lastly, the optimal policy $\pi(s)$ maximizing the action-value function $Q(s, a; \boldsymbol{\Theta})$ is extracted:

$$\pi(s) = \arg\max_a Q(s, a; \boldsymbol{\Theta}) \tag{10.79}$$

10.5.3 Applications

Deep reinforcement learning is mainly used for inverse problems (see [4] for a review within fluid mechanics), where the partial differential equation solver is treated as a black box, and assumed to not be differentiable.

The most prominent application are control problems. One example is discovering swimming strategies for fish—with the goal of efficiently minimizing the distance to a leader fish [525, 526]. The environment is given by the Navier Stokes equation. Another example is balancing rigid bodies with fluid jets while using as little force as possible [527]. Similarly, [528] control jets in order to reduce the drag around a cylinder. Reducing the drag around a cylinder is also achieved by controlling small rotating cylinders in the wake of the flow [529]. A more complex example is controlling unmanned aerial vehicles [530]. The control schemes are learned by interacting with simulations and, subsequently, applied in experiments.

Further applications in connection with inverse problems are learning filters to perturb flows in order to match target flows [531]. Also, constitutive laws can be identified. The individual arithmetic manipulations within a constitutive law can be represented as graphs. An agent constructs the graph in order to best match simulation and measurement [532], which yields an interpretable law.

Topology optimization has also been tackled by reinforcement learning. Specifically, the ability to predict only binary states (material or no material) is desirable—instead of intermediate states, as in solid isotropic material with penalization [248, 533]. This has been shown with binary truss structures, modeled with graphs in order to minimize the total structural volume under stress constraints. In [534], an agent removes trusses from existing structures, and trusses are added in [535]. Similarly, [536] removes finite elements in solid structures to modify the topology. Instead, [537] pursues design diversity. Here a neural network surrogate model predicts near optimal structures from reference designs. The agent then learns to generate reference designs as input, such that the corresponding optimal structures are as diverse as possible.

Also, high-dimensional partial differential equations have been solved with reinforcement learning [538, 539].[23] This is achieved by recasting the partial differential equation into stochastic control problems, thereby solving these with reinforcement learning.

Finally, adaptive mesh refinement algorithms have been learned by reinforcement learning [540]. An agent decides whether an element is to be refined based on the current state, i.e., the mesh and solution. The reward is subsequently defined in terms of the error reduction, which is computed with a ground truth solution. The trained agent can thus be applied to adaptive mesh refinement to previously unseen simulations.

10.5.4 Extensions

Each interaction with the environment requires solving the differential equation, which, due to the many interactions, makes reinforcement learning expensive. The learning can be accelerated through some basic modifications. The learning can be perfectly parallelized by using multiple environments simultaneously [541], or by using multiple agents within the same environment [542]. Another idea is to construct a surrogate model of the environment and thereby exploit model-based approaches [543–546]. The general procedure consists of three steps:

- model learning: learn surrogate of environment,
- behavior learning: learn policy or value function,
- environment interaction: apply learned policy and collect data.

Most approaches construct the surrogate with data-driven modeling (Sect. 10.1.1), but physics-informed approaches have been proposed as well [543, 545] (Sect. 10.2.2).

[23] However, solving high-dimensional optimization problems is to the best of our knowledge not recommended.

References

1. Herrmann, Leon, and Stefan Kollmannsberger. 2024. Deep learning in computational mechanics: A review. *Computational Mechanics*. ISSN: 1432-0924. https://doi.org/10.1007/s00466-023-02434-4.
2. Faller, William E., and Scott J. Schreck. 1996. Neural networks: Applications and opportunities in aeronautics. *Progress in Aerospace Sciences* 32(5): 433–456. ISSN: 0376-0421. https://doi.org/10.1016/0376-0421(95)00011-9.
3. Bock, Frederic E., et al. 2019. A review of the application of machine learning and data mining approaches in continuum materials mechanics. *Frontiers in Materials* 6: 110. ISSN: 2296-8016. https://doi.org/10.3389/fmats.2019.00110.
4. Garnier, Paul, et al. 2019. A review on deep reinforcement learning for fluid mechanics. Physics. arXiv: 1908.04127.
5. Duraisamy, Karthik, Gianluca Iaccarino, and Heng Xiao. 2019. Turbulence Modeling in the Age of Data. *Annual Review of Fluid Mechanics* 51(1): 357–377. ISSN: 0066-4189, 1545-4479. https://doi.org/10.1146/annurev-fluid-010518-040547.
6. Moosavi, Seyed Mohamad, Kevin Maik Jablonka, and Berend Smit. 2020. The Role of machine learning in the understanding and design of materials. *Journal of the American Chemical Society* 142(48): 20273–20287. ISSN: 0002-7863, 1520-5126. https://doi.org/10.1021/jacs.0c09105.
7. Steven Brunton, Bernd Noack, and Petros Koumoutsakos. 2020. Machine Learning for Fluid Mechanics. *Annual Review of Fluid Mechanics* 52(1): 477–508. tex.ids= Brunton2020a. arXiv: 1905.11075, ISSN: 0066-4189, 1545-4479. https://doi.org/10.1146/annurev-fluid-010719-060214.
8. Cai, Shengze, et al. 2021. Physics-informed neural networks (PINNs) for fluid mechanics: a review. *Acta Mechanica Sinica* 37(12): 1727–1738. ISSN: 1614-3116. https://doi.org/10.1007/s10409-021-01148-1.
9. Giovanni Calzolari and Wei Liu. 2021. Deep learning to replace, improve, or aid CFD analysis in built environment applications: A review. *Building and Environment* 206: 108315. ISSN: 0360-1323. https://doi.org/10.1016/j.buildenv.2021.108315.
10. Amir Adler, Mauricio Araya-Polo, and Tomaso Poggio. 2021. Deep learning for seismic inverse problems: Toward the acceleration of geophysical analysis workflows. *IEEE Signal Processing Magazine* 38(2): 89–119. ISSN: 1558-0792. https://doi.org/10.1109/MSP.2020.3037429.
11. Rebekka V. Woldseth et al. 2022. On the use of artificial neural networks in topology optimisation. *Structural and Multidisciplinary Optimization* 65(10): 294. ISSN: 1615-1488. https://doi.org/10.1007/s00158-022-03347-1.
12. Lyle Regenwetter, Amin Heyrani Nobari, and Faez Ahmed. 2022. Deep generative models in engineering design: A review. arXiv:2110.10863 [cs, stat].
13. Seungyeon Shin, Dongju Shin, and Namwoo Kang. 2023. Topology optimization via machine learning and deep learning: a review. *Journal of Computational Design and Engineering* 10(4): 1736–1766. ISSN: 2288-5048. https://doi.org/10.1093/jcde/qwad072.
14. Dana Bishara et al. 2023. A State-of-the-Art Review on Machine Learning-Based Multiscale Modeling, Simulation, Homogenization and Design of Materials. *Archives of Computational Methods in Engineering* 30(1): 191–222. ISSN: 1886-1784. https://doi.org/10.1007/s11831-022-09795-8.
15. Max Rosenkranz et al. 2023. A comparative study on different neural network architectures to model inelasticity. *International Journal for Numerical Methods in Engineering* nme.7319. ISSN: 0029-5981, 1097-0207. https://doi.org/10.1002/nme.7319.
16. Baker, Nathan, et al. 2019. Workshop report on basic research needs for scientific machine learning: Core technologies for artificial intelligence. *Technical Report* 1478744. https://doi.org/10.2172/1478744.

17. Laura von Rueden et al. 2023. Informed machine learning—A taxonomy and survey of integrating prior knowledge into learning systems. *IEEE Transactions on Knowledge and Data Engineering* 35(1): 614–633. ISSN: 1558-2191. https://doi.org/10.1109/TKDE.2021.3079836.
18. Yagawa, G., and H. Okuda. 1996. Neural networks in computational mechanics. *Archives of Computational Methods in Engineering* 3(4): 435–512. ISSN: 1886-1784. https://doi.org/10.1007/BF02818935.
19. Zenon Waszczyszyn and Leonard Ziemiański. 2001. Neural networks in mechanics of structures and materials—new results and prospects of applications. *Computers & Structures* 79(22): 2261–2276. ISSN: 0045-7949. https://doi.org/10.1016/S0045-7949(01)00083-9.
20. Ian Goodfellow, Yoshua Bengio, and Aaron Courville. 2016. *Deep learning*. MIT Press. ISBN: 0-262-03561-8.
21. Richard S. Sutton and Andrew G. Barto. 2018. *Reinforcement learning: An introduction*. 2nd edn. Adaptive computation and machine learning series. Cambridge, Massachusetts: The MIT Press. ISBN: 978-0-262-03924-6.
22. Ethem Alpaydin. 2020. *Introduction to machine learning*. 4th edn. Adaptive computation and machine learning series. Cambridge, Massachusetts: The MIT Press. ISBN: 978-0-262-04379-3.
23. Stuart J. Russell and Peter Norvig. 2022. *Artificial intelligence: A modern approach*. 4th edn. global edition. Pearson series in artificial intelligence. Harlow: Pearson. ISBN: 978-1-292-40113-3.
24. Thomas J.R. Hughes and Gregory M. Hulbert. 1988. Space-time finite element methods for elastodynamics: Formulations and error estimates. *Computer Methods in Applied Mechanics and Engineering* 66(3): 339–363. ISSN: 0045-7825. https://doi.org/10.1016/0045-7825(88)90006-0.
25. Mohamad Alsalman, Brendan Colvert, and Eva Kanso. 2018. Training bioinspired sensors to classify flows. *Bioinspiration & Biomimetics* 14(1): 016009. ISSN: 1748-3190. https://doi.org/10.1088/1748-3190/aaef1d.
26. Brendan Colvert, Mohamad Alsalman, and Eva Kanso. 2018. Classifying vortex wakes using neural networks. *Bioinspiration & Biomimetics* 13(2): 025003. ISSN: 1748-3190. https://doi.org/10.1088/1748-3190/aaa787.
27. Pierret, S., and R. A. Van Den Braembussche. 1999. Turbomachinery blade design using a Navier–Stokes solver and artificial neural network. *Journal of Turbomachinery* 121(2): 326–332. ISSN: 0889-504X, 1528-8900. https://doi.org/10.1115/1.2841318.
28. Poojitha Vurtur Badarinath, Maria Chierichetti, and Fatemeh Davoudi Kakhki. 2021. A machine learning approach as a surrogate for a finite element analysis: status of research and application to one dimensional systems. *Sensors* 21(5): 1654. ISSN: 1424-8220. https://doi.org/10.3390/s21051654.
29. Changhoon Lee et al. 1997. Application of neural networks to turbulence control for drag reduction. *Physics of Fluids* 9(6): 1740–1747. ISSN: 1070-6631, 1089-7666. https://doi.org/10.1063/1.869290.
30. Jambunathan, K., et al. 1996. Evaluating convective heat transfer coefficients using neural networks. *International Journal of Heat and Mass Transfer* 39(11): 2329–2332. ISSN: 0017-9310. https://doi.org/10.1016/0017-9310(95)00332-0.
31. Brendan D. Tracey, Karthikeyan Duraisamy, and Juan J. Alonso. 2015. A machine learning strategy to assist turbulence model development. In *53rd AIAA aerospace sciences meeting*. Kissimmee, Florida: American Institute of Aeronautics and Astronautics. ISBN: 978-1-62410-343-8. https://doi.org/10.2514/6.2015-1287.
32. Ramuhalli, P., L. Udpa, and S.S. Udpa. 2002. Electromagnetic NDE signal inversion by function-approximation neural networks. *IEEE Transactions on Magnetics* 38(6): 3633–3642. ISSN: 1941-0069. https://doi.org/10.1109/TMAG.2002.804817.
33. Mauricio Araya-Polo et al. 2018. Deep-learning tomography. *The Leading Edge* 37(1): 58–66. ISSN: 1070-485X, 1938-3789. https://doi.org/10.1190/tle37010058.1.

34. Yuji Kim and Nori Nakata. 2018. Geophysical inversion versus machine learning in inverse problems. *The Leading Edge* 37(12): 894–901. ISSN: 1070-485X, 1938-3789. https://doi.org/10.1190/tle37120894.1.
35. Van-Nam Hoang et al. 2022. Data-driven geometry-based topology optimization. *Structural and Multidisciplinary Optimization* 65(2): 69. ISSN: 1615-1488. https://doi.org/10.1007/s00158-022-03170-8.
36. Xiaoxuan Zhang and Krishna Garikipati. 2023. Label-free learning of elliptic partial differential equation solvers with generalizability across boundary value problems. *Computer Methods in Applied Mechanics and Engineering* 116214. ISSN: 0045-7825. https://doi.org/10.1016/j.cma.2023.116214.
37. Nils Thuerey et al. 2020. Deep learning methods for Reynolds-Averaged Navier–Stokes simulations of airfoil flows. *AIAA Journal* 58(1): 25–36. ISSN: 0001-1452, 1533-385X. https://doi.org/10.2514/1.J058291.
38. Li-Wei Chen et al. 2021. Numerical investigation of minimum drag profiles in laminar flow using deep learning surrogates. *Journal of Fluid Mechanics* 919: A34. ISSN: 0022-1120, 1469-7645. https://doi.org/10.1017/jfm.2021.398.
39. Xianqi Chen et al. 2021. A deep neural network surrogate modeling benchmark for temperature field prediction of heat source layout. *Science China Physics, Mechanics & Astronomy* 64(11): 1. ISSN: 1869-1927. https://doi.org/10.1007/s11433-021-1755-6.
40. Li-Wei Chen and Nils Thuerey. 2023. Towards high-accuracy deep learning inference of compressible flows over aerofoils. *Computers & Fluids* 250: 105707. ISSN: 0045-7930. https://doi.org/10.1016/j.compfluid.2022.105707.
41. Aditya Khadilkar, Jun Wang, and Rahul Rai. 2019. Deep learning–based stress prediction for bottom-up SLA 3D printing process. *The International Journal of Advanced Manufacturing Technology* 102(5): 2555–2569. ISSN: 1433-3015. https://doi.org/10.1007/s00170-019-03363-4.
42. Zhenguo Nie, Haoliang Jiang, and Levent Burak Kara. 2020. Stress field prediction in cantilevered structures using convolutional neural networks. *Journal of Computing and Information Science in Engineering* 20(1): 011002. ISSN: 1530-9827, 1944-7078. https://doi.org/10.1115/1.4044097.
43. Xiaoxiao Guo, Wei Li, and Francesco Iorio. 2016. Convolutional neural networks for steady flow approximation. In *Proceedings of the 22nd ACM SIGKDD international conference on knowledge discovery and data mining*. San Francisco California USA: ACM, 481–490. ISBN: 978-1-4503-4232-2. https://doi.org/10.1145/2939672.2939738.
44. Zhibo Zhang, Prakhar Jaiswal, and Rahul Rai. 2018. FeatureNet: Machining feature recognition based on 3D convolution neural network. *Computer-Aided Design* 101: 12–22. ISSN: 0010-4485. https://doi.org/10.1016/j.cad.2018.03.006.
45. Glen Williams et al. 2019. Design repository effectiveness for 3D convolutional neural networks: Application to additive manufacturing. *Journal of Mechanical Design* 141(11): 111701. ISSN: 1050-0472, 1528-9001. https://doi.org/10.1115/1.4044199.
46. Yue Wu, Youzuo Lin, and Zheng Zhou. 2018. Inversionet: Accurate and efficient seismic-waveform inversion with convolutional neural networks. In *SEG Technical Program Expanded Abstracts 2018*. Anaheim, California: Society of Exploration Geophysicists, 2096–2100. https://doi.org/10.1190/segam2018-2998603.1.
47. Wenlong Wang, Fangshu Yang, and Jianwei Ma. 2018. Velocity model building with a modified fully convolutional network. In *SEG Technical Program Expanded Abstracts 2018*. Anaheim, California: Society of Exploration Geophysicists, 2086–2090. https://doi.org/10.1190/segam2018-2997566.1.
48. Fangshu Yang and Jianwei Ma. 2019. Deep-learning inversion: A next-generation seismic velocity model building method. *Geophysics* 84(4).
49. York Zheng et al. 2019. Applications of supervised deep learning for seismic interpretation and inversion. *The Leading Edge* 38(7): 526–533. ISSN: 1070-485X, 1938-3789. https://doi.org/10.1190/tle38070526.1.

50. Mauricio Araya-Polo, Stuart Farris, and Manuel Florez. 2019. Deep learning-driven velocity model building workflow. *The Leading Edge* 38(11): 872a1–872a9. ISSN: 1070-485X, 1938-3789. https://doi.org/10.1190/tle38110872a1.1.
51. Vishal Das et al. 2019. Convolutional neural network for seismic impedance inversion. *Geophysics* 84(6): R869–R880. ISSN: 0016-8033, 1942-2156. https://doi.org/10.1190/geo2018-0838.1.
52. Wenlong Wang and Jianwei Ma. 2020. Velocity model building in a crosswell acquisition geometry with image-trained artificial neural networks. *Geophysics* 85(2): U31–U46. ISSN: 0016-8033, 1942-2156. https://doi.org/10.1190/geo2018-0591.1.
53. Shucai Li et al. 2020. Deep-learning inversion of seismic data. *IEEE Transactions on Geoscience and Remote Sensing* 58(3): 2135–2149. ISSN: 0196-2892, 1558-0644. https://doi.org/10.1109/TGRS.2019.2953473.
54. Bangyu Wu et al. 2020. Seismic impedance inversion using fully convolutional residual network and transfer learning. *IEEE Geoscience and Remote Sensing Letters* 17(12): 2140–2144. ISSN: 1558-0571. https://doi.org/10.1109/LGRS.2019.2963106.
55. Min Jun Park and Mauricio D. Sacchi. 2020. Automatic velocity analysis using convolutional neural network and transfer learning. *Geophysics* 85(1): V33–V43. ISSN: 0016-8033, 1942-2156. https://doi.org/10.1190/geo2018-0870.1.
56. Jiaxing Ye and Nobuyuki Toyama. 2022. Automatic defect detection for ultrasonic wave propagation imaging method using spatio-temporal convolution neural networks. *Structural Health Monitoring* 21(6): 2750–2767. ISSN: 1475-9217, 1741-3168. https://doi.org/10.1177/14759217211073503.
57. Jing Rao et al. 2023. Quantitative reconstruction of defects in multi-layered bonded composites using fully convolutional network-based ultrasonic inversion. *Journal of Sound and Vibration* 542: 117418. ISSN: 0022-460X. https://doi.org/10.1016/j.jsv.2022.117418.
58. Qiyin Lin et al. 2018. Investigation into the topology optimization for conductive heat transfer based on deep learning approach. *International Communications in Heat and Mass Transfer* 97: 103–109. ISSN: 0735-1933. https://doi.org/10.1016/j.icheatmasstransfer.2018.07.001.
59. Yonggyun Yu et al. 2019. Deep learning for determining a near-optimal topological design without any iteration. *Structural and Multidisciplinary Optimization* 59(3): 787–799. ISSN: 1615-1488. https://doi.org/10.1007/s00158-018-2101-5.
60. Abueidda, Diab W., Seid Koric, and Nahil A. Sobh. 2020. Topology optimization of 2D structures with nonlinearities using deep learning. *Computers & Structures* 237: 106283. ISSN: 00457949. https://doi.org/10.1016/j.compstruc.2020.106283.
61. Nakamura, Keigo, and Yoshiro Suzuki. 2020. *Deep learning-based topological optimization for representing a user-specified design area.* https://doi.org/10.48550/arXiv.2004.05461.
62. Yiquan, Zhang et al. 2020. *A deep Convolutional Neural Network for topology optimization with strong generalization ability*. arXiv:1901.07761 [cs, stat]. https://doi.org/10.48550/arXiv.1901.07761.
63. Shuai, Zheng, Zhenzhen He, and Honglei Liu. 2021. Generating three-dimensional structural topologies via a U-Net convolutional neural network. *Thin-Walled Structures* 159: 107263. ISSN: 0263-8231. https://doi.org/10.1016/j.tws.2020.107263.
64. Shuai, Zheng et al. 2021. Accurate and real-time structural topology prediction driven by deep learning under moving morphable component-based framework. *Applied Mathematical Modelling* 97: 522–535. ISSN: 0307-904X. https://doi.org/10.1016/j.apm.2021.04.009.
65. Dalei, Wang et al. 2022. A deep convolutional neural network for topology optimization with perceptible generalization ability. *Engineering Optimization* 54(6): 973–988. ISSN: 0305-215X. https://doi.org/10.1080/0305215X.2021.1902998.
66. Jun, Yan et al. 2022. Deep learning driven real time topology optimisation based on initial stress learning. *Advanced Engineering Informatics* 51: 101472. ISSN: 1474-0346. https://doi.org/10.1016/j.aei.2021.101472.
67. Junhyeon, Seo and Rakesh K. Kapania. 2023. Topology optimization with advanced CNN using mapped physics-based data. *Structural and Multidisciplinary Optimization* 66(1): 21. tex.ids= Seo2023. ISSN: 1615-1488. https://doi.org/10.1007/s00158-022-03461-0.

68. Ivan, Sosnovik and Ivan Oseledets. 2019. Neural networks for topology optimization. *Russian Journal of Numerical Analysis and Mathematical Modelling* 34(4): 215–223. ISSN: 1569-3988. https://doi.org/10.1515/rnam-2019-0018.
69. Younghwan, Joo, Yonggyun Yu, and In Gwun Jang. 2021. Unit module-based convergence acceleration for topology optimization using the spatiotemporal deep neural network. *IEEE Access* 9: 149766–149779. ISSN: 2169-3536. https://doi.org/10.1109/ACCESS.2021.3125014.
70. Nikos, Ath. Kallioras, Georgios Kazakis, and Nikos D. Lagaros. 2020. Accelerated topology optimization by means of deep learning. *Structural and Multidisciplinary Optimization* 62(3): 1185–1212. ISSN: 1615-1488. https://doi.org/10.1007/s00158-020-02545-z.
71. Sanchez-Gonzalez, Alvaro, et al. 2020. *Learning to simulate complex physics with graph networks*. https://doi.org/10.48550/arXiv.2002.09405.
72. Pfaff, Tobias, et al. 2021. Learning mesh-based simulation with graph. *Networks*. https://doi.org/10.48550/arXiv.2010.03409.
73. Roberto, Perera, Davide Guzzetti, and Vinamra Agrawal. 2022. Graph neural networks for simulating crack coalescence and propagation in brittle materials. *Computer Methods in Applied Mechanics and Engineering* 395: 115021. ISSN: 0045-7825. https://doi.org/10.1016/j.cma.2022.115021.
74. Charles, R., Qi et al. 2017. *PointNet: Deep learning on point sets for 3D classification and segmentation*. https://doi.org/10.48550/arXiv.1612.00593.
75. Thibault, Groueix et al. 2018. *AtlasNet: A Papier-Mâché approach to learning 3D surface generation*. arXiv:1802.05384 [cs]. https://doi.org/10.48550/arXiv.1802.05384.
76. James, D., Cunningham, Timothy W. Simpson, and Conrad S. Tucker. 2019. An investigation of surrogate models for efficient performance-based decoding of 3D point clouds. *Journal of Mechanical Design* 141(12): 121401. ISSN: 1050-0472, 1528-9001. https://doi.org/10.1115/1.4044597.
77. Peter, W., Battaglia et al. 2018. *Relational inductive biases, deep learning, and graph networks*. arXiv:1806.01261 [cs, stat]. https://doi.org/10.48550/arXiv.1806.01261.
78. Jolliffe, I.T. 2002. *Principal component analysis*. 2nd ed. Springer series in statistics. New York: Springer. ISBN: 978-0-387-95442-4.
79. Tobias, Heimann and Hans-Peter Meinzer. 2009. Statistical shape models for 3D medical image segmentation: A review. *Medical Image Analysis* 13(4): 543–563. ISSN: 1361-8415. https://doi.org/10.1016/j.media.2009.05.004.
80. Kaushik, Bhattacharya et al. 2021. Model reduction and neural networks for parametric PDEs. *The SMAI Journal of Computational Mathematics* 7: 121–157. ISSN: 2426-8399. https://doi.org/10.5802/smai-jcm.74.
81. Berkooz, G., P. Holmes, and J. L. Lumley. 1993. The proper orthogonal decomposition in the analysis of turbulent flows. *Annual Review of Fluid Mechanics* 25(1): 539–575. ISSN: 0066-4189, 1545-4479. https://doi.org/10.1146/annurev.fl.25.010193.002543.
82. Muñoz, D., et al. 2023. Manifold learning for coherent design interpolation based on geometrical and topological descriptors. *Computer Methods in Applied Mechanics and Engineering* 405: 115859. ISSN: 0045-7825. https://doi.org/10.1016/j.cma.2022.115859.
83. Liang, Liang et al. 2018. A deep learning approach to estimate stress distribution: a fast and accurate surrogate of finite-element analysis. *Journal of The Royal Society Interface* 15(138): 20170844. ISSN: 1742-5689, 1742-5662. https://doi.org/10.1098/rsif.2017.0844.
84. Ali, Madani et al. 2019. Bridging finite element and machine learning modeling: Stress prediction of arterial walls in atherosclerosis. *Journal of Biomechanical Engineering* 141(8): 084502. ISSN: 0148-0731, 1528-8951. https://doi.org/10.1115/1.4043290.
85. Muravleva, E., I. Oseledets, and D. Koroteev. 2018. Application of machine learning to viscoplastic flow modeling. *Physics of Fluids* 30(10): 103102. ISSN: 1070-6631. https://doi.org/10.1063/1.5058127.
86. Liang, Liang et al. 2018. A machine learning approach as a surrogate of finite element analysis-based inverse method to estimate the zero-pressure geometry of human thoracic aorta. *International Journal for Numerical Methods in Biomedical Engineering* 34(8): e3103. ISSN: 20407939. https://doi.org/10.1002/cnm.3103.

87. Khouloud, Derouiche et al. 2021. Data-driven modeling for multiphysics parametrized problems-application to induction hardening process. *Metals* 11(5): 738. ISSN: 2075-4701. https://doi.org/10.3390/met11050738.
88. Quercus, Hernández et al. 2023. Thermodynamics-informed neural networks for physically realistic mixed reality. *Computer Methods in Applied Mechanics and Engineering* 407: 115912. tex.ids= Hernandez2023b. ISSN: 0045-7825. https://doi.org/10.1016/j.cma.2023.115912.
89. Hinton, G.E., and R. R. Salakhutdinov. 2006. Reducing the dimensionality of data with neural networks. *Science* 313(5786): 504–507. ISSN: 0036-8075, 1095-9203. https://doi.org/10.1126/science.1127647.
90. Michele Milano and Petros Koumoutsakos. 2002. Neural network modeling for near wall turbulent flow. *Journal of Computational Physics* 182(1): 1–26. tex.ids= Milano2002a. ISSN: 00219991. https://doi.org/10.1006/jcph.2002.7146.
91. Siddharth Nair et al. 2023. GRIDS-Net: Inverse shape design and identification of scatterers via geometric regularization and physics-embedded deep learning. *Computer Methods in Applied Mechanics and Engineering* 414: 116167. ISSN: 0045-7825. https://doi.org/10.1016/j.cma.2023.116167.
92. Ana Fernandez-Navamuel et al. 2022. Supervised deep learning with finite element simulations for damage identification in bridges. *Engineering Structures* 257: 114016. ISSN: 0141-0296. https://doi.org/10.1016/j.engstruct.2022.114016.
93. Olaf Ronneberger, Philipp Fischer, and Thomas Brox. 2015. U-Net: Convolutional networks for biomedical image segmentation. In *Medical Image Computing and Computer-Assisted Intervention – MICCAI 2015*, ed. by Nassir Navab et al. Lecture notes in computer science. Cham: Springer International Publishing, 234–241. ISBN: 978-3-319-24574-4. https://doi.org/10.1007/978-3-319-24574-4sps28.
94. Zongwei Zhou et al. 2018. *UNet++: A nested U-Net architecture for medical image segmentation*. arXiv:1807.10165 [cs, eess, stat]. https://doi.org/10.48550/arXiv.1807.10165.
95. Lu Lu et al. 2022. A comprehensive and fair comparison of two neural operators (with practical extensions) based on FAIR data. *Computer Methods in Applied Mechanics and Engineering* 393: 114778. ISSN: 0045-7825. https://doi.org/10.1016/j.cma.2022.114778.
96. Kurt Hornik, Maxwell Stinchcombe, and Halbert White. 1989. Multilayer feedforward networks are universal approximators. *Neural Networks* 2(5): 359–366. tex.ids= Hornik1989a. ISSN: 0893-6080. https://doi.org/10.1016/0893-6080(89)90020-8.
97. Tianping Chen and Hong Chen. 1995. Universal approximation to nonlinear operators by neural networks with arbitrary activation functions and its application to dynamical systems. *IEEE Transactions on Neural Networks* 6(4): 911–917. ISSN: 1941-0093. https://doi.org/10.1109/72.392253.
98. Lu Lu et al. 2021. Learning nonlinear operators via DeepONet based on the universal approximation theorem of operators. *Nature Machine Intelligence* 3(3): 218–229. tex.ids= Lu2021b. ISSN: 2522-5839. https://doi.org/10.1038/s42256-021-00302-5.
99. Zongyi Li et al. 2021. *Fourier neural operator for parametric partial differential equations*. arXiv:2010.08895 [cs, math].
100. Chensen Lin et al. 2021. A seamless multiscale operator neural network for inferring bubble dynamics. *Journal of Fluid Mechanics* 929: A18. ISSN: 0022-1120, 1469-7645. https://doi.org/10.1017/jfm.2021.866.
101. Aaditya Chandrasekhar and Krishnan Suresh. 2021. Multi-material topology optimization using neural networks. *Computer-Aided Design* 136: 103017. ISSN: 0010-4485. https://doi.org/10.1016/j.cad.2021.103017.
102. Zhiping Mao et al. 2021. DeepM & Mnet for hypersonics: Predicting the coupled flow and finite-rate chemistry behind a normal shock using neural-network approximation of operators. *Journal of Computational Physics* 447: 110698. ISSN: 0021-9991. https://doi.org/10.1016/j.jcp.2021.110698.
103. Clark Di Leoni, P., et al. 2021. *DeepONet prediction of linear instability waves in high-speed boundary layers*. arXiv:2105.08697 [physics]. https://doi.org/10.48550/arXiv.2105.08697.

104. Shengze Cai et al. 2021. DeepM & Mnet: Inferring the electroconvection multiphysics fields based on operator approximation by neural networks. *Journal of Computational Physics* 436: 110296. ISSN: 0021-9991. https://doi.org/10.1016/j.jcp.2021.110296.
105. Chensen Lin et al. 2021. Operator learning for predicting multiscale bubble growth dynamics. *The Journal of Chemical Physics* 154(10): 104118. arXiv:2012.12816 [physics]. ISSN: 0021-9606, 1089-7690. https://doi.org/10.1063/5.0041203.
106. Minglang Yin et al. 2022. Simulating progressive intramural damage leading to aortic dissection using DeepONet: An operator–regression neural network. *Journal of The Royal Society Interface* 19(187): 20210670. ISSN: 1742-5662. https://doi.org/10.1098/rsif.2021.0670.
107. Julian D. Osorio et al. 2022. Forecasting solar-thermal systems performance under transient operation using a data-driven machine learning approach based on the deep operator network architecture. *Energy Conversion and Management* 252: 115063. ISSN: 01968904. https://doi.org/10.1016/j.enconman.2021.115063.
108. Somdatta Goswami et al. 2022. Neural operator learning of heterogeneous mechanobiological insults contributing to aortic aneurysms. *Journal of The Royal Society Interface* 19(193): 20220410. ISSN: 1742-5662. https://doi.org/10.1098/rsif.2022.0410.
109. Koric, Seid, et al. 2023. Deep learning operator network for plastic deformation with variable loads and material properties. *Engineering with Computers. issn* 0177–0667: 1435–5663. https://doi.org/10.1007/s00366-023-01822-x.
110. Seid Koric and Diab W. Abueidda. 2023. Data-driven and physics-informed deep learning operators for solution of heat conduction equation with parametric heat source. *International Journal of Heat and Mass Transfer* 203: 123809. ISSN: 00179310. https://doi.org/10.1016/j.ijheatmasstransfer.2022.123809.
111. Patricio Clark Di Leoni et al. 2023. Neural operator prediction of linear instability waves in high-speed boundary layers. *Journal of Computational Physics* 474: 111793. ISSN: 00219991. https://doi.org/10.1016/j.jcp.2022.111793.
112. Chuang Liu et al. 2023. Operator learning for predicting mechanical response of hierarchical composites with applications of inverse design. *International Journal of Applied Mechanics* 15(04): 2350028. ISSN: 1758-8251, 1758-826X. https://doi.org/10.1142/S175882512350028X.
113. Ahmed, Shady E., and Panos Stinis. 2023. A multifidelity deep operator network approach to closure for multiscale systems. *Computer Methods in Applied Mechanics and Engineering* 414: 116161. ISSN: 0045-7825. https://doi.org/10.1016/j.cma.2023.116161.
114. Sifan Wang, Hanwen Wang, and Paris Perdikaris. 2021. Learning the solution operator of parametric partial differential equations with physics-informed DeepONets. *Science Advances* 7(40): eabi8605. ISSN: 2375-2548. https://doi.org/10.1126/sciadv.abi8605.
115. Somdatta Goswami et al. 2022. *Physics-informed deep neural operator networks*. arXiv:2207.05748 [cs, math]. https://doi.org/10.48550/arXiv.2207.05748.
116. Somdatta Goswami et al. 2022. A physics-informed variational DeepONet for predicting crack path in quasi-brittle materials. *Computer Methods in Applied Mechanics and Engineering* 391: 114587. ISSN: 0045-7825. https://doi.org/10.1016/j.cma.2022.114587.
117. Nikola Kovachki, Samuel Lanthaler, and Siddhartha Mishra. 2021. On universal approximation and error bounds for Fourier neural operators. *The Journal of Machine Learning Research* 22(1): 290:13237–290:13312. ISSN: 1532-4435.
118. Zongyi Li et al. 2020. *Neural operator: Graph kernel network for partial differential equations*. arXiv:2003.03485 [cs, math, stat]. https://doi.org/10.48550/arXiv.2003.03485.
119. Zongyi Li et al. 2020. Multipole graph neural operator for parametric partial differential equations. In *Proceedings of the 34th international conference on neural information processing systems*. NIPS'20. Red Hook, NY, USA: Curran Associates Inc., 6755–6766. ISBN: 978-1-71382-954-6.
120. Qianying Cao, Somdatta Goswami, and George Em Karniadakis. 2023. *LNO: Laplace neural operator for solving differential equations*. arXiv:2303.10528 [cs]. https://doi.org/10.48550/arXiv.2303.10528.

121. Chengke Zhu, Hongxia Ye, and Bin Zhan. 2021. Fast solver of 2D Maxwell's equations based on fourier neural operator. In *2021 Photonics & electromagnetics research symposium (PIERS)*, 1635–1643. Hangzhou, China: IEEE. ISBN: 978-1-72817-247-7. https://doi.org/10.1109/PIERS53385.2021.9695119.
122. Chao Song and Yanghua Wang. 2022. High-frequency wavefield extrapolation using the Fourier neural operator. *Journal of Geophysics and Engineering* 19(2): 269–282. ISSN: 1742-2132, 1742-2140. https://doi.org/10.1093/jge/gxac016.
123. Wei Wei and Li-Yun Fu. Small-data-driven fast seismic simulations for complex media using physics-informed Fourier neural operators. *Geophysics* 87(6): T435–T446. ISSN: 0016-8033, 1942-2156. https://doi.org/10.1190/geo2021-0573.1.
124. Meer Mehran Rashid et al. 2022. Learning the stress-strain fields in digital composites using Fourier neural operator. *iScience* 25(11): 105452. ISSN: 25890042. https://doi.org/10.1016/j.isci.2022.105452.
125. Kai Zhang et al. 2022. Fourier neural operator for solving subsurface oil/water two-phase flow partial differential equation. *SPE Journal* 27(03): 1815–1830. ISSN: 1086-055X, 1930-0220. https://doi.org/10.2118/209223-PA.
126. Bicheng Yan et al. 2022. A robust deep learning workflow to predict multiphase flow behavior during geological C O 2 sequestration injection and Post-Injection periods. *Journal of Hydrology* 607: 127542. ISSN: 00221694. https://doi.org/10.1016/j.jhydrol.2022.127542.
127. Gege Wen et al. 2022. U-FNO—An enhanced Fourier neural operator-based deep-learning model for multiphase flow. *Advances in Water Resources* 163: 104180. ISSN: 03091708. https://doi.org/10.1016/j.advwatres.2022.104180.
128. Wenhui Peng, Zelong Yuan, and Jianchun Wang. 2022. Attention-enhanced neural network models for turbulence simulation. *Physics of Fluids* 34(2): 025111. ISSN: 1070-6631, 1089-7666. https://doi.org/10.1063/5.0079302.
129. Huaiqian You et al. 2022. Learning deep Implicit Fourier Neural Operators (IFNOs) with applications to heterogeneous material modeling. *Computer Methods in Applied Mechanics and Engineering* 398: 115296. ISSN: 0045-7825. https://doi.org/10.1016/j.cma.2022.115296.
130. Tie Kuang et al. 2023. Fast and robust prediction of multiphase flow in complex fractured reservoir using a Fourier neural operator. *Energies* 16(9): 3765. ISSN: 1996-1073. https://doi.org/10.3390/en16093765.
131. Paulo Alexandre Costa Rocha et al. 2023. Deep neural network modeling for CFD simulations: Benchmarking the Fourier neural operator on the Lid-Driven cavity case. *Applied Sciences* 13(5): 3165. ISSN: 2076-3417. https://doi.org/10.3390/app13053165.
132. Ashish Vaswani et al. 2017. Attention is all you need. In *Advances in neural information processing systems*, vol. 30. Curran Associates, Inc.
133. Shuhao Cao. 2021. *Choose a transformer: Fourier or Galerkin*. arXiv:2105.14995 [cs, math]. https://doi.org/10.48550/arXiv.2105.14995.
134. Zongyi Li et al. 2023. *Physics-Informed neural operator for learning partial differential equations*. arXiv:2111.03794 [cs, math]. https://doi.org/10.48550/arXiv.2111.03794.
135. Carlo Marcati et al. 2023. Exponential ReLU neural network approximation rates for point and edge singularities. *Foundations of Computational Mathematics* 23(3): 1043–1127. ISSN: 1615-3383. https://doi.org/10.1007/s10208-022-09565-9.
136. Lukas Gonon and Christoph Schwab. 2023. Deep ReLU neural networks overcome the curse of dimensionality for partial integrodifferential equations. *Analysis and Applications* 21(01): 1–47. ISSN: 0219-5305, 1793-6861. https://doi.org/10.1142/S0219530522500129.
137. Carlo Marcati and Christoph Schwab. 2023. Exponential convergence of deep operator networks for elliptic partial differential equations. *SIAM Journal on Numerical Analysis* 61(3): 1513–1545. ISSN: 0036-1429, 1095-7170. https://doi.org/10.1137/21M1465718.
138. Julen Álvarez-Aramberri Álvarez-Aramberri et al. 2023. Generation of massive databases for deep learning inversion: A goal-oriented hp-adaptive strategy. In *XI International conference on adaptive modeling and simulation (ADMOS 2023)*. Applications of goal-oriented error estimation and adaptivity. https://doi.org/10.23967/admos.2023.027.

139. Erik Lien Bolager et al. 2023. *Sampling weights of deep neural networks*. arXiv:2306.16830 [cs, math]. https://doi.org/10.48550/arXiv.2306.16830.
140. Sepp Hochreiter and Jürgen Schmidhuber. 1997. Long short-term memory. *Neural computation* 9(8): 1735–1780. ISSN: 0899-7667. https://doi.org/10.1162/neco.1997.9.8.1735.
141. Amulya Arun Ballakur and Arti Arya. 2020. Empirical evaluation of gated recurrent neural network architectures in aviation delay prediction. In *2020 5th international conference on computing, communication and security (ICCCS)*, 1–7. https://doi.org/10.1109/ICCCS49678.2020.9276855.
142. Qian Chen et al. 2023. Using the transformer model for physical simulation: An application on transient thermal analysis for 3D printing process simulation.
143. Nicholas Geneva and Nicholas Zabaras. 2020. Modeling the dynamics of PDE systems with physics-constrained deep auto-regressive networks. *Journal of Computational Physics* 403: 109056. tex.ids= Geneva, Geneva2020a arXiv: 1906.05747. ISSN: 00219991. https://doi.org/10.1016/j.jcp.2019.109056.
144. Chang, Michael B., et al. 2017. *A compositional object-based approach to learning physical dynamics*. https://doi.org/10.48550/arXiv.1612.00341.
145. Mrowca, Damian, et al. 2018. *Flexible neural representation for physics prediction*. https://doi.org/10.48550/arXiv.1806.08047.
146. Sanchez-Gonzalez, Alvaro, et al. 2018. *Graph networks as learnable physics engines for inference and control*. https://doi.org/10.48550/arXiv.1806.01242.
147. Li, Yunzhu, et al. 2019. *Propagation networks for model-based control under partial observation*. https://doi.org/10.48550/arXiv.1809.11169.
148. Lino, Mario, et al. 2021. *Simulating continuum mechanics with multi-scale graph neural networks*. https://doi.org/10.48550/arXiv.2106.04900.
149. Motaz Alfarraj and Ghassan AlRegib. 2018. Petrophysical-property estimation from seismic data using recurrent neural networks. In *SEG technical program expanded abstracts 2018*, 2141–2146. Anaheim, California: Society of Exploration Geophysicists. https://doi.org/10.1190/segam2018-2995752.1.
150. Amir Adler, Mauricio Araya-Polo, and Tomaso Poggio. 2019. Deep recurrent architectures for seismic tomography. In *81st EAGE conference and exhibition 2019*. 1–5. https://doi.org/10.3997/2214-4609.201901512.
151. Gabriel Fabien-Ouellet and Rahul Sarkar. 2020. Seismic velocity estimation: A deep recurrent neural-network approach. *Geophysics* 85(1): U21–U29. ISSN: 0016-8033, 1942-2156. https://doi.org/10.1190/geo2018-0786.1.
152. Vlachas, Pantelis R., et al. 2018. Data-driven forecasting of high-dimensional chaotic systems with long short-term memory networks. *Proceedings of the Royal Society A: Mathematical, Physical and Engineering Sciences* 474(2213): 20170844. ISSN: 1364-5021, 1471-2946. https://doi.org/10.1098/rspa.2017.0844.
153. Wei Hou, Darwin Darakananda, and Jeff Eldredge. 2019. Machine learning based detection of flow disturbances using surface pressure measurements. In *AIAA Scitech 2019 Forum*. San Diego, California: American Institute of Aeronautics and Astronautics. ISBN: 978-1-62410-578-4. https://doi.org/10.2514/6.2019-1148.
154. Yagawa , G., and Atsuya Oishi. 2023. *Computational mechanics with deep learning: An introduction*. eng. Cham, Switzerland: Springer. ISBN: 978-3-031-11847-0.
155. Leonhard Heindel, Peter Hantschke, and Markus Kästner. 2021. A virtual sensing approach for approximating nonlinear dynamical systems using LSTM networks. *PAMM* 21(1): e202100119. ISSN: 1617-7061, 1617-7061. https://doi.org/10.1002/pamm.202100119.
156. Heindel, L., P. Hantschke, and M. Kästner. 2022. A data-driven approach for approximating non-linear dynamic systems using LSTM networks. In *Procedia structural integrity. Fatigue design 2021, International conference proceedings*, 9th edn., vol. 38, 159–167. ISSN: 2452-3216. https://doi.org/10.1016/j.prostr.2022.03.017.
157. Freitag, S., et al. 2018. Recurrent neural networks and proper orthogonal decomposition with interval data for real-time predictions of mechanised tunnelling processes. *Computers & Structures*. CIVIL-COMP 2017(207): 258–273. ISSN: 0045-7949. https://doi.org/10.1016/j.compstruc.2017.03.020.

158. Cao, B.T., et al. 2020. Artificial neural network surrogate modelling for real-time predictions and control of building damage during mechanised tunnelling. *Advances in Engineering Software* 149: 102869. ISSN: 0965-9978. https://doi.org/10.1016/j.advengsoft.2020.102869.
159. Ba Trung Cao et al. 2022. Real-time risk assessment of tunneling-induced building damage considering polymorphic uncertainty. *ASCE-ASME Journal of Risk and Uncertainty in Engineering Systems, Part A: Civil Engineering* 8(1): 04021069. ISSN: 2376-7642. https://doi.org/10.1061/AJRUA6.0001192.
160. Anthony Gruber et al. 2022. A comparison of neural network architectures for data-driven reduced-order modeling. *Computer Methods in Applied Mechanics and Engineering* 393: 114764. ISSN: 0045-7825. https://doi.org/10.1016/j.cma.2022.114764.
161. Francisco J. Gonzalez and Maciej Balajewicz. 2018. *Deep convolutional recurrent autoencoders for learning low-dimensional feature dynamics of fluid systems*. arXiv:1808.01346 [physics]. https://doi.org/10.48550/arXiv.1808.01346.
162. Daniel Holden et al. Subspace neural physics: Fast data-driven interactive simulation. In *Proceedings of the 18th annual ACM SIGGRAPH/Eurographics symposium on computer animation*. SCA '19, 1–12. New York, NY, USA: Association for Computing Machinery. ISBN: 978-1-4503-6677-9. https://doi.org/10.1145/3309486.3340245.
163. Stefania Fresca et al. 2020. Deep learning-based reduced order models in cardiac electrophysiology. *PLOS ONE* 15(10): e0239416. ISSN: 1932-6203. https://doi.org/10.1371/journal.pone.0239416.
164. Stefania Fresca, Luca Dede', and Andrea Manzoni. 2021. A Comprehensive deep learning-based approach to reduced order modeling of nonlinear time-dependent parametrized PDEs. *Journal of Scientific Computing* 87(2): 61. ISSN: 1573-7691. https://doi.org/10.1007/s10915-021-01462-7.
165. Stefania Fresca and Andrea Manzoni. 2022. POD-DL-ROM: Enhancing deep learning-based reduced order models for nonlinear parametrized PDEs by proper orthogonal decomposition. *Computer Methods in Applied Mechanics and Engineering* 388: 114181. ISSN: 0045-7825. https://doi.org/10.1016/j.cma.2021.114181.
166. Pu Ren et al. 2022. PhyCRNet: Physics-informed convolutional-recurrent network for solving spatiotemporal PDEs. *Computer Methods in Applied Mechanics and Engineering* 389: 114399. ISSN: 0045-7825. https://doi.org/10.1016/j.cma.2021.114399.
167. Hu, C., S. Martin, and R. Dingreville. 2022. Accelerating phase-field predictions via recurrent neural networks learning the microstructure evolution in latent space. *Computer Methods in Applied Mechanics and Engineering* 397: 115128. ISSN: 0045-7825. https://doi.org/10.1016/j.cma.2022.115128.
168. Kookjin Lee and Kevin T. Carlberg. 2020. Model reduction of dynamical systems on nonlinear manifolds using deep convolutional autoencoders. *Journal of Computational Physics* 404: 108973. ISSN: 0021-9991. https://doi.org/10.1016/j.jcp.2019.108973.
169. Siyuan Shen et al. 2021. *High-order differentiable autoencoder for nonlinear model reduction*. arXiv:2102.11026 [cs]. https://doi.org/10.48550/arXiv.2102.11026.
170. Peter J. Schmid. 2010. Dynamic mode decomposition of numerical and experimental data. *Journal of Fluid Mechanics* 656: 5–28. ISSN: 1469-7645, 0022-1120. https://doi.org/10.1017/S0022112010001217.
171. Jonathan H. Tu et al. 2013. *On dynamic mode decomposition: Theory and applications*. arXiv:1312.0041 [physics]. https://doi.org/10.3934/jcd.2014.1.391.
172. Brunton, Steven L., and Jose Nathan Kutz. 2022. *Data-driven science and engineering: Machine learning, dynamical systems, and control*. Cambridge, United Kingdom, New York, NY: Cambridge University Press 978-1-00-909848-9.
173. Koopman, B.O. 1931. Hamiltonian systems and transformation in hilbert space. *Proceedings of the National Academy of Sciences* 17(5): 315–318. ISSN: 0027-8424, 1091-6490. https://doi.org/10.1073/pnas.17.5.315.
174. Matthew O. Williams, Ioannis G. Kevrekidis, and Clarence W. Rowley. 2015. A data–driven approximation of the koopman operator: Extending dynamic mode decomposition. *Journal of Nonlinear Science* 25(6): 1307–1346. ISSN: 1432-1467. https://doi.org/10.1007/s00332-015-9258-5.

175. Qianxiao Li et al. 2017. Extended dynamic mode decomposition with dictionary learning: A data-driven adaptive spectral decomposition of the Koopman operator. *Chaos: An Interdisciplinary Journal of Nonlinear Science* 27(10): 103111. ISSN: 1054-1500, 1089-7682. https://doi.org/10.1063/1.4993854.
176. Enoch Yeung, Soumya Kundu, and Nathan Hodas. 2019. Learning deep neural network representations for koopman operators of nonlinear dynamical systems. In *2019 American control conference (ACC)*, 4832–4839. ISSN: 2378-5861. https://doi.org/10.23919/ACC.2019.8815339.
177. Naoya Takeishi, Yoshinobu Kawahara, and Takehisa Yairi. 2017. Learning Koopman invariant subspaces for dynamic mode decomposition. In *Proceedings of the 31st international conference on neural information processing systems*, 1130–1140. NIPS'17. Red Hook, NY, USA: Curran Associates Inc. ISBN: 978-1-5108-6096-4.
178. Jeremy Morton et al. 2018. Deep dynamical modeling and control of unsteady fluid flows. In *Proceedings of the 32nd international conference on neural information processing systems*, 9278–9288. NIPS'18. Red Hook, NY, USA: Curran Associates Inc.
179. Bethany Lusch, J. Nathan Kutz, and Steven L. Brunton. 2018. Deep learning for universal linear embeddings of nonlinear dynamics. *Nature Communications* 9(1): 4950. ISSN: 2041-1723. https://doi.org/10.1038/s41467-018-07210-0.
180. Samuel E. Otto and Clarence W. Rowley. 2019. Linearly recurrent autoencoder networks for learning dynamics. *SIAM Journal on Applied Dynamical Systems* 18(1): 558–593. ISSN: 1536-0040. https://doi.org/10.1137/18M1177846.
181. David Cohn, Zoubin Ghahramani, and Michael Jordan. 1994. Active learning with statistical models. In *Advances in neural information processing systems*, vol. 7. MIT Press.
182. Xing Liu et al. 2021. Knowledge extraction and transfer in data-driven fracture mechanics. *Proceedings of the National Academy of Sciences* 118(23): e2104765118. ISSN: 0027-8424, 1091-6490. https://doi.org/10.1073/pnas.2104765118.
183. Bernard Haasdonk et al. 2023. A new certified hierarchical and adaptive RB-ML-ROM surrogate model for parametrized PDEs. *SIAM Journal on Scientific Computing* 45(3): A1039–A1065. ISSN: 1064-8275, 1095-7197. https://doi.org/10.1137/22M1493318.
184. Karl A. Kalina et al. 2023. FE^{ANN}: An efficient data-driven multiscale approach based on physics-constrained neural networks and automated data mining. *Computational Mechanics* 71(5): 827–851. ISSN: 1432-0924. https://doi.org/10.1007/s00466-022-02260-0.
185. Sinno Jialin Pan and Qiang Yang. 2010. A survey on transfer learning. *IEEE Transactions on Knowledge and Data Engineering* 22(10): 1345–1359. ISSN: 1558-2191. https://doi.org/10.1109/TKDE.2009.191.
186. Jason Yosinski et al. 2014. How transferable are features in deep neural networks? In *Proceedings of the 27th international conference on neural information processing systems*, vol. 2, 3320–3328. NIPS'14. Cambridge, MA, USA: MIT Press.
187. Stefan Kollmannsberger, Divya Singh, and Leon Herrmann. 2023. Transfer learning enhanced full waveform inversion. In *2023 IEEE/ASME international conference on advanced intelligent mechatronics (AIM)*, 866–871. Seattle, WA, USA: IEEE. ISBN: 978-1-66547-633-1. https://doi.org/10.1109/AIM46323.2023.10196158.
188. Giada Colella, Volker A. Lange, and Fabian Duddeck. 2024. Transfer learning for crash design. *Discover Mechanical Engineering* 3(1): 26. ISSN: 2731-6564. https://doi.org/10.1007/s44245-024-00059-2.
189. Ziming Liu et al. 2021. *Physics-augmented learning: A new paradigm beyond physics-informed learning*. arXiv:2109.13901 [physics]. https://doi.org/10.48550/arXiv.2109.13901.
190. Yinhao Zhu et al. 2019. Physics-constrained deep learning for high-dimensional surrogate modeling and uncertainty quantification without labeled data. *Journal of Computational Physics* 394: 56–81. arXiv: 1901.06314. ISSN: 00219991. https://doi.org/10.1016/j.jcp.2019.05.024.
191. Jonas Eichelsdörfer, Sebastian Kaltenbach, and Phaedon-Stelios Koutsourelakis. 2021. *Physics-enhanced neural networks in the small data regime*. arXiv:2111.10329 [physics, stat] version: 1. https://doi.org/10.48550/arXiv.2111.10329.

192. Dimitris C. Psichogios and Lyle H. Ungar. 1992. A hybrid neural network-first principles approach to process modeling. *AIChE Journal* 38(10): 1499–1511. ISSN: 0001-1541, 1547-5905. https://doi.org/10.1002/aic.690381003.
193. Dissanayake, M.W.M.G., and N. Phan-Thien. 1994. Neural-network-based approximations for solving partial differential equations. *Communications in Numerical Methods in Engineering* 10(3): 195–201. ISSN: 10698299. https://doi.org/10.1002/cnm.1640100303.
194. Lagaris, I.E., A. Likas, and D.I. Fotiadis. 1998. Artificial neural networks for solving ordinary and partial differential equations. *IEEE Transactions on Neural Networks* 9(5): 987–1000. ISSN: 10459227. https://doi.org/10.1109/72.712178.
195. Maziar Raissi. 2018. Deep hidden physics models: Deep learning of nonlinear partial differential equations.
196. George Em Karniadakis et al. 2021. Physics-informed machine learning. *Nature Reviews Physics* 3(6): 422–440. ISSN: 2522-5820. https://doi.org/10.1038/s42254-021-00314-5.
197. Salvatore Cuomo et al. 2022. Scientific machine learning through physics–informed neural networks: Where we are and what's next. *Journal of Scientific Computing* 92(3): 88. ISSN: 1573-7691. https://doi.org/10.1007/s10915-022-01939-z.
198. Hao, Zhongkai, et al. 2022. Physics-informed machine learning: A survey on problems. *Methods and Applications.* https://doi.org/10.48550/arXiv.2211.08064.
199. Ehsan Haghighat and Ruben Juanes. 2021. SciANN: A Keras/TensorFlow wrapper for scientific computations and physics-informed deep learning using artificial neural networks. *Computer Methods in Applied Mechanics and Engineering* 373: 113552. ISSN: 0045-7825. https://doi.org/10.1016/j.cma.2020.113552.
200. Oliver Hennigh et al. 2021. NVIDIA SimNet$^T M$: An AI-accelerated multi-physics simulation framework. *Computational Science—ICCS 2021*, ed. Maciej Paszynski et al., 447–461. Lecture notes in computer science. Cham: Springer International Publishing. ISBN: 978-3-030-77977-1. https://doi.org/10.1007/978-3-030-77977-1sps36.
201. Lu Lu et al. 2021. DeepXDE: A deep learning library for solving differential equations. *SIAM Review* 63(1): 208–228. ISSN: 0036-1445, 1095-7200. https://doi.org/10.1137/19M1274067.
202. Zhiqiang Cai et al. 2020. Deep least-squares methods: An unsupervised learning-based numerical method for solving elliptic PDEs. *Journal of Computational Physics* 420: 109707. ISSN: 0021-9991. https://doi.org/10.1016/j.jcp.2020.109707.
203. Justin Sirignano and Konstantinos Spiliopoulos. 2018. DGM: A deep learning algorithm for solving partial differential equations. *Journal of Computational Physics* 375: 1339–1364. arXiv: 1708.07469. ISSN: 00219991. https://doi.org/10.1016/j.jcp.2018.08.029.
204. William J. Morokoff and Russel E. Caflisch. 1995. Quasi-Monte Carlo integration. *Journal of Computational Physics* 122(2): 218–230. ISSN: 0021-9991. https://doi.org/10.1006/jcph.1995.1209.
205. 14 - Monte carlo integration I: Basic concepts. *Physically Based Rendering*, ed. by Matt Pharr and Greg Humphreys, 631–660. Burlington: Morgan Kaufmann. ISBN: 978-0-12-553180-1. https://doi.org/10.1016/B978-012553180-1/50016-8.
206. Erich Novak and Klaus Ritter. 1996. High dimensional integration of smooth functions over cubes. *Numerische Mathematik* 75(1): 79–97. ISSN: 0945-3245. https://doi.org/10.1007/s002110050231.
207. Jon A. Rivera et al. 2022. On quadrature rules for solving partial differential equations using neural networks. *Computer Methods in Applied Mechanics and Engineering* 393: 114710. ISSN: 0045-7825. https://doi.org/10.1016/j.cma.2022.114710.
208. Tamara G. Grossmann et al. 2023. *Can physics-informed neural networks beat the finite element method?* https://doi.org/10.48550/arXiv.2302.04107.
209. Ali Kashefi and Tapan Mukerji. 2022. Physics-informed PointNet: A deep learning solver for steady-state incompressible flows and thermal fields on multiple sets of irregular geometries. *Journal of Computational Physics* 468: 111510. ISSN: 00219991. https://doi.org/10.1016/j.jcp.2022.111510.
210. Jens Berg and Kaj Nyström. 2018. A unified deep artificial neural network approach to partial differential equations in complex geometries. *Neurocomputing* 317: 28–41. arXiv: 1711.06464. ISSN: 09252312. https://doi.org/10.1016/j.neucom.2018.06.056.

211. Stefan Kollmannsberger et al. 2021. *Deep learning in computational mechanics: An introductory course*. eng. Studies in computational intelligence, vol. 977. Cham: Springer. ISBN: 978-3-030-76587-3 978-3-030-76586-6.
212. Alexander Henkes, Henning Wessels, and Rolf Mahnken. 2022. Physics informed neural networks for continuum micromechanics. *Computer Methods in Applied Mechanics and Engineering* 393: 114790. ISSN: 0045-7825. https://doi.org/10.1016/j.cma.2022.114790.
213. Lagaris, I.E., A.C. Likas, and D.G. Papageorgiou. 2000. Neural-network methods for boundary value problems with irregular boundaries. *IEEE Transactions on Neural Networks* 11(5): 1041–1049. ISSN: 1941-0093. https://doi.org/10.1109/72.870037.
214. Silvia Ferrari and Mark Jensenius. 2008. A constrained optimization approach to preserving prior knowledge during incremental training. *IEEE Transactions on Neural Networks* 19(6): 996–1009. ISSN: 1941-0093. https://doi.org/10.1109/TNN.2007.915108.
215. Keith Rudd, Gianluca Di Muro, and Silvia Ferrari. 2014. A constrained backpropagation approach for the adaptive solution of partial differential equations. *IEEE transactions on neural networks and learning systems* 25(3): 571–584. ISSN: 2162-2388. https://doi.org/10.1109/TNNLS.2013.2277601.
216. Keith Rudd and Silvia Ferrari. 2015. A constrained integration (CINT) approach to solving partial differential equations using artificial neural networks. *Neurocomputing* 155: 277–285. ISSN: 0925-2312. https://doi.org/10.1016/j.neucom.2014.11.058.
217. Fei Wang et al. 2017. Residual attention network for image classification. In *2017 IEEE conference on computer vision and pattern recognition (CVPR)*, 6450–6458. ISSN: 1063-6919. https://doi.org/10.1109/CVPR.2017.683.
218. Shanshan Zhang, Jian Yang, and Bernt Schiele. 2018. Occluded pedestrian detection through guided attention in CNNs. In *2018 IEEE/CVF conference on computer vision and pattern recognition*, 6995–7003. ISSN: 2575-7075. https://doi.org/10.1109/CVPR.2018.00731.
219. Jim Magiera et al. 2020. Constraint-aware neural networks for Riemann problems. *Journal of Computational Physics* 409: 109345. ISSN: 0021-9991. https://doi.org/10.1016/j.jcp.2020.109345.
220. Yatin Nandwani et al. 2019. A primal dual formulation for deep learning with constraints. In *Advances in neural information processing systems*, vol. 32. Curran Associates, Inc.
221. Levi McClenny and Ulisses Braga-Neto. 2022. *Self-adaptive physics-informed neural networks using a soft attention mechanism*. arXiv:2009.04544 [cs, stat].
222. Lu Lu et al. 2021. Physics-informed neural networks with hard constraints for inverse design. *SIAM Journal on Scientific Computing* 43(6): B1105–B1132. ISSN: 1064-8275, 1095-7197. https://doi.org/10.1137/21M1397908.
223. Qi Zeng et al. 2022. *Competitive physics informed networks*. arXiv:2204.11144 [cs, math]. https://doi.org/10.48550/arXiv.2204.11144.
224. Philipp Moser et al. 2023. Modeling of 3D blood flows with physics-informed neural networks: Comparison of network architectures. *Fluids* 8(2): 46. ISSN: 2311-5521. https://doi.org/10.3390/fluids8020046.
225. Jun Han, Jun Tao, and Chaoli Wang. 2020. FlowNet: A deep learning framework for clustering and selection of streamlines and stream surfaces. *IEEE Transactions on Visualization and Computer Graphics* 26(4): 1732–1744. ISSN: 1941-0506. https://doi.org/10.1109/TVCG.2018.2880207.
226. Saakaar Bhatnagar et al. 2019. Prediction of aerodynamic flow fields using convolutional neural networks. *Computational Mechanics* 64(2): 525–545. ISSN: 1432-0924. https://doi.org/10.1007/s00466-019-01740-0.
227. Han Gao, Luning Sun, and Jian-Xun Wang. 2021. PhyGeoNet: Physics-informed geometry-adaptive convolutional neural networks for solving parameterized steady-state PDEs on irregular domain. *Journal of Computational Physics* 428: 110079. ISSN: 0021-9991. https://doi.org/10.1016/j.jcp.2020.110079.
228. Nils Wandel et al. 2022. *Spline-PINN: Approaching PDEs without data using fast, physics-informed hermite-spline CNNs*. arXiv:2109.07143 [physics]. https://doi.org/10.48550/arXiv.2109.07143.

229. Han Gao, Matthew J. Zahr, and Jian-Xun Wang. 2022. Physics-informed graph neural Galerkin networks: A unified framework for solving PDE-governed forward and inverse problems. *Computer Methods in Applied Mechanics and Engineering* 390: 114502. ISSN: 0045-7825. https://doi.org/10.1016/j.cma.2021.114502.
230. Möller, Matthias, Deepesh Toshniwal, and Frank Van Ruiten. 2021. *Physics-informed machine learning embedded into isogeometric analysis*. In Mathematics: Key enabling technology for scientific machine learning.
231. Hughes, T.J.R., J.A. Cottrell, and Y. Bazilevs. 2005. Isogeometric analysis: CAD, finite elements, NURBS, exact geometry and mesh refinement. *Computer Methods in Applied Mechanics and Engineering* 194(39): 4135–4195. ISSN: 0045-7825. https://doi.org/10.1016/j.cma.2004.10.008.
232. Rishith Ellath Meethal et al. 2022. *Finite element method-enhanced neural network for forward and inverse problems*. arXiv:2205.08321 [cs, math]. https://doi.org/10.48550/arXiv.2205.08321.
233. Hughes, Thomas J. R. 2000. *The finite element method: Linear static and dynamic finite element analysis*. Mineola, NY: Dover Publications 978-0-486-41181-1.
234. Klaus-Jürgen Bathe. 2014. *Finite element procedures*. en. 2nd ed. OCLC: ocn930843107. Englewood Cliffs, N.J: Prentice-Hall. ISBN: 978-0-9790049-5-7.
235. Stefano Berrone, Claudio Canuto, and Moreno Pintore. 2022. Variational physics informed neural networks: The role of quadratures and test functions. *Journal of Scientific Computing* 92(3): 100. ISSN: 1573-7691. https://doi.org/10.1007/s10915-022-01950-4.
236. Santiago Badia, Wei Li, and Alberto F. Martín. 2023. *Finite element interpolated neural networks for solving forward and inverse problems*. arXiv:2306.06304 [cs, math]. https://doi.org/10.48550/arXiv.2306.06304.
237. Alireza Yazdani et al. 2020. Systems biology informed deep learning for inferring parameters and hidden dynamics. *PLOS Computational Biology* 16(11): e1007575. ISSN: 1553-7358. https://doi.org/10.1371/journal.pcbi.1007575.
238. Carlos Uriarte, David Pardo, and Angel Javier Omella. 2022. A finite element based deep learning solver for parametric PDEs. *Computer Methods in Applied Mechanics and Engineering* 391: 114562. ISSN: 0045-7825. https://doi.org/10.1016/j.cma.2021.114562.
239. Ehsan Kharazmi, Zhongqiang Zhang, and George E. M. Karniadakis. 2021. hp-VPINNs: Variational physics-informed neural networks with domain decomposition. *Computer Methods in Applied Mechanics and Engineering* 374: 113547. ISSN: 0045-7825. https://doi.org/10.1016/j.cma.2020.113547.
240. Ameya D. Jagtap, Ehsan Kharazmi, and George Em Karniadakis. 2020. Conservative physics-informed neural networks on discrete domains for conservation laws: Applications to forward and inverse problems. *Computer Methods in Applied Mechanics and Engineering* 365:113028. ISSN: 0045-7825. https://doi.org/10.1016/j.cma.2020.113028.
241. Khemraj Shukla, Ameya D. Jagtap, and George Em Karniadakis. 2021. Parallel physics-informed neural networks via domain decomposition. *Journal of Computational Physics* 447:110683. ISSN: 0021-9991. https://doi.org/10.1016/j.jcp.2021.110683.
242. Ameya D. Jagtap and George Em Karniadakis. 2020. Extended physics-informed neural networks (XPINNs): A generalized space-time domain decomposition based deep learning framework for nonlinear partial differential equations. *Communications in Computational Physics* 28(5): 2002–2041. ISSN: 1815-2406, 1991-7120. https://doi.org/10.4208/cicp.OA-2020-0164.
243. Xinhai Chen et al. 2021. Transfer learning for deep neural network-based partial differential equations solving. *Advances in aerodynamics* 3(1): 36. ISSN: 2524-6992. https://doi.org/10.1186/s42774-021-00094-7.
244. Somdatta Goswami et al. 2019. Transfer learning enhanced physics informed neural network for phase-field modeling of fracture. [cs, stat]. tex.ids= Goswami2019a. arXiv: 1907.02531.
245. Junyan He et al. 2023. Deep energy method in topology optimization applications. *Acta Mechanica* 234(4): 1365–1379. ISSN: 1619-6937. https://doi.org/10.1007/s00707-022-03449-3.

246. Joglekar, Aditya, Hongrui Chen, and Levent Burak Kara. 2023. DMF-TONN: Direct mesh-free topology optimization using neural networks. *Engineering with Computers*. ISSN 0177–0667: 1435–5663. https://doi.org/10.1007/s00366-023-01904-w.
247. Sigmund, O. 2001. A 99 line topology optimization code written in Matlab. *Structural and Multidisciplinary Optimization* 21(2): 120–127. ISSN: 1615-1488. https://doi.org/10.1007/s001580050176.
248. Martin P. Bendsøe and Ole Sigmund. 2004. *Topology optimization*. en. Berlin, Heidelberg: Springer. ISBN: 978-3-642-07698-5 978-3-662-05086-6. https://doi.org/10.1007/978-3-662-05086-6.
249. Mohammad Amin Nabian, Rini Jasmine Gladstone, and Hadi Meidani. 2021. Efficient training of physics-informed neural networks via importance sampling. *Computer-Aided Civil and Infrastructure Engineering* 36(8): 962–977. ISSN: 1093-9687, 1467-8667. https://doi.org/10.1111/mice.12685.
250. John M. Hanna et al. 2022. Residual-based adaptivity for two-phase flow simulation in porous media using physics-informed neural networks. *Computer Methods in Applied Mechanics and Engineering* 396: 115100. ISSN: 0045-7825. https://doi.org/10.1016/j.cma.2022.115100.
251. Stefan Kollmannsberger et al. 2021. Physics-informed neural networks. In *Deep learning in computational mechanics*, vol. 977, 55–84. Cham: Springer International Publishing. ISBN: 978-3-030-76586-6 978-3-030-76587-3. https://doi.org/10.1007/978-3-030-76587-3sps5.
252. David Anton and Henning Wessels. 2021. Identification of material parameters from full-field displacement data using physics-informed neural networks. https://doi.org/10.13140/RG.2.2.24558.89924/1.
253. Yifei Zong, QiZhi He, and Alexandre M. Tartakovsky. 2023. Improved training of physics-informed neural networks for parabolic differential equations with sharply perturbed initial conditions. *Computer Methods in Applied Mechanics and Engineering* 414: 116125. ISSN: 0045-7825. https://doi.org/10.1016/j.cma.2023.116125.
254. Jeremy Yu et al. 2022. Gradient-enhanced physics-informed neural networks for forward and inverse PDE problems. *Computer Methods in Applied Mechanics and Engineering* 393: 114823. ISSN: 0045-7825. https://doi.org/10.1016/j.cma.2022.114823.
255. Jamie M. Taylor, David Pardo, and Ignacio Muga. 2023. A deep fourier residual method for solving PDEs using neural networks. *Computer Methods in Applied Mechanics and Engineering* 405: 115850. ISSN: 0045-7825. https://doi.org/10.1016/j.cma.2022.115850.
256. Pao-Hsiung Chiu et al. 2022. CAN-PINN: A fast physics-informed neural network based on coupled-automatic–numerical differentiation method. *Computer Methods in Applied Mechanics and Engineering* 395: 114909. ISSN: 0045-7825. https://doi.org/10.1016/j.cma.2022.114909.
257. Ameya D. Jagtap and George Em Karniadakis. 2020. Adaptive activation functions accelerate convergence in deep and physics-informed neural networks. *Journal of Computational Physics*. arXiv: 1906.01170, p. 109136. ISSN: 00219991. https://doi.org/10.1016/j.jcp.2019.109136.
258. Guang-Bin Huang, Qin-Yu Zhu, and Chee-Kheong Siew. 2006. Extreme learning machine: Theory and applications. *Neurocomputing Neural Networks* 70(1): 489–501. ISSN: 0925-2312. https://doi.org/10.1016/j.neucom.2005.12.126.
259. Guang-Bin Huang, Dian Hui Wang, and Yuan Lan. 2011. Extreme learning machines: A survey. *International Journal of Machine Learning and Cybernetics* 2(2): 107–122. ISSN: 1868-808X. https://doi.org/10.1007/s13042-011-0019-y.
260. Suchuan Dong and Zongwei Li. 2021. Local extreme learning machines and domain decomposition for solving linear and nonlinear partial differential equations. *Computer Methods in Applied Mechanics and Engineering* 387: 114129. ISSN: 0045-7825. https://doi.org/10.1016/j.cma.2021.114129.
261. Suchuan Dong and Jielin Yang. 2022. Numerical approximation of partial differential equations by a variable projection method with artificial neural networks. *Computer Methods in Applied Mechanics and Engineering* 398: 115284. ISSN: 0045-7825. https://doi.org/10.1016/j.cma.2022.115284.

262. Ehsan Haghighat et al. 2021. A physics-informed deep learning framework for inversion and surrogate modeling in solid mechanics. *Computer Methods in Applied Mechanics and Engineering* 379: 113741. ISSN: 0045-7825. https://doi.org/10.1016/j.cma.2021.113741.
263. Jinshuai Bai et al. 2023. An introduction to programming physics-informed neural network-based computational solid mechanics. *International Journal of Computational Methods* 2350013. ISSN: 0219-8762. https://doi.org/10.1142/S0219876223500135.
264. Georgios Kissas et al. 2020. Machine learning in cardiovascular flows modeling: Predicting arterial blood pressure from non-invasive 4D flow MRI data using physics-informed neural networks. *Computer Methods in Applied Mechanics and Engineering* 358: 112623. ISSN: 0045-7825. https://doi.org/10.1016/j.cma.2019.112623.
265. Maziar Raissi, Alireza Yazdani, and George Em Karniadakis. 2020. Hidden fluid mechanics: Learning velocity and pressure fields from flow visualizations. *Science* 367(6481): 1026–1030. ISSN: 0036-8075, 1095-9203. https://doi.org/10.1126/science.aaw4741.
266. Luning Sun et al. 2020. Surrogate modeling for fluid flows based on physics-constrained deep learning without simulation data. *Computer Methods in Applied Mechanics and Engineering* 361: 112732. ISSN: 0045-7825. https://doi.org/10.1016/j.cma.2019.112732.
267. Xiaowei Jin et al. 2021. NSFnets (Navier-Stokes flow nets): Physics-informed neural networks for the incompressible Navier-Stokes equations. *Journal of Computational Physics* 426: 109951. tex.ids= Jin2021a. ISSN: 0021-9991. https://doi.org/10.1016/j.jcp.2020.109951.
268. Shengze Cai et al. 2021. Flow over an espresso cup: Inferring 3-D velocity and pressure fields from tomographic background oriented Schlieren via physics-informed neural networks. *Journal of Fluid Mechanics* 915: A102. ISSN: 0022-1120, 1469-7645. https://doi.org/10.1017/jfm.2021.135.
269. Fraces, Cedric G., and Hamdi Tchelepi. 2021. Physics informed deep learning for flow and transport in porous media. *OnePetro*. https://doi.org/10.2118/203934-MS.
270. Wenbo Zhang et al. 2022. Simulation of the 3D hyperelastic behavior of ventricular myocardium using a finite-element based neural-network approach. *Computer Methods in Applied Mechanics and Engineering* 394: 114871. ISSN: 0045-7825. https://doi.org/10.1016/j.cma.2022.114871.
271. Jeremy C.-H. Wang and Jean-Pierre Hickey. 2023. FluxNet: A physics-informed learning-based Riemann solver for transcritical flows with non-ideal thermodynamics. *Computer Methods in Applied Mechanics and Engineering* 411: 116070. ISSN: 0045-7825. https://doi.org/10.1016/j.cma.2023.116070.
272. Sina Amini Niaki et al. 2021. Physics-informed neural network for modelling the thermo-chemical curing process of composite-tool systems during manufacture. *Computer Methods in Applied Mechanics and Engineering* 384: 113959. ISSN: 0045-7825. https://doi.org/10.1016/j.cma.2021.113959.
273. Qiming Zhu, Zeliang Liu, and Jinhui Yan. 2021. Machine learning for metal additive manufacturing: predicting temperature and melt pool fluid dynamics using physics-informed neural networks. *Computational Mechanics* 67(2): 619–635. ISSN: 1432-0924. https://doi.org/10.1007/s00466-020-01952-9.
274. Stefano Markidis. 2021. The old and the new: Can physics-informed deep-learning replace traditional linear solvers? *Frontiers in Big Data* 4. ISSN: 2624-909X.
275. Nick McGreivy and Ammar Hakim. 2024. Weak baselines and reporting biases lead to overoptimism in machine learning for fluid-related partial differential equations. *Nature Machine Intelligence* 6(10): 1256–1269. ISSN: 2522-5839. https://doi.org/10.1038/s42256-024-00897-5.
276. Liangliang Li et al. 2022. ReF-nets: Physics-informed neural network for Reynolds equation of gas bearing. *Computer Methods in Applied Mechanics and Engineering* 391: 114524. ISSN: 0045-7825. https://doi.org/10.1016/j.cma.2021.114524.
277. Yuyao Chen et al. 2020. Physics-informed neural networks for inverse problems in nano-optics and metamaterials. *Optics Express* 28(8): 11618. ISSN: 1094-4087. https://doi.org/10.1364/OE.384875.

278. Ruiyang Zhang, Yang Liu, and Hao Sun. 2020. Physics-informed multi-LSTM networks for metamodeling of nonlinear structures. *Computer Methods in Applied Mechanics and Engineering* 369: 113226. arXiv: 2002.10253. ISSN: 00457825. https://doi.org/10.1016/j.cma.2020.113226.
279. Khemraj Shukla et al. 2020. Physics-informed neural network for ultrasound nondestructive quantification of surface breaking cracks. *Journal of Nondestructive Evaluation* 39(3): 61. ISSN: 1573-4862. https://doi.org/10.1007/s10921-020-00705-1.
280. David Anton and Henning Wessels. 2022. *Physics-Informed Neural Networks for Material Model Calibration from Full-Field Displacement Data.*
281. Leon Herrmann et al. 2023. On the use of neural networks for full waveform inversion. *Computer Methods in Applied Mechanics and Engineering* 415: 116278. ISSN: 0045-7825. https://doi.org/10.1016/j.cma.2023.116278.
282. Carlos J. G. Rojas, Marco L. Bitterncourt, and José L. Boldrini. 2021. *Parameter identification for a damage model using a physics informed neural network.* https://doi.org/10.48550/arXiv.2107.08781.
283. Wenjing Li and Kok-Meng Lee. 2021. Physics informed neural network for parameter identification and boundary force estimation of compliant and biomechanical systems. *International Journal of Intelligent Robotics and Applications* 5(3): 313–325. ISSN: 2366-598X. https://doi.org/10.1007/s41315-021-00196-x.
284. Enrui Zhang et al. 2022. Analyses of internal structures and defects in materials using physics-informed neural networks. *Science Advances* 8(7): eabk0644. ISSN: 2375-2548. https://doi.org/10.1126/sciadv.abk0644.
285. Ivan Depina et al. 2022. Application of physics-informed neural networks to inverse problems in unsaturated groundwater flow. *Georisk: Assessment and Management of Risk for Engineered Systems and Geohazards* 16(1): 21–36. ISSN: 1749-9518. https://doi.org/10.1080/17499518.2021.1971251.
286. Chen Xu et al. 2023. Transfer learning based physics-informed neural networks for solving inverse problems in engineering structures under different loading scenarios. *Computer Methods in Applied Mechanics and Engineering* 405: 115852. ISSN: 0045-7825. https://doi.org/10.1016/j.cma.2022.115852.
287. Yubiao Sun, Ushnish Sengupta, and Matthew Juniper. 2023. Physics-informed deep learning for simultaneous surrogate modeling and PDE-constrained optimization of an airfoil geometry. *Computer Methods in Applied Mechanics and Engineering* 411: 116042. ISSN: 0045-7825. https://doi.org/10.1016/j.cma.2023.116042.
288. Majid Rasht-Behesht et al. 2022. Physics-informed neural networks (PINNs) for wave propagation and full waveform inversions. *Journal of Geophysical Research: Solid Earth* 127(5). ISSN: 2169-9313, 2169-9356. https://doi.org/10.1029/2021JB023120.
289. Jonas Zehnder et al. 2021. *NTopo: Mesh-free topology optimization using implicit neural representations.* https://doi.org/10.48550/arXiv.2102.10782.
290. Di Lorenzo, D., et al. 2023. Physics informed and data-based augmented learning in structural health diagnosis. *Computer Methods in Applied Mechanics and Engineering* 414: 116186. ISSN: 0045-7825. https://doi.org/10.1016/j.cma.2023.116186.
291. Jens Berg and Kaj Nyström. 2019. Data-driven discovery of PDEs in complex datasets. *Journal of Computational Physics* 384: 239–252. ISSN: 0021-9991. https://doi.org/10.1016/j.jcp.2019.01.036.
292. Silviu-Marian Udrescu and Max Tegmark. 2020. AI Feynman: A physics-inspired method for symbolic regression. *Science Advances* 6(16): eaay2631. ISSN: 2375-2548. https://doi.org/10.1126/sciadv.aay2631.
293. Richard P. Feynman, Robert B. Leighton, and Matthew L. Sands. 2011. *The Feynman lectures on physics.* New millennium ed. OCLC: ocn671704374. New York: Basic Books. ISBN: 978-0-465-02414-8 978-0-465-02493-3 978-0-465-02416-2 978-0-465-02417-9 978-0-465-02501-5 978-0-465-02382-0.
294. Xuhui Meng et al. 2020. PPINN: Parareal physics-informed neural network for time-dependent PDEs. *Computer Methods in Applied Mechanics and Engineering* 370: 113250. ISSN: 0045-7825. https://doi.org/10.1016/j.cma.2020.113250.

295. Revanth Mattey and Susanta Ghosh. 2022. A novel sequential method to train physics informed neural networks for Allen Cahn and Cahn Hilliard equations. *Computer Methods in Applied Mechanics and Engineering* 390: 114474. ISSN: 0045-7825. https://doi.org/10.1016/j.cma.2021.114474.
296. Arieh Iserles. 2008. *A first course in the numerical analysis of differential equations*. en. Google-Books-ID: 3acgAwAAQBAJ. Cambridge University Press. ISBN: 978-1-139-47376-7.
297. Henning Wessels, Christian Weißenfels, and Peter Wriggers. 2020. The neural particle method—An updated Lagrangian physics informed neural network for computational fluid dynamics. *Computer Methods in Applied Mechanics and Engineering* 368: 113127. ISSN: 0045-7825. https://doi.org/10.1016/j.cma.2020.113127.
298. Jinshuai Bai et al. 2022. A general neural particle method for hydrodynamics modeling. *Computer Methods in Applied Mechanics and Engineering* 393: 114740. ISSN: 0045-7825. https://doi.org/10.1016/j.cma.2022.114740.
299. González-García, R., R. Rico-Martínez, and I. G. Kevrekidis. 1998. Identification of distributed parameter systems: A neural net based approach. *Computers & Chemical Engineering*. In *European symposium on computer aided process engineering-8* 22: S965–S968. ISSN: 0098-1354. https://doi.org/10.1016/S0098-1354(98)00191-4.
300. Zichao Long et al. 2018. PDE-Net: learning PDEs from data. In *Proceedings of the 35th international conference on machine learning*. PMLR, 3208–3216.
301. Zichao Long, Yiping Lu, and Bin Dong. 2019. PDE-Net 2.0: Learning PDEs from data with a numeric-symbolic hybrid deep network. *Journal of Computational Physics* 399: 108925. ISSN: 0021-9991. https://doi.org/10.1016/j.jcp.2019.108925.
302. Binh-Son Hua, Minh-Khoi Tran, and Sai-Kit Yeung. 2018. *Pointwise convolutional neural networks*. arXiv:1712.05245 [cs]. https://doi.org/10.48550/arXiv.1712.05245.
303. Steven L. Brunton, Joshua L. Proctor, and J. Nathan Kutz. 2016. Discovering governing equations from data by sparse identification of nonlinear dynamical systems. *Proceedings of the National Academy of Sciences* 113(15): 3932–3937. ISSN: 0027-8424, 1091-6490. https://doi.org/10.1073/pnas.1517384113.
304. Samuel H. Rudy et al. 2017. Data-driven discovery of partial differential equations. *Science Advances* 3(4): e1602614. ISSN: 2375-2548. https://doi.org/10.1126/sciadv.1602614.
305. Hayden Schaeffer. 2017. Learning partial differential equations via data discovery and sparse optimization. *Proceedings of the Royal Society A: Mathematical, Physical and Engineering Sciences* 473(2197): 20160446. ISSN: 1364-5021, 1471-2946. https://doi.org/10.1098/rspa.2016.0446.
306. Kathleen Champion et al. 2019. Data-driven discovery of coordinates and governing equations. *Proceedings of the National Academy of Sciences* 116(45): 22445–22451. ISSN: 0027-8424, 1091-6490. https://doi.org/10.1073/pnas.1906995116.
307. Paolo Conti et al. 2023. Reduced order modeling of parametrized systems through autoencoders and SINDy approach: continuation of periodic solutions. *Computer Methods in Applied Mechanics and Engineering* 411: 116072. ISSN: 0045-7825. https://doi.org/10.1016/j.cma.2023.116072.
308. Maziar Raissi, Paris Perdikaris, and George Em Karniadakis. 2018. *Multistep neural networks for data-driven discovery of nonlinear dynamical systems*. arXiv:1801.01236 [nlin, physics:physics, stat]. https://doi.org/10.48550/arXiv.1801.01236.
309. Byungsoo Kim et al. 2019. Deep fluids: A generative network for parameterized fluid simulations. *Computer Graphics Forum* 38(2): 59–70. ISSN: 0167-7055, 1467-8659. https://doi.org/10.1111/cgf.13619.
310. Julia Ling, Reese Jones, and Jeremy Templeton. 2016. Machine learning strategies for systems with invariance properties. *Journal of Computational Physics* 318: 22–35. ISSN: 0021-9991. https://doi.org/10.1016/j.jcp.2016.05.003.
311. Julia Ling, Andrew Kurzawski, and Jeremy Templeton. 2016. Reynolds averaged turbulence modelling using deep neural networks with embedded invariance. *Journal of Fluid Mechanics* 807: 155–166. ISSN: 0022-1120, 1469-7645. https://doi.org/10.1017/jfm.2016.615.

312. Smith, G.F. 1965. On isotropic integrity bases. *Archive for rational mechanics and analysis* 18(4): 282–292. ISSN: 1432-0673. https://doi.org/10.1007/BF00251667.
313. Michael Lutter, Kim Listmann, and Jan Peters. 2019. Deep lagrangian networks for end-to-end learning of energy-based control for under-actuated systems. In *2019 IEEE/RSJ international conference on intelligent robots and systems (IROS)*, 7718–7725. ISSN: 2153-0866. https://doi.org/10.1109/IROS40897.2019.8968268.
314. Michael Lutter, Christian Ritter, and Jan Peters. 2019. *Deep lagrangian networks: Using physics as model prior for deep learning*. arXiv:1907.04490 [cs, eess, stat]. https://doi.org/10.48550/arXiv.1907.04490.
315. Miles Cranmer et al. 2020. *Lagrangian neural networks*. arXiv:2003.04630 [physics, stat]. https://doi.org/10.48550/arXiv.2003.04630.
316. Sam Greydanus, Misko Dzamba, and Jason Yosinski. 2019. *Hamiltonian Neural Networks*. arXiv:1906.01563 [cs]. https://doi.org/10.48550/arXiv.1906.01563.
317. Lei Zhang et al. 2016. Road crack detection using deep convolutional neural network. In *2016 IEEE international conference on image processing (ICIP)*, 3708–3712. https://doi.org/10.1109/ICIP.2016.7533052.
318. Fu-Chen Chen and Mohammad R. Jahanshahi. 2018. NB-CNN: Deep learning-based crack detection using convolutional neural network and Naïve Bayes data fusion. *IEEE Transactions on Industrial Electronics* 65(5): 4392–4400. ISSN: 1557-9948. https://doi.org/10.1109/TIE.2017.2764844.
319. Benedict E. Jaeger et al. 2022. Infrared thermal imaging-based turbine blade crack classification using deep learning. *Journal of Nondestructive Evaluation* 41(4): 74. ISSN: 1573-4862. https://doi.org/10.1007/s10921-022-00907-9.
320. Korshunova, N., et al. 2020. Image-based material characterization of complex microarchitectured additively manufactured structures. In *Computers & Mathematics with Applications*. High-order finite element and isogeometric methods 2019 80(11): 2462–2480. ISSN: 0898-1221. https://doi.org/10.1016/j.camwa.2020.07.018.
321. Hall Barbosa, C., et al. 1999. Automation of SQUID nondestructive evaluation of steel plates by neural networks. *IEEE Transactions on Applied Superconductivity* 9(2): 3475–3478. ISSN: 1558-2515. https://doi.org/10.1109/77.783778.
322. Oleg Ovcharenko et al. 2019. Deep learning for low-frequency extrapolation from multioffset seismic data. *Geophysics* 84(6): R989–R1001. ISSN: 0016-8033, 1942-2156. https://doi.org/10.1190/geo2018-0884.1.
323. Hongyu Sun and Laurent Demanet. 2020. Extrapolated full waveform inversion with deep learning. *Geophysics* 85(3): R275–R288. ISSN: 0016-8033, 1942-2156. https://doi.org/10.1190/geo2019-0195.1.
324. Hongyu Sun and Laurent Demanet. 2022. Deep learning for low-frequency extrapolation of multicomponent data in elastic FWI. *IEEE Transactions on Geoscience and Remote Sensing* 60: 1–11. ISSN: 1558-0644. https://doi.org/10.1109/TGRS.2021.3135790.
325. Winston Lewis and Denes Vigh. 2017. Deep learning prior models from seismic images for full-waveform inversion. In *SEG Technical Program Expanded Abstracts 2017*, 1512–1517. Houston, Texas: Society of Exploration Geophysicists. https://doi.org/10.1190/segam2017-17627643.1.
326. Dyck, D.N., D.A. Lowther, and S. McFee. 1992. Determining an approximate finite element mesh density using neural network techniques. *IEEE Transactions on Magnetics* 28(2): 1767–1770. ISSN: 1941-0069. https://doi.org/10.1109/20.124047.
327. Chedid, R., and N. Najjar. 1996. Automatic finite-element mesh generation using artificial neural networks-Part I: Prediction of mesh density. *IEEE Transactions on Magnetics* 32(5): 5173–5178. ISSN: 1941-0069. https://doi.org/10.1109/20.538619.
328. Triantafyllidis, D.G., and D.P. Labridis. 2000. An automatic mesh generator for handling small features in open boundary power transmission line problems using artificial neural networks. *Communications in Numerical Methods in Engineering* 16(3): 177–190. ISSN: 1069-8299, 1099-0887. https://doi.org/10.1002/(SICI)1099-0887(200003)16:3.<177::AID-CNM319spsdoigtsps3.0.CO;2-L.

329. Zheyan Zhang et al. 2020. MeshingNet: A new mesh generation method based on deep learning. In *Computational Science – ICCS 2020*, ed. Valeria V. Krzhizhanovskaya et al., vol. 12139. Lecture notes in computer science, 186–198. Cham: Springer International Publishing. ISBN: 978-3-030-50419-9 978-3-030-50420-5. https://doi.org/10.1007/978-3-030-50420-5sps14.
330. Lock, Callum, et al. 2023. Meshing using neural networks for improving the efficiency of computer modelling. *Engineering with Computers*. ISSN: 1435-5663. https://doi.org/10.1007/s00366-023-01812-z.
331. Bernd Fritzke. 1994. Growing cell structures—A self-organizing network for unsupervised and supervised learning. *Neural Networks* 7(9): 1441–1460. ISSN: 0893-6080. https://doi.org/10.1016/0893-6080(94)90091-4.
332. Alfonzetti, S., et al. 1996. Automatic mesh generation by the let-it-grow neural network. *IEEE Transactions on Magnetics* 32(3): 1349–1352. ISSN: 1941-0069. https://doi.org/10.1109/20.497496.
333. Triantafyllidis, D.G., and D.P. Labridis. 2002. A finite-element mesh generator based on growing neural networks. *IEEE Transactions on Neural Networks* 13(6): 1482–1496. ISSN: 1045-9227. https://doi.org/10.1109/TNN.2002.804223.
334. Jan N. Fuhg et al. 2024. A review on data-driven constitutive laws for solids. *Archives of computational methods in engineering*. ISSN: 1886-1784. https://doi.org/10.1007/s11831-024-10196-2.
335. Lefik, M., and B.A. Schrefler. 2003. Artificial neural network as an incremental non-linear constitutive model for a finite element code. *Computer Methods in Applied Mechanics and Engineering*. Multiscale Computational Mechanics for Materials and Structures 192(28): 3265–3283. ISSN: 0045-7825. https://doi.org/10.1016/S0045-7825(03)00350-5.
336. Dong Phill Jang, Piemaan Fazily, and Jeong Whan Yoon. 2021. Machine learning-based constitutive model for J2- plasticity. *International Journal of Plasticity* 138: 102919. ISSN: 0749-6419. https://doi.org/10.1016/j.ijplas.2020.102919.
337. Lin, Y.C., Jun Zhang, and Jue Zhong. 2008. Application of neural networks to predict the elevated temperature flow behavior of a low alloy steel. *Computational Materials Science* 43(4): 752–758. ISSN: 0927-0256. https://doi.org/10.1016/j.commatsci.2008.01.039.
338. Hong-Ying Li et al. 2012. Artificial neural network and constitutive equations to predict the hot deformation behavior of modified 2.25Cr–1Mo steel. *Materials & Design* 42: 192–197. ISSN: 0261-3069. https://doi.org/10.1016/j.matdes.2012.05.056.
339. Daoping Liu et al. 2022. Mechanistically informed data-driven modeling of cyclic plasticity via artificial neural networks. *Computer Methods in Applied Mechanics and Engineering* 393: 114766. ISSN: 0045-7825. https://doi.org/10.1016/j.cma.2022.114766.
340. Jörg F. Unger and Carsten Könke. 2009. Neural networks as material models within a multiscale approach. *Computers & Structures*. Civil-Comp Special Issue 87(19): 1177–1186. ISSN: 0045-7949. https://doi.org/10.1016/j.compstruc.2008.12.003.
341. Gabriel Hattori and Alberto Luiz Serpa. 2015. Contact stiffness estimation in ANSYS using simplified models and artificial neural networks. *Finite Elements in Analysis and Design* 97: 43–53. ISSN: 0168-874X. https://doi.org/10.1016/j.finel.2015.01.003.
342. Atsuya Oishi and Shinobu Yoshimura. 1970. A new local contact search method using a multi-layer neural network. *Computer Modeling in Engineering & Sciences* 21(2): 93–104. ISSN: 1526-1492, 1526-1492. https://doi.org/10.3970/cmes.2007.021.093.
343. Atsuya Oishi and Genki Yagawa. 2020. A surface-to-surface contact search method enhanced by deep learning. *Computational Mechanics* 65(4): 1125–1147. ISSN: 1432-0924. https://doi.org/10.1007/s00466-019-01811-2.
344. Anand Pratap Singh, Shivaji Medida, and Karthik Duraisamy. 2017. Machine-learning-augmented predictive modeling of turbulent separated flows over airfoils. *AIAA Journal* 55(7): 2215–2227. ISSN: 0001-1452, 1533-385X. https://doi.org/10.2514/1.J055595.
345. Maulik, R., et al. 2019. Subgrid modelling for two-dimensional turbulence using neural networks. *Journal of Fluid Mechanics* 858: 122–144. ISSN: 0022-1120, 1469-7645. https://doi.org/10.1017/jfm.2018.770.

346. Arnau Fabra, Joan Baiges, and Ramon Codina. 2022. Finite element approximation of wave problems with correcting terms based on training artificial neural networks with fine solutions. *Computer Methods in Applied Mechanics and Engineering* 399: 115280. ISSN: 0045-7825. https://doi.org/10.1016/j.cma.2022.115280.
347. Le, B.A., J. Yvonnet, and Q.-C. He. 2015. Computational homogenization of nonlinear elastic materials using neural networks. *International Journal for Numerical Methods in Engineering* 104(12): 1061–1084. ISSN: 0029-5981, 1097-0207. https://doi.org/10.1002/nme.4953.
348. Xiaoxin Lu et al. 2019. A data-driven computational homogenization method based on neural networks for the nonlinear anisotropic electrical response of graphene/polymer nanocomposites. *Computational Mechanics* 64(2): 307–321. ISSN: 1432-0924. https://doi.org/10.1007/s00466-018-1643-0.
349. Daniel Z. Huang et al. 2020. Learning constitutive relations from indirect observations using deep neural networks. *Journal of Computational Physics* 416: 109491. ISSN: 0021-9991. https://doi.org/10.1016/j.jcp.2020.109491.
350. Kun Wang and WaiChing Sun. 2018. A multiscale multi-permeability poroplasticity model linked by recursive homogenizations and deep learning. *Computer Methods in Applied Mechanics and Engineering* 334: 337–380. ISSN: 0045-7825. https://doi.org/10.1016/j.cma.2018.01.036.
351. Bin Li and Xiaoying Zhuang. 2020. Multiscale computation on feedforward neural network and recurrent neural network. *Frontiers of Structural and Civil Engineering* 14(6): 1285–1298. ISSN: 2095-2449. https://doi.org/10.1007/s11709-020-0691-7.
352. Nikolaos N. Vlassis, Ran Ma, and WaiChing Sun. 2020. Geometric deep learning for computational mechanics Part I: Anisotropic hyperelasticity. *Computer Methods in Applied Mechanics and Engineering* 371: 113299. ISSN: 0045-7825. https://doi.org/10.1016/j.cma.2020.113299.
353. Ilona Frankenreiter, Daniele Rosato, and Christian Miehe. 2011. Hybrid micro-macro-modeling of evolving anisotropies and length scales in finite plasticity of polycrystals: Hybrid micro-macro-modeling of evolving anisotropies and length scales in finite plasticity of polycrystals. *PAMM* 11(1): 515–518. ISSN: 16177061. https://doi.org/10.1002/pamm.201110249.
354. Fish, J. 2013. *Practical multiscaling*. Chichester, West Sussex, United Kingdom: Wiley. ISBN: 978-1-118-53485-4 978-1-118-53484-7.
355. Kevin Linka et al. 2021. Constitutive artificial neural networks: A fast and general approach to predictive data-driven constitutive modeling by deep learning. *Journal of Computational Physics* 429: 110010. ISSN: 0021-9991. https://doi.org/10.1016/j.jcp.2020.110010.
356. Mozaffar, M., et al. 2019. Deep learning predicts path-dependent plasticity. *Proceedings of the National Academy of Sciences* 116(52): 26414–26420. ISSN: 0027-8424, 1091-6490. https://doi.org/10.1073/pnas.1911815116.
357. Ling Wu and Ludovic Noels. 2022. Recurrent Neural Networks (RNNs) with dimensionality reduction and break down in computational mechanics; application to multi-scale localization step. *Computer Methods in Applied Mechanics and Engineering* 390: 114476. ISSN: 0045-7825. https://doi.org/10.1016/j.cma.2021.114476.
358. Diab W. Abueidda et al. 2021. Deep learning for plasticity and thermo-viscoplasticity. *International Journal of Plasticity* 136: 102852. ISSN: 0749-6419. https://doi.org/10.1016/j.ijplas.2020.102852.
359. Yu-Chuan Hsu, Chi-Hua Yu, and Markus J. Buehler. 2020. Using deep learning to predict fracture patterns in crystalline solids. *Matter* 3(1): 197–211. ISSN: 2590-2385. https://doi.org/10.1016/j.matt.2020.04.019.
360. Andrew J. Lew et al. 2021. Deep learning model to predict fracture mechanisms of graphene. *npj 2D Materials and Applications* 5(1): 1–8. ISSN: 2397-7132. https://doi.org/10.1038/s41699-021-00228-x.
361. Minliang Liu, Liang Liang, and Wei Sun. 2020. A generic physics-informed neural network-based constitutive model for soft biological tissues. *Computer Methods in Applied Mechanics and Engineering* 372: 113402. ISSN: 0045-7825. https://doi.org/10.1016/j.cma.2020.113402.

362. Patrick Weber, Jeremy Geiger, and Werner Wagner. 2021. Constrained neural network training and its application to hyperelastic material modeling. *Computational Mechanics* 68(5): 1179–1204. ISSN: 1432-0924. https://doi.org/10.1007/s00466-021-02064-8.
363. Yue Leng et al. 2021. Predicting the mechanical properties of biopolymer gels using neural networks trained on discrete fiber network data. *Computer Methods in Applied Mechanics and Engineering* 387: 114160. arXiv:2101.11712 [cs, q-bio]. ISSN: 00457825. https://doi.org/10.1016/j.cma.2021.114160.
364. Vahidullah Tac, Francisco Sahli Costabal, and Adrian B. Tepole.2022. Data-driven tissue mechanics with polyconvex neural ordinary differential equations. *Computer Methods in Applied Mechanics and Engineering* 398: 115248. ISSN: 0045-7825. https://doi.org/10.1016/j.cma.2022.115248.
365. Lennart Linden et al. 2023. *Neural networks meet hyperelasticity: A guide to enforcing physics*. arXiv:2302.02403 [cs]. https://doi.org/10.48550/arXiv.2302.02403.
366. Dominik K. Klein et al. 2022. Finite electro-elasticity with physics-augmented neural networks. *Computer Methods in Applied Mechanics and Engineering* 400: 115501. ISSN: 0045-7825. https://doi.org/10.1016/j.cma.2022.115501.
367. Dominik K. Klein et al. 2022. Polyconvex anisotropic hyperelasticity with neural networks. *Journal of the Mechanics and Physics of Solids* 159: 104703. ISSN: 0022-5096. https://doi.org/10.1016/j.jmps.2021.104703.
368. Faisal As'ad and Charbel Farhat. 2023. A mechanics-informed neural network framework for data-driven nonlinear viscoelasticity. In *AIAA SCITECH 2023 Forum*. National Harbor, MD & Online: American Institute of Aeronautics and Astronautics. ISBN: 978-1-62410-699-6. https://doi.org/10.2514/6.2023-0949.
369. Vahidullah Taç et al. 2023. Data-driven anisotropic finite viscoelasticity using neural ordinary differential equations. *Computer Methods in Applied Mechanics and Engineering* 411: 116046. ISSN: 0045-7825. https://doi.org/10.1016/j.cma.2023.116046.
370. Brandon Amos, Lei Xu, and J. Zico Kolter. 2017. Input convex neural networks. In *Proceedings of the 34th international conference on machine learning*, 146–155. PMLR.
371. Ricky, T., Q. Chen, et al. 2019. Neural ordinary. *Differential Equations*. https://doi.org/10.48550/arXiv.1806.07366.
372. Peiyi Chen and Johann Guilleminot. 2022. Polyconvex neural networks for hyperelastic constitutive models: A rectification approach. *Mechanics Research Communications* 125: 103993. ISSN: 0093-6413. https://doi.org/10.1016/j.mechrescom.2022.103993.
373. Filippo Masi et al. 2021. Thermodynamics-based artificial neural networks for constitutive modeling. *Journal of the Mechanics and Physics of Solids* 147: 104277. ISSN: 0022-5096. https://doi.org/10.1016/j.jmps.2020.104277.
374. Filippo Masi et al. 2021. Material modeling via thermodynamics-based artificial neural networks. *Geometric Structures of Statistical Physics, Information Geometry, and Learning*, ed. Frédéric Barbaresco and Frank Nielsen. Springer Proceedings in Mathematics & Statistics, 308–329. Cham: Springer International Publishing. ISBN: 978-3-030-77957-3. https://doi.org/10.1007/978-3-030-77957-3sps16.
375. Filippo Masi and Ioannis Stefanou. 2022. Multiscale modeling of inelastic materials with Thermodynamics-based Artificial Neural Networks (TANN). *Computer Methods in Applied Mechanics and Engineering* 398: 115190. ISSN: 0045-7825. https://doi.org/10.1016/j.cma.2022.115190.
376. Ladeveze, P., D. Nedjar, and M. Reynier. 1994. Updating of finite element models using vibration tests. *AIAA Journal* 32(7): 1485–1491. ISSN: 0001-1452, 1533-385X. https://doi.org/10.2514/3.12219.
377. Basile Marchand, Ludovic Chamoin, and Christian Rey. 2019. Parameter identification and model updating in the context of nonlinear mechanical behaviors using a unified formulation of the modified constitutive relation error concept. *Computer Methods in Applied Mechanics and Engineering* 345: 1094–1113. ISSN: 0045-7825. https://doi.org/10.1016/j.cma.2018.09.008.

378. Hai Nam Nguyen, Ludovic Chamoin, and Cuong Ha Minh. 2022. mCRE-based parameter identification from full-field measurements: Consistent framework, integrated version, and extension to nonlinear material behaviors. *Computer Methods in Applied Mechanics and Engineering* 400: 115461. ISSN: 0045-7825. https://doi.org/10.1016/j.cma.2022.115461.
379. Antoine Benady, Emmanuel Baranger, and Ludovic Chamoin. 2023. NN-mCRE: a modified constitutive relation error framework for unsupervised learning of nonlinear state laws with physics-augmented neural networks. https://doi.org/10.13140/RG.2.2.32171.00804.
380. Antoine Benady Benady, Ludovic Chamoin Chamoin, and Emmanuel Baranger Baranger. 2023. A modified Constitutive Relation Error (mCRE) framework to learn nonlinear constitutive models from strain measurements with thermodynamics-consistent neural networks. In *XI International Conference on Adaptive Modeling and Simulation (ADMOS 2023)* Advanced Techniques for Data Assimilation, Inverse Analysis, and Data-based Enrichment of Simulation Models. https://doi.org/10.23967/admos.2023.020.
381. Xueyang Li, Christian C. Roth, and Dirk Mohr. 2019. Machine-learning based temperature- and rate-dependent plasticity model: Application to analysis of fracture experiments on DP steel. *International Journal of Plasticity* 118: 320–344. ISSN: 0749-6419. https://doi.org/10.1016/j.ijplas.2019.02.012.
382. Prakash Thakolkaran et al. 2022. NN-EUCLID: Deep-learning hyperelasticity without stress data. *Journal of the Mechanics and Physics of Solids* 169: 105076. ISSN: 0022-5096. https://doi.org/10.1016/j.jmps.2022.105076.
383. Xiang Li et al. 2019. Predicting the effective mechanical property of heterogeneous materials by image based modeling and deep learning. *Computer Methods in Applied Mechanics and Engineering* 347: 735–753. ISSN: 0045-7825. https://doi.org/10.1016/j.cma.2019.01.005.
384. Alexander Henkes, Ismail Caylak, and Rolf Mahnken. 2021. A deep learning driven pseudospectral PCE based FFT homogenization algorithm for complex microstructures. *Computer Methods in Applied Mechanics and Engineering* 385: 114070. ISSN: 00457825. https://doi.org/10.1016/j.cma.2021.114070.
385. Minliang Liu, Liang Liang, and Wei Sun. 2019. Estimation of in vivo constitutive parameters of the aortic wall using a machine learning approach. *Computer Methods in Applied Mechanics and Engineering* 347: 201–217. ISSN: 0045-7825. https://doi.org/10.1016/j.cma.2018.12.030.
386. Lu Lu et al. 2020. Extraction of mechanical properties of materials through deep learning from instrumented indentation. *Proceedings of the National Academy of Sciences* 117(13): 7052–7062. ISSN: 0027-8424, 1091-6490. https://doi.org/10.1073/pnas.1922210117.
387. Xuhui Meng and George Em Karniadakis. 2020. A composite neural network that learns from multi-fidelity data: Application to function approximation and inverse PDE problems. *Journal of Computational Physics* 401: 109020. ISSN: 0021-9991. https://doi.org/10.1016/j.jcp.2019.109020.
388. Xing Liu et al. 2020. A machine learning approach to fracture mechanics problems. *Acta Materialia* 190: 105–112. ISSN: 13596454. https://doi.org/10.1016/j.actamat.2020.03.016.
389. Ridha Hambli, Houda Katerchi, and Claude-Laurent Benhamou. 2011. Multiscale methodology for bone remodelling simulation using coupled finite element and neural network computation. *Biomechanics and Modeling in Mechanobiology* 10(1): 133–145. ISSN: 1617-7940. https://doi.org/10.1007/s10237-010-0222-x.
390. Moritz Flaschel, Siddhant Kumar, and Laura De Lorenzis. 2021. Unsupervised discovery of interpretable hyperelastic constitutive laws. *Computer Methods in Applied Mechanics and Engineering* 381: 113852. ISSN: 0045-7825. https://doi.org/10.1016/j.cma.2021.113852.
391. Robert Tibshirani. 1996. Regression Shrinkage and Selection via the Lasso. *Journal of the Royal Statistical Society. Series B (Methodological)* 58(1): 267–288. ISSN: 0035-9246.
392. Moritz Flaschel, Siddhant Kumar, and Laura De Lorenzis. 2022. Discovering plasticity models without stress data. *npj Computational Materials* 8(1): 91. arXiv:2202.04916 [cs]. ISSN: 2057-3960. https://doi.org/10.1038/s41524-022-00752-4.
393. Enzo Marino et al. 2023. Automated identification of linear viscoelastic constitutive laws with EUCLID. *Mechanics of Materials* 181: 104643. ISSN: 0167-6636. https://doi.org/10.1016/j.mechmat.2023.104643.

394. Moritz Flaschel, Siddhant Kumar, and Laura De Lorenzis. 2023. Automated discovery of generalized standard material models with EUCLID. *Computer Methods in Applied Mechanics and Engineering* 405: 115867. ISSN: 0045-7825. https://doi.org/10.1016/j.cma.2022.115867.
395. Akshay Joshi et al. 2022. Bayesian-EUCLID: Discovering hyperelastic material laws with uncertainties. *Computer Methods in Applied Mechanics and Engineering* 398: 115225. ISSN: 0045-7825. https://doi.org/10.1016/j.cma.2022.115225.
396. Kevin Linka, Sarah R. St. Pierre, and Ellen Kuhl. 2023. Automated model discovery for human brain using constitutive artificial neural networks. *Acta Biomaterialia* 160: 134–151. ISSN: 1742-7061. https://doi.org/10.1016/j.actbio.2023.01.055.
397. Kevin Linka and Ellen Kuhl. 2023. A new family of constitutive Artificial Neural Networks towards automated model discovery. *Computer Methods in Applied Mechanics and Engineering* 403: 115731. ISSN: 0045-7825. https://doi.org/10.1016/j.cma.2022.115731.
398. Atsuya Oishi and Genki Yagawa. 2017. Computational mechanics enhanced by deep learning. *Computer Methods in Applied Mechanics and Engineering* 327: 327–351. ISSN: 00457825. https://doi.org/10.1016/j.cma.2017.08.040.
399. Jaeho Jung, Kyungho Yoon, and Phill-Seung Lee. 2020. Deep learned finite elements. *Computer Methods in Applied Mechanics and Engineering* 372: 113401. ISSN: 0045-7825. https://doi.org/10.1016/j.cma.2020.113401.
400. Yohai Bar-Sinai et al. 2019. Learning data-driven discretizations for partial differential equations. *Proceedings of the National Academy of Sciences* 116(31): 15344–15349. tex.ids= Bar-Sinai2019a. ISSN: 0027-8424, 1091-6490. https://doi.org/10.1073/pnas.1814058116.
401. Panos Pantidis and Mostafa E. Mobasher. 2023. Integrated Finite Element Neural Network (I-FENN) for non-local continuum damage mechanics. *Computer Methods in Applied Mechanics and Engineering* 404: 115766. ISSN: 0045-7825. https://doi.org/10.1016/j.cma.2022.115766.
402. Daniel Andrés Arcones et al. 2022. Neural network-based surrogate models applied to fluid-structure interaction problems. In *WCCM-APCOM 2022* 1700 Data Science, Machine Learning and Artificial Intelligence. https://doi.org/10.23967/wccm-apcom.2022.080.
403. Changnian Han et al. 2021. Artificial intelligence for accelerating time integrations in multiscale modeling. *Journal of Computational Physics* 427: 110053. ISSN: 0021-9991. https://doi.org/10.1016/j.jcp.2020.110053.
404. Tomasz Służalec et al. 2023. Automatic stabilization of finite-element simulations using neural networks and hierarchical matrices. *Computer Methods in Applied Mechanics and Engineering* 411: 116073. ISSN: 0045-7825. https://doi.org/10.1016/j.cma.2023.116073.
405. Mariusz Bujny et al. 2023. Learning hyperparameter predictors for similarity-based multidisciplinary topology optimization. *Scientific Reports* 13(1): 14856. ISSN: 2045-2322. https://doi.org/10.1038/s41598-023-42009-0.
406. Casadei, F., J.J. Rimoli, and M. Ruzzene. 2013. A geometric multiscale finite element method for the dynamic analysis of heterogeneous solids. *Computer Methods in Applied Mechanics and Engineering* 263: 56–70. ISSN: 0045-7825. https://doi.org/10.1016/j.cma.2013.05.009.
407. Oguz Oztoprak et al. 2023. Two-scale analysis and design of spaceframes with complex additive manufactured nodes. https://doi.org/10.48550/ARXIV.2301.06921.
408. Arnd Koeppe, Franz Bamer, and Bernd Markert. 2020. An intelligent nonlinear meta element for elastoplastic continua: deep learning using a new Time-distributed Residual U-Net architecture. *Computer Methods in Applied Mechanics and Engineering* 366: 113088. ISSN: 0045-7825. https://doi.org/10.1016/j.cma.2020.113088.
409. German Capuano and Julian J. Rimoli. 2019. Smart finite elements: A novel machine learning application. *Computer Methods in Applied Mechanics and Engineering* 345: 363–381. ISSN: 0045-7825. https://doi.org/10.1016/j.cma.2018.10.046.
410. Taichi Yamaguchi and Hiroshi Okuda. 2021. Zooming method for FEA using a neural network. *Computers & Structures* 247: 106480. ISSN: 0045-7949. https://doi.org/10.1016/j.compstruc.2021.106480.
411. Minglang Yin et al. 2022. Interfacing finite elements with deep neural operators for fast multiscale modeling of mechanics problems. *Computer Methods in Applied Mechanics and Engineering. A Special Issue in Honor of the Lifetime Achievements of J. Tinsley Oden* 402: 115027. ISSN: 0045-7825. https://doi.org/10.1016/j.cma.2022.115027.

412. Ole Sigmund. 2011.On the usefulness of non-gradient approaches in topology optimization. *Structural and Multidisciplinary Optimization* 43(5): 589–596. ISSN: 1615-1488. https://doi.org/10.1007/s00158-011-0638-7.
413. Nils Thuerey et al. 2021. *Physics-based Deep Learning*. WWW.
414. Philipp Holl, Vladlen Koltun, and Nils Thuerey. 2020. *Learning to Control PDEs with Differentiable Physics*. arXiv:2001.07457 [physics, stat]. https://doi.org/10.48550/arXiv.2001.07457.
415. Kiwon Um et al. 2020. Solver-in-the-loop: learning from differentiable physics to interact with iterative PDE-solvers. In *Proceedings of the 34th International Conference on Neural Information Processing Systems*. NIPS'20, 6111–6122. Red Hook, NY, USA: Curran Associates Inc. ISBN: 978-1-71382-954-6.
416. Kiwon Um et al. 2021. *Solver-in-the-loop: Learning from differentiable physics to interact with Iterative PDE-Solvers*. arXiv:2007.00016 [physics]. https://doi.org/10.48550/arXiv.2007.00016.
417. Jensen, C.A., et al. 1999. Inversion of feedforward neural networks: Algorithms and applications. *Proceedings of the IEEE* 87(9): 1536–1549. ISSN: 1558-2256. https://doi.org/10.1109/5.784232.
418. Chi-Hua Yu, Zhao Qin, and Markus J Buehler. 2019. Artificial intelligence design algorithm for nanocomposites optimized for shear crack resistance. *Nano Futures* 3(3): 035001. ISSN: 2399-1984. https://doi.org/10.1088/2399-1984/ab36f0.
419. Chun-Teh Chen and Grace X. Gu. 2020. Generative deep neural networks for inverse materials design using backpropagation and active learning. *Advanced Science* 7(5): 1902607. ISSN: 2198-3844, 2198-3844. https://doi.org/10.1002/advs.201902607.
420. Derick Nganyu Tanyu et al. 2022. *Deep learning methods for partial differential equations and related parameter identification problems*. https://doi.org/10.48550/arXiv.2212.03130.
421. Zohdi, T.I. 2023. A machine-learning digital-twin for rapid large-scale solar-thermal energy system design. *Computer Methods in Applied Mechanics and Engineering* 412: 115991. ISSN: 0045-7825. https://doi.org/10.1016/j.cma.2023.115991.
422. Plessix, R.-E. 2006. A review of the adjoint-state method for computing the gradient of a functional with geophysical applications. *Geophysical Journal International* 167(2): 495–503. ISSN: 0956540X, 1365246X. https://doi.org/10.1111/j.1365-246X.2006.02978.x.
423. Dan Givoli. 2021. A tutorial on the adjoint method for inverse problems. *Computer Methods in Applied Mechanics and Engineering* 380: 113810. ISSN: 00457825. https://doi.org/10.1016/j.cma.2021.113810.
424. Vahid Keshavarzzadeh, Robert M. Kirby, and Akil Narayan. 2021. Robust topology optimization with low rank approximation using artificial neural networks. *Computational Mechanics* 68(6): 1297–1323. ISSN: 1432-0924. https://doi.org/10.1007/s00466-021-02069-3.
425. Chao Qian and Wenjing Ye. 2021. Accelerating gradient-based topology optimization design with dual-model artificial neural networks. *Structural and Multidisciplinary Optimization* 63(4): 1687–1707. ISSN: 1615-1488. https://doi.org/10.1007/s00158-020-02770-6.
426. Heng Chi et al. 2021. Universal machine learning for topology optimization. *Computer Methods in Applied Mechanics and Engineering* 375: 112739. tex.ids= Chi2021a, Chi2021b. ISSN: 0045-7825. https://doi.org/10.1016/j.cma.2019.112739.
427. Nikola Aulig and Markus Olhofer. 2013. Evolutionary generation of neural network update signals for the topology optimization of structures. In *Proceedings of the 15th annual conference companion on Genetic and evolutionary computation*. GECCO '13 Companion, 213–214. New York, NY, USA: Association for Computing Machinery. ISBN: 978-1-4503-1964-5. https://doi.org/10.1145/2464576.2464685.
428. Nikola Aulig and Markus Olhofer. 2014. Topology Optimization by predicting sensitivities based on local state features.
429. Nikola Aulig and Markus Olhofer. 2015. Neuro-evolutionary topology optimization with adaptive improvement threshold. In *Applications of evolutionary computation*, ed. Antonio M. Mora and Giovanni Squillero. Lecture notes in computer science, 655–666. Cham: Springer International Publishing. ISBN: 978-3-319-16549-3. https://doi.org/10.1007/978-3-319-16549-3sps53.

430. Zhang, Yuyu, et al. 2021. *Speeding up computational morphogenesis with online neural synthetic gradients.* https://doi.org/10.48550/arXiv.2104.12282.
431. Thomas Hunter Hunter, Steven Hulsoff Hulsoff, and Arnish Sitaram Sitaram. 2023. Super-Adjoint: Super-resolution neural networks in adjoint-based output error estimation. In *XI international conference on adaptive modeling and simulation (ADMOS 2023)* Recent Developments in Methods and Applications for Mesh Adaptation. https://doi.org/10.23967/admos.2023.058.
432. Kai Fukami, Koji Fukagata, and Kunihiko Taira. 2021. Machine-learning-based spatio-temporal super resolution reconstruction of turbulent flows. *Journal of Fluid Mechanics* 909:A9. ISSN: 0022-1120, 1469-7645. https://doi.org/10.1017/jfm.2020.948.
433. Fernando V. Senhora et al. 2022. Machine learning for topology optimization: Physics-based learning through an independent training strategy. *Computer Methods in Applied Mechanics and Engineering* 398: 115116. ISSN: 0045-7825. https://doi.org/10.1016/j.cma.2022.115116.
434. Jun-Ting Hsieh et al. 2019. *Learning neural PDE solvers with convergence guarantees.* arXiv:1906.01200 [cs, stat]. https://doi.org/10.48550/arXiv.1906.01200.
435. Hong-Ling Ye et al. 2021. Acceleration design for continuum topology optimization by using Pix2pix neural network. *International Journal of Applied Mechanics* 13(04): 2150042. ISSN: 1758-8251. https://doi.org/10.1142/S1758825121500423.
436. Hoyer, Stephan, Jascha Sohl-Dickstein, and Sam Greydanus. 2019. *Neural reparameterization improves structural optimization.* https://doi.org/10.48550/arXiv.1909.04240.
437. Kailai, Xu., and Eric Darve. 2019. *The neural network approach to inverse problems in differential equations.* https://doi.org/10.48550/arXiv.1901.07758.
438. Jens Berg and Kaj Nyström. 2021. Neural networks as smooth priors for inverse problems for PDEs. *Journal of Computational Mathematics and Data Science* 1: 100008. tex.ids= Berg2021a. ISSN: 27724158. https://doi.org/10.1016/j.jcmds.2021.100008.
439. Liang Chen and Mo-How Herman Shen. 2021. A new topology optimization approach by physics-informed deep learning process. *Advances in Science, Technology and Engineering Systems Journal* 6(4): 233–240. ISSN: 24156698, 24156698. https://doi.org/10.25046/aj060427.
440. Alex Halle, Lucio Flavio Campanile, and Alexander Hasse. 2021. An Artificial intelligence–Assisted design method for topology optimization without pre-optimized training data. *Applied Sciences* 11(19): 9041. ISSN: 2076-3417. https://doi.org/10.3390/app11199041.
441. Hao Deng and Albert C. 2020. To. Topology optimization based on deep representation learning (DRL) for compliance and stress-constrained design. *Computational Mechanics* 66(2): 449–469. ISSN: 1432-0924. https://doi.org/10.1007/s00466-020-01859-5.
442. Aaditya Chandrasekhar and Krishnan Suresh. 2021. TOuNN: Topology optimization using neural networks. *Structural and Multidisciplinary Optimization* 63(3): 1135–1149. ISSN: 1615-1488. https://doi.org/10.1007/s00158-020-02748-4.
443. Chandrasekhar, Aaditya, and Krishnan Suresh. 2021. *Length scale control in topology optimization using fourier enhanced neural networks.* https://doi.org/10.48550/arXiv.2109.01861.
444. Leon Herrmann et al. 2024. *Neural networks for generating better local optima in topology optimization [Software].* https://doi.org/10.5281/zenodo.12806412.
445. Jeong Joon Park et al. 2019. DeepSDF: Learning continuous signed distance functions for shape representation. In *2019 IEEE/CVF conference on computer vision and pattern recognition (CVPR)*, 165–174. ISSN: 2575-7075. https://doi.org/10.1109/CVPR.2019.00025.
446. Mateusz Michalkiewicz et al. 2019. Implicit surface representations as layers in neural networks. In *2019 IEEE/CVF international conference on computer vision (ICCV)*, 4742–4751. ISSN: 2380-7504. https://doi.org/10.1109/ICCV.2019.00484.
447. Amos Gropp et al. 2020. Implicit geometric regularization for learning shapes. In *Proceedings of the 37th international conference on machine learning*, vol. 119, 3789–3799. ICML'20. JMLR.org.

448. Vincent Sitzmann et al. 2020. Implicit neural representations with periodic activation functions. arXiv:2006.09661 [cs, eess].
449. Zhichun Huang, Shaojie Bai, and J. Zico Kolter. 2021. (Implicit)^2: Implicit layers for implicit representations. In *Advances in neural information processing systems*, vol. 34, 9639–9650. Curran Associates, Inc.
450. Deng, Hao, and C. To Albert. 2021. *A parametric level set method for topology optimization based on Deep Neural Network (DNN)*. https://doi.org/10.48550/arXiv.2101.03286.
451. Zeyu Zhang et al. 2021. TONR: An exploration for a novel way combining neural network with topology optimization. *Computer Methods in Applied Mechanics and Engineering* 386: 114083. ISSN: 00457825. https://doi.org/10.1016/j.cma.2021.114083.
452. Reetam Biswas et al. 2019. Prestack and poststack inversion using a physics-guided convolutional neural network. *Interpretation* 7(3): SE161–SE174. ISSN: 2324-8858, 2324-8866. https://doi.org/10.1190/INT-2018-0236.1.
453. Motaz Alfarraj and Ghassan AlRegib. 2019. Semi-supervised learning for acoustic impedance inversion. In *SEG technical program expanded abstracts 2019*, 2298–2302. San Antonio, Texas: Society of Exploration Geophysicists. https://doi.org/10.1190/segam2019-3215902.1.
454. Divya Shyam Singh et al. 2024. *Accelerating full waveform inversion by transfer learning*. arXiv:2408.00695 [cs]. https://doi.org/10.48550/arXiv.2408.00695.
455. Chao Dong et al. 2014. Learning a deep convolutional network for image super-resolution. In *Computer vision—ECCV 2014*, ed. David Fleet et al. Lecture notes in computer science, 184–199. Cham: Springer International Publishing. ISBN: 978-3-319-10593-2. https://doi.org/10.1007/978-3-319-10593-2sps13.
456. Chao Dong et al. 2015. *Image super-resolution using deep convolutional networks*. arXiv:1501.00092 [cs]. https://doi.org/10.48550/arXiv.1501.00092.
457. Kai Fukami, Koji Fukagata, and Kunihiko Taira. 2019. Super-resolution reconstruction of turbulent flows with machine learning. *Journal of Fluid Mechanics* 870: 106–120. ISSN: 0022-1120, 1469-7645. https://doi.org/10.1017/jfm.2019.238.
458. Nicholas Napier et al. 2020. An artificial neural network approach for generating high-resolution designs from low-resolution input in topology optimization. *Journal of Mechanical Design* 142(1): 011402. ISSN: 1050-0472, 1528-9001. https://doi.org/10.1115/1.4044332.
459. Chunpeng Wang et al. 2021. Deep super-resolution neural network for structural topology optimization. *Engineering Optimization* 53(12): 2108–2121. ISSN: 0305-215X. https://doi.org/10.1080/0305215X.2020.1846031.
460. Liang Xue et al. 2021. Efficient, high-resolution topology optimization method based on convolutional neural networks. *Frontiers of Mechanical Engineering* 16(1): 80–96. ISSN: 2095-0241. https://doi.org/10.1007/s11465-020-0614-2.
461. Atsuya Oishi and Genki Yagawa. 2021. Finite elements using neural networks and a posteriori error. *Archives of Computational Methods in Engineering* 28(5): 3433–3456. ISSN: 1886-1784. https://doi.org/10.1007/s11831-020-09507-0.
462. Martin Ohrt Elingaard et al. 2022. De-homogenization using convolutional neural networks. *Computer Methods in Applied Mechanics and Engineering* 388: 114197. ISSN: 0045-7825. https://doi.org/10.1016/j.cma.2021.114197.
463. Zhong Yi Wan et al. 2018. Data-assisted reduced-order modeling of extreme events in complex dynamical systems. *PLOS ONE* 13(5): e0197704, ed. Daniel Durstewitz. ISSN: 1932-6203. https://doi.org/10.1371/journal.pone.0197704.
464. Syuhei Sato et al. 2018. Example-based turbulence style transfer. *ACM Transactions on Graphics* 37(4): 84:1–84:9. ISSN: 0730-0301. https://doi.org/10.1145/3197517.3201398.
465. Mengyu Chu and Nils Thuerey. 2017. Data-driven synthesis of smoke flows with CNN-based feature descriptors. *ACM Transactions on Graphics* 36(4): 69:1–69:14. ISSN: 0730-0301. https://doi.org/10.1145/3072959.3073643.
466. Yildiz, A.R., et al. 2003. Integrated optimal topology design and shape optimization using neural networks. *Structural and Multidisciplinary Optimization* 25(4): 251–260. ISSN: 1615-1488. https://doi.org/10.1007/s00158-003-0300-0.

467. Chyi-Yeu Lin and Shin-Hong Lin. 2005. Artificial neural network based hole image interpretation techniques for integrated topology and shape optimization. *Computer Methods in Applied Mechanics and Engineering* 194(36): 3817–3837. ISSN: 0045-7825. https://doi.org/10.1016/j.cma.2004.09.005.
468. Guodong Chen and Krzysztof Fidkowski. 2020. Output-based error estimation and mesh adaptation using convolutional neural networks: Application to a scalar advection-diffusion problem. In *AIAA Scitech 2020 Forum*. Orlando, FL: American Institute of Aeronautics and Astronautics. ISBN: 978-1-62410-595-1. https://doi.org/10.2514/6.2020-1143.
469. Jun Takeuchi and Yukio Kosugi. 1994. Neural network representation of finite element method. *Neural Networks* 7(2): 389–395. ISSN: 0893-6080. https://doi.org/10.1016/0893-6080(94)90031-0.
470. Ramuhalli, P., L. Udpa, and S.S. Udpa. 2005. Finite-element neural networks for solving differential equations. *IEEE Transactions on Neural Networks* 16(6): 1381–1392. ISSN: 1941-0093. https://doi.org/10.1109/TNN.2005.857945.
471. Sikora, R., et al. 1999. Artificial neural network application for material evaluation by electromagnetic methods. In *IJCNN'99. International joint conference on neural networks. Proceedings (Cat. No.99CH36339)*, vol. 6, 4027–4032. https://doi.org/10.1109/IJCNN.1999.830804.
472. Xu, G., et al. 1999. Application of FE-based neural networks to dynamic problems. In *ICONIP'99. ANZIIS'99 & ANNES'99 & ACNN'99. 6th international conference on neural information processing. Proceedings (Cat. No.99EX378)*, vol. 3, 1039–1044. https://doi.org/10.1109/ICONIP.1999.844679.
473. Fei Guo et al. 1999. Finite element analysis based Hopfield neural network model for solving nonlinear electromagnetic field problems. In *IJCNN'99. International joint conference on neural networks. Proceedings (Cat. No.99CH36339)*, vol. 6, 4399–4403. https://doi.org/10.1109/IJCNN.1999.830877.
474. Hyuk Lee and In Seok Kang. 1990. Neural algorithm for solving differential equations. *Journal of Computational Physics* 91(1), 110–131. ISSN: 0021-9991. https://doi.org/10.1016/0021-9991(90)90007-N.
475. Kalkkuhl, J., K.J. Hunt, and H. Fritz. 1999. FEM-based neural-network approach to nonlinear modeling with application to longitudinal vehicle dynamics control. *IEEE Transactions on Neural Networks* 10(4): 885–897. ISSN: 1941-0093. https://doi.org/10.1109/72.774241.
476. Chao Xu et al. 2012. Finite-element neural network-based solving 3-D differential equations in MFL. *IEEE Transactions on Magnetics* 48(12): 4747–4756. ISSN: 1941-0069. https://doi.org/10.1109/TMAG.2012.2207732.
477. Yang, Z., et al. 2012. An efficient integration technique for the voxel-based finite cell method: Efficient Integration Technique For Finite Cells. *International Journal for Numerical Methods in Engineering* 91(5): 457–471. ISSN: 00295981. https://doi.org/10.1002/nme.4269.
478. Lei Zhang et al. 2021. Hierarchical deep-learning neural networks: finite elements and beyond. *Computational Mechanics* 67(1): 207–230. ISSN: 1432-0924. https://doi.org/10.1007/s00466-020-01928-9.
479. Sourav Saha et al. 2021. Hierarchical Deep Learning Neural Network (HiDeNN): An artificial intelligence (AI) framework for computational science and engineering. *Computer Methods in Applied Mechanics and Engineering* 373: 113452. ISSN: 00457825. https://doi.org/10.1016/j.cma.2020.113452.
480. Lei Zhang et al. 2022. HiDeNN-TD: Reduced-order hierarchical deep learning neural networks. *Computer Methods in Applied Mechanics and Engineering* 389: 114414. ISSN: 0045-7825. https://doi.org/10.1016/j.cma.2021.114414.
481. Yingjian Liu et al. 2023. HiDeNN-FEM: a seamless machine learning approach to nonlinear finite element analysis. *Computational Mechanics* 72(1): 173–194. ISSN: 1432-0924. https://doi.org/10.1007/s00466-023-02293-z.
482. Ye Lu et al. 2023. Convolution Hierarchical Deep-learning Neural Networks (C-HiDeNN): finite elements, isogeometric analysis, tensor decomposition, and beyond. *Computational Mechanics* 72(2): 333–362. ISSN: 1432-0924. https://doi.org/10.1007/s00466-023-02336-5.

483. Chanwook Park et al. 2023. Convolution Hierarchical Deep-learning Neural Networks (C-HiDeNN) with Graphics Processing Unit (GPU) acceleration. *Computational Mechanics* 72(2): 383–409. ISSN: 1432-0924. https://doi.org/10.1007/s00466-023-02329-4.
484. Hengyang Li et al. 2023. Convolution Hierarchical Deep-Learning Neural Network Tensor Decomposition (C-HiDeNN-TD) for high-resolution topology optimization. *Computational Mechanics* 72(2): 363–382. ISSN: 1432-0924. https://doi.org/10.1007/s00466-023-02333-8.
485. Grosse, I.R., P. Katragadda, and J. Benoit. 1992. An adaptive accuracy-based a posteriori error estimator. *Finite Elements in Analysis and Design* 12(1): 75–90. ISSN: 0168-874X. https://doi.org/10.1016/0168-874X(92)90008-Z.
486. Zhu, J.Z., and O.C. Zienkiewicz. 1997. A posteriori error estimation and three-dimensional automatic mesh generation. *Finite Elements in Analysis and Design*. Adaptive Meshing, Part 1 25(1): 167–184. ISSN: 0168-874X. https://doi.org/10.1016/S0168-874X(96)00037-6.
487. Möller, M., and D. Kuzmin. 2006. Adaptive mesh refinement for high-resolution finite element schemes. *International Journal for Numerical Methods in Fluids* 52(5): 545–569. ISSN: 0271-2091, 1097-0363. https://doi.org/10.1002/fld.1183.
488. Houpu Yao, Yi Ren, and Yongming Liu. 2019. FEA-Net: A deep convolutional neural network with physics prior for efficient data driven PDE learning. In *AIAA Scitech 2019 Forum*. San Diego, California: American Institute of Aeronautics and Astronautics. ISBN: 978-1-62410-578-4. https://doi.org/10.2514/6.2019-0680.
489. Houpu Yao, Yi Gao, and Yongming Liu. 2020. FEA-Net: A physics-guided data-driven model for efficient mechanical response prediction. *Computer Methods in Applied Mechanics and Engineering* 363: 112892. ISSN: 0045-7825. https://doi.org/10.1016/j.cma.2020.112892.
490. Mishra, R.K., and P.S. Hall. 2005. NFDTD concept. *IEEE Transactions on Neural Networks* 16(2): 484–490. ISSN: 1941-0093. https://doi.org/10.1109/TNN.2004.841799.
491. Richardson, Alan. 2018. *Seismic full-waveform inversion using deep learning tools and techniques*. https://doi.org/10.48550/arXiv.1801.07232.
492. Jian Sun et al. 2020. A theory-guided deep-learning formulation and optimization of seismic waveform inversion. *Geophysics* 85(2): R87–R99. ISSN: 0016-8033, 1942-2156. https://doi.org/10.1190/geo2019-0138.1.
493. Tyler W. Hughes et al. 2019. Wave physics as an analog recurrent neural network. *Science Advances* 5(12): eaay6946. ISSN: 2375-2548. https://doi.org/10.1126/sciadv.aay6946.
494. Zeliang Liu, C. T. Wu, and M. Koishi. 2019. A deep material network for multiscale topology learning and accelerated nonlinear modeling of heterogeneous materials. *Computer Methods in Applied Mechanics and Engineering* 345: 1138–1168. ISSN: 0045-7825. https://doi.org/10.1016/j.cma.2018.09.020.
495. Zeliang Liu and C. T. Wu. 2019. Exploring the 3D architectures of deep material network in data-driven multiscale mechanics. *Journal of the Mechanics and Physics of Solids* 127: 20–46. ISSN: 0022-5096. https://doi.org/10.1016/j.jmps.2019.03.004.
496. Eldad Haber and Lars Ruthotto. 2018. Stable architectures for deep neural networks. *Inverse Problems* 34(1): 014004. arXiv:1705.03341 [cs, math]. ISSN: 0266-5611, 1361-6420. https://doi.org/10.1088/1361-6420/aa9a90.
497. Lars Ruthotto and Eldad Haber. 2018. *Deep neural networks motivated by partial differential equations*. arXiv:1804.04272 [cs, math, stat]. https://doi.org/10.48550/arXiv.1804.04272.
498. Yiping Lu et al. 2020. *Beyond finite layer neural networks: Bridging deep architectures and numerical differential equations*. arXiv:1710.10121 [cs, stat]. https://doi.org/10.48550/arXiv.1710.10121.
499. Pontriagin, L.S., Lucien W. Neustadt, and L.S. Pontriagin. 1986. *The mathematical theory of optimal processes*. eng. English ed. Classics of Soviet mathematics, vol. 1. New York: Gordon and Breach Science Publishers, 1986. ISBN: 978-2-88124-077-5.
500. Yang Yu, Houpu Yao, and Yongming Liu. 2018. Physics-based learning for aircraft dynamics simulation. In *Annual conference of the PHM society*, vol. 10, issue 1. ISSN: 2325-0178. https://doi.org/10.36001/phmconf.2018.v10i1.513.

501. Rishikesh Ranade, Chris Hill, and Jay Pathak. 2021. DiscretizationNet: A machine-learning based solver for Navier–Stokes equations using finite volume discretization. *Computer Methods in Applied Mechanics and Engineering* 378: 113722. ISSN: 0045-7825. https://doi.org/10.1016/j.cma.2021.113722.
502. Richard S. Sutton. 1991. Dyna, an integrated architecture for learning, planning, and reacting. *ACM SIGART Bulletin* 2(4): 160–163. ISSN: 0163-5719. https://doi.org/10.1145/122344.122377.
503. Michael Janner et al. 2019. When to trust your model: Model-based policy optimization. In *Proceedings of the 33rd International Conference on Neural Information Processing Systems*, vol. 1122, 12519–12530. Red Hook, NY, USA: Curran Associates Inc.
504. Lukasz Kaiser et al. 2020. *Model-based reinforcement learning for Atari*. arXiv:1903.00374 [cs, stat]. https://doi.org/10.48550/arXiv.1903.00374.
505. Yuping Luo et al. 2021. *Algorithmic framework for model-based deep reinforcement learning with theoretical guarantees*. arXiv:1807.03858 [cs, stat]. https://doi.org/10.48550/arXiv.1807.03858.
506. Marc Peter Deisenroth and Carl Edward Rasmussen. 2011. PILCO: A model-based and data-efficient approach to policy search. In *Proceedings of the 28th international conference on international conference on machine learning*. ICML'11, 465–472. Madison, WI, USA: Omnipress. ISBN: 978-1-4503-0619-5.
507. Sergey Levine and Pieter Abbeel. 2014. Learning neural network policies with guided policy search under unknown dynamics. In *Advances in neural information processing systems*, vol. 27. Curran Associates, Inc.
508. Nicolas Heess et al. 2015. Learning continuous control policies by stochastic value gradients. In *Advances in neural information processing systems*, vol. 28. Curran Associates, Inc.
509. Ignasi Clavera, Violet Fu, and Pieter Abbeel. 2020. *Model-augmented actor-critic: Backpropagating through Paths*. arXiv:2005.08068 [cs, stat]. https://doi.org/10.48550/arXiv.2005.08068.
510. Danijar Hafner et al. 2020. *Dream to control: Learning behaviors by latent imagination*. arXiv:1912.01603 [cs]. https://doi.org/10.48550/arXiv.1912.01603.
511. Danijar Hafner et al. 2022. *Mastering atari with discrete world models*. arXiv:2010.02193 [cs, stat]. https://doi.org/10.48550/arXiv.2010.02193.
512. Ronald J. Williams. 1992. Simple statistical gradient-following algorithms for connectionist reinforcement learning. *Machine Learning* 8(3): 229–256. ISSN: 1573-0565. https://doi.org/10.1007/BF00992696.
513. Richard S. Sutton et al. 1999. Policy gradient methods for reinforcement learning with function approximation. In *Advances in Neural Information Processing Systems*, vol. 12. MIT Press
514. Sham M. Kakade. 2001. A natural policy gradient. In *Advances in neural information processing systems*, vol. 14. MIT Press.
515. David Silver et al. 2014. Deterministic policy gradient algorithms. In *Proceedings of the 31st international conference on machine learning*, 387–395. PMLR.
516. John Schulman et al. 2015. Trust region policy optimization. In *Proceedings of the 32nd international conference on machine learning*, 1889–1897. PMLR.
517. Christopher J. C. H. Watkins and Peter Dayan. 1992. Q-learning. *Machine Learning* 8(3): 279–292. ISSN: 1573-0565. https://doi.org/10.1007/BF00992698.
518. Hado van Hasselt, Arthur Guez, and David Silver. 2016. Deep reinforcement learning with double Q-Learning. In *Proceedings of the thirtieth AAAI conference on artificial intelligence*, 2094–2100. AAAI'16. Phoenix, Arizona: AAAI Press.
519. Ziyu Wang et al. 2016. Dueling network architectures for deep reinforcement learning. In *Proceedings of the 33rd international conference on international conference on machine learning*, vol. 48, 1995–2003. ICML'16. New York, NY, USA: JMLR.org.
520. John Schulman et al. 2017. *Proximal policy optimization algorithms*. arXiv:1707.06347 [cs]. https://doi.org/10.48550/arXiv.1707.06347.
521. Richard Bellman. 1957. A markovian decision process. *Journal of Mathematics and Mechanics* 6(5): 679–684. ISSN: 0095-9057.

522. Capuzzo Dolcetta, I., and H. Ishii. 1984. Approximate solutions of the bellman equation of deterministic control theory. *Applied Mathematics & Optimization* 11(1): 161–181. ISSN: 0095-4616, 1432-0606. https://doi.org/10.1007/BF01442176.
523. Richard S. Sutton. 1988. Learning to predict by the methods of temporal differences. *Machine Learning* 3(1): 9–44. ISSN: 1573-0565. https://doi.org/10.1007/BF00115009.
524. Steven J. Bradtke and Andrew G. Barto. 1996. Linear Least-Squares algorithms for temporal difference learning. *Machine Learning* 22(1): 33–57. ISSN: 1573-0565. https://doi.org/10.1007/BF00114723.
525. Guido Novati et al. 2017. Synchronised swimming of two fish. *Bioinspiration & Biomimetics* 12(3): 036001. arXiv:1610.04248 [physics]. ISSN: 1748-3190. https://doi.org/10.1088/1748-3190/aa6311.
526. Siddhartha Verma, Guido Novati, and Petros Koumoutsakos. 2018. Efficient collective swimming by harnessing vortices through deep reinforcement learning. *Proceedings of the National Academy of Sciences* 115(23): 5849–5854. ISSN: 0027-8424, 1091-6490. https://doi.org/10.1073/pnas.1800923115.
527. Pingchuan Ma et al. 2018. Fluid directed rigid body control using deep reinforcement learning. *ACM Transactions on Graphics* 37(4): 96:1–96:11. ISSN: 0730-0301. https://doi.org/10.1145/3197517.3201334.
528. Jean Rabault et al. 2019. Artificial neural networks trained through deep reinforcement learning discover control strategies for active flow control. *Journal of Fluid Mechanics* 865: 281–302. ISSN: 0022-1120, 1469-7645. https://doi.org/10.1017/jfm.2019.62.
529. Dixia Fan et al. 2020. Reinforcement learning for bluff body active flow control in experiments and simulations. *Proceedings of the National Academy of Sciences* 117(42): 26091–26098. ISSN: 0027-8424, 1091-6490. https://doi.org/10.1073/pnas.2004939117.
530. Jie Xu et al. 2019. Learning to fly: Computational controller design for hybrid UAVs with reinforcement learning. *ACM Transactions on Graphics* 38(4): 42:1–42:12. ISSN: 0730-0301. https://doi.org/10.1145/3306346.3322940.
531. Xian Yeow Lee et al. 2018. *Flow shape design for microfluidic devices using deep reinforcement learning*. arXiv:1811.12444 [cs, stat]. https://doi.org/10.48550/arXiv.1811.12444.
532. Kun Wang and WaiChing Sun. 2019. Meta-modeling game for deriving theory-consistent, microstructure-based traction–separation laws via deep reinforcement learning. *Computer Methods in Applied Mechanics and Engineering* 346: 216–241. ISSN: 0045-7825. https://doi.org/10.1016/j.cma.2018.11.026.
533. Bendsøe, M.P. 1989. Optimal shape design as a material distribution problem. *Structural Optimization* 1(4): 193–202. ISSN: 1615-1488. https://doi.org/10.1007/BF01650949.
534. Kazuki Hayashi and Makoto Ohsaki. 2020. Reinforcement learning and graph embedding for binary truss topology optimization under stress and displacement constraints. *Frontiers in Built Environment* 6. ISSN: 2297-3362.
535. Shaojun Zhu et al. 2021. Machine-specified ground structures for topology optimization of binary trusses using graph embedding policy network. *Advances in Engineering Software* 159: 103032. ISSN: 0965-9978. https://doi.org/10.1016/j.advengsoft.2021.103032.
536. Hongbo Sun and Ling Ma. 2020. Generative design by using exploration approaches of reinforcement learning in density-based structural topology optimization. *Designs* 4(2): 10. ISSN: 2411-9660. https://doi.org/10.3390/designs4020010.
537. Seowoo Jang, Soyoung Yoo, and Namwoo Kang. 2022. Generative design by reinforcement learning: enhancing the diversity of topology optimization designs. *Computer-Aided Design* 146: 103225. ISSN: 0010-4485. https://doi.org/10.1016/j.cad.2022.103225.
538. Jiequn Han, Arnulf Jentzen, and Weinan E. 2018. Solving high-dimensional partial differential equations using deep learning. *Proceedings of the National Academy of Sciences* 115(34): 8505–8510. ISSN: 0027-8424, 1091-6490. https://doi.org/10.1073/pnas.1718942115.
539. Weinan, E., and Bing Yu. 2017. The Deep Ritz method: A deep learning-based numerical algorithm for solving variational problems. arXiv:1710.00211 [cs, stat]. tex.ids= E2018.

540. Jiachen Yang et al. 2023. Reinforcement learning for adaptive mesh refinement. In *Proceedings of The 26th international conference on artificial intelligence and statistics*, 5997–6014. PMLR.
541. Jean Rabault and Alexander Kuhnle. 2019. Accelerating deep reinforcement learning strategies of Flow Control through a multi-environment approach. *Physics of Fluids* 31(9): 094105. arXiv:1906.10382 [physics]. ISSN: 1070-6631, 1089-7666. https://doi.org/10.1063/1.5116415.
542. Guido Novati, Hugues Lascombes de Laroussilhe, and Petros Koumoutsakos. 2020. *Automating turbulence modeling by multi-agent reinforcement learning*. arXiv:2005.09023 [physics]. https://doi.org/10.48550/arXiv.2005.09023.
543. Xin-Yang Liu and Jian-Xun Wang. 2021. Physics-informed Dyna-style model-based deep reinforcement learning for dynamic control. *Proceedings of the Royal Society A: Mathematical, Physical and Engineering Sciences* 477(2255): 20210618. ISSN: 1364-5021, 1471-2946. https://doi.org/10.1098/rspa.2021.0618.
544. Haotian Shi et al. 2023. Physics-informed deep reinforcement learning-based integrated two-dimensional car-following control strategy for connected automated vehicles. *Knowledge-Based Systems* 269: 110485. ISSN: 0950-7051. https://doi.org/10.1016/j.knosys.2023.110485.
545. Adithya Ramesh and Balaraman Ravindran. 2023. *Physics-informed model-based reinforcement learning*. arXiv:2212.02179 [cs]. https://doi.org/10.48550/arXiv.2212.02179.
546. Colin Rodwell and Phanindra Tallapragada. 2023. Physics-informed reinforcement learning for motion control of a fish-like swimming robot. *Scientific Reports* 13(1): 10754. ISSN: 2045-2322. https://doi.org/10.1038/s41598-023-36399-4.

Chapter 11
The Future of Deep Learning in Computational Mechanics

Abstract The rapid rise of scientific literature on deep learning in computational mechanics illustrates its potential but also its pitfalls due to its hyped-up over-optimism. This chapter summarizes the methodological overview of the preceding chapter, including assessments. These pinpoint the most promising research directions and highlight which to avoid. In addition, clear guidelines within deep learning for simulation are provided. In particular, the well-known but under-utilized principles of good scientific practice are reiterated in the context of deep learning in computational mechanics. Following these ensures much-needed transparency and integrity, thereby accelerating future improvements in this field—creating a pathway for deep learning to contribute meaningfully to computational mechanics.

In order to structure the state-of-the-art, an overview of the most prominent deep learning methods employed in computational mechanics was presented in Chap. 10. Five main categories were identified: simulation substitution, simulation enhancement, discretizations as neural networks, generative approaches, and deep reinforcement learning. The potentials and limitations of each category are assessed in Sect. 11.1. Finally, general recommendations and guidelines for research in deep learning and computational mechanics are provided in Sect. 11.2. Following these not only ensures meaningful research but also enables more realistic (and critical) assessments of deep learning technologies (i.e., with a realistic perspective on their true capabilities in relation to state-of-the-art techniques).

11.1 Outlook

Despite the variety and abundance of the literature, few approaches are competitive in comparison to classical methods, as similarly seen throughout this book. In the literature, this manifests itself in the lack of comparisons of neural network-based methods

---

This chapter contains modified excerpts of the publications *Deep Learning in Computational Mechanics: a review* [1] and *Towards a Meaningful Integration of Neural Networks in Computational Solid Mechanics* [2].

L. Herrmann et al., *Deep Learning in Computational Mechanics*,
https://doi.org/10.1007/978-3-031-89529-6_11

to classical methods. We have found little evidence that neural network-based methods truly outperform classical methods in computational mechanics. However, with only few exceptions, current research is still in its early stages, with a focus on showcasing possibilities without focusing too much attention on accuracy and efficiency. Future research must, nevertheless, shift its focus to incorporate more in-depth investigations into the performance of the developed methods—including thorough and meaningful comparisons to performant classical methods dedicated to the task under investigation. This is in agreement with recent critical review articles on deep learning in simulation [3, 4], where honest and fair assessments are requested. This includes the determination of generalization capabilities, greater transparency by including, e.g., worst case performances to illustrate reliability, and computation times without disregarding the training time.

In line with this, and to the best of our knowledge, we provide a final overview outlining the potentials and limitations of the discussed methods.

- **Simulation substitution** has potential for surrogate modeling of parameterized models that need to be evaluated many times. However, currently this is only realizable for small parameter spaces, due to the amount of data required and unlikely to replace established methods, as also stated in [5]. Complex problems can still be tackled by neural network surrogates if they are first reduced to a low-dimensional space through model order reduction techniques. Physics-informed learning further reduces the amount of required data and improves the generalization capabilities. However, enforcing physics through penalty terms increases the computational effort, where the solutions still do not necessarily satisfy the corresponding physical laws. Instead, enforcing physical laws by construction guarantees that they are obeyed, which is more favorable to adding constraints through penalty terms.
- **Simulation enhancement** is currently one of the most promising areas of investigation. It is in particular beneficial for tasks where classical methods show difficulties. An excellent example for this is the formulation of constitutive laws, which are inherently phenomenological and thereby well-suited to be identified from data using tools such as deep learning. In addition, simulation enhancement, makes it possible to draw on insights gained from classical methods developed since the inception of computational mechanics. Furthermore, it is currently more realistic to learn smaller components of the simulation chain with neural networks rather than the entire model. These components should ideally be expensive and have limited requirements regarding accuracy and reliability. Lastly, it is also easier to assess whether a method enhanced by deep learning outperforms the classical method, as direct and fair comparisons are readily possible.
- An interesting research direction is to employ **discretizations as neural networks**, as this offers the potential to discover neural networks tailored to computational mechanics tasks, such as convolutional neural networks for computer vision or recurrent neural networks and transformers for natural language processing. In computational mechanics, their main benefit seems to stem from being able to exploit the computational benefits of tools and hardware that were created for the

wider community of deep learning—such as neural network libraries programmed for GPUs which enable an efficient, yet effortless massive parallelization. In our assessment, none of the methods encountered in this review were shown to be able to consistently outperform classical approaches using a comparable amount of computational resources.

- **Generative approaches** have been shown to be highly versatile in applications of computational mechanics since the accuracy of a specific instance under investigation is less of a concern here. They have been used to generate statistically equivalent data to train other machine learning models, to incorporate vague constraints based on data within optimization frameworks, and to detect anomalies.
- **Deep reinforcement learning** has already shown encouraging results—for example in controlling unmanned vehicles in complex physics environments. It is mainly applicable for problems where efficient differentiable physics solvers are unavailable, which is why it is popular in control problems for turbulence. In the presence of differentiable solvers, gradient-based methods are, however, still the state-of-the-art [6] and, thus, preferred.

To truly harvest these potentials and ensure meaningful research, good scientific principles have to be upheld (or rather reinstated) when delving into the field of deep learning in computational mechanics. To this end, this book closes with essential recommendations on general challenges and good scientific practice in the context of deep learning and, more specifically, deep learning in computational mechanics.

11.2 Recommendations

A major reason for deep learning's attractiveness, beyond its novelty and successes, is often the simplicity of formulating and implementing new methods through neural network frameworks, such as TensorFlow [7] and PyTorch [8]. Thus, without deep knowledge of either computational mechanics or deep learning, novelty can easily be generated. The majority of effort is spent on mindlessly tuning hyperparameters, for which expert domain knowledge is not required. Moreover, most insights are thereby derived from empirical observations rather than from fundamental theories or first principles. As many empirical studies are not easily reproduced, such a shift in research focus is particularly problematic when good scientific principles are neglected—which is unfortunately often the case. To this end, the following two sections outline general challenges of deep learning and, more concretely, what needs to be considered when integrating neural networks in computational mechanics.

11.2.1 General Challenges in the Assessment of Deep Learning Methods

The three biggest challenges of deep learning are:

- Neural networks break in unpredictable ways [9] and can thereby be consistently fooled [10–16].
- Deep learning is not robust due to its sensitivity to hyperparameters and thus requires extensive tuning by the practitioner [17, 18].
- Neural networks are uninterpretable and have limited explainability [19, 20], which compromises their reliability.

Regarding the first point, it seems that the only robust aspect of neural networks is the possibility of fooling them. Adversarial attacks [10, 11, 15], for example, can achieve this by using small perturbations (as small as a single pixel in an image) to provoke incorrect predictions by the model. Some works emphasize how this can be achieved in robust ways [12, 14, 16]. However, [9] highlights that neural networks break not only when attacked but also under naturally occurring circumstances, highlighting the insufficient reliability of neural networks for critical tasks.

Regarding the second point, deep learning is highly sensitive to hyperparameters. This is especially problematic when comparing different neural networks. Besides the tedious tuning, whose time is often ignored in performance assessments, the robustness issue means that one can never be sure that a method has been sufficiently tuned. Therefore, [21] argue that in most cases, a researcher is more familiar with one method (often his/her own), which biases a comparison that he/she conducts towards another. In addition, the need for extensive tuning to find specific hyperparameters [22], which are often problem-dependent and, in some cases, even machine-dependent, often kills reproducibility [23].

The superiority of neural networks over other simpler approaches is not guaranteed [24], as no single method dominates over all the others [25]. Therefore, simpler (and more interpretable) methods should not be neglected. Often, the superiority of sophisticated methods is only illusory [26]. For many neural networks, 90% of the predictive power can be reached by simple methods [21]. The additional accuracy may be swamped by uncertainties due to violations of assumptions, such as applying the model to samples outside the training distribution, errors in the labels, changes in the data over time, or an unsuitable choice of cost function. Without a clear quantitative advantage, simpler methods are preferable due to their interpretability [24]. However, novel and more sophisticated methods are often harder to criticize [27] since it is much easier to show why a simple method does not work than why a complicated approach, such as neural networks, does not work.

Before shifting the focus to computational mechanics, some of the main differences between deep learning in general and deep learning in computational mechanics are highlighted. First, the amount of training data is smaller by multiple orders of magnitude. In 2012, AlexNet [28] relied on 1.2 million labeled data samples to recognize 1 000 different objects. Today, 12 years later, large language models, such as ChatGPT [29] rely on hundreds of gigabytes to train models with billions of trainable parameters. This stands in stark contrast to the data available in computational mechanics, where some of the largest models rely on tens to hundreds of thousands of datapoints [30, 31]. Furthermore, in many deep learning tasks, such as image classification, small changes in the input typically do not lead to different outputs. By contrast, modifying the boundary conditions in problems such as wave propagation or topology optimization leads to very different outcomes, as demonstrated in [3] by changing a load direction by $0.01°$ in topology optimization. This implies that patterns in computational mechanics can be much more complex—making current pattern recognition algorithms less effective for some tasks in computational mechanics.

11.2.2 Good Scientific Practice for Deep Learning in Computational Mechanics

As mentioned for deep learning in general, but equally as critical for deep learning in computational mechanics, reproducibility issues exist for neural networks. Statements such as the following are troubling: $\sim$70% of researchers have at some point encountered issues in reproducing another scientist's results [32], or that a claim found in the scientific literature is more likely to be false than true [33]. These statistics are likely to be worse in experimental sciences but, through deep learning, become more important in computational mechanics. The reason is that claims in deep learning are mostly based on empirical observations rather than sound theoretical foundations. Unpublished codes and the sensitivity to the hyperparameters have led to a reproducibility crisis in machine learning in general [23, 26, 34–36], from which computational mechanics is not unaffected.

Beyond the technicalities, a major cause is the systematic reporting bias toward positive results. This bias mainly consists of a publication bias, i.e., publishers accepting only positive results, and an outcome reporting bias, i.e., scientists selectively disregarding negative results or relying on outcome switching. Only publishing good results with highly sensitive methods leads to a wrong impression in science [37].[1] Moreover, the proportion of positive results in a field is thought to correlate with the strength of reporting biases [38, 39]. In computational fluid mechanics [4], $\sim$95% of the 232 considered articles present positive results, and $\sim$5% present mixed results, while none present purely negative results. This is an unusually high rate of positive results compared to other fields [38]. Related is the inflated share of deep learning-related publications in computational mechanics, seen in Fig. 1.3. The share is more than twice as large as the average in all other fields. This inflation and reporting bias leads to an overly optimistic and unrealistic view of the state-of-the-art of deep learning in computational mechanics [4] and slows progress in the field by wasting other

[1] Consider, for example, a method that does not work in nine out of ten independently conducted experiments. If only the one successful trial is reported, the method is falsely promoted as favorable.

scientists' resources with investigating methods, which turn out to be inefficient, inappropriate, or do not generalize well, once examined closely.

Structural reforms in scientific publishing to encourage the publication of bad results (if well documented) could act as a partial remedy [4]. Rather than evaluating the results, the research should be evaluated on the quality of the hypothesis and its investigation. This would ensure awareness of the true capabilities of methods, which is currently lacking in deep learning in computational mechanics.

However, some basic ground rules should nevertheless be followed when reporting results. Computational resources must be presented transparently, including training time, data generation, and, ideally, hyperparameter tuning. For data-driven surrogate models, the breakeven threshold [3, 40] (see Eq. 3.48) is a valuable measure that quantifies the number of model evaluations to make a training worthwhile. This tool also counteracts the argument that a method's performance increases with more data. More data would simultaneously also lead to a higher breakeven threshold. With the limited applicability of some models in simulation, it is impossible to reach current breakeven thresholds [3].

Furthermore, comparisons with classical methods must be performed. When comparing computation times with classical methods, state-of-the-art methods must be considered instead of inferior approaches. This is unfortunately common practice, as seen in [4] for forward problems in fluid mechanics, where $\sim$79% considered weak baselines. Similarly, only 14/111 considered machine learning publications [3] on structural optimization were published in the *Structural and Multidisciplinary Optimization* journal. The majority were published in non-optimization journals, where the current structural optimization state-of-the-art is unknown. To overcome this problem, [4] emphasizes that state-of-the-art solution techniques with either the same accuracy or the same computational effort must be considered. This contrasts with many works that compare solution times to the high-resolution solver used during data generation while imposing looser standards on the accuracy of the approximation.

The solution quality must be evaluated in a fair and quantifiable way, i.e., using meaningful error measures [3, 41]. A bad example is pixel-wise differences between optimized topologies, which severely underestimate the effect of disconnections. In optimization scenarios, the mechanical quantity of interest is a better metric. Similarly, for the solution of differential equations, norms such as the energy norm [42] should be utilized instead of pixel-wise L^2 errors common in machine learning. Additionally, statistical assessments are particularly useful because the quality of results from a single deep learning model varies widely. Average performances or errors across a dataset are misleading. Instead, worst-case performances provide more insight into the reliability and applicability of a method.

As mentioned in Sect. 11.2.1, reliability and robustness are major issues in deep learning. Reliability can be reported through statistical analyses. However, robustness with respect to tunable model parameters is more challenging. Currently, the best option is full transparency in terms of qualitative descriptions and estimates of the time required for hyperparameter tuning. During method development, an important criterion should be ensuring a higher robustness and decreasing the number of tunable parameters.

In addition, problems should be selected more carefully when working with deep learning. Besides assessing the viability, problem scopes should be restricted to more specialized tasks, as opposed to physics-informed neural networks or general-purpose surrogate models tackling any kind of problem. These general-purpose deep learning-based approaches are unlikely to become universal solutions in scientific computing. Therefore, specific problems in which improvements seem viable or where no serious competition from established methods exists should be selected, as emphasized in this chapter. However, the range of viable applications should be large enough, and the method should be adaptable enough (e.g., to changes in geometry or material parameters) to make training worthwhile [3].

The three main takeaways can be summarized as follows:

- Honest assessments and transparent explanations of the developed method, taking into account the state-of-the-art, are essential.
- The proposed methods should be robust with respect to hyperparameters and modifications of the problem. In other words, extensive tuning by the user for similar problems should not be required.
- A more careful and narrower selection of problem types. Neural networks are unlikely to be universal tools that replace all kinds of specialized discretization techniques. Yet, some flexibility should be maintained in the application domain.

Finally, although deep learning-based approaches have been critically assessed throughout this book and the works mentioned in the preceding Chap. 10, the goal is not to discourage the exploration of neural networks for simulation. As long as good scientific principles are followed, meaningful and useful future research directions can be ensured. This, in turn, will enable deep learning to benefit and contribute to the field of computational mechanics.

References

1. Herrmann, Leon, and Stefan Kollmannsberger. 2024. Deep learning in computational mechanics: A review. en. *Computational Mechanics*. ISSN: 1432-0924. https://doi.org/10.1007/s00466-023-02434-4.
2. Herrmann, Leon. 2025. Towards A Meaningful Integration of Neural Networks in Computational Solid Mechanics. PhD thesis. Technische Universität München.
3. Woldseth, Rebekka V., et al. 2022. On the use of artificial neural networks in topology optimisation. en. *Structural and Multidisciplinary Optimization* 65(10): 294. ISSN: 1615-1488. https://doi.org/10.1007/s00158-022-03347-1.
4. McGreivy, Nick, and Ammar Hakim. 2024. Weak baselines and reporting biases lead to overoptimism in machine learning for fluid-related partial differential equations. en. *Nature Machine Intelligence* 6(10): 1256–1269. ISSN: 2522-5839. https://doi.org/10.1038/s42256-024-00897-5.
5. Baker, Nathan, et al. 2019. *Workshop Report on Basic Research Needs for Scientific Machine Learning: Core Technologies for Artificial Intelligence*. en. Tech. rep. 1478744. Feb. 2019, 1478744. https://doi.org/10.2172/1478744.

6. Sigmund, Ole. 2011. On the usefulness of non-gradient approaches in topology optimization. en. *Structural and Multidisciplinary Optimization* 43(5): 589–596. ISSN: 1615-1488. https://doi.org/10.1007/s00158-011-0638-7.
7. Abadi, Martín, et al. 2016. *TensorFlow: Large-Scale Machine Learning on Heterogeneous Distributed Systems*. arXiv:1603.04467 [cs]. Mar. 2016.
8. Paszke, Adam, et al. 2019. PyTorch: An Imperative Style, High-Performance Deep Learning Library. arXiv:1912.01703 [cs, stat] (Dec. 2019). arXiv: 1912.01703.
9. Heaven, Douglas. 2019. Why deep-learning AIs are so easy to fool. en. *Nature* 574(7777): 163–166. ISSN: 0028-0836, 1476-4687. https://doi.org/10.1038/d41586-019-03013-5.
10. Szegedy, Christian, et al. 2014. *Intriguing properties of neural networks*. arXiv:1312.6199 [cs]. Feb. 2014. https://doi.org/10.48550/arXiv.1312.6199.
11. Goodfellow, Ian J., Jonathon Shlens, and Christian Szegedy. 2015. *Explaining and Harnessing Adversarial Examples*. arXiv:1412.6572 [cs, stat]. Mar. 2015. https://doi.org/10.48550/arXiv.1412.6572.
12. Moosavi-Dezfooli, Seyed-Mohsen, Alhussein Fawzi, and Pascal Frossard. 2016. DeepFool: A simple and accurate method to fool deep neural networks. In *2016 IEEE Conference on Computer Vision and Pattern Recognition (CVPR)*. ISSN: 1063-6919. June 2016, 2574–2582. https://doi.org/10.1109/CVPR.2016.282.
13. Kurakin, Alexey, Ian Goodfellow, and Samy Bengio. 2017. *Adversarial Machine Learning at Scale*. arXiv:1611.01236 [cs, stat]. Feb. 2017. https://doi.org/10.48550/arXiv.1611.01236.
14. Eykholt, Kevin, et al. 2018. Robust physical-world attacks on deep learning visual classification. In *2018 IEEE/CVF Conference on Computer Vision and Pattern Recognition*. ISSN: 2575-7075. June 2018, 1625–1634. https://doi.org/10.1109/CVPR.2018.00175.
15. Finlayson, Samuel G., et al. 2019. Adversarial attacks on medical machine learning. eng. *Science (New York, N.Y.)* 363(6433): 1287–1289. ISSN: 1095-9203. https://doi.org/10.1126/science.aaw4399.
16. Zhang, Ming, Yongkang Chen, and Cheng Qian. 2023. Fooling examples: Another intriguing property of neural networks. en. *Sensors* 23(14): 6378. ISSN: 1424-8220. https://doi.org/10.3390/s23146378.
17. Smith, Leslie N. 2018. *A disciplined approach to neural network hyper-parameters: Part 1 – learning rate, batch size, momentum, and weight decay*. arXiv:1803.09820 [cs, stat]. Apr. 2018. https://doi.org/10.48550/arXiv.1803.09820.
18. Géron, Aurélien. 2023. *Hands-on machine learning with Scikit-Learn, Keras and TensorFlow: Concepts, tools, and techniques to build intelligent systems*. Third edition. OCLC: on1346503549. Beijing: O'Reilly Media, Inc, 2023. ISBN: 978-1-09-812597-4.
19. Arrieta, Alejandro Barredo, et al. 2020. Explainable Artificial Intelligence (XAI): Concepts, taxonomies, opportunities and challenges toward responsible AI. *Information Fusion* 58: 82–115. ISSN: 1566-2535. https://doi.org/10.1016/j.inffus.2019.12.012.
20. Zhong, Xiaoting, et al. 2022. Explainable machine learning in materials science. en. *npj Computational Materials* 8(1): 204. ISSN: 2057-3960. https://doi.org/10.1038/s41524-022-00884-7.
21. Hand, David J. 2006. Classifier technology and the illusion of progress. *Statistical Science* 21(1). ISSN: 0883-4237. https://doi.org/10.1214/088342306000000060.
22. Sanu, Suryanarayanan Manoj, Alejandro M. Aragon, and Miguel A. Bessa. 2024. *Neural topology optimization: The good, the bad, and the ugly*. arXiv:2407.13954 [cs, math]. July 2024. https://doi.org/10.48550/arXiv.2407.13954.
23. Gundersen, Odd Erik, et al. 2023. *Sources of Irreproducibility in Machine Learning: A Review*. arXiv:2204.07610 [cs]. Apr. 2023. https://doi.org/10.48550/arXiv.2204.07610.
24. Schmidl, Sebastian, Phillip Wenig, and Thorsten Papenbrock. 2022. Anomaly detection in time series: A comprehensive evaluation. en. In *Proceedings of the VLDB Endowment* 15(9): 1779–1797. ISSN: 2150-8097. https://doi.org/10.14778/3538598.3538602.
25. Wolpert, D.H., and W.G. Macready. 1997. No free lunch theorems for optimization. *IEEE Transactions on Evolutionary Computation* 1(1): 67–82. ISSN: 1941-0026. https://doi.org/10.1109/4235.585893.

26. Kapoor, Sayash, and Arvind Narayanan. 2023. Leakage and the reproducibility crisis in machine-learning-based science. *Patterns* 4(9): 100804. ISSN: 2666-3899. https://doi.org/10.1016/j.patter.2023.100804.
27. Breiman, Leo. 2001. Statistical modeling: The two cultures (with comments and a rejoinder by the author). *Statistical Science* 16(3): 199–231. ISSN: 0883-4237, 2168-8745. https://doi.org/10.1214/ss/1009213726.
28. Krizhevsky, Alex, Ilya Sutskever, and Geoffrey E. Hinton. 2017. ImageNet classification with deep convolutional neural networks. *Communications of the ACM* 60(6): 84–90. ISSN: 0001-0782. https://doi.org/10.1145/3065386.
29. OpenAI. 2023. *GPT-4 Technical Report*. arXiv:2303.08774 [cs]. Mar. 2023. https://doi.org/10.48550/arXiv.2303.08774.
30. Bielecki, Dustin, et al. 2021. Multi-stage deep neural network accelerated topology optimization. en. *Structural and Multidisciplinary Optimization* 64(6): 3473–3487. ISSN: 1615-1488. https://doi.org/10.1007/s00158-021-03028-5.
31. Bastek, Jan-Hendrik, and Dennis M. Kochmann. 2023. Inverse design of nonlinear mechanical metamaterials via video denoising diffusion models. en. *Nature Machine Intelligence* 5(12): 1466–1475. ISSN: 2522-5839. https://doi.org/10.1038/s42256-023-00762-x.
32. Baker, Monya. 2016. 1,500 scientists lift the lid on reproducibility. en. *Nature* 533(7604): 452–454. ISSN: 1476-4687. https://doi.org/10.1038/533452a.
33. Ioannidis, John P. A. 2005. Why most published research findings are false. en. *PLOS Medicine* 2(8): e124. ISSN: 1549-1676. https://doi.org/10.1371/journal.pmed.0020124.
34. Hutson, Matthew. 2018. Artificial intelligence faces reproducibility crisis. en. *Science* 359(6377): 725–726. ISSN: 0036-8075, 1095-9203. https://doi.org/10.1126/science.359.6377.725.
35. Sculley, D., et al. 2018. Winner's Curse? On Pace, Progress, and Empirical Rigor. en. In: (June 2018).
36. Kapoor, Sayash, et al. 2024. REFORMS: Consensus-based recommendations for machine-learning-based science. en. *Science Advances* 10(18): eadk3452. ISSN: 2375-2548. https://doi.org/10.1126/sciadv.adk3452.
37. Bergstrom, Carl T., and Jevin D. West. 2021. *Calling bullshit: The art of skepticism in a data-driven world*. eng. First published by Allen Lane. London: Penguin Random House. ISBN: 978-0-14-198705-7.
38. Fanelli, Daniele. 2010. "Positive" Results Increase Down the Hierarchy of the Sciences. en. *PLOS ONE* 5(4): e10068. ISSN: 1932-6203. https://doi.org/10.1371/journal.pone.0010068.
39. Ritchie, Stuart. 2020. *Science fictions: How fraud, bias, negligence, and hype undermine the search for truth*. First edition. New York: Metropolitan Books; Henry Holt and Company. ISBN: 978-1-250-22269-5.
40. Keshavarzzadeh, Vahid, et al. 2021. Image-based multiresolution topology optimization using deep disjunctive normal shape model. en. *Computer-Aided Design* 130: 102947. ISSN: 00104485. https://doi.org/10.1016/j.cad.2020.102947.
41. Most, Thomas, Lars Gräning, and Sebastian Wolff. 2024. *Robustness investigation of quality measures for the assessment of machine learning models*. arXiv:2408.04391 [cs, stat] version: 1. Aug. 2024. https://doi.org/10.48550/arXiv.2408.04391.
42. Hughes, Thomas J. R. 2000. *The finite element method: Linear static and dynamic finite element analysis*. Mineola, NY: Dover Publications. ISBN: 978-0-486-41181-1.

Solutions to the Exercises

Selected solutions are provided with a focus on the pen-and-paper exercises. The remaining solutions to the coding exercises are provided at www.deeplearningincomputationalmechanics.com.

Chapter 1

S.2 *Computing Stress Fields with Tensors*

2. The material's density is $\rho = 8 \cdot 10^{-6}$.

Chapter 2

S.4 *Linear Regression*

1. Using Eq. (2.3), the cost function can be computed.

$$\begin{aligned}
&\frac{1}{3}\sum_{i=1}^{3}(\tilde{y}_i - (w\tilde{x}_i + b)^2) \\
&= \frac{1}{3}\big((1-(1\cdot 0+1))^2 + (6-(1\cdot 2+1))^2 + (7-(1\cdot 3+1))^2\big) \\
&= 6
\end{aligned}$$

L. Herrmann et al., *Deep Learning in Computational Mechanics*,
https://doi.org/10.1007/978-3-031-89529-6

2. The gradients are found by differentiating Eq. (2.3).

$$\frac{\partial C}{\partial w} = \frac{1}{3}\sum_{i=1}^{3}\big(-2\tilde{x}_i(\tilde{y}_i - (w\tilde{x}_i + b))\big)$$

$$\frac{\partial C}{\partial b} = \frac{1}{3}\sum_{i=1}^{3}\big(-2(\tilde{y}_i - (w\tilde{x}_i + b))\big)$$

3. The gradients are equal to zero at the minimum. Thereby, the minimum of the cost function can be found.

$$\frac{\partial C}{\partial w} = \frac{1}{3}(26w + 10b - 66) \overset{!}{=} 0$$
$$\frac{\partial C}{\partial b} = \frac{1}{3}(10w + 6b - 28) \overset{!}{=} 0$$

The equations are solved for the weight and bias.

$$w = \frac{29}{14}$$
$$b = \frac{17}{14}$$

The resulting linear regression is then given as $\hat{y} = \frac{29}{14}x + \frac{17}{14}$.

4. The matrix and corresponding vectors can be set up as

$$\boldsymbol{X} = \begin{pmatrix} 0 & 1 \\ 2 & 1 \\ 3 & 1 \end{pmatrix}, \boldsymbol{\Theta} = \begin{pmatrix} w \\ b \end{pmatrix}, \boldsymbol{y} = \begin{pmatrix} 1 \\ 6 \\ 7 \end{pmatrix}.$$

This can now be solved using the linear regression formula given in the script

$$\boldsymbol{\Theta} = (\boldsymbol{X}^\top \boldsymbol{X})^{-1}\boldsymbol{X}^\top \boldsymbol{y} = \begin{pmatrix} \frac{29}{14} \\ \frac{17}{14} \end{pmatrix}.$$

The linear regression is identical to the one found in the previous exercise.

5. Using the gradient expressions derived in the previous exercises, the gradients can be found for each iteration.

- iteration 1

$$\frac{\partial C}{\partial w} = -10$$
$$\frac{\partial C}{\partial b} = -4$$
$$w = w - \alpha \frac{\partial C}{\partial w} = 2$$
$$b = b - \alpha \frac{\partial C}{\partial b} = 1.4$$

- iteration 2

$$\frac{\partial C}{\partial w} = 0$$
$$\frac{\partial C}{\partial b} = 0.133$$
$$w = w - \alpha \frac{\partial C}{\partial w} = 2$$
$$b = b - \alpha \frac{\partial C}{\partial b} = 1.387$$

- iteration 3

$$\frac{\partial C}{\partial w} = -0.044$$
$$\frac{\partial C}{\partial b} = 0.107$$
$$w = w - \alpha \frac{\partial C}{\partial w} = 2.004$$
$$b = b - \alpha \frac{\partial C}{\partial b} = 1.376$$

It is observed that there is a tendency towards $w = 2.071$ and $b = 1.214$. If more iterations are conducted, a clear convergence is seen.

S.5 Higher-Order Regression

The solution implementation is provided in `exercise5_solution.ipynb`. Minor explanations are provided below

1. The cost function is given as

$$C = \frac{1}{m_{\mathcal{D}}} \sum_{i=1}^{m_{\mathcal{D}}} \big(\tilde{y}_i - \sum_{j=1}^{p} (w_j \tilde{x}_i^j) + b\big)^2.$$

2. The gradients of the cost function with respect to the weights and the bias are given as

$$\frac{\partial C}{\partial w_j} = \frac{1}{m_{\mathcal{D}}} \sum_{i=1}^{m_{\mathcal{D}}} 2\tilde{x}_i^j \big(\tilde{y}_i - \sum_{j=1}^{p} (w_j \tilde{x}_i^j) + b\big),$$

$$\frac{\partial C}{\partial b} = \frac{1}{m_{\mathcal{D}}} \sum_{i=1}^{m_{\mathcal{D}}} 2\big(\tilde{y}_i - \sum_{j=1}^{p} (w_j \tilde{x}_i^j) + b\big).$$

5. For linear polynomials, underfitting can be observed. This might be seen in the MSE during learning, as the MSE is relatively high. For a polynomial of degree 2, one observes ideal fitting. This is to be expected, as the data generated is a polynomial of degree 2. For the very large polynomial of degree 9, overfitting is observed. This is illustrated by the large difference between the training and testing data. Therefore, it can often already be spotted during learning, where the MSE of the training data decreases, whereas the MSE of the testing data does not change significantly or even increases.
   Underfitting can be avoided by using a model with a higher complexity, so an increase of the polynomial degree and overfitting can be avoided by using a regularization.
6. The cost function with the regularization parameter λ is now given as

$$C = \frac{1}{m_{\mathcal{D}}} \sum_{i=1}^{m_{\mathcal{D}}} \big(\tilde{y}_i - \sum_{j=1}^{p} (w_j \tilde{x}_i^j) + b\big)^2 + \lambda \sum_{j=1}^{p} w_j^2$$

   and the gradients of the cost function with respect to the weights and the bias are given as

$$\frac{\partial C}{\partial w_j} = \frac{1}{m_{\mathcal{D}}} \sum_{i=1}^{m_{\mathcal{D}}} 2\tilde{x}_i^j \big(\tilde{y}_i - \sum_{j=1}^{p} (w_j \tilde{x}_i^j) + b\big) + 2\lambda w_j,$$

$$\frac{\partial C}{\partial b} = \frac{1}{m_{\mathcal{D}}} \sum_{i=1}^{m_{\mathcal{D}}} 2\big(\tilde{y}_i - \sum_{j=1}^{p} (w_j \tilde{x}_i^j) + b\big).$$

S.6 Adam Optimizer

1. The implementation follows Algorithm 2.
2. The Adam algorithm consistently reaches better optima in comparison to the gradient descent algorithm. This is valid for both different initial guesses and different learning rates. Thus, Adam is more robust, as it is less sensitive to the choice of hyperparameters.

Chapter 3

S.7 Approximating the Sine Function

1. Underfitting
2. Overfitting
3. Reduce the capacity of the model (fewer layers/neurons), use a smaller learning rate, choose fewer epochs, or apply regularization, e.g., early stopping, weight decay (L^1-/L^2-regularization), dropout, etc.
9. The neural network did not receive training data beyond the boundaries and is not able to extrapolate this information. Neural networks are only useful for interpolation. To improve the result beyond the boundaries, training data would have to be supplied in this regime.

S.8 Neural Network Derivatives

1. The function describing the neural network is:

$$\hat{y} = w_1^{(2)} \sigma(w_{11}^{(1)} x_1 + w_{12}^{(1)} x_2 + b_1^{(1)}) + w_2^{(2)} \sigma(w_{21}^{(1)} x_1 + w_{22}^{(1)} x_2 + b_2^{(1)}) + b^{(2)}$$

2. The partial derivatives with respect to the weights are given as

$$\frac{\partial \hat{y}}{\partial w_{11}^{(1)}} = w_1^{(2)} 2(w_{11}^{(1)} x_1 + w_{12}^{(1)} x_2 + b_1^{(1)}) x_1,$$

$$\frac{\partial \hat{y}}{\partial w_{12}^{(1)}} = w_1^{(2)} 2(w_{11}^{(1)} x_1 + w_{12}^{(1)} x_2 + b_1^{(1)}) x_2,$$

$$\frac{\partial \hat{y}}{\partial w_{21}^{(1)}} = w_2^{(2)} 2(w_{21}^{(1)} x_1 + w_{22}^{(1)} x_2 + b_2^{(1)}) x_1,$$

$$\frac{\partial \hat{y}}{\partial w_{22}^{(1)}} = w_2^{(2)} 2(w_{21}^{(1)} x_1 + w_{22}^{(1)} x_2 + b_2^{(1)}) x_2,$$

$$\frac{\partial \hat{y}}{\partial w_1^{(2)}} = (w_{11}^{(1)} x_1 + w_{12}^{(1)} x_2 + b_1^{(1)})^2,$$

$$\frac{\partial \hat{y}}{\partial w_2^{(2)}} = (w_{21}^{(1)} x_1 + w_{22}^{(1)} x_2 + b_2^{(1)})^2.$$

The partial derivatives with respect to the biases are given as

$$\frac{\partial \hat{y}}{\partial b_1^{(1)}} = w_1^{(2)} 2(w_{11}^{(1)} x_1 + w_{12}^{(1)} x_2 + b_1^{(1)}),$$

$$\frac{\partial \hat{y}}{\partial b_2^{(1)}} = w_2^{(2)} 2(w_{21}^{(1)} x_1 + w_{22}^{(1)} x_2 + b_2^{(1)}),$$

$$\frac{\partial \hat{y}}{\partial b^{(2)}} = 1.$$

3. The partial derivatives of the cost function C with respect to the network output $\hat{y}$ is given as

$$\frac{\partial C}{\partial \hat{y}} = -2(\tilde{y} - \hat{y})$$

The partial derivatives with respect to the weights are given as

$$\frac{\partial C}{\partial w_{11}^{(1)}} = \frac{\partial C}{\partial \hat{y}} \frac{\partial \hat{y}}{\partial w_{11}^{(1)}} = -2(\tilde{y} - \hat{y}) w_1^{(2)} 2(w_{11}^{(1)} x_1 + w_{12}^{(1)} x_2 + b_1^{(1)}) x_1,$$

$$\frac{\partial C}{\partial w_{12}^{(1)}} = \frac{\partial C}{\partial \hat{y}} \frac{\partial \hat{y}}{\partial w_{12}^{(1)}} = -2(\tilde{y} - \hat{y}) w_1^{(2)} 2(w_{11}^{(1)} x_1 + w_{12}^{(1)} x_2 + b_1^{(1)}) x_2,$$

$$\frac{\partial C}{\partial w_{21}^{(1)}} = \frac{\partial C}{\partial \hat{y}} \frac{\partial \hat{y}}{\partial w_{21}^{(1)}} = -2(\tilde{y} - \hat{y}) w_2^{(2)} 2(w_{21}^{(1)} x_1 + w_{22}^{(1)} x_2 + b_2^{(1)}) x_1,$$

$$\frac{\partial C}{\partial w_{22}^{(1)}} = \frac{\partial C}{\partial \hat{y}} \frac{\partial \hat{y}}{\partial w_{22}^{(1)}} = -2(\tilde{y} - \hat{y}) w_2^{(2)} 2(w_{21}^{(1)} x_1 + w_{22}^{(1)} x_2 + b_2^{(1)}) x_2,$$

$$\frac{\partial C}{\partial w_1^{(2)}} = \frac{\partial C}{\partial \hat{y}} \frac{\partial \hat{y}}{\partial w_1^{(2)}} = -2(\tilde{y} - \hat{y})(w_{11}^{(1)} x_1 + w_{12}^{(1)} x_2 + b_1^{(1)})^2,$$

$$\frac{\partial C}{\partial w_2^{(2)}} = \frac{\partial C}{\partial \hat{y}} \frac{\partial \hat{y}}{\partial w_2^{(2)}} = -2(\tilde{y} - \hat{y})(w_{21}^{(1)} x_1 + w_{22}^{(1)} x_2 + b_2^{(1)})^2.$$

The partial derivatives with respect to the biases are given as

$$\frac{\partial C}{\partial b_1^{(1)}} = \frac{\partial C}{\partial \hat{y}} \frac{\partial \hat{y}}{\partial b_1^{(1)}} = -2(\tilde{y} - \hat{y}) w_1^{(2)} 2(w_{11}^{(1)} x_1 + w_{12}^{(1)} x_2 + b_1^{(1)}),$$

$$\frac{\partial C}{\partial b_2^{(1)}} = \frac{\partial C}{\partial \hat{y}} \frac{\partial \hat{y}}{\partial b_2^{(1)}} = -2(\tilde{y} - \hat{y}) w_2^{(2)} 2(w_{21}^{(1)} x_1 + w_{22}^{(1)} x_2 + b_2^{(1)}),$$

$$\frac{\partial C}{\partial b^{(2)}} = \frac{\partial C}{\partial \hat{y}} \frac{\partial \hat{y}}{\partial b^{(2)}} = -2(\tilde{y} - \hat{y}).$$

S.10 Neural Network Representational Capacity

The number of biases N_b and the number of weights N_w can be computed with

$$N_b = 0 + \sum_{i=1}^{N_d} N_w + 1,$$

$$N_w = 2N_w + \sum_{i=1}^{N_d - 1} N_w^2 + N_w \cdot 1.$$

This problem can be solved visually by considering a neural network representation like Fig. 3.1. The number of biases is simply the number of neurons (excluding the input neurons). The number of weights is obtained by counting the number of connections, i.e., multiplying the number of neurons of adjacent layers for each layer.

S.11 Convolutional Neural Network

1. The one-dimensional output (using a stride of $s = 1$) can be computed as follows.

$$\begin{aligned}\hat{y}_1 &= [1 \cdot 1 + 4 \cdot 0 + 6 \cdot 1,\\ &\quad 4 \cdot 1 + 6 \cdot 0 + 8 \cdot 1,\\ &\quad 6 \cdot 1 + 8 \cdot 0 + 2 \cdot 1]\\ &= [7, 12, 8]\\ \hat{y}_2 &= [1 \cdot 0 + 4 \cdot 0 + 6 \cdot 2,\\ &\quad 4 \cdot 0 + 6 \cdot 0 + 8 \cdot 2,\\ &\quad 6 \cdot 0 + 8 \cdot 0 + 2 \cdot 2]\\ &= [12, 16, 4]\end{aligned}$$

2. The one-dimensional output (using a stride of $s = 2$) can be computed as follows.

$$\begin{aligned}\hat{y}_1 &= [1 \cdot 1 + 4 \cdot 0 + 6 \cdot 1,\\ &\quad 6 \cdot 1 + 8 \cdot 0 + 2 \cdot 1]\\ &= [7, 8]\\ \hat{y}_2 &= [1 \cdot 0 + 4 \cdot 0 + 6 \cdot 2,\\ &\quad 6 \cdot 0 + 8 \cdot 0 + 2 \cdot 2]\\ &= [12, 4]\end{aligned}$$

3. The one-dimensional output from two inputs (using a stride of $s = 2$) can be computed as follows.

$$\begin{aligned}\hat{y}_1 &= [(3 \cdot 2 + 2 \cdot 3 + 1 \cdot 0) + (5 \cdot 1 + 2 \cdot 1 + 0 \cdot 1),\\ &\quad (1 \cdot 2 + 5 \cdot 3 + 2 \cdot 0) + (0 \cdot 1 + 8 \cdot 1 + 0 \cdot 1)]\\ &= [19, 25]\\ \hat{y}_2 &= [(3 \cdot 0 + 2 \cdot 2 + 1 \cdot 0) + (5 \cdot 0 + 2 \cdot 2 + 0 \cdot 2),\\ &\quad (1 \cdot 0 + 5 \cdot 2 + 2 \cdot 0) + (0 \cdot 0 + 8 \cdot 2 + 0 \cdot 2)]\\ &= [8, 26]\end{aligned}$$

Chapter 4

S.14 Physics-Informed Neural Network for a Static Bar

1. Using Eq. (4.4), the distributed load $p(x)$ is computed by insertion of the displacement $u(x)$ and cross-sectional properties $EA(x)$.

$$
\begin{aligned}
p(x) &= -\frac{d}{dx}(EA(x)\frac{du}{dx}) \\
&= -6x\pi \sin(3\pi x) - 9\pi^2(x^2+1)\cos(3\pi x)
\end{aligned}
$$

The homogeneous boundary conditions can be determined by plotting the displacements

$$
\begin{aligned}
u(0) &= 0, \\
\left.\frac{du}{dx}\right|_{x=0} &= 0.
\end{aligned}
$$

8. The following code snippet defines the problem from Sect. 4.2.1.

```
# analytial solution
uAnalytic = lambda x: np.sin(2 * np.pi * x)

# problem data
L = 1
E = lambda x: 1
A = lambda x: 1

uB = [[0, 0, 0], [0, 0, L]]

distLoad = lambda x: 4 * np.pi ** 2 * torch.sin(2 * np.pi * x)
```

S.15 Data-Driven Identification using Physics-Informed Neural Networks for a Static Bar

1. The axial stiffness $\lambda = EA(x)$ depends on the solution $u(x)$.
2. The following code snippet defines the inverse problem.

```
# analytial solution
EAAnalytic = lambda x: np.sqrt(2 * np.sin(1) * x - 2 * np.sin(x) +
    1)
```

```

# problem data
L = 1.
uAnalytic = lambda x: 1 - torch.sqrt(2 * torch.sin(torch.tensor
    ([1])) * x
                                             - 2 * torch.sin(x) + 1)
distLoad = lambda x: torch.sin(x)
```

5. The following code snippet reproduces the example from Sect. 4.3.

```
# analytic solution
EAAnalytic = lambda x: x ** 3 - x ** 2 + 1

# problem data
L = 1.
uAnalytic = lambda x: torch.sin(2 * np.pi * x)

distLoad = lambda x: (-2 * (3 * x ** 2 - 2 * x) * np.pi * torch.cos
    (2 * np.pi * x)
                     + 4 * (x ** 3 - x ** 2 + 1) * np.pi ** 2 * torch.sin(2
    * np.pi * x))
```

S.17 Physics-Informed Neural Network for a Plate in Membrane Action

4. Case 2 is more challenging due to the weak singularity, i.e., the stress singularity. The physics-informed neural network acting on the strong form of the partial differential equation is unable to learn the solution. A remedy is to consider the weak form, as is done in Chap. 5.

Chapter 5

S.18 Variations of Physics-Informed Neural Networks

2. The problem setup is defined by the code snippet below.

```
# analytical solution
uAnalytic = lambda x: -x ** 0.65 + 0.65 * x

# problem data
E = lambda x: 1. + x * 0
```

```
A = lambda x: 1. + x * 0
L = 1
uB = [[0, 0, 0], [0, 1, L]]
distLoad = lambda x: -0.2275 * x ** -1.35
```

3. The issue is, that the distributed load $p(x)$ and the strain $\varepsilon(x)$ have a singularity at $x = 0$. This causes a problem for integration techniques evaluating the parameters at the boundaries, such as the trapezoidal integration. The midpoint integration circumvents this issue. Note that this issue cannot be resolved by physics-informed neural networks, where the cost is usually evaluated at the boundaries. Avoiding an evaluation at the boundary leads to a worse estimate of $u(x)$.

S.19 Deep Energy Method for a Plate in Membrane Action

4. The indicator `alpha` can be modified as follows to incorporate a circular hole.

```
radius = 0.2
index = (x1 - 0.5) ** 2 + (x2 - 0.5) ** 2 < radius ** 2
alpha[index] = 0
```

6. Only incorporating the indicator function in the strong form would not be sufficient. In addition, the homogeneous Neumann boundary condition at the additional boundaries would have to be imposed. An alternative to the indicator function would be to simply not allocate collocation points in the fictitious domain, i.e., only sampling in the real domain. The challenging part remains, however, the imposition of the corresponding Neumann boundary conditions, which are naturally imposed when considering the weak form as in the deep energy method.

Chapter 6

S.22 Singular Value Decomposition

1. Determine the two matrix products

$$
\boldsymbol{A}\boldsymbol{A}^{\intercal} = \begin{pmatrix} 10 & 4 \\ 4 & 16 \end{pmatrix},
$$

$$
\boldsymbol{A}^{\intercal}\boldsymbol{A} = \begin{pmatrix} 9 & 3 \\ 3 & 17 \end{pmatrix}.
$$

Estimate the eigenvalues of either $\boldsymbol{A}\boldsymbol{A}^\mathsf{T}$ or $\boldsymbol{A}^\mathsf{T}\boldsymbol{A}$.

$$\det(\boldsymbol{A}\boldsymbol{A}^\mathsf{T} - \boldsymbol{I}\lambda) = \det\begin{pmatrix} 10-\lambda & 4 \\ 4 & 16-\lambda \end{pmatrix} = \lambda^2 - 26\lambda + 144$$

$$\lambda_{1,2} = 13 \pm 5$$
$$\lambda_1 = 18$$
$$\lambda_2 = 8$$

Thus $\boldsymbol{\Sigma}$ is given by

$$\boldsymbol{\Sigma} \approx \begin{pmatrix} \sqrt{18} & 0 \\ 0 & \sqrt{8} \end{pmatrix}.$$

Determine the normalized eigenvectors $\boldsymbol{u_1}, \boldsymbol{u_2}$ of $\boldsymbol{A}\boldsymbol{A}^\mathsf{T}$.

$$\lambda = 18, \begin{pmatrix} 2 & 4 \\ 4 & 8 \end{pmatrix} \boldsymbol{u_1} = \begin{pmatrix} 0 \\ 0 \end{pmatrix} \rightarrow \boldsymbol{u_1} \approx \begin{pmatrix} 0.447 \\ 0.894 \end{pmatrix}$$
$$\lambda = 8, \begin{pmatrix} -8 & 4 \\ 4 & -2 \end{pmatrix} \boldsymbol{u_2} = \begin{pmatrix} 0 \\ 0 \end{pmatrix} \rightarrow \boldsymbol{u_2} \approx \begin{pmatrix} 0.894 \\ -0.447 \end{pmatrix}$$

Thus $\boldsymbol{U}$ is given by

$$\boldsymbol{U} \approx \begin{pmatrix} 0.447 & 0.894 \\ 0.894 & -0.447 \end{pmatrix}.$$

Determine the normalized eigenvectors $\boldsymbol{v_1}, \boldsymbol{v_2}$ of $\boldsymbol{A}^\mathsf{T}\boldsymbol{A}$.

$$\lambda = 18, \begin{pmatrix} -9 & 3 \\ 3 & -1 \end{pmatrix} \boldsymbol{v_1} = \begin{pmatrix} 0 \\ 0 \end{pmatrix} \rightarrow \boldsymbol{v_1} \approx \begin{pmatrix} 0.316 \\ 0.949 \end{pmatrix}$$
$$\lambda = 8, \begin{pmatrix} 1 & 3 \\ 3 & 9 \end{pmatrix} \boldsymbol{v_2} = \begin{pmatrix} 0 \\ 0 \end{pmatrix} \rightarrow \boldsymbol{v_2} \approx \begin{pmatrix} 0.316 \\ 0.949 \end{pmatrix}$$

Thus $\boldsymbol{V}$ is given by

$$\boldsymbol{V} \approx \begin{pmatrix} 0.316 & 0.949 \\ 0.949 & -0.316 \end{pmatrix}.$$

The decomposition can be checked with

$$U\Sigma V^\top = \begin{pmatrix} 3 & 1 \\ 0 & 4 \end{pmatrix}$$

3. The lowest possible truncation without loss of information is $r = 3$. The reason is the linear dependency of the first and fourth rows and the second and last rows, respectively.
5. The lowest possible truncation level is the smallest image dimension, i.e., 3024. The main difference to Figs. 6.3 and 6.4 is simply that in Fig. 6.3, the second dimension is smaller, i.e., m, while in Fig. 6.23 the first is smaller, i.e., n. This order of the aspect ratio does, however, not impact the lowest possible truncation level. The corresponding compression ratios for $r = 1, 2, 10, 20, 100, 200, (3024)$ are $0.0579\%, 0.116\%, 0.580\%, 1.16\%, 5.87\%,$
$11.9\%, (250\%)$. Despite the truncation with $r = 3024$, representing an image with the economy singular value decomposition with the matrices of size $n \times r, r \times r, r \times m$, no compression is achieved over the original matrix of size $n \times m$. (The maximum compression ratio is reached for $n = m$, yielding 300%.)

S.25 Introduction to SINDy

1. The snapshot matrix is given as

$$X = \begin{pmatrix} x_1(0) & x_2(0) \\ x_1(\pi/4) & x_1(\pi/4) \\ x_1(\pi/2) & x_1(\pi/2) \end{pmatrix} = \begin{pmatrix} 0 & 2 \\ 1 & -1 \\ 0 & -2 \end{pmatrix}.$$

The temporal derivative of the snapshot matrix is given as

$$\dot{X} = (2\cos(2t), -2\cos(2t) - 4\sin(2t))$$

$$\dot{X} = \begin{pmatrix} \dot{x}_1(0) & \dot{x}_2(0) \\ \dot{x}_1(\pi/4) & \dot{x}_1(\pi/4) \\ \dot{x}_1(\pi/2) & \dot{x}_1(\pi/2) \end{pmatrix} = \begin{pmatrix} 2 & -2 \\ 0 & -4 \\ -2 & 2 \end{pmatrix}.$$

2. The library of the candidate functions is given as

$$\Theta(X) = \begin{pmatrix} 1 & x_1(0) & x_2(0) \\ 1 & x_1(\pi/4) & x_2(\pi/4) \\ 1 & x_1(\pi/2) & x_2(\pi/2) \end{pmatrix} = \begin{pmatrix} 1 & 0 & 2 \\ 1 & 1 & -1 \\ 1 & 0 & -2 \end{pmatrix}.$$

Perform an initial least squares regression.

$$\Xi = (\Theta^\top \Theta)^{-1} \Theta^\top \dot{X} = \begin{pmatrix} 0.5 & 0 & 0.5 \\ -0.25 & 1 & -0.75 \\ 0.25 & 0 & -0.25 \end{pmatrix} \dot{X} = \begin{pmatrix} 0 & 0 \\ 1 & -5 \\ 1 & -1 \end{pmatrix}$$

When convergence is reached, the recovered system of differential equation is

$$\dot{x}_1 = x_1 + x_2,$$
$$\dot{x}_2 = -5x_1 - x_2.$$

S.26 SINDy for Structural Dynamics

1. Introduce the substitution $y_1 = x_1$, $y_2 = \dot{x}_1$, $y_3 = x_2$, $y_4 = \dot{x}_2$.

$$\dot{y}_1 = y_2$$
$$\dot{y}_2 = \frac{-K_1 y_1 - D_1 y_2 - K_2(y_1 - y_3)}{m_1}$$
$$\dot{y}_3 = y_4$$
$$\dot{y}_4 = \frac{-k_3(y_3) - K_2(y_3 - y_1)}{m_2}$$

2. The system can be incorporated with the following code snippet.

```
# multi-dof parameters
k1 = 1
k2 = 2
k3 = lambda x: x ** 3
m1 = 5
m2 = 2
d1 = 1

# generation of artificial data
def rhs(t, y):
    return [y[1],
            (-k1 * y[0] - d1 * y[1] - k2 * (y[0] - y[2])) /
     m1,
            y[3],
            (-k3(y[2]) - k2 * (y[2] - y[0])) / m2]
```

Chapter 9

S.33 Iterative Forward Solver

The peculiarities in the training loop are listed below.

- The boundary values are always set to c_0 and thereby not learned.
- The gradient computation is special, as the sensitivites $\partial C/\partial c(x)$ are passed to the backward propagation call `.backward()`. Instead of fully relying on automatic differentiation, the adjoint method is partially employed to reduce the computational effort.
- In the piece-wise constant ansatz, an activation function is not used to ensure a scaling between zero and one, as it would make the ansatz non-linear. Instead, the parameters are always clamped in the range of zero and one subsequent to the optimization update.

S.34 Data-Driven Solver

3. The fully connected neural network performs better due to the model order reduction through singular value decomposition in the input and the simple parametrization of the wave velocity in the output.

S.35 Finite Difference Method with ρ-Scaling

Using the finite difference stencils from Eqs. (9.16) and (9.18), and the harmonic mean expressions from Eqs. (9.23) and (9.24) (applied to γ instead of c^2), the modified scalar wave equation form Eq. (9.43) can be written in its discretized form as

$$\gamma_i \rho_0 \frac{u_i^{n+1} - 2u_i^n + u_i^{n-1}}{\Delta t^2} - 2\rho_0 c_0^2 \frac{\left(\frac{1}{\gamma_i} + \frac{1}{\gamma_{i+1}}\right)^{-1}(u_{i+1}^n - u_i^n) - \left(\frac{1}{\gamma_{i-1}} + \frac{1}{\gamma_i}\right)^{-1}(u_i^n - u_{i-1}^n)}{\Delta x^2} = p_i^n.$$

Rewriting the equation with respect to u_i^{n+1}, an explicit stencil is established with

$$\begin{aligned} u_i^{n+1} &= -u_i^{n-1} + 2u_i^n + \frac{\Delta t^2 p_i^n}{\rho_0 \gamma_i} \\ &+ \frac{2}{\gamma_i}\left(\frac{\Delta t c_0}{\Delta x}\right)^2 \left[\left(\frac{1}{\gamma_i} + \frac{1}{\gamma_{i+1}}\right)^{-1}(u_{i+1}^n - u_i^n) - \left(\frac{1}{\gamma_{i-1}} + \frac{1}{\gamma_i}\right)^{-1}(u_i^n - u_{i-1}^n)\right]. \end{aligned}$$

Index

L. Herrmann et al., *Deep Learning in Computational Mechanics*,
https://doi.org/10.1007/978-3-031-89529-6

P

Q

R